PLEASE STAMP DATE DUE, BOTH BELOW AND ON CARD

DATE DUE	DATE DUE	DATE DUE	DATE DUE
SEP 3 0 1996			

GL-15

Early Precambrian Processes

Geological Society Special Publications
Series Editor A. J. FLEET

Professor John Sutton FRS

GEOLOGICAL SOCIETY SPECIAL PUBLICATION NO. 95

Early Precambrian Processes

EDITED BY

M. P. COWARD

Geology Department
Imperial College London, UK

and

A. C. RIES

Ries–Coward Associates Ltd
Reading, UK

1995

Published by
The Geological Society
London

THE GEOLOGICAL SOCIETY

The Society was founded in 1807 as the Geological Society of London and is the oldest geological society in the world. It received its Royal Charter in 1825 for the purpose of 'investigating the mineral structure of the Earth'. The Society is Britain's national society for geology with a membership of around 7500. It has countrywide coverage and approximately 1000 members reside overseas. The Society is responsible for all aspects of the geological sciences including professional matters. The Society has its own publishing house, which produces the Society's international journals, books and maps, and which acts as the European distributor for publications of the American Association of Petroleum Geologists, SEPM and the Geological Society of America.

Fellowship is open to those holding a recognized honours degree in geology or cognate subject and who have at least two years' relevant postgraduate experience, or who have not less than six years' experience in geology or a cognate subject. A Fellow who has not less than five years' relevant postgraduate experience in the practice of geology may apply for validation and, subject to approval, may be able to use the designatory letters C Geol (Chartered Geologist).

Further information about the Society is available from the Membership Manager, The Geological Society, Burlington House, Piccadilly, London W1V 0JU, UK. The Society is a Registered Charity, No. 210161

Published by the Geological Society from:
The Geological Society Publishing House
Unit 7
Brassmill Enterprise Centre
Brassmill Lane
Bath BA1 3JN
UK
(*Orders*:   Tel 01225 445046
          Fax 01225 442836)

First published 1995

The publisher makes no representation, express or implied, with regard to the accuracy of the information contained in this book and cannot accept any legal responsibility for any errors or omissions that may be made.

British Library Cataloguing in Publication Data

A catalogue record for this book is available from the British Library.

ISBN 1-897799-36-5

Typeset by EJS Chemical Composition,
Midsomer Norton, Bath, Avon

Printed by The Alden Press, Osney Mead, Oxford, UK.

Distributors

USA
  AAPG Bookstore
  PO Box 979
  Tulsa
  OK 74101-0979
  USA
(*Orders*:   Tel (918) 584-2555
          Fax (918) 548-0469)

Australia
  Australian Mineral Foundation
  63 Conyngham Street
  Glenside
  South Australia 5075
  Australia
(*Orders*:   Tel (08) 379-0444
          Fax (08) 379-4634)

India
  Affiliated East-West Press PVT Ltd
  G-1/16 Ansari Road
  New Delhi 110 002
  India
(*Orders:*   Tel (11) 327-9113
          Fax (11) 326-0538)

Japan
  Kanda Book Trading Co.
  Tanikawa Building
  3–2 Kanda Surugadai
  Chiyoda-Ku
  Tokyo 101
  Japan
(*Orders*:   Tel (03) 3255-3497
          Fax (03) 3255-3495)

Contents

Preface

The papers in this volume are grouped together, following particular themes or models of Early Precambrian processes.

In the first paper **Maarten de Wit & Andrew Hynes** discuss the onset of interaction between the hydrosphere and oceanic crust. According to these authors the Jamestown Ophiolite Complex, part of the Barberton greenstone belt of South Africa, represents very early ocean-like lithosphere, obducted onto a volcanic arc terrain by processes very similar to those which have emplaced modern ophiolites. The early metamorphism and degree of hydration are similar to modern sea-floor hydrothermal processes. **Ken Eriksson** discusses the relationships between crustal growth, surface processes and atmospheric evolution in the Early Precambrian. Models for the evolution of the Precambrian atmosphere range from those proposing little change through time to those advocating a stepwise increase in oxygen content. According to Ken Eriksson the evolution of the atmosphere to an oxidizing state, necessary for the formation of red beds and red palaeosols, was only achieved by 2.0 Ga following the escape of oxygen to the atmosphere from the oceans. Prior to 3.0 Ga, oxygen was consumed mainly in oceanic crust alteration. From 3.0 to 2.0 Ga it was consumed in the precipitation of Fe and Mn supplied to the oceans by fumerolic activity. **Euan Nisbet** reviews Archaean ecology and speculates on the development of a global scale biosphere. The earliest living communities may have consisted of chemotrophic bacteria that existed around hydrothermal systems.

Three papers discuss the tectonic evolution of greenstone belts. **Robert Shackleton** divides greenstone belt structures into pre-diapiric, diapiric and post-diapiric and discusses their origin in terms of regional tectonic processes. **Pierre Choukroune** notes the rapid variations in foliation trends, fold and thrust vergence and strain state in the Archaean greenstone belts compared to more modern orogenic belts. He argues that the Archaean cratons lacked a rigid upper plate and that continental crust formed and was deformed during episodes of enhanced mantle plume activity. **Pete Treloar & Tom Blenkinsop** examine Archaean deformation patterns in Zimbabwe and the adjacent Limpopo Belt and discuss the relationship between major shear zones and crustal shortening and thickening. They discuss how far the Archaean orogenic and greenstone belts can be compared with modern collisional mountain ranges such as the Himalayas and Tibet.

Stratigraphic and tectono-stratigraphic problems in Zimbabwe and Western Australia are described in the next group of papers. **Jimmy Wilson** and others produce new isotopic age data on zircon populations from Zimbabwean Archaean felsic volcanic sequences. The data demonstrate a Late Archaean volcanicity spanning 250 Ma from 2.9 Ga to 2.65 Ga. **Alec Trendall** discusses the tectono-stratigraphic components of the Pilbara craton in Western Australia. He notes that there has been a tendancy to interpret the geology of this and other cratons in terms of paradigms, involving a simplistic comparison with other areas thought to be better understood. This tendency has impeded the development of independant hypotheses. **John Myers** shows how the Yilgarn craton has been generated and assembled from diverse crustal fragments, including volcanic arcs, back-arc basins and microcontinents.

Two papers discuss aspects of the economic geology of Archaean rocks. **Dave Groves** and others discuss the lode-gold deposits of the Yilgarn Block, in particular the source of the gold-bearing fluids. **Richard Herrington** describes the Late Archaean structure and gold mineralization in the Kadoma region of the Zimbabwe Midlands.

Three papers describe the structures of the gneissic terranes of the North Atlantic craton. **Dave Bridgwater** examines the character of the basic dyke swarms across the craton and their reltionship to different tectonic events. **Graham Park** reviews the tectonics of the different components which make up the craton and discusses a speculative plate tectonic history of the region between *c.* 2.6 Ga and *c.* 1.5 Ga. **Tim Wynn** re-examines the shear zones and deformed Scourie dykes of the Lewisian Complex of Northwest Scotland. He notes that a map of the Scourie–Laxford area can be regarded as a section through the mid to lower crust of Laxfordian age (*c.* 1.8 Ga). The geometry of the Laxfordian shear zones may be analogous to the geometry of deep extensional faults in Phanerozoic terrains.

Mike Coward and others show that Archaean basins, such as the gold-bearing Witwatersrand basin in South Africa, formed on already consolidated crust, by processes analogous to those of modern destructive plate margins. The Witwatersrand basin has been analysed by techniques similar to those used for Phanerozoic basins. The tectonic history comprises phases of compressional and extensional tectonics related to events to the northwest at the edge of the South African plate.

Rod Graham discusses present-day analogues for different regimes of Archaean tectonics. He uses the Black Sea as an analogue for tonalite–greenstone terrains and the deep levels of the Tarim basin for reworked granulite terrains. The discussion is supported by new structural studies and cross sections showing the evolution of the central Asian basins.

This collection of papers therefore provides a broad spectrum, covering many different Early Precambrian processes, together with suggestions for more recent analogues.

Mike Coward
July 1995

Dedication

This Memorial Volume for Professor John Sutton FRS is concerned with Early Precambrian processes because that was the field in which, with Janet Watson, he made his scientific reputation.

John Sutton was one of the most eminent geologists of his time, interested in every aspect of the Earth Sciences. His scientific passion was for Precambrian rocks throughout the world. It was however in the Northwest Highlands of Scotland that he established his reputation. In their joint PhD researches on the formidable problems of the Lewisian Complex in NW Scotland, he and Janet demonstrated that the Lewisian is not a single system but reflects two quite distinct sets of events, their Scourian and Laxfordian, separated by a vast period of time. This new work, initially inspired and supervised by H. H. Read, provoked a strong but expected outburst from the conservative authorities in geology, one of the first of a series of noteworthy, if at times vitriolic, debates on Lewisian geology at the Geological Society.

John Sutton and Janet Watson turned their attention to a range of subjects in Scottish geology, first the Lewisian, then the Moine, the Dalradian and finally the Torridonian. John in particular applied the techniques developed in the Lewisian and Moine rocks to basement analysis worldwide, in South Africa, Australia and the Canadian Shield. He was the inspiration behind much of the exciting and innovative work done in Greenland, showing how techniques of basement analysis could be used to generate a tectono-stratigraphy over more than 1500 Ma of Early Precambrian time. Although never a laboratory geologist, he encouraged collaboration with geochemists, geophysicists and geochronologists and was delighted by the way the new radiometric data from the Precambrian supported his ideas based solely on field data. He was once heard to announce at a Royal Society meeting that he would personally offer a crate of the finest champagne to the first scientist to obtain an age of over 4000 Ma from Precambrian rocks.

John Sutton will be remembered for his enthusiasm and far-sightedness. He had tremendous insight into the potential for future research in geology and science as a whole. He was more concerned with the possibilities of science than with everyday administration. Nevertheless he became Head of the Department of Geology at Imperial College in 1964, a post he held for ten years during a time of great developments in geology. John Sutton had the vision and capacity to make the Department into a cohesive entity, uniting the separate (disparate) specialist research groups. He presided over a major expansion in the Department, making it the largest and most influential in Europe, an achievement that some people never forgave him for. He became one of the world's leading, if not larger than life, professors in geology, with a central role in guiding the future of research and geological teaching internationally. For all his influence, John always remained an approachable and impartial person. He was interested in all aspects of the subject, in all the people, staff and students, and in their work. He would love to leave the administration for day and spend time discussing ideas and techniques with students and colleagues, generally on a one to one basis.

In the mid 1970s, after John retired as Head of Department in Geology, he recognized the importance of scientific work related to the environment and established the highly successful interdisciplinary Centre for Environmental Technology and then a Centre for Remote Sensing at Imperial College. John Sutton, who had been Dean of the Royal School of Mines at Imperial College in 1965–68, then became Dean for a second term in 1974–77. In 1979 he went on to become Pro-Rector of Imperial College until his retirement in 1983.

Outside Imperial College, he was secretary of the Geological Society (1966–68), President of the Geologists' Association (1966–67) and Vice-President of the Royal Society (1975–77).

Following the death of Janet Watson, he married Betty Middleton-Sandford, and together they pursued the second of John's great enthusiasms, that of gardening. With his family connections it was natural that he should be passionately interested in gardens, flowers and plants. He designed the delightful garden at Chobham where he and Janet lived after they married. He was fascinated by plants and their environment no matter where he was, be it in the South African low-veld, the remote moorlands of Tibet or the rocky coasts of the Outer Hebrides. The restoration of the garden in Dorset, where he retired with Betty, was certainly a major achievement.

We should appreciate then, in reading this Memorial Volume, that there is much more for which to remember John Sutton than his geological work on the Precambrian, important though that was.

Mike Coward, Alison Ries & Robert Shackleton
with help from colleagues and fellow students of John Sutton.

The onset of interaction between the hydrosphere and oceanic crust, and the origin of the first continental lithosphere

MAARTEN J. DE WIT[1] & ANDREW HYNES[2]

[1]Department of Geological Sciences, University of Cape Town,
Rondebosch 7700, South Africa
[2]Department of Earth and Planetary Sciences, McGill University,
3450, University Street, Montreal, Quebec H3A 2A7, Canada

Abstract: New continental crust forms above subduction zones through the recycling of hydrated oceanic lithosphere. The most efficient process known for oceanic lithosphere hydration takes place at the submerged mid-ocean ridges where the lithosphere is young and warm, and cools through hydrothermal convection. Such mid-ocean ridge hydrothermal interactions were operative at least as far back as 3.5–3.8 Ga. The apparent absence of preserved continental crust older than 4.0 Ga may reflect the absence of hydrothermal interaction before that time. This model requires that prior to about 4.0 Ga mid-ocean ridges stood above sea level. Our calculations show, however, that on a plate-tectonic early Earth with substantially less continent, realistic higher heat flow, and a volume of sea water similar to that of today's ocean, Archaean mid-ocean ridges would have remained below sea level. Only with a substantial reduction of surface water would Earth have been able to recycle dry oceanic lithosphere, and thus prevent the present day style of continental crust formation.

A 30% reduction of surface water is required to elevate early Earth's ridges above sea level. This excess water may have been stored in nominally anhydrous minerals of the mantle. Early Earth's mantle may have released a significant proportion of its initial water only gradually through convective overturn of the oceanic floor. Given realistic ocean-floor creation rates, it would have taken roughly 500 Ma to process the early Earth's mantle through a MORB generation event if only the upper mantle was involved and considerably longer if whole mantle convection was involved. The inefficiency of water extraction during this process is illustrated by the amount of water apparently present in the source regions for present-day MORB. In this scenario, the Hadean–Archaean transition may mark the time when Earth changed its style of cooling from one dominated by heat exchange directly to the atmosphere to one dominated by heat exchange with the hydrosphere, which still buffers Earth's heat loss today.

The oldest continental rocks discovered to date are the Acasta gneisses from the Slave Province in Canada (Bowring *et al.* 1989, 1990). Precise U/Pb zircon dating has shown that these rocks are about 4.0 Ga old. Small areas of such old rocks (> 3.8 Ga) have been found in only a few other places around the world: over very limited areas of Antarctica, Australia, China and Greenland (Bennett *et al.* 1993). One of the major conclusions of the work on the Acasta gneisses is that there is nothing unusual about these rocks. In fact they chemically so resemble rocks that are forming above subduction zones today, that Bowring and his colleagues feel that plate tectonics might well have operated as long ago as 4.0 Ga.

Continental rocks between 3.8 and 3.5 Ga old have now been found on all continents, and two sizable continental fragments, the Pilbara and Kaapvaal cratons, have preserved continental areas of $> 0.5 \times 10^6$ km^2 which had stabilised by 3.0 Ga.

There is circumstantial evidence that these two fragments were part of a much greater volume of continental crust that must have appeared between 4.0 and 3.0 Ga (de Wit & Hart 1993; Myers this volume).

A search in Archaean sediments has identified single minerals older than 4.0 Ga in one place (Frouche *et al.* 1983). A small percentage of zircon grains from a 3.0 Ga sandstone in Western Australia have yielded ages of *c.* 4.2 Ga. The chemical compositions of these zircons are similar to those found in continental rocks of all ages. It is not clear yet if these tiny zircons indicate that there were substantial continents in existence by 4.2 Ga. If so, they have not yet been discovered or they have been destroyed, perhaps by meteorite impacts or by subduction. Hf isotope data from zircons in Archaean sediments, however, provide no evidence for an abundance of continental crust prior to 4.0 Ga (Stevenson & Patchett 1990; McCulloch &

From COWARD, M. P. & RIES, A. C. (eds), 1995, *Early Precambrian Processes*,
Geological Society Special Publication No. 95, pp. 1–9.

1

Bennett 1993; Bennett *et al.* 1993). Moreover, the general lack of inherited zircons does not provide support for the presence of large volumes of very ancient continental crust (> 4.0 Ga) preserved in the old cratons such as in Greenland (Nutman *et al.* 1993).

In contrast to the continents, rocks of the present ocean basins are always young (< 0.2 Ga). This is because oceanic basalts are continuously recycled into the asthenosphere at subduction zones. In exceptional cases, old pieces of oceanic crust (ophiolites) are preserved because they are emplaced on the continents during mountain building. These fragments offer the only clues to ancient oceanic environments.

Continental rocks are 'second-hand' rocks (cf. Taylor 1989) formed at subduction zones during the recycling of the oceanic lithosphere. The precise genesis of continental rocks above subduction zones is still a matter of debate, but almost certainly devolatilization and dehydration of the descending oceanic lithosphere plays a first-order role in the production of these continental rocks (Grove & Kinzler 1986, Ellam & Hawkesworth 1988, Baker *et al.* 1994). Without the involvement of fluids from the descending slab it is unlikely that efficient formation of granitoid rocks, which predominate on the continents, would occur (Campbell & Taylor 1983). Thus, a significant water content in the oceanic lithosphere is a prerequisite for the formation of continental rocks in today's plate tectonic scheme. Where does the oceanic lithosphere obtain this water?

Mid-oceanic ridge hydrothermal processes

The most efficient process known through which oceanic material becomes hydrated takes place at mid-ocean ridges during the formation of oceanic lithosphere. Submarine volcanism at mid-ocean ridges is the most important source of molten rock of Earth's crust (Sclater *et al.* 1981). The heat from igneous rocks, cooling within the oceanic crust, drives the convection of cold seawater which hydrothermally cools and chemically alters the crust (Hart 1973; Spooner & Fyfe 1973; Lister 1977, de Wit & Stern 1976; Basaltic Volcanism Study Project 1981; Seyfried 1987; Edmond 1992). Some 25% of Earth's total heat budget is believed to be lost to the hydrosphere through this hydrothermal heat exchange. Oxygen and strontium isotopes used as tracers of hydrothermal activity indicate that ophiolite sections up to 7 km thick have similarly interacted with seawater during their formation at spreading centres (Gregory & Taylor 1981; Alt *et al.* 1986; Schiffman & Smith 1988; Alexander *et al.* 1993).

Recently, similar processes have been recognized to have affected ophiolite-like sequences in the oldest preserved greenstone belts. The most detailed studies of this type have been carried out on *c.* 3.5 Ga mafic–ultramafic rock sequences of the Barberton greenstone belt in South Africa. The comagmatic mafic–ultramafic rocks of the southern Barberton terrain (referred to as the Jamestown ophiolite complex) are interpreted to represent a remnant of *c.* 3.5 Ga ocean-like lithosphere (de Wit *et al.* 1987) which was obducted approximately 45 Ma after its formation (de Wit *et al.* 1992). Evidence for submarine hydrothermal metamorphism is extensive (de Wit & Hart 1993; de Ronde *et al.* 1994). The igneous and hydration ages (3.48–3.49 Ga) of these mafic–ultramafic rocks are synchronous; this provides the most compelling evidence for mid-ocean ridge-like hydrothermal interaction (de Wit & Hart 1993). The extent of hydration and the chemical reactions of meta-igneous rocks in the Jamestown ophiolite complex are similar to those of hydrothermally altered igneous rocks from the present-day oceanic-crust. Water contents of up to 16% have been documented in the ultramafic rocks (de Wit & Hart 1993).

Work to date indicates that mid-ocean ridge-like hydrothermal interactions were operative at least as far back as 3.5 Ga. More circumstantial evidence (viz. the presence of pillow lavas and associated BIF) from the oldest mafic–ultramafic rocks, such as those in the deformed 3.8 Ga Isua supracrustal sequence in Greenland, suggests that major interactions between the hydrosphere and ocean floor were also active as far back as 3.8 Ga (Maruyama *et al.* 1992; Nutman & Collerson 1991; Maruyama pers. comm., 1992). The circumstances for generation of continental crust were therefore present at least as early as 3.8 Ga, in conformity to the presence of continental crust that old. The apparent absence of preserved continental crust, older than 4.0 Ga, has led to the suggestion that the formation of the first second-hand continental crust may have been coincident with the onset of hydrothermal alteration of oceanic crust (de Wit *et al.* 1992; de Wit & Hart 1993). In this model, the first production of substantial continental crust is directly related to the onset of mid-ocean ridge-like hydrothermal processes. The earliest hydrothermal interaction must, therefore, have started by 4.0 Ga and perhaps as early as 4.2 Ga.

If this model is correct, then prior to *c.* 4.0 Ga the average mid-ocean ridge must have stood above sea level. Under these circumstances the degree of hydrothermal alteration of the newly formed crust is highly dependent on the amount of precipitation (i.e. rain etc.) the region experiences. There is, for example more hydrothermal alteration in subaerial spreading centres on Iceland than in the Afar (both of which are dominated by meteoric water (SIO 1977; Sveinbjornsdottir et al. 1986; Lonker et al. 1993), but even on Iceland alteration is less pervasive by a factor of 3–10 and probably reaches shallower depths than the alteration in ophiolite complexes and ocean-floor rocks (Stefannson 1983). Furthermore, since most of the cooling of oceanic crust occurs at very young ages, even if such cooling is solely by conduction, the probability of substantial hydrothermal circulation after submergence declines markedly with increase in submergence age, and is probably neglible by an age of 50 Ma. With emergent spreading ridges, therefore, Earth's oceanic lithosphere would have been recycled with much less contained water than it has today, and there would have been no significant formation of continental crust. Such a model is compatible with the geochemical observations that the earliest preserved mafic–ultramafic rocks on Earth were derived from mantle sources which already had a long history of relative chemical depletion and extraction of possibly earlier, but recycled, oceanic crust (Hamilton et al. 1983; DePaolo 1988; Hoffman 1988; Chase & Patchett 1988; Carlson & Silver 1988; Bowring et al. 1990; Galer & Goldstein 1991; McCulloch & Bennett 1993). Assuming a closed-system for argon and helium in the mantle, crust and atmosphere, the model presented here is also compatable with the high $^{40}Ar/^{36}Ar$-low $^3He/^4He$ ratios of the present day mid-ocean ridge mantle source, and with exhaustive Early Archaean depletion of ^{36}Ar relative to ^{40}K, and ^3He relative to U–Th, from this mantle source (Schwartzman 1973, Hart et al. 1979, 1985); such depletion would have accompanied the onset of mid-ocean ridge hydrothermal activity and continental crust formation.

Mid-ocean ridges of present Earth stand 2–2.5 km below present sea level. Only in a few places do these spreading ridges become shallow and subaerial (viz. Iceland and the Afar). Shallow-water spreading activity has been documented during the formation of the Jamestown ophiolite complex in Barberton (de Wit & Hart 1993; de Ronde et al. 1994) but the setting of this ophiolite might have corresponded to an unusual area like the Afar. It is pertinent to ask, therefore, what the average elevation of spreading ridges might have been relative to sea level on an Archaean Earth with little or no continental material.

Below, we calculate the depths of mid-oceanic ridges at 4.0 Ga, given a plate-tectonic Earth with substantially less continent, a realistically higher heat flow and a volume of seawater similar to that of today's oceans.

Water depths of Archaean mid-oceanic ridges

On present Earth, there is a direct relationship between the elevation of oceanic lithosphere and its age (Parsons & Sclater 1977) which is thought to be a direct consequence of conductive cooling following formation of the oceanic crust at ridges. Following Parsons & Sclater (1977), an approximate expression for this relationship is:

$$e(t) = \frac{\rho_0 \, a\alpha T_1}{2(\rho_0 - \rho_w)} - \frac{2\rho_0 \, \alpha T_1}{(\rho_0 - \rho_w)} \left(\frac{\kappa t}{\pi} \right)^{1/2} \quad (1)$$

where $e(t)$ is the elevation of ocean floor of age t, ρ_0 is the density of the lithosphere, α is the coefficient of thermal expansion for the ocean floor, ρ_w is the density of water and κ is the thermal diffusivity. T_1 is the temperature at the base of the plate using the 'plate' model of McKenzie (1967) for the thermal structure of the oceans. In the plate model, the plate is considered unable to convect throughout its depth, and has a constant temperature at its base. Since the thermal boundary layer of Earth is probably able to convect at depths considerably lower than those corresponding to the plate base in the plate model (e.g. Parsons & McKenzie 1978), the basal temperature used in the model is not a true indicator of the temperature at some specific depth; it is simply a convenient device with which to parameterize oceanic thermal evolution. It is, however, very similar in magnitude to the temperature of the mantle beneath Earth's thermal boundary layer (e.g. Parsons & Slater 1978, fig. 7) and may be treated as such for the purposes of this paper.

There is, furthermore, a simple relationship between the area of ocean floor younger than a specified age and that specified age (Parsons 1982):

$$A(t) = C_0 \, t \left(1 - \frac{t}{2t_m} \right) \quad (2)$$

where $A(t)$ is the area of ocean floor of age less than or equal to age t, C_0 is the rate of generation of new oceanic crust, and t_m is the age of the oldest oceanic crust preserved in situ. Parsons (1982) showed that a relationship of this form would be a direct consequence of the consumption of oceanic crust (by subduction) at uniform rates regardless of its age, and that this uniform consumption rate is approximately true on present Earth.

From equation 2:

$$t_m = \frac{2 \, A(t_m)}{C_0} \quad (3)$$

where t_m is the age of the oldest extant oceanic lithosphere and $A(t_m)$ is the total area of ocean floor on the planet. The cumulative conductive heat loss from ocean floor of age greater than or equal to t, $Q(t)$, is approximately given by (Parsons 1982):

$$Q(t) = \frac{2kT_1 C_0}{\kappa} \left(\frac{\kappa t_m}{\pi}\right)^{\frac{1}{2}} \left[\left(\frac{t}{t_m}\right)^{\frac{1}{2}} - \frac{1}{3}\left(\frac{t}{t_m}\right)^{1\frac{1}{2}}\right]$$

(4)

where k is the thermal conductivity. Equations 3 and 4 with $t = t_m$ may be solved for t_m and C_0 if $Q(t_m)$, the total heat flow from the oceans, T_1 and $A(t_m)$, the total area of ocean floor, are known. After this, Equations 1 and 2 may be used to calculate the total volume of the oceans beneath the ridge crests and, if the total volume of oceanic water is known, the height of the ridge crest with respect to sea level.

These relationships depend only on the assumptions of a 'plate' cooling model for the ocean floor and a subduction rate that is independent of the age of the ocean floor being subducted. They are not affected by the thickness of the oceanic crust, which may have been thicker or thinner in the Archaean than at present (cf. Sleep & Windley 1982, or Nisbet & Fowler 1983, respectively). This is because any continental crust present would have adjusted itself to stable thicknesses giving rise to minimal freeboard as it does today, through a combination of orogenic and erosional processes. It is unlikely that any significant proportion of the continental crust would have been deeply submerged for long time intervals.

These relationships have been used to calculate the elevation of the mid-ocean ridges for a range of assumed values of total heat loss from the oceans, plate basal temperature and total oceanic area. Other parameters in Equations 1 to 4 were assigned the values given in Table 1.

On modern Earth the temperature at the base of the plates is about 1350°C (Parsons & Sclater 1977). Spreading-ridge heights are calculated

Table 1. *Values of constants used*

Constant	Value
ρ_0	3330 kg m^{-3}
α	3.28×10^{-5} °C^{-1}
κ	7×10^{-7} m^2 s^{-1}
ρ_w	1000 kg m^{-3}
k	3.14 W m^{-1} °C^{-1}
a	1.25×10^5 m

Values of constants used in equations 1 to 4, after Parsons & Sclater (1977) & Parsons (1982).

relative to sea level, assuming this value for the basal temperature T_1, but allowing total heat flow $Q(t_m)$ to increase incrementally to up to four times its present value, and with oceanic surfaces ($A(t_m)$) ranging up to 100% of the total planetary surface (Fig. 1a). It is apparent that ocean ridges would have stood below sea level for most combinations of heat flow and oceanic area. Only those for which the total heat flow out of the oceans is less than roughly one and a half times present give ridges that emerge, and these do so only for oceanic areas considerably larger than at present, so that the total heat loss from Earth would be less than that indicated by past/present total heat-flow ratio. These models are clearly unrealistic. Total Earth's heat flow in the Archaean must have been at least 2–3 times that of present heat loss (Burke *et al.* 1976; Basalt Volcanism Study Project 1981).

If the total heat flow through the oceans was twice that today, ridges could have emerged provided there was very little continental crust, and the basal temperature was considerably higher than at present. For example, with a basal temperature of 1550°C, ridges would have been emergent if oceanic crust covered 95% of Earth's surface (Fig. 1b). For this value of total heat flow, however, the ridges would never have been emergent with basal temperatures of only 1450°C (Fig. 1b). Twice the present total heat flow through the oceans is still an improbably low figure for the Archaean if total oceanic area was larger than at present, and it can be seen from Fig. 1a that raising the heat flow further would only result in submergence of the ridges, even for basal temperatures as high as 1550°C. This modelling indicates, therefore, that mid-ocean ridges are unlikely to have been emergent in the Archaean if there was as much oceanic water on the surface as there is today. In Fig. 2 and Table 2 we illustrate, however, using what we believe to be a realistic thermal model for the Archaean planet, a total oceanic heat-flow three times the present one and basal temperatures of 1550°C, that a reduction of the volume of the oceans by 30% would have caused the ridges to remain above sea level.

In conclusion, it appears from our calculations that with the present volume of seawater, the mid-ocean-ridges would have remained below sea level throughout the Archaean. Although increase in basal temperature tends to favour the emergence of ridges, its effect is far outweighed by that of the accompanying increase in total heat loss from the oceans. With constant oceanic volume, interaction between oceanic lithosphere and the hydrosphere would thus have been operative before the time of the first preserved continental fragments. A relationship between the onset of hydrothermal processes and continental crust formation therefore

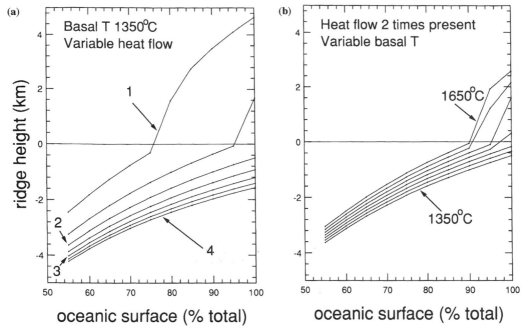

Fig. 1. (a) Height of the mid-ocean ridge above sea level as a function of the percentage of the Earth's surface covered by oceanic crust, assuming a basal temperature (approximately equivalent to the mantle temperature below the thermal boundary-layer; see text) of 1350°C. The curves are for different assumed values for the total heat flow out of the ocean floor, expressed as multiples of the present total heat flow out of the oceans (1, 1.5, 2, 2.5, 3, 3.5 and 4 times; every second curve is labelled), which is taken to be 29.0×10^{12} W, after Parsons (1982). The total volume of oceanic water used to calculate the height of the ridges is the present volume of the oceans (1.349×10^{18} m^3; Emiliani 1987).(b) as for (a), but assuming a total heat flow out of the oceans of twice the present value. The curves are for different assumed values of the basal temperature (1350, 1400, 1450, 1500, 1550, 1600 and 1650°C; the lowest- and highest-temperature curves are labelled).

seems unlikely unless the volume of the seawater in the oceans was substantially smaller than today. Below, we explore the likelihood of such a reduced volume of ocean water on Archaean Earth. We do not, however, consider the potential addition of exogenous water via cometary impact about that time (cf. Cogley & Henderson-Sellers 1984; Chyba & Sagan 1992).

Water content of the mantle

Recent work has indicated that the Earth's mantle may contain more water than has previously been suspected (Smyth 1987; Smyth *et al.* 1991; Bell & Rossman 1992*a, b;* Thompson 1992). Nominally anhydrous minerals believed to be present in the mantle are potential reservoirs for substantial amounts of water (see Bell & Rossman 1992*a*, and Thompson 1992 for comprehensive lists of these minerals).

Measured concentrations of water in MORB from the Juan de Fuca ridge and the Pacific–Nazca ridge have led to estimates of water concentrations in their mantle source regions of 250–300 ppm and 100–180 ppm, respectively (Dixon *et al.* 1988; Michael 1988) and incompatible-enriched basalts from Pacific–Nazca region require H_2O contents in the source region of 250–450 ppm (Michael 1988). Even higher values have been reported by Jambon & Zimmerman (1990). The infrared spectroscopic data of Bell & Rossman (1992*a*) indicate that olivine, orthopyroxene, clinopyroxene and garnet could accommodate most or all of this water.

The potential importance of the mantle as a reservoir for water is illustrated by the fact that 100 ppm of H_2O distributed uniformly through the upper mantle (shallower than 670 km) would be equivalent to 7.5% of the present oceanic volume (for an oceanic volume of 1.349×10^{18} m^3; Emiliani 1987) and this figure rises to 26.5% if the entire mantle is considered to have H_2O concentrations of this magnitude. If undepleted or enriched mantle is considered these figures become correspondingly higher again.

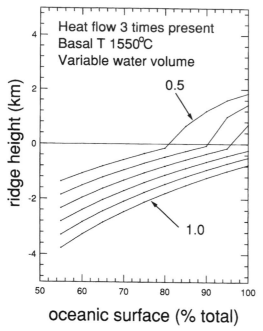

Fig. 2. As for Fig. 1a, but assuming a total heat flow out of the oceans of three times the present value and a basal temperature of 1550°C. The curves are for different assumed total volumes of oceanic water, expressed as multiples of the present oceanic volume (0.5, 0.6, 0.7, 0.8, 0.9 and 1.0 times present volume; the lowest- and highest-volume curves are labelled).

With the maximum measured OH contents in nominally anhydrous minerals, an amount of water equivalent to c. 85% of the current ocean mass can be accommodated in the whole mantle. Alternatively, if the oceans were derived from the mantle (which is not established) and originated from that portion modelled to have degassed to yield Earth's atmosphere (c. 46%), then the maximum observed OH concentrations suggest that up to 40% of the present oceans could originally have been stored in this form (Bell & Rossman 1992a).

It is thus possible that during Earth's early history a greater volume of water was stored in the mantle. Given such a scenario, we investigate first if Earth could have retained a realistic volume of water in the mantle for a long enough period to have allowed any mid-ocean ridge to have remained above sea levels for a realistic period of time; and secondly, how long it would have taken to dehydrate the entire mantle.

Mantle dehydration

This problem is approached using the results of our earlier calculations, since these calculations yield a creation rate for the ocean floor, expressed as area per unit time (e.g. Table 2). Using these creation rates, together with estimates of the crustal thickness created and the amount of mantle involved in (depleted by) the creation of the crust, it is then possible to determine how long it would

Table 2. *Ridge heights for Archaean Earth model and variable ocean volume*

Area of ocean floor (% total planet area)	Ocean-basin volume (multiple of present)	Ocean floor creation rate (km² a⁻¹)	Oldest extant ocean floor (Ma)	Ocean volume above ridge crests (10⁹ km³)	Ocean volume below ridge crests (10⁹ km³)	Ridge crest height (km)	Age of ocean floor at sea level (Ma)
60	1.0	23.35	26.21	0.346	1.003	−3.277	
70	1.0	20.02	35.67	0.471	0.878	−2.459	
80	1.0	17.51	46.59	0.615	0.734	−1.798	
90	1.0	15.57	58.97	0.779	0.570	−1.242	
100	1.0	14.01	72.80	0.962	0.387	−0.760	
60	0.7	23.35	26.21	0.346	0.598	−1.955	
70	0.7	20.02	35.67	0.471	0.473	−1.325	
80	0.7	17.51	46.59	0.615	0.329	−0.806	
90	0.7	15.57	58.97	0.779	0.165	−0.360	
100	0.7	14.01	72.80	0.962	0	0.699	2.844
60	0.5	23.35	26.21	0.346	0.328	−1.073	
70	0.5	20.02	35.67	0.471	0.203	−0.570	
80	0.5	17.51	46.59	0.615	0.059	−0.145	
90	0.5	15.57	58.97	0.779	0	1.193	8.292
100	0.5	14.01	72.80	0.962	0	1.849	19.906

Results of representative calculations of mid-ocean-ridge height, for a basal temperature of 1550°C and a total oceanic heat flow of three times present, for variously reduced volumes of oceanic water. These results are some of those plotted on Fig. 2.

take to deplete the mantle through the creation of the ocean floor. Using an ocean-floor creation-rate of 14.01×10^6 m^2 a^{-1} (the last entry from Column 3, Table 2) as an example, for a crustal thickness of 15 km, extracted from a mantle column initially three times as thick (33% melting by volume), 6.30×10^{11} m^3 of mantle is depleted each year. At this rate, the mantle above the 670 km discontinuity would be overturned only in 483 Ma, and the entire mantle, if it is involved, only in 1703 Ma.

Although there is uncertainty in these figures, they serve to illustrate that the time necessary to overturn the upper mantle is of at least the same order as the time between Earth formation and the age of the oldest preserved continent fragments (4.5 Ga minus 4.0 Ga). It is therefore quite possible that oceanic volume rose to its present value only gradually in the first 0.5 Ga of its history.

Discussion and conclusions

The calculations presented here show that with the present ocean volume, ocean ridges would have remained below sea level given realistic Archaean heat content/flux. Only with a substantial reduction of surface water would Earth have been able to recycle dry oceanic lithosphere. It has also been shown that, given our present understanding of the water content of Earth's mantle, a realistic reduction of early Earth surface water could have been attained for a period long enough to allow a gradual rise in sea levels of Hadean Earth. It is possible, therefore, that the higher mantle temperatures of Hadean Earth, and lower ocean volumes, may have kept mid-oceanic ridges above sea level. The gradual dehydration of Earth's mantle may then have drowned the mid-oceanic ridges by the onset of the Archaean, resulting in hydrothermal alteration of the ocean floor and, in turn, the formation of continental crust above subduction zones. For the first time in the history of Earth there was henceforth efficient and continuous interaction of our planet's hydrosphere and mantle. Although this may have been a gradual process, it would nevertheless have initiated a major change in the efficiency of Earth's chemical differentiation, and would have resulted in a change from planetary cooling dominated by heat exchange directly to the atmosphere (perhaps with a Venus-style mantle convection, cf. Solomon 1993) to one dominated by heat exchange with the hydrosphere, which still buffers Earth's heat loss today.

The initial rate at which continental crust formation took place is not known, but this must have been related to prevailing rates of oceanic lithosphere formation, which is believed to have been at least 3 to 10 times greater in the early Archaean

than today (Burke *et al.* 1976; Basaltic Volcanism Study 1981; Abbott & Hoffman; 1984, de Wit & Hart 1993; Table 2). Furthermore, there are compelling arguments for the continuous recycling of continental crust into the mantle (Armstrong 1990). The geological record indicates, however, that by 3.5 Ga the net rate of continental crust formation was in the order of 0.05 km^3 a^{-1} (de Wit & Hart 1993), more than sufficient to produce the granite–greenstone nuclei of the presently preserved old cratons, which, together with their depleted dunitic–harzburgitic keels, provided continental kernels for further (plate) tectonic accretion and collisions (de Wit *et al.* 1992; de Wit & Hart 1993).

A critical feature of the evolution model proposed in this paper is that Earth's mantle released a significant proportion of its initial water only gradually, through convective overturn of the ocean floor, over the first 500 Ma if the upper mantle only was involved, and considerably longer, if whole mantle convection was involved. This feature may appear inconsistent with inferences that Earth underwent essentially complete melting after core formation (e.g. Nisbet & Walker 1982). If this inference is correct, our model requires that this whole-Earth melting did not lead to complete extraction of the water from the mantle. While it is beyond the scope of this paper to discuss the reasons for this, it is consistent with the compelling evidence for significant amounts of water in the source regions for MORB. Although it is possible that some of this water has been recycled from the hydrosphere through subduction processes (e.g. Jambon & Zimmerman 1990) the depleted character and low ^{87}Sr/^{86}Sr of N-MORB make it unlikely that all the water is derived from this source. The Sr ratios would be particularly difficult to explain, since any water transported into the mantle would probably carry sufficient Sr to dominate the mantle signature.

This study serves to illustrate the importance of a greater understanding of the involvement of water in mantle melting-processes. We need to know more about the influence of water on this melting and on the efficiency of water extraction during melting. An implication of the model is that melting in the early Archaean may have occurred in the presence of larger amounts of water than at present. There is the potential to test this through the detailed examination of water-bearing phases such as pyroxene and amphibole in Archaean mafic–ultramafic rocks.

M. J. de W. acknowledges Foundation for Research and Development (South Africa) support, and A.H. was supported by Natural Sciences and Engineering Research Council (Canada). Two reviewers provided helpful suggestions to sharpen our arguments.

References

ABBOTT, D. H. & HOFFMAN, S. E. 1984. Archaean plate tectonics revisited. Part 1. Heat flow, spreading rate and the age of the subducting oceanic lithosphere and their effects on the origin and evolutions of continents. *Tectonics*, **3**, 429–448.

ALEXANDER, R. J., HARPER, G. D. & BOWMAN, J. R. 1993. Oceanic faulting and fault-controlled subseafloor hydrothermal alteration in the sheeted dike complex of the Josephine ophiolite. *Journal of Geophysical Research*, **98**, 9731–9759.

ιLT, J. C., MEUHLENBACHS, K. & HONNOREZ, J. 1986. An oxygen isotope profile through the upper kilometer of oceanic crust, DSPP hole 504-B. *Earth and Planetary Science Letters*, **80**, 217–229.

ARMSTRONG, R. L. 1990. The persistent myth of crustal growth. *Australian Journal of Earth Sciences*, **38**, 613–630.

BAKER, M. B., GROVE, T. L. & PRICE, R. 1994. Primitive basalts and andesites from the Mt Shaska region, N. California: products of varying melt fraction and water contents. *Contributions to Mineralogy and Petrology*, **118**, 111–129.

BASALTIC VOLCANISM STUDY PROJECT, 1981. *Basaltic volcanism on the Terrestrial Planet*. Pergamon, New York.

BELL, D. R. & ROSSMAN, G. R. 1992a. Water in the Earth's mantle: the role of nominally anhydrous minerals. *Science*, **255**, 1391–1397.

—— & —— 1992b. The distributions of hydroxyl in garnets from the subcontinental mantle of southern Africa. *Contributions to Mineralogy and Petrology*, **111**, 161–178.

BENNETT, V. C., NUTMAN, A. P. & McCULLOCH, M. T. 1993. Nd isotopic evidence for transcient, highly depleted mantle reservoirs in the early history of the Earth. *Earth and Planetary Science Letters*, **119**, 299–317.

BOWRING, S. A., WILLIAMS, I. S. & COMPSTON, W. 1989. 3.96 Ga gneisses from the Slave Province, NW Territories, Canada. *Geology*, **17**, 971–975.

——, HOUSH, T. B. & ISACHSEN, C. E. 1990. The Acasta gneisses; remnant of Earth's early crust. *In*: NEWSOM, H. E.& JONES, J. H. (eds) *Origin of the Earth*. Oxford University Press, New York, 319–344.

BURKE, K., DELANEY, J. F. & KIDD, W. S. F. 1976. Dominance of horizontal movements and micro-continental collisions during the later permobile regime. *In*: WINDLEY, B. F. (ed.) *The Early History of the Earth*. J. Wiley and Sons, London, 113–130.

CAMPBELL, I. H. & TAYLOR, S. R. 1983. No water, no granites, no oceans, no continents. *Geophysical Research Letters*, **10**, 1061–1064.

CARLSON, R. W. & SILVER, P. G. 1988. Incompatible element enriched and depleted reservoirs in the Earth: the possible importance of the lower mantle. *EOS*, **69**, 494.

CHASE, C. G. & PATCHETT, P. J. 1988. Stored mafic/ultramafic crust and early Archean mantle depletion. *Earth and Planetary Science Letters*, **91**, 66–72.

CHYBA, C. & SAGAN, C. 1992. Endogenous production,

exogenous delivery and impact-shock synthesis of organic molecules: an inventory for the origins of life. *Nature*, **355**, 125–132.

COGLEY, J. G. & HENDERSON-SELLERS, A. H. 1984. The origin and earliest state of the earth's hydrosphere. *Reviews of Geophysics and Space Physics*, **22**, 131–175.

DE WIT, M. J & HART, R. A. 1993. Earth's earliest continental lithosphere, hydrothermal flux and crustal recycling. *Lithos*, **30**, 309–336.

—— & STERN, C. R 1976. Ocean floor metamorphism, seismic layering and magnetism. *Nature*, **274**, 615–619.

——, HART, R. A. & HART, R. J. 1987. The Jamestown Ophiolite Complex, Barberton greenstone belt: a section through 3.5 Ga oceanic crust. *Journal African Earth Sciences*, **5**, 681–730.

——, ROERING, C., HART, R. J., ARMSTRONG, R. A., DE RONDE, R. E. J., GREEN, R. W. E., TREDOUX, M., PERBERDY, E. & HART, R. A. 1992. Formation of an Archaean continent. *Nature*, **357**, 553–562.

DEPAOLO, D. J. 1988. *Neodymium isotope geochemistry: an introduction*. Springer-Verlag, Berlin.

DE RONDE, C. E. J., DE WIT, M. J. & SPOONER, E. T. C 1994. Early Archaean (> 3.2 Ga) Fe-oxide-rich, hydrothermal discharge vents in the Barberton Greenstone Belt, South Africa. *Geological Society of America Bulletin*, **106**, 86–104.

DIXON, J. E., STOLPER, E. & DELANEY, J. R. 1988. Infrared spectroscopic measurements of CO_2 and H_2O in Juan de Fuca Ridge basaltic glasses. *Earth and Planetary Science Letters*, **90**, 87–107.

EDMOND, J. M. 1992. Results of water rock interactions in mid-ocean ridges. *In*: KHARAKA, Y. K. & MAEST, A. S. (eds.) *Water–Rock Interaction*. Balkema, Rotterdam, 885–889.

ELLAM, R. M. & HAWKESWORTH, C. J. 1988. Is average continental crust generated at subduction zones? *Geology*, **16**, 314–317.

EMILIANI, C. 1987. *Dictionary of the Physical Sciences*. Oxford University Press, Oxford.

FROUCHE, D. O., IRELAND, T. R., KINNY, P. D., WILLIAMS, I. S., COMPSTON, W., WILLIAMS, I. R & MEYERS, J. S. 1983. Ion microprobe identification of 4100–4200 Ma-old terrestrial zircons. *Nature*, **304**, 616–618.

GALER, S. J. G. & GOLDSTEIN, S. L. 1991. Early mantle differentiation and its thermal consequences. *Geochimica Cosmochimica Acta*, **55**, 227–239.

GREGORY, R. T. & TAYLOR, H. P. 1981. An oxygen isotope profile in a section of Cretaceous oceanic crust, Semail ophiolite, Oman: evidence for $\partial^{18}O$ buffering of the oceans by deep (> 5 km) seawater hydrothermal circulation at mid ocean ridges. *Journal Geophysical Research*, **86**, 2737–2755.

GROVE, T. L. & KINZLER, R. 1986. Petrogenesis of andesites. *Annual Reviews of Earth and Planetary Science*, **14**, 417–454.

HAMILTON, P. J., O'NIONS, R. K., BRIDGEWATER, D. & NUTMAN, A. 1983. Sm–Nd studies of Archean metasediments and metavolcanics from West Greenland and their implication for the Earth's early history. *Earth and Planetary Science Letters*, **62**, 263–272.

HART, R. A. 1973. Geochemical and geophysical implications of the reactions between seawater and the oceanic crust. *Nature*, **243**, 76–78.

HART, R. A., DYMOND, J. & HOGAN, L. 1979. Preferential formation of the atmosphere-sialic crust system from the upper mantle. *Nature*, **278**, 156–159.

——, HOGAN, L. & DYMOND, J. 1985. The closed system approximation for the evolution of argon and helium in the mantle, crust and atmosphere. *Chemical Geology (Isotope Geoscience Section)*, **52**, 45–73.

HOFFMAN, A. W. 1988. Chemical differentiation of the earth: the relationship between mantle, continental crust and oceanic crust. *Earth and Planetary Science Letter*, **90**, 297–314.

JAMBON, A. & ZIMMERMANN, J. L. 1990. Water in oceanic basalts: evidence for dehydration of recycled crust. *Earth and Planetary Science Letters*, **101**, 323–331.

LISTER, C. R. B. 1977. Qualitative models of spreading center processes, including hydrothermal penetration. *Tectonophysics*, **37**, 203–218.

LONKER, S. W., FRANZSON, H. & KRISTMANNSDÓTIIR, H. 1993. Mineral–fluid interactions in the Reykjanes and Svartsengi geothermal systems, Iceland. *American Journal of Science*, **293**, 605–670.

MARUYAMA, S., MASUDA, T., NOHDA, S., APPEL, P., OTOFUJY, Y., MIKI, M., SHIBATA, T. & HAGIYA, H. 1992. The 3.9–3.8 Ga plate tectonics on the earth; evidence from Isua, Greenland. *In: Evolving Earth Symposium, abstract volume, Okazaki, Japan*, 113.

MCCULLOCH, M. T. & BENNETT, V. C. 1993. Evolution of the earth: constraints from ^{143}Nd–^{142}Nd isotope systematics. *Lithos*, **30**, 237–256.

MCKENZIE, D. P. 1967. Some remarks on hcat flow and gravity anomalies. *Journal of Geophysical Research*. **72**, 6261–6273.

MYERS, J. S. 1995. The generation and assembly of an Archaean supercontinent: evidence from the Yilgarn craton, Western Australia. *This volume*.

MICHEAL, J. T. 1988. The concentration, behaviour and storage of H_2O in the suboceanic upper mantle: implications for mantle metasomatism. *Geochimica Cosmochimica Acta*, **52**, 555–556.

NISBET, E. G. & FOWLER, C. M. R. 1983. Model for Archean plate tectonics. *Geology*, **19**, 376–379.

—— & WALKER, D. 1982. Komatiites and the structure of the Archean mantle. *Earth and Planetary Science Letters*, **60**, 105–113.

NUTMAN, A. P. & COLLERSON, K. D. 1991. Very early Archean crustal accretion complexes preserved in the North Atlantic craton. *Geology*, **19**, 791–794.

——, FRIEND, C. R. L., KINNEY, P. K. & MCGREGOR, V. R. 1993. Anatomy of an early Archaean gneiss complex: 3900–3600 Ma crust in southern West Greenland, *Geology*, **21**, 415–418.

PARSONS, B. 1982. Causes and consequences of the relation between area and age of the ocean floor. *Journal of Geophysical Research*, **82**, 289–302.

—— & MCKENZIE, D. 1978. Mantle convection and the thermal structure of the plates. *Journal Geophysical Research*, **83**, 4485–4496.

—— & SCLATER, J. G. 1977. An analysis of the variation of ocean floor bathymetry and heat flow with age. *Journal Geophysical Research*, **82**, 802–827.

SCHIFFMAN, P. & SMITH, B. M. 1988. Petrology and oxygen isotope geochemistry of a fossil seawater hydrothermal system within the Solea Graben, northern Troodos ophiolite, Cyprus. *Journal of Geophysical Research*, **93**, 4612–4624.

SCHWARTZMAN, D. W. 1973. Ar degassing and the origin of sialic crust. *Geochimica et Cosmochimica Acta*, **37**, 2495–2497.

SCLATER, J. G., PARSON, B. & JAUPART, C. 1981. Oceans and continents: similarities and differences in the mechanism of heat loss. *Journal of Geophysical Research*, **86**, 11535–11552.

SEYFRIED, W. R. 1987. Experimental and theoretical constraints on hydrothermal alteration processes at mid-ocean ridges. *Annual Reviews of Earth and Planetary Science*, 15, 317–335.

SIO 1977. *Isotopic geochemistry and hydrology of geothermal waters in the Ethiopian rift valley*. Scrips Institution of Oceanography, La Jolla, California, Technical Report **77-14**.

SLEEP, N. H. & WINDLEY, B. F. 1982. Archean plate tectonics: constraints and inferences. *Journal of Geology*, **90**, 363–379.

SMYTH, J. R. 1987. ß-Mg$_2$SiO$_4$: a potential host for water in the mantle? *American Mineralogist*, **72**, 1051–1055.

——, BELL, D. R. & ROSSMAN, G. R. 1991. Incorporation of hydroxyl in upper-mantle clinopyroxene. *Nature*, **351**, 732–735.

SOLOMON, S. C. 1993. The geophysics of Venus. *Physics Today*, **46**, 48–55.

SPOONER, E. T. C & FYFE, W. S. 1973. Subseafloor metamorphism, heat and mass transfer. *Contributions to Mineralogy and Petrology*,. 42, 287–304.

STEFANNSON, V. 1983. Physical environment of hydrothermal systems in Iceland and on submerged oceanic ridges. *In*: RONA, P. A., BOSTRÖM, K., LAUBIER, L. & SMITH, K. L. Jr., (eds) *Hydrothermal processes at seafloor spreading centres*. Plenum Press, New York, 321-360.

STEVENSON, R. K. & PATCHETT, P. J. 1990. Implications for the evolution of continental crust from Hf isotope systematics of Archean detrital zircons. *Geochimica et Cosmochimica Acta*, **54**, 1683–1697.

SVEINBJORNSDOTTIR, A. E., COLEMAN, M. L. & YARDLEY, B. W. D. 1986. Origin and history of hydrothermal fluids of the Reykjanes and Krafla geothermal fields, Iceland. *Contributions to Mineralogy and Petrology*, **94**, 99–109.

TAYLOR, S. R. 1989. Growth of planetary crust. *Tectonophysics*, **161**, 147–156.

THOMPSON, A. B. 1992. Water in the Earth's upper Mantle. *Nature*, **358**, 295–302.

Crustal growth, surface processes, and atmospheric evolution on the early Earth

KENNETH A. ERIKSSON

Department of Geological Sciences, Virginia Polytechnic Institute and State University, Blacksburg, Virginia 24061, USA

Abstract: Evolution of the atmosphere between 3.9 and 1.8 Ga can be related to changes in physical, chemical and biological surface processes. Changes in surface processes, in turn, took place in response to growth of continents. Growth of stable continents favoured expansion of epeiric seas and resultant increase in oxygen productivity. Prior to 2.6 Ga, continents probably comprised less than 10% of the Earth's crust and, as a consequence, physical sedimentation and volcanism, and chemical and biological sedimentation took place mainly in oceanic and island-arc settings. Following major crustal growth at 2.6 Ga, surface processes became dominated by widespread stromatolite growth and iron and manganese precipitation in vast epicontinental seas.

Prior to 2.6 Ga, oxygen was consumed mainly in ocean–crust alteration. Following growth of large continents, the ocean crust oxygen reservoir decreased in scale freeing oxygen for the second major reservoir, namely reduced iron and manganese. Only after oceanic sinks were utilized did oxygen escape to the atmosphere. This was taking place on a large scale by 1.8 Ga at which time the atmosphere had an oxidation state necessary for formation of continental red beds and palaeosols.

Evolution of the Precambrian atmosphere is a subject of considerable debate. Models range from those proposing that the atmosphere has changed little through time (e.g. Hallam 1981, p. 205; Clemmey & Badham 1982) to those advocating a stepwise increase in oxygen content through the Precambrian Era (e.g. Cloud 1988; Hsü 1992; Kasting 1993). Much of the controversy is related to different interpretations of the same evidence. This contribution expands on the hypothesis advocated by Cloud and other proponents by developing the thesis that evolution of the atmosphere can be related to changing physical, chemical and biological processes through the time period 4.0 to 2.0 Ga. Implicit in this thesis is that evolution of surface processes, in turn, took place in response to growth of stable continental crust.

Evidence for growth of continents

The rate at which continental crust has grown is controversial. Models range from rapid Early Archean growth (Fyfe 1978; Armstrong 1981), to growth at a fairly constant rate (Hurley & Rand 1969). Other models envisage rapid growth around the Archaean–Proterozoic boundary or episodic growth (Fig. 1; Veizer & Jansen 1979; McLennan & Taylor 1982; Nelson & DePaolo 1985); these are now the most widely accepted models.

Veizer & Jansen (1979) based their model on Sr isotope ratios of marine carbonates. $^{87}Sr/^{86}Sr$ ratios display a pronounced increase at c. 2.5 ±

0.3 Ga (Fig. 2). Archaean values typically are in the range of the coeval mantle, whereas Proterozoic samples are distinctly more radiogenic. Veizer *et al.* (1982) attribute this increase to an exponential decrease in heat flux from the mantle, associated with cooling of the Earth, and an accompanying growth of continents that resulted in greater continental flux of more radiogenic strontium after 2.5 Ga.

The model of McLennan & Taylor (1982) is based on the recognition of two pelite geochemical populations and their relative abundances through time. The first population is characterized by flat to

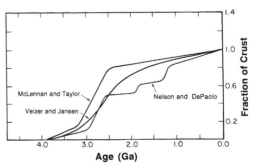

Fig. 1. Crustal growth models visualizing rapid growth at the Archaean-Proterozoic boundary (Veizer & Jansen 1979; McLennan & Taylor 1982) with episodic growth continuing through the Early and Middle Proterozoic (Nelson & DePaolo 1985).

From COWARD, M. P. & RIES, A. C. (eds), 1995, *Early Precambrian Processes*, Geological Society Special Publication No. 95, pp. 11–25.

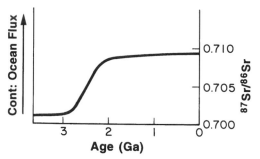

Fig. 2. Secular trend in $^{87}Sr/^{86}Sr$ ratios of marine carbonates reflecting a decrease in mantle flux associated with growth of continental crust at the Archaean–Proterozoic boundary (from Veizer & Jansen 1979).

relatively steep rare earth element (REE) patterns, low to moderate REE abundances and lack of prominent negative Eu anomalies (Fig. 3a). This population also has low Th/Sc, La/Sc and Th/U ratios and low Th and U abundances. These pelites

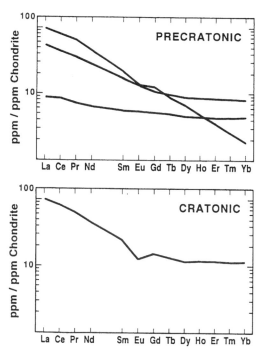

Fig. 3. Schematic rare earth element patterns for 'pre-cratonic' and 'cratonic' pelites (based on Taylor & McLennan 1985). Rare earth element patterns for 'pre-cratonic' mudstones range from flat to relatively steep. Intermediate patterns are most typical of 'pre-cratonic' mudstones and represent mixtures of mafic and felsic igneous end members.

are characterized as 'pre-cratonic' derived from undifferentiated granitoids such as tonalites, and felsic and mafic volcanic rocks in oceanic island arc and accretionary prism settings (Fig. 4). The second population has moderately steep REE patterns, higher REE abundances, and prominent negative Eu anomalies (Fig. 3b). In addition, Th/Sc, La/Sc and Th/U ratios and Th and U contents are higher than in the first population. These signatures imply derivation from continental crust that consisted of differentiated granitoids and are typically 'cratonic' (Fig. 4). The negative Eu anomalies reflect Eu retention in plagioclase as a result of intracrustal melting at lower crustal levels (Taylor & McLennan 1985). 'Precratonic' signatures typify Archaean greenstone belts, whereas 'cratonic' dominate the post-Archaean record but are developed locally in Archaean terrains (Taylor *et al.* 1986). These secular trends are interpreted by McLennan & Taylor (1982) to indicate major crustal growth at the Archaean–Proterozoic boundary. However, the presence locally of 'cratonic' pelites in Archaean terrains imply the existence of cratonic nucleii prior to 3.0 Ga (Taylor *et al.* 1986).

Nd-isotopic data of Nelson & De Paolo (1985) support the crustal growth of Veizer & Jansen (1979) and McLennan & Taylor (1982) but indicate more episodic crustal growth in the southwestern USA. T_{DM} model ages indicate addition of new crust to Archaean nucleii at 1.8 Ga and 1.3 Ga rather than recycling of older crust. Also, each of the models discussed (Fig. 1) indicate greater average ages for the continental crust than implied by the constant growth rate model of Hurley & Rand (1969).

Surface processes

Evolution of surface processes on the early Earth is investigated with respect to five time periods: pre-3.9 Ga, 3.9–3.3.2 Ga, 3.2–2.6 Ga, 2.6–2.0 Ga and post-2.0 Ga. Evidence for physical, chemical and biological processes is reviewed for each period.

Pre-3.9 Ga: asteroid bombardment

By analogy with the Moon, it is likely that during this time period surface processes on the Earth were dominated by intense meteorite bombardment. Grieve (1980) argues that following formation of a mafic proto-crust, subsidence of major impact-related basins resulted in partial melting of large volumes of basaltic volcanics leading to production of sialic crustal components. Ion microprobe zircon dating has revealed evidence for early sialic crust in gneisses and in detrital zircons from siliciclastic

Cratonic   Precratonic

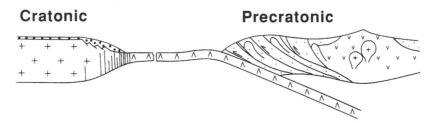

Fig. 4. Plate-tectonic cross section distinguishing between 'pre-cratonic' island arc–accretionary prism–ocean crust and 'cratonic' settings.

sedimentary sequences. Gneisses from the western Slave Province of Canada and Enderby land in Antarctica are dated at c. 3.95 Ga (Black et al. 1986; Bowring et al. 1989). Detrital zircons from the Wyoming Province in the USA are up to 3.96 Ga (Mueller et al. 1992) and from Mount Narryer and Jack Hills in the western Yilgarn Block, Australia, a small percentage of 4.10– 4.27 Ga detrital zircons have been identified (Froude et al. 1983; Compston & Pidgeon 1986).

3.9 to 3.2 Ga: small continents, large oceans

Isotopic data discussed previously are compatible with continents making up 5–10% of the Earth's crust during this period of geological time. Examples of this pre-3.0 Ga continental crust include: 3.65–3.5 Ga gneisses of the Bimodal gneiss suite and Tsawela gneiss in Swaziland (Compston & Kröner 1988; Hunter & Wilson 1988), the 3.3–3.2 Ga Sand River gneisses in the Limpopo province of South Africa (Retiet et al. 1990), 3.6–3.3 Ga gneisses of the Tokwe segment in Zimbabwe (Taylor et al. 1991), the 3.7–3.35 Ga Narryer gneiss complex in the western Yilgarn Block, Australia (Nutman et al. 1991), 3.8 Ga gneisses from the Singhbhum craton and 3.3– 3.0 Ga gneisses from the Dharwar craton of India (Beckinsale et al. 1980; Basu et al. 1981), the 3.8 Ga Amîtsoq gneiss of Greenland (Kinny 1986), the 3.9–3.4 Ga Uivak gneisses of Labrador (Nutman et al. 1989), and 3.8–3.3 Ga gneisses in the Wyoming province (Mueller et al. 1985; Mogk et al. 1992).

Shelf sediments, > 3.2 Ga, that were deposited on these continental nucleii include the c. 3.4 Ga Mkhondo and Mahamba sequences in Swaziland. These arenites (now metaquartzites), iron-formations and pelites (now paragneisses) un-conformably overlie the Bimodal gneiss suite (Hunter & Wilson 1988).

Remnants of the widespread oceanic crust surrounding the cratonic nucleii may be represented by the c. 3.55–3.3 Ga Onverwacht Group in the Barberton greenstone belt and the Warrawoona Group in the Pilbara block (Eriksson et al. 1994; de Wit 1991) although Bickle et al. (1994) dispute that these successions are of oceanic affinity. Mafic and ultramafic volcanics in the Barberton greenstone belt, some of which are pillowed, are interpreted as an ophiolite sequence and are considered to display evidence for seafloor-type metamorphism, metasomatism and black-smoker mineralization (de Wit 1991; de Ronde et al. 1994). Subordinate felsic volcanic rocks are present in the Barberton and Pilbara belts; these are interpreted as arc related and locally contain small massive sulphide deposits (Barley & Groves 1990a; de Wit 1991). If the mafic–ultramafic volcanics represent remnants of oceanic crust, those associated with felsic volcanics indicate proximity to an arc. A back-arc setting is favoured by Eriksson et al. (1994) for these mafic–ultramafic volcanics.

Sedimentary rocks are a minor but important component of the Onverwacht and Warrawoona groups insofar as providing information of water depth and chemistry. Lowe (1982) recognized three types of sedimentary rocks that record physical, chemical and biochemical processes. Volcaniclastic sediments, associated with felsic volcanic centres, accumulated in sub-aerial and shallow subaqueous environments. Orthochemical sediments are represented by white chert, carbonate and, most importantly, gypsum. Black cherts, algal laminites and stromatolites record biochemical processes (Lowe 1980; Walter et al. 1980; Byerley et al. 1986). Microfossils include filamentous, sheath-like, spheroidal and elliptical structures representing the remains of prokaryotic cyanobacteria or bacteria (Walsh & Lowe 1985; Schopf & Packer 1987; Walsh 1992; Schopf 1993); at least some of these microorganisms probably were oxygen-producing photoautotrophs (Schopf 1993).

Sedimentary rocks indicate that the preserved remnants of 'oceanic' crust in the Barberton greenstone belt and Pilbara block developed in shallow water (Lowe 1982). Detritus is exclusively of intra-basinal derivation indicating environments far removed from a continental influence. Felsic volcanic and volcaniclastic rocks probably were derived from the flanking island arcs, suggesting that mafic–ultramafic volcanism took place in back-arc or marginal basins (de Wit 1991; Krapez 1993).

Volcanic successions in the Barberton greenstone belt and Pilbara block are overlain by c. 3.2 Ga siliciclastic sedimentary sequences that record a dramatic change in tectonic setting associated with ocean closure. The Fig Tree and Moodies groups in the Barberton greenstone belt represent flysch and molasse facies of a foreland basin that developed in response to thrust loading from the south (Jackson et al. 1987). Turbidites, tidal and beach sandstones and braided-alluvial conglomerates and sandstones record comparable physical processes to those operating today. Similar syn-orogenic foreland-basin deposits are also present in the Pilbara block (Krapez 1993; Eriksson et al. 1995). Foreland-basin deposits in both the Barberton greenstone belt and Pilbara block contain substantial thicknesses of mudstone. These imply subaerial weathering and thus refute the contention of Moores (1993) that continental crust was emergent only from the Neoproterozoic Era onwards. In addition, the mudstones have characteristic 'pre-cratonic' geo-chemical signatures indicating derivation from undifferentiated provenances (McLennan et al. 1983a, b).

3.2–2.6 Ga: continental growth

Continental growth on a global scale was diachronous as recorded by different ages of continental growth on different continents and by ages of syn-rift and stable-shelf cover rocks. By 3.1 Ga extensive portions of the Kaapvaal province had been cratonized (e.g., Anhaeusser & Robb 1981; Tankard et al. 1982) and became the sites of syn-rift volcanism and sedimentation and post-rift stable-shelf sedimentation at c. 3.0 Ga. On the Kaapvaal province these cover successions include the syn-rift Dominion and Nsuze groups and over-lying, widespread quartz arenites, mudstones and iron formations of the Mozaan Group and lower-most Witwatersrand Supergroup (Eriksson et al. 1981; Bickle & Eriksson 1982; Burke et al. 1985; Beukes & Cairncross 1991; Stanistreet & McCarthy 1991). The Dominion and Nsuze groups are 3.1–3.0 Ga old (Hegner et al. 1984; Armstrong et al. 1991), whereas minimum ages of detrital zircons indicate that the lowermost Witwatersrand Supergroup was deposited after 3.05 Ga (Barton et al. 1989; Robb et al. 1990). REE signatures of pelites in these successions (Wronkiewiez & Condie 1987, 1989) imply that large parts of the Kaapvaal province consisted of differentiated

continental crust by 3.1–3.0 Ga. Other c. 3.0 Ga stable-shelf cover sequences have a more limited distribution and appear to have accumulated on small cratonic nucleii. Shelf sediments, including carbonates, were deposited on the Tokwe gneisses in southern Zimbabwe and on the Sand River gneisses in the central Limpopo province; these are represented, respectively, by the Buhwa succession and the Beit Bridge complex (Eriksson et al. 1988; Fedo & Eriksson 1993). Detrital zircons indicate that these two cover successions were deposited after 3.2 Ga (Dodson et al. 1988; Brandl & Barton 1991). REE data point to mixed provenance for the Beit Bridge complex but including a significant component of differentiated continental crust (Taylor et al. 1986; Boryta & Condie 1990). Conglomerates, arenites, pelites and iron formation in the Narryer and Jack Hills areas of the western Yilgarn block have a maximum age of 3.1 Ga (Maas & McCulloch 1991). Pelites have typical 'cratonic' geochemical signatures (Taylor et al. 1986; Maas & McCulloch 1991). Other examples of cover sequences include 3.3–2.8 Ga quartz arenites (now quartzites) and iron formations in the Wyoming province (Mueller et al. 1992), the > 3.05 Ga Malene supracrustals including quartzites, metapelites and marbles in the North Atlantic province of Greenland (Chadwick 1990), 3.2–3.1 Ga conglomerates, sandstones, iron formations containing stromatolites, pelites and volcanics of the Iron Ore Group in the Singhbhum craton of eastern India (Naqvi & Rogers 1987), and the c. 3.2–3.0 Ga Bababudan Group in the Dharwar craton of southwestern India (Taylor et al. 1984; Srinivasan & Ojakangas 1986; Bhaskar Rao et al. 1992).

Initial cratonization of the Zimbabwe province took place at c. 2.9 Ga followed at c. 2.7 Ga by deposition of the thin but laterally persistent Manjeri Formation (Wilson et al. 1978). This formation contains alluvial conglomerates and arkoses but consists mainly of shallow-shelf quartz arenites, mudstones, iron formations and stromatolitic carbonates (Nisbet et al. 1993). Final cratonization of the province was accomplished by emplacement of the potassic Chilimanzi suite of granites at c. 2.6 Ga (Wilson et al. 1978).

Following final tectonic assembly including late-stage, collision-related, lateral-escape tectonics and pull-apart basin development (Krapez & Barley 1987; Krapez 1993), the Pilbara block was cratonized by 2.85 Ga. Basalts, conglomerates and arkoses of the Fortescue Group were deposited between 2.77 and 2.69 Ga in syn-rift basins on differentiated continental crust (Blake & Groves 1987; Barley et al. 1992).

Coincident with the existence of widespread continental crust in inter alia South Africa and northwestern Australia, arc-related volcanism and sedimentation was taking place in regions that later developed into the stable Superior and Slave provinces and Norseman–Wiluna belt of Yilgarn block (Kusky 1989; Barley & Groves 1990; Thurston & Chivers 1990). Ferric oxide crusts are developed locally on pillowed basalts (Dimroth & Lichtblau 1978). Sedimentary rocks include reworked pyroclastics and epiclastics, deposited in fluvial and submarine environments, as well as 'Algoman-type' iron formation (e.g. Barrett & Fralick 1989; Devaney & Williams 1989). Massive sulphide deposits are widely developed within these arc-related successions (e.g. Barley & Groves 1990; Thurston & Chivers 1990). Local cratonic nucleii in the Superior and Slave provinces were the sites of accumulation of stable-shelf conglomerate quartz arenites, pelites, iron formations and stromatolitic carbonates (Wilks & Nisbet 1988; Thurston & Chivers 1990), but final cratonization of these provinces in response to island-arc accretion and emplacement of granitoids was achieved only by 2.65–2.6 Ga (Percival & Williams 1988; Kusky 1989; Hoffman 1990). Similarly, the Yilgarn block, the Guiana, Central Brazilian and Atlantic shields in Brazil and the Baltic shield of Russia and Finland were cratonized around 2.6 Ga.

Surface processes in cratonic settings during the time period 3.2–2.6 Ga were dominated by volcanism and alluvial and lacustrine sedimentation in rift basins and shelf sedimentation on cratons. Stromatolites are developed locally within the shelf assemblages including the Steep Rock, Pongola and Manjeri successions (Wilks & Nisbet 1988; Beukes & Lowe 1989; Nisbet *et al.* 1993) and also are present within syn-rift lacustrine facies of the *c.* 2.7 Ga Fortescue Group in Australia and Ventersdorp and Wolkberg groups in South Africa (Button 1973; Buck 1980; Walter 1983). Iron formations also are an important component of the *c.* 3.0 Ga shelf successions. Alluvial conglomerates and arenites are characteristically dull, tan coloured; the *only* reported example of red beds of this age is from Canada (Shegelski 1980) but these may be a result of younger (?Cretaceous) weathering.

Fluvial conglomerates in the Dominion, Fortescue and Bababudan groups, and in the Mozaan and lowermost Witwatersrand successions host placer gold and pyrite ± uraninite (e.g. Saager *et al.* 1986; Barley *et al.* 1992). More important occurrences of these minerals are in fluvial sediments of the syn-orogenic upper Witwatersrand Supergroup, and Uitkyk Formation in the Pietersburg greenstone belt in South Africa (Muff & Saager 1979; Stanistreet & McCarthy 1991; de Wit *et al.* 1992). MacLean & Fleet (1990) and

Robb & Meyer (1991) have presented convincing arguments in favour of a detrital rather than epigenetic replacement origin for most pyrite in the Witwatersrand basin as proposed by Phillips & Myers (1989). In particular, pyrite grains display random truncation of arsenic growth bands and have U–Pb older ages than the host rocks. Heavy minerals are concentrated mainly in thin, well-sorted conglomerates above low-angle unconformities rather than in poorly sorted, pebble-to-cobble conglomerates. Concentration took under conditions of prolonged footwall degradation associated with a lowering of base level (Button & Adams 1980).

Sandstones in the Witwatersrand supergroup, in particular, are enriched in quartz; rock fragments are absent, and feldspar is only a minor constituent (Sutton *et al.* 1990). Pelites from the Witwatersrand and Pongola successions display normal weathering histories and are strongly enriched in Al_2O_3 (Fig. 5). These pelites have CIA values (Nesbitt & Young 1982) similar to, or higher than, residual clays and Amazon Cone muds (Wronkiewicz & Condie 1987). Palaeosols developed beneath the Pongola, Dominion and

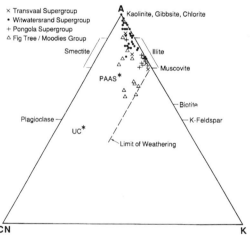

Fig. 5. Major element geochemistry of samples from the Pongola and Witwatersrand successions on an Al_2O_3–CaO* + Na_2O–K_2O plot of Nesbitt & Young (1989). CaO* is CaO in silicate minerals only. Also shown for comparison are samples from the older Fig Tree, Moodies and younger Transvaal successions. Note that all samples display normal weathering histories. Average shale (PAAS) and Upper Crust (UC) data from Taylor & McLennan (1985). Witwatersrand data from Fuller *et al.* (1981) and Wronkiewicz & Condie (1987), Pongola data from Wronkiewicz & Condie (1989), Fig Tree and Moodies from McLennan *et al.* (1983*b*), and Transvaal data from Wronkiewicz & Condie (1990).

Witwatersrand basins are depleted in CaO, Na_2O, MgO and total iron and enriched in SiO_2, Al_2O_3 and K_2O (Button & Tyler 1981). The high degree of chemical maturity of these sandstones, pelites and palaeosols implies intense weathering (Eriksson & Soegaard 1985; Maynard *et al.* 1991).

2.6–2.0 Ga: vast epicontinental seas

By the end of the Archaean Era at 2.5 Ga, continental land masses had attained areal extents of up to $2.6 \times 10^6 \, km^2$ (Fig. 6; Condie 1981). Dimensions of continental land masses shown on Fig. 6 are minimum only because recycling of older crust is known to have been widespread. For instance, Devonian granites in New England contain Archaean xenocrystic zircons (Eriksson *et al.* 1989). These cratons became the sites of stable-shelf sedimentation. Important differences between the *c.* 3.0 Ga and these post-2.5 Ga shelf sediments is the areal extent of and presence of thick stromatolitic carbonates in the latter. Examples reviewed below are from the Kaapvaal province, Pilbara block, Superior province, São Francisco craton, Baltic shield and India shield (Fig. 6). Available data indicate that pelites within cover sequences of this age have distinctive 'cratonic' geochemical signatures (e.g. Wronkiewicz & Condie 1990).

The Campbellrand and Malmani subgroups were deposited between 2.56 and 2.43 Ga (Jahn *et al.* 1990; Trendall *et al.* 1990; Barton *et al.* 1994) and

consist of thin, basal, siliciclastic-dominated intervals containing quartz arenites and pelites, overlain by carbonate successions with preserved thicknesses in excess of 1.5 km (Beukes 1987). Carbonate platform facies comprise tidal-flat algal mats and small-scale stromatolites, desiccation breccias, and shallow-subtidal, large-scale stromatolitic mounds (Truswell & Eriksson 1973; Eriksson & Truswell 1974). These mainly stromatolitic facies are preserved in two structural basins but probably extended originally across the entire Kaapvaal province, an area in excess of 500 000 km^2 (Fig. 7). Platform facies pass westward into platform-edge columnar stromatolites, oolites and oncolites, slope breccias and slump deposits, and basinal ferruginous dolomites and limestones containing carbonate turbidites and tuffs (Beukes 1987).

The lowermost Hamersley Group on the southern and eastern margins of the Pilbara block overlies the syn-rift Fortescue Group and also represents a stable-shelf succession (Eriksson 1988). Deposition took place between 2.69 and 2.47 Ga (Trendall *et al.* 1990). Platform facies are represented by tidal-flat carbonates of the Carawine Dolomite in the eastern Pilbara block (Walter 1983; Simonsin *et al.* 1993). Unlike the succession in the Kaapvaal province, basinal facies predominate and consist of iron formation, ferruginous dolomite and limestone containing carbonate and chert turbidites, and tuffs (Simonsin & Goode 1988; Hassler 1993; Simonson *et al.* 1993). Turbidites were derived

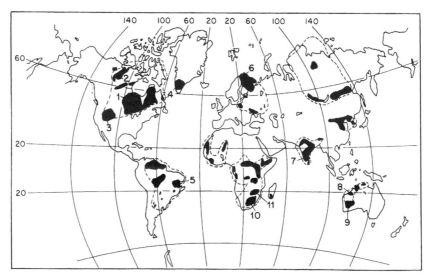

Fig. 6. Distribution of Archaean crustal provinces (from Condie 1981). Dashed lines demarcate regions that probably consist of Archaean crust. Provinces referred to in the text are: 1, Superior; 2, Slave; 3, Wyoming; 4, North Atlantic, Labrador and Greenland; 5, São Francisco; 6, Baltic; 7, Indian; 8, Pilbara; 9, Yilgarn; 10, Kaapvaal; 11, Zimbabwe.

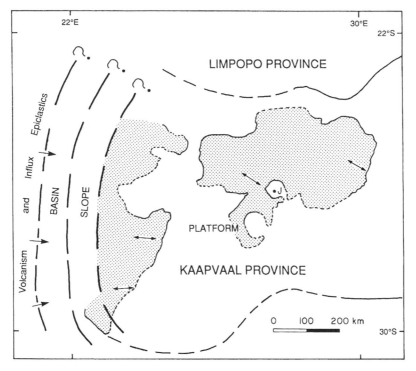

Fig. 7. Palaeogeographic map for the Campbellrand and Malmani Subgroups on the Kaapvaal Province (from Beukes 1987). Stippled areas are present outcrop/suboutcrops belts but the two subgroups probably were deposited across the entire Kaapvaal Province.

from the north, and thus it is likely that platform facies originally extended across all or much of the Pilbara block.

Early Proterozoic stratigraphic successions preserved on the margins of the Superior province display a similar stratigraphic evolution to the Fortescue–Hamersley basin. In the northern Great Lakes region, the c. 2.45 Ga, lower Huronian synrift sediments and volcanics are overlain by 2.3–2.2 Ga sedimentary units attributed to regional downwarp (Young 1983) or deposition along a divergent continental margin (Fralick & Miall 1989). The latter sedimentary units contain quartz arenites, pelites with gypsum and anhydrite, and stromatolitic carbonates. The Kaniapiskau Supergroup in the Labrador Trough displays a similar transition from syn-rift, immature fluvial sediments and mafic volcanics to stable-shelf quartz arenites and stromatolitic carbonates (Hoffman 1987). These dominantly sedimentary successions are preserved on the margins of the Superior province, but the shallow-shelf facies, in particular, may have extended across all, or part, of the province.

Other Lower Proterozoic, stable-shelf sequences

are represented by the c. 2125 Ma lower Minas Supergroup in the São Francisco craton of Brazil (Babinski et al. 1991a, b), the 2.3–2.0 Ga Jatulian Group in the Baltic shield (Gaal & Gorbatschev 1987), and the Aravalli Supergroup in the north-west Indian shield (Roy & Paliwal 1981). In addition to typical shelf assemblages of sediments, the Jatulian and Aravalli successions also contain phosphate sediments.

Concentrations of placer minerals are present locally in Lower Proterozoic fluvial sediments. Where unconformably overlying the Witwatersrand Supergroup, the Black Reef Formation, at the base of the Malmani Subgroup, contains detrital pyrite, gold and uraninite (Papenfus 1964). Similar placer minerals are present in the lower but not in the upper Huronian Supergroup (Roscoe 1973). In the > 2.1 Ga, amphibolite-grade Jacobina belt conglomerates and arenites in Brazil also contain placer concentrations of gold, pyrite and uraninite (Mascarenhas & da Silva 1982).

Major occurrences of laminated and clastic-textured Superior-type iron formations are associated with stable-shelf successions or with over-lying, > 2.0 Ga foredeep deposits. The Malmani and Campbellrand subgroups are overlain conformably by laminated, basinal and coeval, clastic-textured shelf iron formations (Beukes 1983). Oolitic ironstones are intercalated locally within the overlying Pretoria Group whereas laminated iron formations and manganiferous deposits are developed in the coeval Postmasburg Group to the southwest (Nel *et al.* 1986).

Laminated iron formations in the Hamersley basin represent basinal facies to the platform to the north and also comprise the base of the overlying foredeep succession that consists mainly of Turee Creek turbidites (Eriksson 1988; Barley *et al.* 1992). Iron formations in the circum-Superior basins also make up the base of foredeep successions (Hoffman 1987). Clastic-textured varieties containing stromatolites developed under shallow-water conditions on the foreland ramp, whereas laminated iron formations represent deeper-water deposits of the adjoining trough (Ojakangas 1983; Simonsin 1985). Ferruginous sediments are overlain by thick sequences of syn-orogenic flysch and molasse deposits.

Similar transitions from stable-shelf to foredeep sedimentation are recorded in the Minas Supergroup, Jatulian Group and Aravalli Supergroup. In the Minas Supergroup, extensive laminated iron formations represent basinal equivalents of stable-shelf sediments, whereas in the Jatulian Group ferruginous and manganiferous sediments are clastic textured and appear to represent a ramp facies of the foredeep. Manganese deposits and laminated iron formations overlie the auriferous conglomerates and arenites in the Jacobina belt, but their tectonic setting is not known.

Palaeosols developed beneath the Transvaal basin that has a maximum age of *c.* 2.6 Ga, are depleted in total iron (Button & Tyler 1981). In contrast, some hematitic alteration profiles are present within the 2.2–2.0 Ga uppermost stratigraphic interval in the Transvaal basin (Wiggering & Beukes 1990). Arenites in this same stratigraphic interval have matrices of iron-stained clay but lack hematitic grain coatings (Eriksson & Cheney 1992). Fluvial sediments in the *c.* 2.2 Ga upper Huronian Supergroup locally display hematitic staining (Roscoe 1973).

Post-2.0 Ga: continental red beds

Texturally immature fluvial sandstones < 2.0 Ga old typically display a red pigmentation. The oldest stratigraphic successions containing sandstones with hematitic coatings on grains are the *c.* 2.0 Ga Deweras Group in the Magondi mobile belt of Zimbabwe and sandstone units in the Kheis belt of South Africa (Tankard *et al.* 1982; Treloar 1988). Similar sandstones are widespread in the *c.* 1.9 Ga,

molasse-phase Takiyuak Formation in the Wopmay orogen, Canada (Hoffman 1980) and in the *c.* 1.8 Ga Waterberg, Soutpansberg and Matsap successions in South Africa, Umkondo Group in Zimbabwe, Haslingden Group in the Mount Isa inlier in Australia, and Dubawnt Group in northern Canada (Blake 1980; Tankard *et al.* 1982; Eriksson *et al.* 1993). Metamorphic grade in the Haslingden Group ranges from sub-greenschist to amphibolite facies (Blake 1987). However, even at highest grades, hematite is preserved and has not re-crystallized to specularite (cf. Clemmey & Badham 1982). Palaeosols associated with some of the above units, most notably the 1.8–1.75 Ga Haslingden Group (Simpson *et al.* 1993; Driese *et al.* in press), are characteristically red coloured. Unlike most pre-2.0 Ga palaeosols, these are enriched in ferric iron.

A model for atmospheric evolution

It is likely that free oxygen has been produced photosynthetically since 3.5 Ga (Schidlowski *et al.* 1983; Schopf & Walter 1983; Schopf 1993) but it is probable that actual productivity lagged behind potential productivity until expansion of epicontinental seas coincident with growth of continents (Fig. 8). If so, the concentration of free oxygen in any reservoir is then a consequence of its rate of supply versus consumption (Veizer 1983). Early life forms apparently occupied shallow marine environments, and it is thus likely that the first oxygen reservoir was the ocean. If modern oceans can be used as counterparts of ancient oceans, then oxygen will have decreased rapidly with depth due to its consumption in deep waters (Kasting 1993).

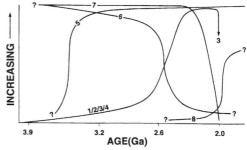

Fig. 8. Secular changes in parameters incorporated into the model. 1, Growth of continental crust. 2, Expansion of epicontinental seas. 3, Iron and manganese in oceanic chemical sediments. 4, Actual oxygen productivity. 5, Potential oxygen productivity. 6, Ferrous iron dissolved in ocean. 7, Detrital pyrite and uranite. 8, Atmospheric pO_2, continental red beds, red palaeosols.

Evolution of the early ocean-atmosphere system can be viewed in terms of different oxygen sinks through time (Veizer 1983; Table 1). During the time period 3.9–3.2 Ga, prolific mantle flux supplied divalent Fe and Mn as well as reduced gases to the ocean. These gases, as well as divalent Fe released by serpentinization, acted as sinks for oxygen during rapid turnover of ocean water through oceanic crust (Veizer 1983). If ophiolites are analogues for Archaean ocean crust, chemical data from the Troodos complex in Cyprus illustrate the importance of the ferrous iron sink. Magmatic phases at Troodos are depleted in Fe_2O_3, whereas altered pillow lavas and sheeted dykes have Fe_2O_3:FeO ratios as high as 4:1. Ferric iron in the altered volcanic rocks is present in epidote and magnetite (Richardson et al. 1987). Black smoker-type deposits interbedded within the volcanic succession in the Barberton greenstone belt represent a sea-floor oxygen sink, but extensive iron formations are not preserved in the 3.9–3.2 Ga rock record. Thus, the ocean became enriched progressively in divalent Fe and Mn. Gypsum associated with shallow-water facies in the Barberton and Pilbara greenstone belts record another, but minor, oxygen sink above the pycnocline. The paucity of gypsum in the Archaean, as well as the Early Proterozoic rock record, is ascribed by Grotzinger & Kasting (1993) to a critically low concentration of sulphate and a high bicarbonate : calcium ratio that resulted upon evaporation in removal of calcium from seawater before the gypsum field

was reached. Any sulphur below the pycnocline was utilized in the precipitation of Fe and other sulphides present locally within 3.5 Ga volcanic successions. Carbonate precipitation was rare during this time period and, as a consequence, carbon was not fixed in limestone or dolomite but was available as another oxygen sink for conversion to carbon dioxide. All siliciclastic alluvial sediments of this age, whether mineralogically mature or not, are devoid of red pigmentation indicating a dearth of atmospheric oxygen.

The most important oxygen sink during the 3.2–2.6 Ga time period was iron. Algoman-type iron formations in greenstone belts and Superior-type iron formations associated with shallow-shelf successions developed in response to upwelling of reduced bottom waters and precipitation of iron above the pycnocline (Button et al. 1982). Massive sulphides in greenstone belts developed below the pycnocline. Thin carbonate units fixed some shallow-water carbon but most carbon above the pycnocline was still available for oxidation. The lack of extensive carbonates within this time period probably is a function of the small size of stable cratons but may also reflect an unfavourable atmosphere. Garrels & Perry (1973) argue that oxidation of atmospheric hydrogen, ammonia and carbon monoxide preceded development of extensive carbonates. The presence of detrital pyrite and uraninite in alluvial sediments and absence of red arenites and palaeosols indicates that the atmosphere was still anoxygenic. Weathering was intense, probably due to a chemically aggressive atmosphere. The evidence for intense weathering coupled with the concentration of pyrite and uraninite in winnowed conglomerates exposed to the atmosphere for long periods of time, negates the contention of Clemmey & Badham (1982) that the presence of detrital pyrite and uraninite in the rapidly aggrading Indus River and of detrital pyrite in the Atacama Desert are evidence for an oxygenic atmosphere 3.0 Ga ago.

At the end of the 3.2–2.6 Ga time period, ocean waters remained enriched in divalent Fe and Mn because insufficient oxygen was available in the pre-2.5 Ga oceans to precipitate all of the Fe and much if any of the more soluble Mn. The 2.6–2.0 Ga time period was characterized by voluminous precipitation of these metals to produce the large Superior-type iron formations and economically important Mn deposits (Fig. 8). Many of these iron formations are finely laminated and lack any evidence for wave reworking; relatively deep waters are implied. Following growth of continents by 2.6 Ga, less oxygen was being consumed in ocean crust alteration probably resulting in a deeper pycnocline than before 2.6 Ga. Prolific algal growth resulted in abundant oxygen

Table 1. *Summary of major oxygen sinks during the time period 3.8–1.8 Ga*

Pre-3.2 Ga
- Ocean crust alteration including oxidation of reduced volvanic gases
- Gypsum precipitation
- Oxidation of carbon

3.2–2.6 Ga
- Precipitation of some iron
- Oxidation of carbon
- Intense weathering
- (Detrital pyrite and uraninite in fluvial sediments)

2.6–2.0 Ga
- Precipitation of Fe and Mn
- (Extensive shelf carbonates)
- (Detrital pyrite and uraninite in fluvial sediments)

Post-2.0 Ga
- Red beds and red palaeosols

Detrital pyrite and uraninite in fluvial sediments indicate an anoxygenic atmosphere until at least 2.3 Ga ago. From 2.5 Ga onwards carbon was fixed in carbonates rather than being available for oxidation to carbon dioxide.

production above the photic zone and it is likely that iron and manganese were precipitated in this zone in response to upwelling of reduced bottom waters (Button *et al.* 1982). Light rare-earth-element depletion and positive Eu anomalies in ~~these~~ as well as older iron formations support a ~~sub~~marine hydrothermal origin for the iron (Derry Jacobsen 1991). Laterally extensive and thick stromatolitic carbonate successions that developed between 2.6 and 2.0 Ga on the large stable cratons fixed large quantities of carbon in carbonates. Thus, less carbon dioxide was escaping to the atmosphere (Walker 1990). The presence of detrital pyrite and uraninite in rocks of this time period indicate that the atmosphere remained anoxygenic until at least 2.3 Ga (Fig. 8).

Roscoe (1973) has used the absence of detrital pyrite and uraninite in the upper Huronian Supergroup as evidence for a transition to an oxygenic atmosphere by 2.3 Ga. However, red pigmented arenites and hematitic alteration profiles are developed only locally, suggesting that atmospheric oxygen was not widespread. Laterally persistent and thick red bed successions as well as hematitic palaeosols are confined to rock units less than 2.0 Ga old. These observations indicate that abundant, free atmospheric oxygen appeared only after oceanic oxygen sinks were utilized prior to 2.0 Ga.

Summary

The preceding discussion has attempted to demonstrate that surface processes during the time period from before 3.9 to after 2.0 Ga changed in response to growth of stable continental crust. Continents comprised less than 10% of the Earth's crust prior to 2.6 Ga and as a consequence physical sedimentation and volcanism, and chemical and biochemical sedimentation took place mainly in oceanic-island arc environments. The small continents were sites of subordinate syn-rift volcanism and sedimentation, and stable shelf physical, chemical and biochemical sedimentation. Following major continental growth at *c.* 2.6 Ga, surface processes were dominated by widespread stromatolite growth, and iron and manganese precipitation in vast, shallow epicontinental seas.

Evolution of the atmosphere was closely related to changing surfaces processes as influenced by continental growth. During the Archaean Era, oxygen produced by photosynthetic organisms was consumed mainly in alteration of ocean crust and locally in gypsum and iron oxide precipitation. Following growth of continents at 2.6 Ga, the ocean crust reservoir decreased in scale and freed oxygen for the second major sink, namely reduced iron and manganese. The extensive stable continents also favoured prolific carbonate precipitation thereby fixing carbon and diminishing another Archaean oxygen sink. Only after oceanic sinks were utilized could oxygen escape to the atmosphere. This may have commenced as early as 2.3 Ga but was taking place on a large scale only by 1.8 Ga.

I thank M. Coward for the opportunity to participate in the meeting in memory of John Sutton. This paper was prepared during my tenure at the University of Zimbabwe as a Fulbright Scholar in 1992. J. F. Wilson and other staff members in the Geology Department provided valuable comments. C. Fedo commented on an earlier version of this paper. The paper benefitted from reviews by M. J. Bickle and A. F. Trendall.

References

ANHAEUSSER, C. R. & ROBB, L. J. 1981. Magmatic cycles and the evolution of Archean crust in the eastern Transvaal and Swaziland. *In*: GLOVER, J. E. & GROVES, D. I. (eds) *Archaean Geology: Second International Symposium.* Geological Society Australia Special Publications, **7**, 457–467.

ARMSTRONG, R. A., COMPSTON, W., RETIEF, E. A., WILLIAMS, I. S. & WELKE, H. J. 1991. Zircon ion microprobe studies on the age and evolution of the Witwatersrand triad. *Precambrian Research*, **53**, 243–266.

ARMSTRONG, R. L. 1981. Radiogenic isotopes: The case for crustal recycling on a near steady-state no-continental-growth Earth. *Philosophical Transactions of the Royal Society, London*, **A301**, 43–472.

BABINSKI, M., CHEMALE, F. & VAN SCHMUS, W. R. 1991a. Geocronologia Pb/Pb em rochas carbonaticas do Supergrupo Minas, Quadrilatero Ferrifero, Minas Gerais, Brazil. *An. III Congresso Brasileira de Geoquimica, Sao Paulo, Soceidade Brasiliera Geociencias*, 628–631.

——, —— & —— 1991b. Pb/Pb geochronology of carbonate rocks of Minas Supergroup, Quadrilatero Ferrifero, Minas Gerais. *AGU Fall Meeting, San Francisco, USA. EOS Transactions Supplement*, 531.

BARLEY, M. E. & GROVES, D. I. 1990. Deciphering the tectonic evolution of Archaean greenstone belts: the importance of contrasting histories to the distribution of mineralization in the Yilgarn Craton, Western Australia. *Precambrian Research*, **46**, 3–20.

——, BLAKE, T. S. & GROVES, D. I. 1992. The Mount Bruce Megasequence Set and eastern Yilgarn Craton: examples of late Archaean to early Proterozoic divergent and convergent craton margins and controls on mineralization. *Precambrian Research*, **58**, 55–70.

BARRETT, T. J. & FRALICK, P. W. 1985. Turbidites and

iron formations, Beardmore-Geraldton, Ontario: Application of a combined ramp fan model to Archean clastic and chemical sedimentation. *Sedimentology*, **36**, 221–234.

BARTON, E. S., COMPSTON, W., WILLIAMS, I. S., BRISTOW, J. W., HALLBAUER, D. K. & SMITH, C. B. 1989. Provenance ages for the Witwatersrand Supergroup and the Ventersdorp Contact Reef: constraints from ion microprobe U-Pb ages of detrital zircons. Economic *Geology*, **84**, 2012–2019.

BASU, A. R., RAY, S. L., SAHA, A. K. & SARKAR, S. N. 1981. Eastern India 3800 million-year-old crust and early mantle differentiation. *Science*, **212**, 1502–1506.

BECKINSALE, R. D., DRURY, S. A. & HOLT, R. W. 1980. 3360 Myr old gneisses from the South India Craton. *Nature*, **283**, 469–470.

BEUKES, N. J. 1983. Palaeoenvironmental setting of iron-formations in the depositional basin of the Transvaal Supergroup, South Africa. *In*: TRENDALL, A. F. & MORRIS, R. C. (eds) *Iron-Formation: Facts and Problems*. Elsevier, Amsterdam, 131–209.

—— 1987. Facies relations, depositional environments and diagenesis in a major early Proterozoic stromatolitic carbonate platform to basinal sequence, Campbellrand Subgroup, Transvaal Supergroup, Southern Africa. *Sedimentary Geology*, **54**, 1–46.

—— & CAIRNCROSS, B. 1991. A lithostratigraphic-sedimentological reference profile for the Late Archaean Mozaan Group, Pongola Sequence: application to sequence stratigraphy and correlation with the Witwatersrand Supergroup. *South African Journal of Geology*, **94**, 44–69.

—— & LOWE, D. R. 1989. Environmental control on diverse stromatolite morphologies in the 3000 Myr Pongola Supergroup, South Africa. *Sedimentology*, **36**, 383–397.

BHASKAR RAO, Y. J., SIVARAMAN, T. V., PANTULU, G. V. C., GOPALAN, K. & NAQVI, S. M. 1992. Rb–Sr ages of Late Archean metavolcanics and granites, Dharwar Craton, South India and evidence for Early Proterozoic thermotectonic event(s). *Precambrian Research*, **59**, 145–170.

BICKLE, M. J. & ERIKSSON, K. A. 1982. Evolution and subsidence of early Precambrian sedimentary basins. *Philosophical Transactions of the Royal Society, London*, **A305**, 225–247.

——, NISBET, E. G. & MARTIN, A. 1994. Archean greenstone belts are not oceanic crust. *Journal of Geology*, **102**, 121–138.

BLACK, L. P., WILLIAMS, I. S. & COMPSTON, W. 1986. Four zircon ages from one rock: the complex history of a 3930 Ma old granulite from Mount Sones, Enderby Land, Antarctica. *Contributions Mineralogy and Petrology*, **94**, 427–437.

BLAKE, D. H. 1980. *Volcanics rocks of the Paleohelikian Dubawnt Group in the Baker Lake–Angikuni Lake area. District of Keewatin, NWT*. Geological Survey of Canada Bulletin **309**.

—— 1987. *Geology of the Mount Isa Inlier and environs, Queensland and Northern Territory*. Australia Bureau Mineral Resources Bulletin **225**.

BLAKE, T. S. & GROVES, D. I. 1987. Continental rifting

and the Archean-Proterozoic transition. *Geology*, **15**, 229–232.

BORYTA, M. & CONDIE, K. C. 1990. Geochemistry and origin of the Archaean Beit Bridge complex, Limpopo Belt, South Africa. *Journal of the Geological Society, London*, **147**, 229–239.

BOWRING, S. A., WILLIAMS, I. S. & COMPSTON, W. 1989. 3.96 Ga gneisses from the Slave Province, Northwest Territories, Canada. *Geology*, **17**, 971–975.

BRANDL, G. & BARTON, J. M. 1991. Pre-Limpopo geology of the Central Zone. *In*: ASHWAL, L. D. (ed.) *Two Cratons and an Orogen-Excursion Guidebook and Review Articles for a Field Workshop through Selected Archaean Terranes of Swaziland, South Africa and Zimbabwe*. IGCP Project 280, Dept. of Geology, Univ. Witwatersrand, Johannesburg, 235–249.

BUCK, S. G. 1980. Stromatolite and ooid deposits within the fluvial and lacustrine sediments of the Precambrian Ventersdorp Supergroup of South Africa. *Precambrian Research*, **12**, 311–330.

BURKE, K., KIDD, W. S. F. & KUSKY, T. M. 1985. The Pongola structure of Southeastern Africa: The World's oldest preserved rift? *Journal of Geodynamics*, **2**, 35–49.

BUTTON, A. 1973. Algal stromatolites in the Early Proterozoic Wolkberg Group, Transvaal Sequence. *Journal of Sedimentary Petrology*, **43**, 160–167.

—— & ADAMS, S. S. 1981. *Geology and recognition criteria for uranium deposits of the quartz-pebble conglomerate type*. National Uranium Resource Evaluation Report **GJBX-3**(81).

—— & TYLER, N. 1981. The characteristics and economic significance of Precambrian paleo-weathering and erosional surfaces in Southern Africa. *In*: SKINNER, B. J. (ed.) *Economic Geology, 75th Anniversary Volume*, 686–709.

——, BROCK, T. D., COOK, P. J., EUGSTER, H. P., GOODWIN, A. M., JAMES, H. L., MARGULIS, L., NEALSON, K. H., NRIAGU, J. O., TRENDALL, A. F. & WALTER, M. R.1982. Sedimentary iron deposits, evaporites and phosphorites. *In*: HOLLAND, H. D. & SCHIDLOWSKI, M. (eds) *Mineral Deposits and Evolution of the Biosphere*. Dahlem Konferenzen-Dahlem Workshop Report, **16**, Berlin, Springer, 259–273.

BYERLEY, G. R., LOWE, D. R. & WALSH, M. M. 1986. Stromatolites from the 3300–3500-Myr Swaziland Supergroup, Barberton Mountain Land, South Africa. *Nature*, **319**, 489–491.

CHADWICK, B. 1990. The stratigraphy of a sheet of supracrustal rocks within orthogneisses and its bearing on Late Archaean structure in southern West Greenland. *Journal of the Geological Society, London*, **147**, 639–652.

CLEMMEY, H. & BADHAM, N. 1982. Oxygen in the Precambrian atmosphere: An evaluation of the geological evidence. *Geology*, **10**, 141–146.

CLOUD, P. E. 1988. *Oasis in Space: Earth History from the Beginning*. W.W. Norton and Co., New York.

COMPSTON, W. & KRÖNER, A. 1988. Multiple zircon growth within early Archean tonalitic gneiss from the Ancient Gneiss Complex, Swaziland. *Earth and Planetary Science Letters*, **87**, 13–28.

——— & PIDGEON, R. T. 1986. Jack Hills, evidence of more very old detrital zircons in Western Australia. *Nature*, **321**, 766–769

CONDIE, K. C. 1981. Archean Greenstone Belts. *Developments in Precambrian Geology* 3, Elsevier, Amsterdam.

DE WIT, M. J. 1991, Archaean greenstone belt tectonism and basin development: some insights from the Barberton and Pietersburg greenstone belts, Kaapvaal Craton, South Africa. *Journal of African Earth Sciences*, **13**, 45–63.

———, JONES, M. G. & BUCHANAN, D. L. 1992. The geology and tectonic evolution of the Pietersburg Greenstone Belt, South Africa. *Precambrian Research*, **55**, 123–153.

DE RONDE, C. E. J., DE WIT, M. J. & SPOONER, E. T. C. 1994. Early Archean (>3.2 Ga) Fe-oxide-rich, hydrothermal discharge vents in the Barberton greenstone belt, South Africa. *Geological Society of America Bulletin*, **106**, 86–104.

DERRY, L. A. & JACOBSEN, S. B. 1990. The chemical evolution of Precambrian seawater: Evidence from REE's in banded iron formations. *Geochimica et Cosmochimica Acta*, **54**, 2965–2977.

DEVANEY, J. R. & WILLIAMS, H. R. 1989. Evolution of an Archean subprovince boundary: A sedimentological and structural study of part of the Wabigoon-Quetico boundary in northern Ontario. *Canadian Journal of Earth Sciences*, **26**, 1013–1026.

DIMROTH, E. & LICHTBLAU, A. P. 1978. Oxygen in the Archean ocean: Comparison of ferric oxide crusts on Archean and Cenozoic pillow basalts. *Neues Jahrbuch für Mineralogie Abhandlungen*, **133**, 1–22.

DODSON, M. H., COMPSTON, W., WILLIAMS, I. S. & WILSON, J. F. 1988. A search for ancient zircons in Zimbabwean sediments. *Journal of the Geological Society, London*, **145**, 977–983.

DRIESE, S. G., SIMPSON, E. L. & ERIKSSON, K. A. 1.8 Ga paleosols from the Lochness Formation, Mount Isa, Australia: Paleoclimatic implications. *Journal Sedimentary Research*, in press.

ERIKSSON, K. A. 1988. Precambrian basin formation and evolution: comparison with possible Phanerozoic counterparts. *Ninth Australian Geological Convention*, 24.

——— & SOEGAARD, K. 1985. Petrography and geochemistry of Precambrian sediments: implications for weathering and crustal evolution. *Geological Survey of Finland Bulletin*, **331**, 7–32.

——— & TRUSWELL, J. F. 1974. Tidal flat associations from a Lower Proterozoic carbonate sequence in South Africa. *Sedimentology*, **21**, 293–309.

———, KIDD, W. S. F. & KRAPEZ, B. 1988. Basin analysis in regionally metamorphosed and deformed Early Archean terrains: examples from southern Africa and Western Australia, *In*: KLEINSPEHN, K. L. & PAOLA, C. (eds) *New Perspectives in Basin Analysis*. Springer-Verlag, New York, 371–404.

———, KRAPEZ, B. & FRALICK, P. W. 1994. Sedimentology of Archaean greenstone belts: signatures of

tectonic evolution. *Earth-Science Reviews*, **37**, 1–88.

———, SIMPSON, E. L. & JACKSON, M. J. (1993). Stratigraphic evolution of a Proterozoic rift to thermal-relaxation basin, Mount Isa Inlier, Australia: Constraints on nature of lithospheric extension. *In*: FROSTICK, L. E. & STEEL, R. J. (eds) *Tectonic Controls and Signatures in Sedimentary Successions*. International Association of Sedimentologists Special Publication, **20**, 203–221.

———, TURNER, B. R. & VOS, R. G. 1981. Evidence for tidal processes from the lower part of the Witwatersrand Supergroup. *Sedimentary Geology*, **29**, 309–325.

ERIKSSON, P. G. & CHENEY, E. S. 1992. Evidence for the transition to an oxygen-rich atmosphere during the evolution of red beds in the lower Proterozoic sequences of southern Africa. *Precambrian Research*, **54**, 257–269.

ERIKSSON, S. C., HOGAN, J. P. & WILLIAMS, I. S. 1989. Ion microprobe resolution of age details in heterogeneous sources of plutonic rocks from a comagmatic province. *Geological Society of America, Abstracts with Programs*, **21**, A361.

FEDO, C. M. & ERIKSSON, K. A. 1993. Chronological evolution of the Archean (~3.0 Ga) Buhwa Greenstone Belt. *16th International Colloquium of African Geology*, Ezulweni, Swaziland, 125–126.

FRALICK, W. & MIALL, A. D. 1989. Sedimentology of the Lower Huronian Supergroup (Early Proterozoic), Elliot Lake area, Ontario, Canada. *Sedimentary Geology*, **63**, 127–153.

FROUDE, D. O., IRELAND, T. R., KINNY, P. D., WILLIAMS, I. S., COMPSTON, W., WILLIAMS, I. R. & MYERS, J. S. 1983. Ion microprobe identification of 4,100–4,200 Myr-old terrestrial zircons. *Nature*, **304**, 616–618.

FULLER, A. O., CAMDEN-SMITH, P., SPRAGUE, A. R. G., WATERS, D. J. & WILLIS, J. P. 1981. Geochemical signature of shales from the Witwatersrand Supergroup. *South African Journal of Science*, **77**, 379–381.

FYFE, W. S. 1978. The evolution of the Earth's crust: Modern plate tectonics to ancient hot spot tectonics? *Chemical Geology*, **23**, 89–114.

GAAL, G. & GORBATSCHEV, R. 1987. An outline of the Precambrian evolution of the Baltic Shield. *Precambrian Research*, **35**, 15–52.

GARRELS, R. M. & PERRY, E. A. 1973. Cycling of carbon, sulfur and oxygen through geologic time. *In*: GOLDBERG, E. D. (ed.) *The Seas: Ideas and Observations*, **Vol. V**. Wiley International, New York, 303–336.

GRIEVE, R. A. F. 1990. Impact bombardment and its role in protocontinental growth on the early Earth. *Precambrian Research*, **42**, 63–75.

GROTZINGER, J. P. & KASTING, J. F. 1993. New constraints on Precambrian ocean composition. *Journal of Geology*, **101**, 235–243.

HALLAM, A. 1981. *Facies Interpretation and the Stratigraphic Record*. W. H. Freeman and Co.

HASSLER, S. W. 1993. Depositional history of the Main Tuff Interval of the Wittenoom Formation, late Archean–early Proterozoic Hamersley Group,

Western Australia. *Precambrian Research*, **60**, 337–359.

HEGNER, V. E., KRÖNER, A. & HOFMANN, A. W. 1984. Age and isotope geochemistry of the Archean Pongola and Usushwana suites, southern Africa: a case for crustal contamination of the mantle derived magma. *Earth and Planetary Science Letters*, **70**, 267–279.

HOFFMAN, P. F. 1980. Wopmay Orogen: a Wilson cycle of Early Proterozoic age in the northwest of the Canadian Shield. *In*: STANGWAY, D. W. (ed.) *The Continental Crust and its Mineral Deposits*. Geological Association Canada Special Papers, **20**, 523–549.

—— 1987. Early Proterozoic foredeeps, foredeep magmatism, and Superior-type iron-formations of the Canadian Shield. *In*: KRÖNEF, A. (ed.) *Proterozoic Lithospheric Evolution*. AGU Geodynamics Series, **17**, 85–98.

—— 1990. On accretion of granite-greenstone terranes. *In*: SHEAMAN, R. F. & GREEN, S. B. (eds) *Greenstone, Gold and Crustal Evolution*. Geological Association of Canada, St John's, Newfoundland, 32–45.

HSÜ, K. J. 1992. Is Gaia endothermic? *Geological Magazine*, **129**, 129–141.

HUNTER, D. R. & WILSON, A. H. 1988. A continuous record of Archean evolution from 3.5 Ga to 2.6 Ga in Swaziland and northern Natal. *South African Journal of Geology*, **91**, 57–74.

HURLEY, P. M. & RAND, J. R. 1969. Pre-drift continental nuclei. *Science*, **164**, 1229–1242.

JACKSON, M. P. A., ERIKSSON, K. A. & HARRIS, C. W. 1987 Early Archean foredeep sedimentation related to crustal shortening: A reinterpretation of the Barberton sequence, southern Africa. *Tectonophysics*, **136**, 197–221.

JAHN, B., BERTRAND-SARFATI, J., MORIN, N. & MACE, J. 1990. Direct dating of stromatolitic carbonates from the Schmidtsdrif Formation (Transvaal Dolomite), South Africa, with implications on the age of the Ventersdorp Supergroup. *Geology*, **18**, 2111–1214.

KASTING, J. F. 1993. Earth's early atmosphere. *Nature*, **259**, 920–926.

KINNY, D. D. 1986. 3820 Ma zircons from a tonalitic Amitsoq gneiss in the Godthab district of southern West Greenland. *Earth and Planetary Science Letters*, **79**, 337–347.

KRAPEZ, B. 1993 Sequence stratigraphy of the Archaean supracrustal-belts of the Pilbara Block, Western Australia. *Precambrian Research*, **60**, 1–45.

—— & BARLEY, M. E. 1987. Archaean strike-slip faulting and related ensialic basins: evidence from the Pilbara Block, Australia. *Geological Magazine*, **124**, 555–567.

KUSKY, T. M. 1989 Accretion of the Archean Slave Province. *Geology*, **17**, 63–67.

LOWE, D. R. 1980. Stromatolites 3,400-Myr old from the Archean of Western Australia. *Nature*, **284**, 441–443.

—— 1982. Comparative sedimentology of the principle volcanic sequence of Archean greenstone belts in South Africa, Western Australia and Canada:

implications for crustal evolution. *Precambrian Research*, **8**, 145–167.

MAAS, R. & McCULLOCH, M. T. 1991. The provenance of Archean clastic metasediments in the Narryer Gneiss Complex, Western Australia: Trace element geochemistry, Nd isotopes, and U-Pb ages for detrital zircons. *Geochimica et Cosmochimica Acta*, **55**, 1915–1932.

MACLEAN, P. J. & FLEET, M. E. 1990. Detrital pyrite in the Witwatersrand goldfields of South Africa: Evidence from truncated growth banding. *Economic Geology*, **84**, 2008–2011.

MASCARENHAS, J. F. & DA SILVA, SA. J. H. 1892. Geological and metallogenic patterns in the Archean and Early Proterozoic of Bãhia State, eastern Brazil. *Revista Brasileira Geociencias*, **12**, 193–214.

MAYNARD, J. B., RITGER, S. D, & SUTTON, S. J. 1991. Chemistry of sands from the modern Indus River and the Archean Witwatersrand basin: Implications for the composition of the Archean atmosphere. *Geology*, **19**, 265–268.

McLENNAN, S. M. & TAYLOR, S. R. 1982. Geochemical constraints on the growth of the continental crust. *Journal of Geology*, **90**, 342–361.

——, TAYLOR, S. R. & ERIKSSON, K. A. 1983*a*. Geochemistry of Archean shales from the Pilbara Supergroup, Western Australia. *Geochimica et Cosmochimica Acta*, **47**, 1211–1222.

——, —— & KRÖNER, A. 1983*b*. Geochemical evolution of Archean shales from South Africa. I. The Swaziland and Pongola Supergroups. *Precambrian Geology*, **22**, 93–124.

MOGK, D. W., MUELLER, P. A. & WOODEN, J. L. 1992. The nature of Archean terrane boundaries: an example from the northern Wyoming Province. *Precambrian Research*, **55**, 155–168.

MOORES. E. M. 1993. Neoproterozoic oceanic crustal thinning, emergence of continents, and origin of the Phanerozoic ecosystem: A model. *Geology*, **21**, 5–8.

MUELLER, P. A., WOODEN, J. L. & NUTMAN, A. P. 1992. 3.96 Ga zircons from an Archean quartzite, Beartooth Mountains, Montana. *Geology*, **20**, 327–330.

——, ——, HENRY, D. J. & BOWES, D. R. 1985. Archean crustal evolution of the eastern Beartooth Mountains, Montana and Wyoming. *In*: CSAMANSKE, G. K. & ZIENTEK, M. I. (eds) *The Stillwater Complex, Montana: Geology and Guide*. Montana Bureau of Mines Special Publications, **92**, 9–20.

MUFF, R. & SAAGER, R. 1979. Petrographic and mineragraphic of the Archaean gold placer at Mount Robert, Pietersburg, N. Transvaal. *In*: ANDERSON, A. M. & VAN BILTON, W. J. (eds) *Some Sedimentary Basins and Associated Ore Deposits of South Africa*. Geological Society of South Africa Special Publications, **6**, 23–31.

NAQVI, S. M. & ROGERS, J. J. W. 1987. *The Precambrian Geology of India*. Clarendon Press, New York.

NEL, C. J., BEUKES, N. J. & DEVILLIERS, J. P. R. 1986. The Mamatwan manganese mine of the Kalahari manganese field. *In*: ANHAEUSSER, C. R. & MASKE, S. (eds) *Mineral Deposits of Southern Africa*,

Geological Society of South Africa, Johannesburg, 963–978,

NELSON, B. K. & DEPAOLO, D. J. 1985. Rapid production of continental crust 1.7 to 1.9 b.y. ago: Nd isotopic evidence from the basement of the North American mid-continent. *Geological Society of America Bulletin*, **96**, 746–754.

NESBITT, H. W. & YOUNG, G. M. 1982. Early Proterozoic climates and plate motions inferred from major element chemistry of lutites. *Nature*, **299**, 715–717.

—— & YOUNG, G. M. 1989. Formation and diagenesis of weathering profiles. *Journal of Geology*, **97**, 129–147.

NISBET, E. G., MARTIN, A., BICKLE, M. J. & ORPEN, J. L. 1993. The Ngezi Group: Komatiites, basalts and stromatolites on continental crust. *In*: BICKLE, M. J. & NISBET, E. G. (eds) *The Geology of the Belingwe Greenstone Belt: a Study of the Evolution of Continental Crust*. Geological Society of Zimbabwe Special Publications, **2**, 121–165.

NUTMAN, A. P., FRYER, B. J. & BRIDGWATER, D. 1989. The early Archean Nulliak (supracrustal) assemblage, northern *Labrador*. *Canadian Journal Earth Sciences*, **26**, 2159–2168.

——, KINNY, P. D., COMPSTON, W. & WILLIAMS, I. S. 1991. SHRIMP U-Pb zircon geochronology of the Narryer Gneiss Complex, Western Australia. *Precambrian Research*, **52**, 275–300.

OJAKANGAS, R. W. 1983. Tidal deposits in the early Proterozoic basin of the Lake Superior region-the Palms and Pokegama Formations: evidence for subtidal-shelf deposition of Superior-type banded iron-formation. *In*: MEDARIS, L. G. Jr. (ed.) *Early Proterozoic Geology of the Great Lakes Region*. Geological Society of America, Memoirs, **160**, 49–66.

PAPENFUS, J. A. 1964. The Black Reef Series in the Witwatersrand basin with special reference to its occurrence at Government Gold Mining Areas. *In*: HAUGHTON, S. H. (ed.) *The Geology of Some Ore Deposits in Southern Africa*, Geological Society of South Africa, Johannesburg, 191–218.

PERCIVAL, J. A. & WILLIAMS, H. R. 1989. Late Archean Quetico accretionary complex, Superior Province, Canada. *Geology*, **17**, 23–25.

PHILLIPS, G. N. & MYERS, R. E. 1989. The Witwatersrand gold fields: Part 2. An origin for Witwatersrand gold during metamorphism and associated alteration. *In*: KEAYS, R. R., RAMSAY, W. R. H. & GROVES, D. I. (eds) *The Geology of Gold Deposits: the perspectives in 1988*. Economic Geology Monograph, **6**, 598–608.

RETIEF, E. A., COMPSTON, W., ARMSTRONG, R. A. & WILLIAMS, I. S. 1990. Characteristics and preliminary U-Pb ages from Limpopo Belt lithologies. *In*: VAN REENEN, D. D. & ROERING, C. D. (eds) *The Limpopo Belt: A Field Workshop on Granulites and Deep Crustal Tectonics*. Rand Afrikaans University Publications, 289.

RICHARDSON, C. J., CANN, J. R., RICHARDS, H. G. & COWAN, J. G. 1987. Metal-depleted root zones of the Troodos ore-forming hydrothermal systems. *Earth and Planetary Science Letters*, **84**, 243–253.

ROBB, L. J. & MEYER, F. M. 1991. A contribution to recent debate concerning epigenetic versus syngenetic mineralization processes in the Witwatersrand Basin. *Economic Geology*, 86, 396–401.

——, DAVIS, D. W. & KAMO, S. L. 1990. U-Pb ages on single detrital zircon grains from the Witwatersrand Basin, South Africa: constraints on the age of sedimentation and on the evolution of granites adjacent to the basin. *Journal of Geology*, **98**, 311–328.

ROSCOE, S. M. 1973. The Huronian Supergroup, a Paleoaphebian succession showing evidence of atmospheric evolution. *In*: YOUNG, G. M. (ed.) *Huronian Stratigraphy and Sedimentation*. Geological Society of Canada, Special Paper **12**, 31–47.

ROY, A. B. & PALIWAL, B. S. 1981. Evolution of Lower Proterozoic epicontinental deposits: stromatolite-bearing Aravalli rocks of Udaipur, Rajasthan, India. *Precambrian Research*, **14**, 49–74.

SAAGER, R., STUPP, H. D., UTTER, T. & MATTHEY, H. O. 1986. Geological and mineralogical notes on placer occurrences in some conglomerates of the Pongola Sequence. *In*: ANHAEUSSER, C. R. & MASKE, S. (eds) *Mineral Deposits of Southern Africa*, Geological Society South Africa, Johannesburg, 473–487.

SCHIDLOWSKI, M., HAYES, J. M. & KAPLAN, I. R. 1983. Isotopic inferences of ancient biochemistries: carbon, sulfur, hydrogen, and nitrogen. *In*: SCHOPF, J. W. (ed.) *Earth's Earliest Biosphere: Its Origin and Evolution*. Princeton University Press, Princeton, 149–186.

SCHOPF, J. W. 1993. Microfossils of the Early Archean Apex Chert: New evidence of the antiquity of life. *Science*, **260**, 640–646.

—— & PACKER, B. M. 1987. Early Archean (3.3-billion to 3.5-billion-year-old) microfossils from Warrawoona Group, Australia. *Science*, **237**, 70–73.

—— & WALTER, M. R. 1983. Ancient microfossils: new evidence of ancient microbes. *In*: SCHOPF, J. W. (ed.) *Earth's Earliest Biosphere: Its Origin and Evolution*. Princeton University Press, Princeton, 214–239.

SHEGELSKI, R. J. 1980. Archean cratonization, emergence and red-bed development, Lake Shebandowan area, Canada. *Precambrian Research*, **12**, 331–347.

SIMONSON, B. M. 1985. Sedimentological constraints on the origins of Precambrian iron-formations. *Geological Society of America Bulletin*, **96**, 244–252

—— & GOODE, A. D. T. 1988. First discovery of ferruginous chert arenites in the Early Precambrian Hamersley Group of Western Australia. *Geological Society of America, Abstracts with Programs*, **20**, A206.

——, SCHUBEL, K. A. & HASSLER, S. W. 1993. Carbonate sedimentology of the early Precambrian Hamersley Group of Western Australia. *Precambrian Research*, **60**, 287–335.

SIMPSON, E. L., DRIESE, S. G. & ERIKSSON, K. A. 1993. 1.8 Ga paleosols from the Lochness Formation, Mount Isa Inlier, Australia: Paleoclimatic implications. *Geological Soceity America, Abstracts with Program*, **25**, A69.

SRINIVASAN, R. & OJAKANGAS, R. W. 1986. Sedimentology of quartz-pebble conglomerates and quartzites of the Archean Babudan Group, Dharwar Craton, South India: Evidencve for early crustal stability. *Journal of Geology*, **94**, 199–214.

STANTISTREET, I. G. & McCARTHY, T. S. 1991. Changing tectono-sedimentary scenarios relevant to the development of the Late Archaean Witwatersrand Basin. *Journal of African Earth Sciences*, **13**, 65–81.

SUTTON, S. J., RITGER, S. D. & MAYNARD, J. B. 1990. Stratigraphic control of chemistry and mineralogy in metamorphosed Witwatersrand quartzites. *Journal of Geology*, **98**, 329–341.

TANKARD, A. J., JACKSON, M. P. A., ERIKSSON, K. A., HOBDAY, D. K., HUNTER, D. R. & MINTER, W. E. L. 1982. *Crustal Evolution of Southern Africa: 3.8 Billion Years of Earth History*. Springer-Verlag, New York.

TAYLOR, P. N., CHADWICK, B., MOORBATH, S., RAMIKRISHNAN, M. & VISWANATHA, M. N. 1984. Petrology, chemistry and isotopic ages of Peninsula gneiss, Dharwar acid volcanics and the Chitradurga granite with special reference to the late Archean evolution of the Karnataka craton, Southern India. *Precambrian Research*, **23**, 349–375.

——, KRAMERS, J. D. MOORBATH, S., WILSON, J. F., ORPEN, J. L. & MARTIN, A. 1991. Pb/Pb, Sm-Nd and Rb-Sr geochronology in the Archean Craton of Zimbabwe. *Chemical Geology*, **87**, 175–196.

TAYLOR, S. R. & McLENNAN, S. M. 1985. *The Continental Crust: Its Composition and Evolution*. Blackwell Scientific Publications.

——, RUDNICK, R. L., McLENNAN, S. M. & ERIKSSON, K. A. 1986. Rare earth element patterns in Archean high- grade metasediments and their tectonic significance. *Geochimica et Cosmochimica Acta*, **50**, 2267–2279.

THURSTON, P. C. & CHIVERS, K. M. 1990. Secular variation in greenstone sequence development emphasizing Superior Province, Canada. *Precambrian Research*, **46**, 21–58.

TRELOAR, P. J. 1988. The geological evolution of the Magondi Mobile Belt. *Precambrian Research*, **38**, 55–73.

TRENDALL, A. F., COMPSTON, W., WILLIAMS, I. S., ARMSTRONG, R. A., ARNDT, N. T., McNAUGHTON, N. J., NELSON, D. R., BARLEY, M. E., BEUKES, N. J., deLAETER, J. R., RETIEF, E. A. & THORNE, A. M. 1990. Precise zircon U-Pb chronological comparison of the volcano-sedimentary sequences of the Kaapvaal and Pilbara Cratons between about 3.1 and 2.4 Ga. *Third International Archaean Symposium*, Perth, 81–83.

TRUSWELL, J. F. & ERIKSSON, K. A. 1973. Stromatolitic associations and their palaeoenvironmental significance: a re-appraisal of a Lower Proterozoic locality from the northern Cape Province, South Africa. *Sedimentary Geology*, **10**, 1–23.

VEIZER, J. 1983. Geologic evolution of the Archean-Early Proterozoic Earth. *In*: SCHOPF, J. W. (ed.) *Earth's Earliest Biosphere: Its Origin and Evolution*. Princeton University Press, Princeton, 240–259.

—— & JANSEN, S. L. 1979. Basement and sedimentary recycling and continental evolution. *Journal of Geology*, **87**, 341–370.

——, COMPSTON, W., HOEFS, J. & NIELSEN, H. 1982. Mantle buffering of the early oceans. *Naturwissenschaften*, **69**, 173–180.

WALKER, J. C. G. 1990. Precambrian evolution of the climate system. *Palaeogeography, Palaeoclimatology, Palaeoecology*, **82**, 261–289.

WALSH, M. M. 1992. Microfossils and possible microfossils from the early Archean Onverwacht Group, Barberton Mountain Land, South Africa. *Precambrian Research*, **54**, 271–294.

—— & LOWE, D. R. 1985. Filamentous microfossils from the 3500 Myr-old Onverwacht Group, Barberton Mountain Land, South Africa. *Nature*, **314**, 530–532.

WALTER, M. R. 1983. Archean stromatolites: evidence of the Earth's earliest benthos. *In*: SCHOPF, J. W. (ed.) *Earth's Earliest Biosphere: Its Origin and Evolution*. Princeton University Press, Princeton, 187–213.

——, BUICK, R. & DUNLOP, J. S. R. 1980. Stromatolites 3400–3500 Myr old from the North pole area, Western Australia. *Nature*, **284**, 443–445.

WIGGERING, H. & BEUKES, N. J. 1990. Petrography and geochemistry of a 2000–2200 Ma old hematitic paleoweathering alteration profile on Ongeluk Basalt of the Transvaal Supergroup, Griqualand West, South Africa. *Precambrian Research*, **46**, 241–258.

WILKS, M. E. & NISBETT, E. G. 1988. Stratigraphy of the Steep Rock Group, northwest Ontario: a major Archean unconformity and Archean Stromatolites. *Canadian Journal of Earth Sciences*, **25**, 370–391.

WILSON, J. F., BICKLE, M. J., HAWKESWORTH, C. J., MARTIN, A., NISBETT, E. & ORPEN, J. L. 1978. Granite-greenstone terrains of the Rhodesian Archaean craton. *Nature*, **271**, 23–27.

WRONKIEWICZ, D. J. & CONDIE, K. C. 1987. Geochemistry of Archean shales from the Witwatersrand Supergroup, South Africa: Source-area weathering and provenance. *Geochimica et Cosmochimica Acta*, **51**, 2401–2416.

—— & —— 1989. Geochemistry and provenance of sediments from the Pongola Supergroup, South Africa: Evidence for a 3.0-Ga-old continental craton. *Geochimica et Cosmochimica Acta*, **53**, 1537–1549.

—— & —— 1990. Geochemistry and mineralogy of sediments from the Ventersdorp and Transvaal Supergroups, South Africa: Cratonic evolution during the early Proterozoic. *Geochimica et Cosmochimica Acta*, **54**, 343–354.

YOUNG, G. M. 1983. Tectono-sedimentary history of early Proterozoic rocks of the northern Great Lakes region. *In*: YOUNG, G. M. (ed.) *Huronian Stratigraphy and Sedimentation*. Geological Society of America, Memoirs **160**, 15–32.

Archaean ecology: a review of evidence for the early development of bacterial biomes, and speculations on the development of a global-scale biosphere

E. G. NISBET

Department of Geology, Royal Holloway, University of London, Egham, Surrey TW20 0EX, UK

Abstract: The antiquity of the carbon cycle, which began prior to 3.5 Ga ago, implies that a complex and diverse biosphere has existed for most of the Earth's history. The earliest living community may have consisted of a variety of chemotrophic bacteria, including both archaea and eubacteria, that existed around hydrothermal systems. It is possible that photosynthesis began in thermophilic bacteria that originally developed infra-red thermotaxis for detecting hot vents, maximizing the chance of survival in the close vicinity of hydrothermal systems. The post-photosynthetic biosphere may have inhabited a set of distinct habitats, including marine littoral, open ocean, terrestrial and hydrothermal biomes. Sequestration of oxidation power, by partitioning into zones of oxidation and reduction, may have allowed an inflationary biosphere to develop, limited ultimately by the availability of crucial nutrients such as phosphorus. Biomes develop and collapse, and, in consequence, selective pressures over time would have created a cooperative, Gaian biosphere.

Discussions of Hadean (pre-life) and Archaean (post-life) geology tend to concentrate on the evolution of the Earth's crust and mantle: the Archaean biological record is often regarded as the product of a veneer on the surface of the planet, historically interesting but not important in the long-term physical evolution of the Earth. In this review, an attempt is made to show that the biological impact on the early evolution of the planet was massive, and that a truly global biosphere existed from the earliest time, which played a major role in determining the physical and chemical evolution of the planet.

The review first examines the home of the earliest life, and how life may have been shaped by its home. Then the origin of photosynthesis is examined, and the subsequent colonisation of the planet by communities of life. Finally, the impact of those communities on the planet itself is examined, with a discussion of the extent to which the communites could have been interactive.

The modern biosphere is global in scope, consisting of a set of distinct though interconnected biomes, such as the boreal forest, the tundra, or the tropical rainforest. Collectively, these biomes play a major role in the management of the environment, helping to regulate the atmosphere and the climate. The productivity of the modern biosphere is mainly from land, and the ocean plays a subsidiary role. Moreover, today the productivity is based mainly on plants: eukaryotes. The pro-

karyotic substratum exists in all biomes, from the anaerobic bacteria in submarine ooze to the methanogens of boreal peat bogs, but, although remaining fundamental to the operation of the biosphere, does not today directly dominate its productivity.

In the Archaean, in contrast, the biological community was mainly marine and mainly prokaryotic. Eukarya may have been present, but single-celled. In the discussion that follows, an attempt is made to imagine the scope of this Archaean bacterial biosphere. It is argued that the Archaean bacterial biosphere, though less productive than the modern system, was also global in scope, and was also capable of managing the atmosphere and climate of the Earth. By so doing, it helped to set the surface temperature, managed the atmosphere, and may have helped to stabilize the presence of liquid water on the surface.

The home of the earliest life

The pre-biotic Earth's atmosphere was clearly very different from that today, but there is little consensus on exactly how different it was. The classical picture of a reducing atmosphere rich in methane and ammonia gained much support from the Miller–Urey experiment to synthesize pre-biotic molecules, but is photochemically unlikely, although low but significant concentrations of reducing species may have existed.

From COWARD, M. P. & RIES, A. C. (eds), 1995, *Early Precambrian Processes*, Geological Society Special Publication No. 95, pp. 27–51.

Earth has had oceans throughout the time of the geological record, and it is likely that the total inventory of water on the planet's surface has been of the order of 350 bars (= 3.5 km depth) or more (perhaps 500 bars in early time) since the Hadean. The D/H ratio of hydrogen in the atmosphere on Venus, Earth and Mars strongly implies that significant H loss occurred on all three bodies. If the original ratio were cosmic, the current D/H enrichments are a factor of 10 for Earth, 60 for Mars and 1500 for Venus (e.g. see Hunten 1993). On Venus, this implies that a substantial ocean was lost (Donahue et al. 1982; Watson et al. 1984). Earth too may have undergone lesser but still significant loss of H, leaving a surplus of O_2. The next most abundant volatile is carbon, as CO_2, CO and CH_4. The total carbon inventory on the Earth's surface today, around 60–80 bars if expressed as CO_2, is comparable to the carbon inventory on the surface of Venus. How much was on or in the Earth's crust and how much in the mantle in Hadean time is a matter of much debate. Given that water and noble gases clearly degassed quickly, it is arguable that, although chemically different, significant carbon degassing is likely also to have taken place. On the Earth, by the end of the Hadean an atmosphere of, say, 10 bars CO_2 + CO is not unlikely (Kasting 1993). Nitrogen must have degassed from the interior, and would have been present as ammonium minerals in lavas. In an atmosphere with abundant lightning and with water present, N_2 may have been fixed out of the air to become dissolved NO_3^- in the water and may have been recycled back via hydrothermal systems (Summers & Chang 1993) and also by photochemical reduction (Kasting et al. 1993b). The dominant species in the early atmosphere were probably H_2O and CO_2. If temperatures were low and H_2O was present mainly as ocean, the air would have been mainly CO_2.

Other species in the air would have included O_2 formed after loss to space of H_2, and possibly minor photochemical CO and CH_4 (Wen et al. 1989). The amount of oxygen present in the atmospheric burden is difficult to assess. It is likely that at least trace amounts were present, photolytically made, but whether the early air was oxidising (as opposed to neutral) is unknown. If significant H loss to space occurred, as implied by the D/H ratio, the build-up of more than trace amounts of oxygen (and, conversely, of reduced species such as methane and ammonia) would depend on how fast the oxygen was taken up by atmospheric and surface oxidation, and by interaction with the mantle, via volcanism and subduction. If oxygen removal were rapid, as is likely (i.e. active tectonic and volcanic exposure of new rock and magma), the atmospheric oxygen content would have been

very low; however, if removal were slow, trace atmospheric oxygen content might have been sustained over tens and hundreds of millions of years.

A hot mantle would have degassed water quickly, rapidly forming oceans and partitioning water outside the mantle, and would have stabilised with the mantle geotherm and melting regime involving voluminous deep melting in a near-dry mantle. Komatiites flowing as lava, for instance, would have carried little water (Green 1975) although they may have conveyed it to the surface in explosive eruptions.

The water content of the early mantle is not known. It can be argued that the mantle is essentially self-fluxing over time. In this argument, in the Hadean Earth, melting in a hot mantle would have taken place easily and rapidly, dissipating heat to space. In the Archaean, with a well-developed solid lithosphere (Nisbet 1987) and some system of plate tectonics developing as a result of the cold surface (itself thermally managed by the atmosphere), the ridges would express melt generated from hot, fairly dry mantle. If anomalously water-rich patches of mantle existed for some reason, abundant melt would occur, removing the water to the surface. The re-entry of water to the mantle down subduction zones would have been limited thermally by the temperature regime of the overlying mantle wedge. In the hot Archaean geothermal regime when surface water descended the Archaean subduction zones, it would have been almost completely driven off and consequently would have acted as a flux to promote melting in the overlying wedge, returning the water to the surface. The same process still takes place today. However, on the modern slightly cooler Earth, slightly more water may now remain with the descending slab and return to the deep mantle. If the mantle has cooled slowly as the Earth aged, it may have progressively accepted more water down the subduction system: this water entered and mixed into the mantle, steadily slightly rehydrating it and also oxidizing it as H_2 outgassed (Kasting et al. 1993a). In this way, gradual rehydration of the mantle helps to flux the sustained operation of the mid-ocean ridges. In future, more water will enter the mantle, and continuing ridge operation in a cooler mantle will depend on increasing water content of the bulk mantle as the ocean is returned to the interior. To quote a remark by W.S. Fyfe: 'degas when hot, engas when cool'.

These are speculations: the facts are few. In the following discussion, the following assumptions, based on best current knowledge, are made.

(1) Life began roughly 4000 ± 200 Ma ago.

(2) The first fully organic life was a simple organism, probably little more than RNA in an

enclosing bag, that used the kinetics of chemical exchange with its surrounding environment to maintain a fitful disequilibrium with its setting. Chemotrophy came first. Photosynthesis, much more difficult, came later.

(3) When life began, the Earth possessed deep oceans (3.5–5 km deep).

(4) The atmosphere was predominantly CO_2, possibly at 1–10 bars pressure.

(5) The Earth was tectonically active, and had become sufficiently thermally organised in the late Hadean to sustain a form of plate tectonics with lengthy mid-ocean ridges, as opposed to a more chaotic system of heat loss in individual plumes. Subduction zones existed. Shallow water and subaerial (meteoric water) hydrothermal systems existed around arc and plume volcanoes.

(6) The Moon was present, and tides occurred.

(7) Land areas existed, with weathering and run-off contributing chemical species such as Ca^{++} to the ocean water.

(8) The chemistry of the seas, controlled by hydrothermal circulation at ridges and by erosion from land, was broadly similar to today. As a long-lived moist greenhouse was avoided, sea temperature, as a global average, was unlikely to have been above 40–50°C by 4 Ga ago.

(9) There was no pre-biotic soup. The bulk chemistry of the sea was geochemically controlled by hydrothermal exchange, and perhaps also by erosional input. The vigour of the late Hadean and early Archaean hydrothermal systems was probably very great, enough to cycle a volume of water equal to the total volume of the oceans in a period of the order of, say, a million years. In such circumstances it is unlikely any significant pre-biotic soup developed.

The grounds for these assumptions are complex and too detailed to justify here: a more lengthy discussion is given in Nisbet (1987, 1991).

Environmental resources of pre-photosynthetic life

There is excellent molecular evidence that the first life forms were not photosynthetic (e.g. see Achenbach-Richter *et al*. 1987, 1988. Kandler 1992), but rather depended on chemical reactions for their energy. Photosynthesis is complex and the product of much evolutionary accident; early organisms instead were chemotrophic. To live, they needed to use local chemical disequilibrium, most probably at hydrothermal systems where the mantle interacted with the ocean–atmosphere system. Here, lavas from the interior would have been in chemical contact and oxidation contrast with water

and gases that were exposed to photochemistry and suffered H loss to space.

What resources did the first life have? Before photosynthesis, all life would have depended directly on the ability to use chemical oxidation. For this to be possible, there must have been an oxidation contrast. The best sites would have been where there was contrast between the mantle, the ultimate source of magma, and the atmosphere/hydrosphere, the source of fluids interacting with the crust and providing oxidation potential. Most probably the supply of reductants was mediated via hydrothermal systems (see Tunnicliffe 1991 for a review of the operation of hydrothermal systems), themselves supplied with seawater oxidized by contact with the atmosphere (Fig. 1). Formation of pyrite and other sulphides in hydrothermal settings may have provided sources of reducing power.

It is possible that other more distant habitats (e.g. tidal waters) could have been supplied with chemical species transported in the sea and could have provided settings for the first life. However, the availability of substrates for life in the near-hydrothermal environment must be a persuasive argument in favour of the hypothesis that the first life (the RNA world), was in the proximity of a hydrothermal field (e.g. see Nisbet 1986, 1987). A likely setting would have been at shallow or sub-aerial level in a hydrothermal system, possibly in shallow water around a large komatitite plume volcano, with near-neutral pH (as opposed to acid mid-ocean ridge systems), temperatures in large parts of the system around 35–40°C, but locally ranging up to boiling, with regions around 80°C, and perhaps bicarbonate-based chemistry. Another potential site is in tidal muds, but here the resources of oxidation contrast would have been much more limited.

The other prerequisite resources are the chemicals of life: most are common, but some, especially phosphorus, sulphur, nickel, magnesium and usable nitrogen are rarer and their availability would have limited the total early Archaean biomass.

First consider the supply of oxidants. Prior to photosynthesis, the first biotic communities must have depended on the natural supply of oxidation power in the pre-biotic world. Various models of planetary accretion give varying heterogeneity, but the likelihood of a late giant impact makes it plausible that, after precipitation of the iron-rich core, the earliest Earth's mantle would have been well homogenised in terms of internal oxidation state. The contrast comes from the air. In the atmosphere, early hydrogen and deuterium loss may have been considerable. In the first 1 Ga of the planet's history, Yung *et al*. (1989) estimate from the

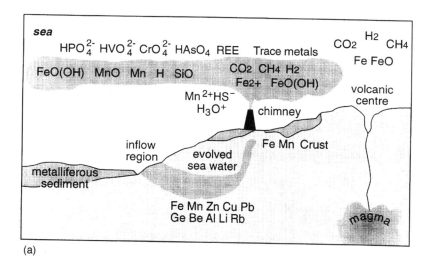

(a)

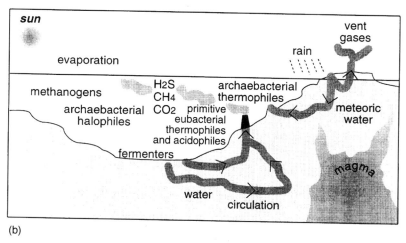

(b)

Fig. 1. Operation of a hydrothermal system. (a) The chemical environment. Modified from various sources, especially from Tunnicliffe (1991). Seawater percolating into the system heats and leaches a wide variety of elements by reaction with host hot rock. The hot acid solution exits from vents. Chimneys form at exit vents; sulphates precipitate first from seawater, then sulphide deposits rich in copper or zinc. Plumes form in the water above the vent, spreading out several hundred metres over the bottom. This illustration is based on processes at a mid-ocean ridge, but similar processes occur in and around shallow water vents. (b) Possible setting of an early pre-photosynthetic biological community in a shallow hydrothermal pool near an andesitic volcano, subject to evaporation: hot, acid and salty. Hyperthermophiles may have inhabited regions close to vents and surface fumaroles; further away from direct volcanic heat, methanogens and halophiles could have existed. Eubacterial fermenters may have exploited available resources. Heat shock would have been a common danger. Other environments would also have existed. In the Archaean, komatiite volcanism would have been widespread, forming very wide shield volcanoes and islands similar to Hawaii but wider and lower. Hydrothermal circulation around komatiite flows could locally also have produced Ni-rich fluids. Under 3–5 km of water, vent temperatures exceed 350°C. Because thermal gradients near the hydrothermal exit vents are very sharp, bacteria that are thermotactic have a selective evolutionary advantage. Thermotaxis by infra-red detection evolves, preadapting bacteria for optional photosynthesis at higher levels in the sea, using H_2S or H_2, from the hydrothermal gases.

present oceanic D/H ratio that the equivalent of up to 36% of the present ocean (up to more than 1 km of water) might have been lost in the past, or very roughly 100 bars of oxygen. This figure is, however, subject to many unknowns. A lower limit of the amount of H_2O lost can be calculated assuming Deuterium did not escape: if so, the amount of water lost is 8.3% of the present ocean (about 300 m). This escape of hydrogen, although considerable, is within the limits of the input flux of hydrogen to the early atmosphere. Thus it is likely, though not certain, that there was a significant source of atmospheric oxygen derived from hydrogen loss to space.

In the modern air,

$$O_3 \xrightarrow{hv} O(^1D) + O_2$$

$$O(^1D) + H_2O \longrightarrow 2\ OH\cdot$$

The distribution of OH in the modern atmosphere depends on the availability of O derived from O_3, as well as on many other species such as CH_4. If oxygen and hence ozone were only trace species, and UV penetrated into the moist troposphere, it is possible that photochemical OH

$$H_2O \xrightarrow{hv} H + OH\cdot$$

was a trace but important factor in the Archaean tropospheric air (Wen *et al.* 1989); alternatively, if the partial pressure of oxygen were not minimal, OH may have been produced by the modern ozone-mediated chain.

Even if it is assumed that by loss of hydrogen to space about 100 bars of oxygen were liberated, in total, in the atmosphere through the Hadean (i.e. loss of about a third of the present oceanic hydrogen content; loss of about 1 km of water), this does not mean that the air was necessarily rich in oxygen. The standing atmospheric burden of oxygen would have depended on the balance between the creation of oxygen by H loss and rate of removal of oxygen, via seawater and hydrothermal systems into new oceanic crust and hence via subduction into the mantle. Nevertheless, it is highly likely that the dominant carbon gas was carbon dioxide, not CO or CH_4. This is an important conclusion. In pre-photosynthetic times, at the end of the Hadean and the start of the Archaean, dissolved CO_2 atmospheric oxygen and other oxidised species are the most likely supply of oxidation power in the seas, and contrast between this relatively oxidised atmosphere and relatively reduced magmas from the mantle would have provided an adequate basis on which to build a biosphere. Most likely, the atmosphere was neither reducing (barite evaporites are known from the Pilbara: see Schidlowski 1989 and discussion

below) nor strongly oxidizing. It is possible that there was a low but nevertheless significant disequilibrium trace component of oxygen.

The needs of the biosphere are complex, and even in the most fruitful location, the hydrothermal systems, the earliest biosphere would soon have been resource-limited, very shortly after reproducing biological activity began. Life uses resources to their limits. The early chemotrophic organisms had various options, depending on oxidation of NH_3, NO_2^-, H_2, reduced sulphur, CO or ferrous iron. Pyrite formation may have been the first reaction to sustain life (the 'Wachtershauser reactions' Wachtershauser 1990; see discussion in Kandler 1992).

The nitrogen and sulphur cycles must be of great antiquity. Lovelock (1988) has pointed out the constraints on the biosphere prior to the development of the nitrogen cycle. Nitrogen is fixed by lightning, and it is reasonable to infer that the early atmosphere, over a large ocean, sustained thunderstorms: if so, the pre-biotic atmosphere over some millions of years may have lost much of any initial complement of N_2, and nitrogen would have been present in the surface system perhaps as NH_4^+ in the oceans. Magma from the mantle carries nitrogen, which would be added via seawater interaction to the oceans, or precipitated as ammonium minerals in zeolite-facies metamorphism of mafic rock. Lovelock (1988) suggested that, had the atmosphere been CO_2 rich, as is likely, the upper levels of the sea would have been somewhat more acid than today, and perhaps rich in ferrous iron (depending on the supply of oxidant). If so, the ferrous iron may have sequestered the ammonium, leading to a global nitrogen shortage except in the proximity of hydrothermal fluids. Only when the biological nitrogen cycle had fully evolved, with denitrifying bacteria, would N_2 begin to build up in the air, raising air pressure, with greenhouse implications for the carbon gases.

The sulphur cycle is equally important. Schidlowski (1989) has argued persuasively for the antiquity of dissimilatory sulphate reduction, at least predating 2.8 Ga, and has pointed out that the existence of a marine sulphate reservoir as early as 3.5 Ga ago is indisputable. Schidlowski (1989) comments that the evidence is 'consistent with the onset of a biologically mediated sulphur oxidation in the surficial exchange reservoir and the build-up of oceanic sulphate as from at least 3.5 Ga ago'.

Life needs phosphorus, which is mobilised in hydrothermal systems, but does not typically travel far, being reprecipitated. It is also present in continental alkaline volcanic rocks, and a small river-borne supply may have been contributed by the land, even in the early Archaean. But in general, phosphorus must have been difficult to obtain across much of the planet. This is perhaps an argument in favour of the hypothesis that the earliest biotic community must have been physically close to available phosphorus (possibly at the vent of a hydrothermal system, which, though poor in P, would have had more available than other settings); only later, as a complex biota developed could phosphorous be recycled and transported biologically.

'A warm little pond': the first, pre-photosynthetic, biome?

The site where life began is a matter for a different debate, not to be considered in detail here (for discussion see Nisbet 1986, 1987). However, there are strong molecular grounds (e.g. see Achenbach-Richter *et al.* 1987, 1988) for concluding that hydrothermal systems must have been the host sites of the first, pre-photosynthetic biome, where I define an Archaean biome as a complex ecological community, of a scale that had global impact. This hypothesis is not new: it dates back to the voyage of *HMS Challenger* and to Charles Darwin's famous letter to Hooker (published in 1959) 'But if (and oh, what a big if) we could conceive in some warm little pond, with all sorts of ammonia and phosphoric salts, light, heat, electricity, etc., present …'.

In no other pre-photosynthetic environment is it likely that there was sufficient flux of oxidants and chemicals, and adequate temporal stability to sustain the evolution of a complex community. It is possible that tidal muds would have a flux of varied chemical species, moved by the tides, and adequate temporal stability, but in contrast to the hydrothermal setting the likelihood of the tidal environment hosting a complex pre-photosynthetic community is small (see later).

It is possible to imagine an early hydrothermal 'world', in which the ambient environment supplies seawater containing species such as dissolved CO_2 and SO_4^{2-} to hot rock (Fig. 1). At depth, by reaction with basalt, the resulting hydrothermal liquid is enriched in a variety of species such as S^{2-} (from the reduction of sulphate and by leaching sulphur from the rocks) and in various leaching processes Fe^{2+}, Mn^{2+}, Ni^{2+}, H^+ are formed. This varied hydrothermal fluid rises to sustain a bacterial community in the upper levels of the volcanic system, based on chemotrophy. Such processes can occur both in deep water mid-ocean ridge settings and also in shallower and subaerial conditions around andesite volcanoes. Initially, the population in this 'world' may have been of a single ancestral progenotic cell type, accidentally spread around by the vagaries of currents and volcanic explosions. Later, accidental mutation and adaptation to habitat may have given rise to two, possibly three distinct

bacterial types, ancestors of the eubacteria and the archaea, and perhaps the eukarya. Then the archaea diverged into two archaebacterial 'phyla' (Woese 1985). One of these would have been sulphur dependent and thermophilic, the other occupying hot salty environments and giving rise to the methanogens. The ten or so eubacterial phyla descended from the ancestral eubacterium, probably a thermophile, may also have evolved at this stage (Woese 1985).

From these beginnings evolved a diverse array of prokaryotes, both archaea and eubacteria (Fig. 2). Both domains of the bacterial community are today very varied. Apart from those bacteria oxidizing hydrogen, sulphur, iron and manganese, nitrifying and denitrifying, sulphur-oxidizing bacteria and sulphur and sulphate reducers occur, as well as methanogens and halobacteria. In modern hydrothermal communities, nitrate respiration occurs (Hentschel & Felbeck 1993). In the early Archaean, on the organic substrates provided by the activities of the first bacteria, other fermenting bacteria would have developed. Bacteria such as *Desulfuromonas acetoxidans* can grow on acetate and elemental sulphur as the sole source of energy (Thauer & Morris 1984):

$$\text{acetate}^- + 4S° + 2H_2O + H^+ \longrightarrow 2CO_2 + 4H_2S.$$

Collectively, a diverse community is built upon fermentation.

The origin of the eubacteria took place very early in life's history. Achenbach-Richter *et al.* (1987), investigating eubacterial ribosomal RNA, found that in the eubacteria the deepest divergence is between the hyperthermophilic species *Aquifex, and Thermotoga* and all other eubacteria. The implication is that the most primitive eubacteria were hyperthermophiles. Interestingly, *Aquifex* and *Hydrogenobacter*, the most deeply rooted eubacteria, need free O_2 (see Kandler 1992 for a discussion of this). Possibly this is not a primitive characteristic, but rather a product of later evolution. Nevertheless, these hyperthermophiles may indeed preserve early Archaean metabolism, and may have evolved in an environment with free O_2 generated by H loss to space (see discussion above in this work). Green non-sulphur bacteria may also be very old. It is intriguing to speculate on the origin of the streamlined eubacterial genome, which has either removed introns from an originally noisy DNA, or possibly is of the greatest antiquity, if introns came later as archaea evolved. However, there is a strong argument for the first hypothesis, that introns date back to an original RNA world, likely to have been much noisier. If so, it is possible that the eubacterial deletion of introns may have evolved as a more rapid and less accident-prone method of reproduction in response

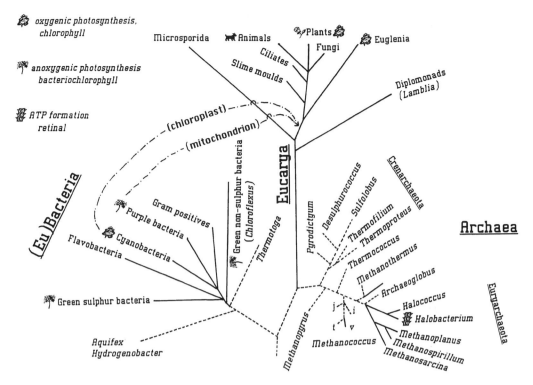

Fig. 2. The domains of life. Phylogenetic tree, showing the three domains of life (Woese *et al.* 1990), with branch order and lengths based on 16S rRNA sequence comparisons (Kandler 1994). *Dashed lines* show the descent of hyperthermophiles (living at temperatures close to 100°C). *Solid lines* show descent of organisms that live at lower temperatures. In both the archaea and the (eu)bacteria the most deeply rooted organisms are hyperthermophilic. In the archaea, deeply rooted branches include those leading to organisms such as *Sulfolbus* in the Crenarchaeota, and in the Euryarchaeota to *Thermococcus* and *Methanopyrus*. Halobacteria such as *Halobacterium* as well as other methanobacteria that are not hyperthermophiles appear to have evolved later. In the (eu)bacteria also, the hyperthermophiles such as *Thermotoga* and *Hydrogenobacter* appear to lie on the most deeply rooted branches. The root of the Eucarya branch appears to lie also in the hyperthermophiles. Source of tree: Kandler (1994) and references cited therein.

either to assault by heat or chemicals in a hydrothermal setting, or to the UV flux in shallow water away from an early rock-hosted setting.

Archaea, otherwise known as archaebacteria, are uniquely suited to inhabit an imaginary early hydrothermal setting. The thermophilic archaea are diverse, including some extreme thermophiles that have optimum growth at over 80°C (Stetter 1986) Of the thermoacidophiles (Sundarm 1986), which live in the pH range 1.5–4, *Sulfolobus,* for instance, can grow heterotrophically or autotrophically by oxidizing elemental sulphur or iron, and *Thermoproteus* is anaerobic and sulphur respiring. An important archaebacterial group that can occur in a hydrothermal setting is the methanogens, which convert CO_2, H_2 or fermentation products formed by other anaerobes such as formate or acetate to methane or methane and CO_2. Methanogens (Stanier *et al.*, 1987) seem to be universally capable of using elemental sulphur as electron acceptor, respiring by sulphur reduction: some cease methanogenesis when grown in the presence of sulphur while others can continue. Autotrophs, most grow well with CO_2 as the sole carbon source. They occur in highly reducing habitats.

On a global scale, the impact of this biological activity at volcanic centres may have been considerable. Direct removal of carbon from the water/atmosphere system is possible, with the carbon eventually incorporated into the oceanic crust and returned into the mantle via subduction. Nisbet *et al.* (1994) have suggested that this may be the source of some diamonds with isotopically very light carbon and 'ophiolite-like' mineral inclusions, which may have begun life as hydrothermal clumps of bacteria at mid-ocean ridges, and then descended down subduction zones to become clumps of carbon attached to the base of the lithosphere.

Could a hydrothermally-based biosphere have had a large enough impact on the atmosphere to change global climate? This depends on the balance of other processes in the atmosphere, as well as on the limitations imposed on biological activity by the availability of oxidation power and of phosphorus (probably more easily accessible in the hydrothermal setting than anywhere else) and other essential trace chemicals. Walker (1977) took the rate of burial of organic carbon to be equal to the hydrogen production rate from volcanoes: the size of the biosphere is limited by the supply of reduced compounds as electron donors.

Even a limited biosphere can influence the global setting. The latest Hadean CO_2 budget in the atmosphere–ocean system would have been controlled by mainly inorganic processes: CO_2 would have been emitted by andesitic volcanoes, and consumed by inorganic precipitation on the sea floor. However, if, for example, in the early Archaean the consumption of CO_2 by a hydrothermal community tilted the global balance of supply/consumption, the atmospheric CO_2 burden would change until a new balance was attained, as the concentration in seawater fell. This suggests that biological control could have been important, by returning reduced carbon as dead bacterial bodies (stripped of nutrients such as phosphorus) to the sea floor and hence by subduction to the mantle. However, it should also be noted that there is good evidence for the existence of continents (zircons; Compston & Pidgeon et al. 1986; and a locally thick continental lithosphere; Nisbet 1987) in the early Archaean. CO_2 would have been consumed by inorganic processes of weathering and $CaCO_3$ precipitation, which would have overwhelmed the impact of pre-photosynthetic biology on the global carbon budget .

Methane is also important. It may have been emitted by the methanogens in a hydrothermally-based biosphere. The emission of methane from bacterial communities of methanogens may have had important global consequences, as this is a major greenhouse gas, and would also have affected the OH content of the air and consequently a variety of other gases. Although methane may not have been of major significance in the carbon balance, and would have been consumed by other bacteria its climatic impact may have been considerable. If oxygen were present in the atmosphere, its total atmospheric burden would have been limited by the availability of OH, and in a low oxygen atmosphere, methane may

have been a much more significant greenhouse gas than today. Furthermore, Lovelock (1988) has made the interesting suggestion that the release of CH_4 and H_2S to air would produce an atmospheric smog capable of absorbing ultra-violet light from the Sun in the way the ozone layer does today (but did not then, at low levels of O_2). Thus, although the hydrothermal biome may have been small in total net biological productivity, limited by the availability of habitat and oxidants, it is possible that it could nevertheless have had a global impact on climate, even as early as, say, 4 Ga ago.

The colonization of other habitats

An early community based on volcanic hydrothermal systems (Figs 1 & 2) would inevitably colonise outward, accidentally from the first volcano to nearby systems, forming new communities and then cross-exchanging newly evolved attributes and bacterial genera between communities. Whether the eubacteria pre- or post-dated the first archaea may never be known for certain, but it is possible that the eubacteria and archaea originally diverged from an extremely thermophilic ancestor (Achenbach-Richter et al. 1988). The diversification of the archaebacterial phyla may have proceeded with the colonization outward from an original hot, acid thermal setting, of shallow bicarbonate pools and hot shallow-water evaporative salty pools, giving rise to methanogens and halobacteria (e.g. Zillig et al. 1985; Bock et al. 1985). From the original ancestral line evolved a varied community of descendants that could generate energy by producing methane, or could occupy lower temperature or more oxidizing environments, or could survive in very salty water or carry out disimilatory sulphate reduction (Stackebrandt 1985).

Given an active hydrothermal biome, life would have soon spread to develop communities in other settings such as tidal mud flats, either based on local chemotropy (e.g. iron oxidising bacteria, exploiting the oxidation contrast between the atmosphere and plumes of reducing water containing ferrous iron liberated from volcanoes), or by utilizing other possibilities (e.g. nitrification, using NH_4^+), or by using as a substrate the organic debris from bacteria that grew at volcanic systems but were washed away by currents or eruption. One can imagine tidal communities of scattered bacteria using the tidal flux to provide themselves with nutrient from more distant sources.

Using light: the origins of photosynthesis

From chemolithoautotrophy, the early community of bacteria widened, using fermentation to increase diversity. Respiration and photosynthesis most probably came later. One form of photosynthesis is a simple light-driven proton pump such as in the archaebacterium *Halobacterium halobium*, which uses a protein known as bacteriorhodopsin. *Halobacterium* can generate ATP anaerobically under light, but cannot grow: the photophosphorylation is only used to prolong viability of non-growing cells (Stanier *et al.* 1987). The process occurs with deficient oxygen, but not if oxygen is absent: whether this is of palaeoenvironmental significance, and indeed the antiquity of the process, are moot points, but suggests early evolution in saline pools with low oxygen partial pressure. Bacteriorhodopsin is similar to rhodopsin, the visual pigment used by vertebrates, and it is possible to speculate (without evidence) that all eukaryote eyes share a common ancestry, inherited from an ancient ancestor which also passed the use of retinal to early archaea. In the eukaryotes, the succession passed down to the metazoa, including vertebrates and insects with highly conserved genes for both rhodopsin and for switching on eye growth. The problem of Darwin's eye falls away.

Photosynthesis that supports life, rather than simply prolonging it, is a different matter, and comes through the eubacterial domain. The forms of photosynthesis (Fig. 2) that are presumably the more primitive are anoxygenic. In the purple sulphur bacteria,

$$2H_2O + H_2S + 2CO_2 \xrightarrow{hv} 2(CH_2O) + H_2SO_4$$

occuring in two stages:

$$H_2S + \tfrac{1}{2}CO_2 \xrightarrow{hv} S^0 + \tfrac{1}{2}CH_2O) + \tfrac{1}{2}H_2O$$
$$S^0 + 2\tfrac{1}{2}H_2O + 1\tfrac{1}{2}CO_2 \xrightarrow{hv}$$
$$SO_4{}^{2-} + 1\tfrac{1}{2}(CH_2O) + 2H^+.$$

During this process, there is a massive transient build-up of sulphur, often deposited in the cell as globules, or excreted and later reabsorbed. Some purple sulphur bacteria are capable of chemoautotrophic growth under low oxygen partial pressure, with reduced sulphur compounds as electron donor (Stanier *et al.* 1987). It is thus possible to imagine an early environment in a shallow hydrothermal pool, with abundant H_2S, where chemoautotrophic bacteria evolved primitive photosynthesis (Fig. 3).

The steps in evolution producing respiration and photosynthesis are most likely explained in terms of pre-adaptation, both of chemical equipment (in the development eventually of chlorophyll) and of function (e.g. from an earlier use that could be adapted to light harvesting). Iron, sulphur, nickel, molybdenum, magnesium and manganese are all significant; interestingly, iron and sulphur would have been readily available in hydrothermal settings, where nickel and molybdenum would also have been present. Jones (1985) discussed the various possible stages from fermentation through respiration to photosynthesis. Fumarate reduction may have been the link between early fermentation

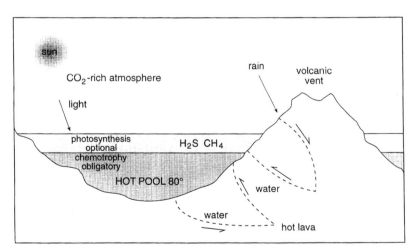

Fig. 3. Cartoon to show possible setting in which an early hydrothermal biome could have developed. The diagram shows a small andesitic volcanic centre (e.g. in an island arc) surrounded by sea or pools. Circulation of seawater or meteoric water around the vent provides a supply to the pool of hot (80–100°C) fluid, enriched in leached chemicals such as sulphur and minor phosphorus, Cu, Zn, Mo, Se, and H_2S, H_2, etc. A primitive chemotrophic community of bacteria exists in the pool. Those bacteria in shallow water which develop pigments as shields against solar ultra-violet radiation are further selectively advantaged, and thereby preadapted for the use of these pigments in more sophisticated photosynthesis.

and anaerobic respiration. Possibly fumarate reductase was originally a simple iron–sulphur flavoprotein. Hydrogenase (containing iron, nickel–iron and nickel–iron–selenium and 4Fe–4S clusters: Volbeda *et al.* 1995) and formate dehydrogenase (containing molybdenum) may then have been developed.

Jones (1985) pointed out that an organism capable both of synthesising porphyrin and of catalysing a primitive form of anaerobic respiration (probably fumarate respiration) would be preadapted to initiate photosynthesis as a supplementary energy source. Replacement of iron by magnesium in the centre of the porphyrin molecule, together with small changes in the structure of the ring system would allow the formation of a primitive bacteriochlorophyll, capable of undergoing a photochemical charge separation to release an electron (Jones 1985). However, there would also be the problem of preventing back reaction, recombining the released electron with the bacteriochlorophyll again. Jones suggested that, in the primitive respirer, the initially soluble bacteriochlorophyll-protein complex became embedded in the respiratory membrane so that the electron was transferred to an acceptor on one side of the membrane, and a donor was located on the other side of the membrane, to give an electron back to the bacteriochlorophyll. The acceptor and donor may have been respectively an iron-sulphur protein and cytochrome <u>b</u> or a high redox potential protein. Later developments would include the evolution of cytochrome <u>c</u> and then development of more sophisticated bacteriochlorophylls able to use more of the electromagnetic spectrum, and evolution of quinones to act as proton carriers (see Barber & Andersson 1994 for a review of the blueprint of photosynthesis).

This is a possible chemical story; what was the physical setting? The need for iron, sulphur, magnesium, nickel, molybdenum, copper, etc is intriguing. Copper and molybdenum imply conditions around andesitic vents, above porphyry deposits (e.g. see Hedenquist & Lowenstern 1994). Nickel immediately suggests to a geologist the environment around a cooling komatiite flow in shallow water (e.g. those of the Reliance Formation; Nisbet *et al.* 1993), with water rich in many of these metals emerging hot from hydrothermal circulation in the flow. Indeed, the broad flat surface of a large komatiite plume volcano, rising to shallow water or above sea-level, offers what is geologically a very attractive substrate for the first home of life.

The purple sulphur bacteria and green bacteria use bacteriochlorophyll, and the reaction centre absorbs in the infra-red region between 870–960 nm (Lawlor 1993). One form of bacteriochlorophyll, bacteriochlorophyll *a*, is common to both green and purple bacteria, in the photochemical reaction centres, and is arguably primitive. It absorbs at 773 nm in organic solvents, and at 805-910 nm in cells (Stanier *et al.* 1987), in the infra-red. Bacteriochlorophyll *b*, in purple bacteria, absorbs at even longer wavelengths, at 1020–1035 nm. This ability to absorb infra-red light is geologically intriguing, as it suggests an evolutionary pathway by which photosynthesis may have developed. If the existence of an early hydrothermal biome is accepted, it is possible that early hydrothermal bacteria, living by fermentation or respiration, may have survived better if they were thermotactic, able to detect hot vents by sensing the infra-red radiation emitted from hydrothermal vents, just as modern shrimps around submarine vents appear to be equipped with infra-red sensors ('eyes')(van Dover *et al.* 1988). Since the initial submission of this paper, the infra-red-thermotaxis-leads-to-photosynthesis hypothesis has been explored further by Nisbet *et al.* (1995), to which readers are referred.

The argument that photosynthesis developed from infra-red thermotaxis (Nisbet *et al.* 1995) is based on two main points: one is consideration of the phylogeny of bacteria, with photosynthetic bacteria being descended from hyperthermophilic bacteria likely to have been living in hydrothermal settings (e.g. see Kandler 1994); the other point is the similarity of the light wavelengths used by bacteriochlorophylls to the infra-red light transmitted through the water around hot vents. However, it is of supporting interest to the thermotaxis hypothesis that the assembly of rubisco, which is the catalyst at the core of photosynthesis, underpinning the modern biosphere (e.g. Hartman & Harpel 1994), involves chaperonin (Hemmingsen *et al.* 1988; Gething & Sambrook 1992; Ellis 1994). Possibly this process may have developed accidentally by exploitation of the properties of previously existing heat shock proteins. Where is precaution against heat shock important? — Around hydrothermal vents. Furthermore, rubisco incorporates at its heart FeS, a definitively hydrothermal chemical species.

A scenario can be imagined in which mobile bacteria (e.g. like *Chloroflexus,* a gliding green bacterium, deeply rooted in the eubacterial line of descent; Achenbach-Richter *et al.* 1987) gain advantage either by their ability to sense warmth and hence nutrient, or to avoid great heat and hence death. Hydrothermal systems can sustain sharp temperature gradients in the water surrounding vents, and infra-red light travels poorly underwater. For a bacterium, the difference of a centimetre in position can mean the difference between being cooked, on one side, or moving into water so cool that the bacterial chemical reactions shut down, on the other side. In between is the narrowly bounded optimum location, where growth and reproduction are favoured by the ambient temperature. Thus it would be greatly to the benefit of a bacterium if it could sense infra-red and move accordingly: the sense would be for survival, not for energy gathering. Such an ability, if evolved,would have preadapted bacteria to be able to photosynthesise using volcanogenic H_2S in shallow water. The reaction centre that had initially evolved to sense heat (850 nm with bacteriochlorophyll a) may have been modified with the addition of the antenna complexes to harvest red light at shorter wavelengths (750–810nm). This sequence, depending on specific local chance and preadaptation, was most probably monophyletic: it only happened to one line of bacteria, and was then acquired by others from the first line.

This photosynthesis uses light at long wavelength. Consequently, bacteriochlorophyll cannot generate a sufficiently strong oxidant to remove electrons from water. For the biosphere to have escaped from a volcanic habitat rich in H_2S or other reduced species, a photosynthetic process capable of using shorter-wavelength light was needed. The relationship between heat shock proteins and rubisco assembly has already been mentioned: rubisco may have been put together with the help of evolutionary tinkering with heat shock proteins originally developed as a protective device in the hydrothermal setting. The prochlorobacteria and cyanobacteria use chlorophyll, not bacteriochlorophyll. Prochlorobacteria differ from other cyanobacteria in that accessory pigments, phycobilisomes, are present in the cyanobacteria but undetectable in the prochlorobacteria, though 16S RNA evidence suggests the two are related (Stanier et al. 1987). Thermophilic cyanobacteria occur abundantly in neutral to alkaline warm springs, characteristic of subaerial springs around andesitic volcanoes above subduction zones. The cyanobacteria, using chlorophyll at the reaction centre, are able to harvest light at 670 nm (orange-red). In addition, their antenna complexes, using phycobilisomes, are able to absorb light at wavelengths as short as 570 nm (green). The consequence is that water splitting is possible, and hence the modern biosphere can begin. Cyanobacteria may perhaps have first evolved around hydrothermal systems, but their abilities allow them to colonize the planet, from the open ocean to the desert interior of the continents (see discussion in Knoll & Bauld 1989).

The role of the carotenoids in this process is also interesting, as they play an essential role in detoxifying reactive forms of O_2 and must have pre-dated widespread oxygenic photosynthesis. They share the early part of the synthetic pathway with chlorophyll, and may have functioned as light-harvesting pigments in early photosynthesis, later becoming important in regulating the energy state of the reaction centre (Lawlor 1993). Cyanobacterial photosynthesis would immediately cause the problem of oxygen accumulation in the local environment, and carotenoids may have played a protective role from this stage. Carotenoids absorb light in the wavelengths 380–520 nm (violet-blue), and would have helped to protect bacteria in shallow water from this energetic radiation. They may have evolved first as early protective pigments, prior to their utilisation in photosynthesis.

Aerobic respiration may be polyphyletic (Jones 1985), a widespread response to the challenges of an environment in which photosynthesis is common. One type may have evolved from early fumarate respirers, others from phototrophic ancestors via the loss of bacteriochlorophyll, yet others from chemolithotropes.

Carbon isotopic fractionation in the rubisco pathway is, depending on various factors, around –29‰ (Summons & Hayes 1992). This signature was stamped on the planet, as the biosphere built via rubisco took over. The effects of this are clearly expressed in the geological record back to about 3.5 Ga and perhaps back to 3.8 Ga ago, through the fractionation between organic carbon ($c.$ –26 ± 7‰) and carbon in carbonate (0‰), both derived from original mantle carbon at $c.$ –7‰. Schidlowski (1991 and papers cited therein) has argued persuasively that the continuity through the geological record of the isotopic distinction between organic and inorganic carbon demonstrates that the Calvin cycle has operated at least since 3.5 Ga time and most probably since prior to 3.8 Ga ago, although it should be remembered that other bacterial processes (e.g. methanogenesis) also fractionate carbon strongly, and have in part contributed to this record. Throughout this time, biological carbon must have represented roughly a fifth of total carbon precipitated (both cumulatively and, for the most part, instantaneously). The modern carbon cycle, therefore, has been in operation virtually since the beginning of the geological record, and throughout this time biological productivity has been adequate to sequester as organic matter a constant proportion of the carbon dioxide that has been degassed to the atmosphere/ocean system from volcanoes at ridges, plumes and subduction zones.

Nitrogen, sulphur, phosphorus and iron

Consider nitrogen again. On the late Hadean Earth, a scenario can be imagined in which nitrogen is present in the sea as NH_4^+, and in hydrothermal systems, and only trace amounts of N_2, NO_x, and NH_3 are available in air. Nitrification by an evolving biosphere would rapidly deplete these resources to nitrate, and the system would soon face a nitrogen crisis, that would limit total biological productivity to the level set by the availability of freshly erupted ammonia. A biosphere, in which cyanobacteria were spreading widely, actively photosynthesising and building up locally oxic environments (Kasting 1992), would soon be sharply limited in its productivity by the availability of nitrogen. Unless, that is, the biosphere evolves a way of returning the usable nitrogen to the environment. This occurs when denitrifying bacteria that can remove nitrogen from the dead bodies of their sisters evolve, and the productivity of the system rises. However, the end of the denitrification cascade is N_2 gas, released to atmosphere, and the aqueous system loses available nitrogen. Nitrogen fixation, by methanogens, green and purple bacteria evolved, and then more widely in cyanobacteria. The first nitrogen fixation must have taken place very early, well prior to 3.5 Ga, to allow the level of sustained productivity recorded by the C isotopes.

The presence of molybdenum and iron-sulphur centres in both nitrate reductase and nitrite oxidoreductase suggests (Jones 1985) that the former may have evolved from the latter, in a way similar to the postulated evolution of succinate dehydrogenase from fumarate reductase. Manganese appearance in cytochrome $\underline{c}$ may have helped in the catalytic oxidation of NO to NO_2^- (Rao *et al.* 1985). Copper proteins are ubiquitous in denitrification. These requirements of the nitrogen cycle for iron, sulphur, manganese, copper and molybdenum suggest an early hydrothermal parentage.

Similar arguments may be made in the case of the sulphur cycle, but with the difference that the volcanogenic availability of sulphur would have been greater, and resource limitation may have been less crucial than in the case of nitrogen or phosphorus. It has been suggested from the available isotopic evidence that a major change took place around the late Archaean or early Proterozoic, marking the beginning of dissimilatory sulphate reduction (e.g. see Runnegar 1992), although this inference may fall as more data are obtained.

Phosphate is obviously crucial to the biosphere, which recycles it readily. The balance between uptake from the vicinity of hydrothermal systems and from seawater fed by continental run-off from granites and carbonatites, and loss to sea floor sediment, may have limited the total productivity of the biosphere then and now.

Iron availability is another puzzle. In an oxygen-poor world, ferrous iron could be transported in solution away from the proximity of volcanoes, to supply newly evolved cyanobacterial photosynthesisers in the open ocean or shallow coastal waters. Any build-up of oxygen as a local result of cyanobacterial activity would limit this transport of dissolved iron, and control productivity. Again, however, biology comes to the rescue. Even in a highly oxidising environment such as the present-day, the vicinity of active hydrothermal systems can remain acid and reducing. In the late Archaean and Proterozoic, capture of iron in volcanic settings by bacteria may have introduced iron into the biosphere: from there, the iron may have been recycled widely across the oceans by bacteria eating other bacteria, floating iron, like phosphorus, worldwide through the food chains. Fe^{3+} reduction by bacteria may also have been an important iron source in oceans near dusty continents (Knoll, pers. comm.).

The hydrothermal fingerprint on life

If the hydrothermal-first hypothesis is correct, then hydrothermal systems should have left their fingerprint on all life. Each feature of deep antiquity in the genetic heritage of the three domains (eubacteria, archaea, and eukarya) should bear witness to a hydrothermal source.

It can be argued that this is exactly what is seen: many of the deeply conserved processes in life have characteristics that seem to imply hydrothermal origin. First, it is worth recapitulating the list of key chemicals. Nitrogen, sulphur, iron, zinc and especially phosphorus are common to many life processes. Although they are presently common in the environment, much of that distribution is now biologically mediated (e.g. N, Fe, S, P). Abiologically, species such as phosphorus, and iron and zinc sulphides are most available around volcanoes. Phosphorus in particular is immobile and only locally accessible unless biologically held. Nitrogen in the late Hadean may have been best accessed as ammonium minerals around volcanic systems and lava flows. Other trace components such as cobalt, molybdenum (nitrogenase), managanese and selenium are characterisic of porphyry deposits around volcanoes. Another interesting element is nickel, which is essential in the Ni–Fe hydrogenase of *Desulfovibrio gigas* (Volbeda *et al.* 1995) critical for oxidising molecular hydrogen. Early life must have been able to access nickel. Possibly this would have been available near basalt lava flows, but the most likely site would have been around submarine komatiite flows, from which hydrothermal liquids rich in Ni and S would have issued. Magnesium (e.g. in ATPase) is another of the essentials of an interactive biosphere, for instance in eukaryotic mitochondria. While Mg is commonly available in seawater, nevertheless here too the komatiite (high-Mg lavas) association seems likely.

Indeed, although all elements are present in sea water, it is difficult to imagine how nickel, molybdenum, zinc or selenium could have been sufficiently accessible to be used by the first, primitive life if they had not been easily available. Early life would not have had sophisticated biological techniques capable of extracting scarce minerals and then hoarding and recycling them to sustain an ecosystem. There must have been an easy availability of the metals that are incorporated in enzymes. This would only have been possible if the living community had been in the close proximity of a hydrothermal system. Molybdenum, zinc, selenium and copper are widely characteristic of hydrothermal systems, though especially of andesitic volcanism; nickel sulphide is a feature of komatiitic volcanism. Magnesium is more widely available in seawater, but magnesian solutions characteristically circulate or are taken up in the alteration of komatiites and high magnesium basalts. Collectively, these elements suggest a greenstone belt setting, with komatiitic plume volcanism in close proximity to andesitic activity, as in some Zimbabwean belts (e.g. see Nisbet 1987). Later throughout geological time, some of these elements, such as magnesium, have probably been widely available. However, the early biosphere may well have faced supply crises with many trace elements.

Apart from the possibility that photosynthesis itself has hydrothermal roots in thermotaxis, many of the deepest-rooted component structures of life may also have hydrothermal origins. As mentioned, the heat shock proteins (Ellis 1994) are the most obvious of these, perhaps evolved by a very ancient ancestor in response to the thermal dangers of the environment: possibly the chaperonins in mitochondria (Gething & Sambrook 1992), in rubisco-binding protein, and in archaea may all have a common distant ancestor in mechanisms originally designed to contain heat damage.

The carbon isotopic evidence that a prolific global biosphere, based on rubisco, has existed since prior to 3.5 Ga ago suggests that these crises were resolved very early on by biological recycling of essentials such as phosphorus, accessible iron, nitrogen and sulphur. The biosphere may have left its putative hydrothermal home, but the hydrothermal stamp may remain on all life.

The post-photosynthetic biomes

On the modern world, a biome is a collective patch-work of living systems, jointly adapted to a common climate, sharing habitat, recycling nutrients, and in which the various species must accommodate to each other (Colinvaux 1993). The hydrothermal biome fits this concept well, but away from the hydrothermal setting, for instance in the open ocean, the idea of a bacterial biome must refer to an assemblage of organisms that has temporal continuity, but not necessarily spatial stability. Bacteria move with the wind and the current; as the habitat moves, so do the blooms, yet over time there is great stability to the bacterial world.

We can imagine a set of bacterial biomes in the Archaean biosphere (Fig. 4). Around the hydrothermal systems of the mid-ocean ridges and the island arcs would be the hydrothermal biome, including both shallow-water habitats (mainly above subduction zones, but also where plume volcanism took place), and deep water ridge communities. The shallow-water calc-alkaline hydrothermal communities would be rich in thermophilic photosynthesisers, but the ridge community would depend on an influx of chemical species such as SO_4^{2-} from neutral or oxidised ambient seawater, ultimately from the surface photosynthetic communities. The hydrothermal environment would host a complex community of chemotrophs and heterotrophs, and would be responsible for the capture from the lava and rock into the biosphere of chemicals such as iron, sulphur, nitrogen, phosphorus and cobalt. Currents would export these chemicals, in bacterial matter, from the hydrothermal vicinity to the other Archaean biomes.

Colonization of non-hydrothermal shallow coastal waters, the marine littoral biome (Whitton & Potts 1982), would be likely as soon as cyanobacterial photosynthesis developed. Shallow water environments would provide nutrients in tidal currents, and accessible mineral substrates in muds. Coastal upwelling may have provided metals from deep water hydrothermal systems, or they may have been brought by currents from shallower calc-alkaline volcanoes. Near river mouths, supply of mineral nutrients (especially calcium and phosphorus) from continental run-off would have been significant. Active photosynthesis in tidal lagoons would have led to a complex array of local habitats, ranging from local oxic shallow water (e.g. see Kasting 1992), to anoxic lagoonal floor muds where organic debris collected. In this setting a very varied bacterial community could develop.

In most cases where Archaean stromatolites have been described in the geological record, volcanic rocks are stratigraphically very close by. For instance, in the 2.7 Ga old Ngezi Group of the

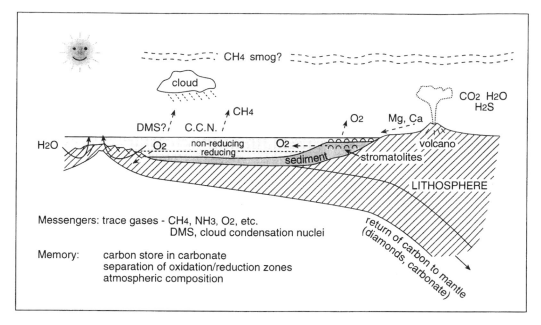

Fig. 4. A model of the Archaean environment. At mid-ocean ridges, in island arcs and above plumes, volcanism drives the circulation of hydrothermal systems. Incoming water, containing species such as SO_4^{2-}, enters the system and in the hot rock H_2S etc. are produced. There is also mobilisation (to varying degrees) of other essentials of life, such as Fe, P, Ni, Mg, Zn, Cu, Mo, Co, etc. Exiting hot water sustains a hydrothermal biome. In the marine littoral setting and also floating in the open oceans, photosynthetic cyanobacteria flourish, producing oxygen, and allowing the build-up of large-scale zones of oxic and anoxic habitat and complex picoplanktonic communities of eubacteria and archaea. In cooler water, communities of archaea spread. The atmosphere is influenced by both inorganic and organic processes, including emission of volcanic gases, and emission of O_2 and possibly methane compounds (such as dimethyl sulphide and other chemicals helping as cloud condensation nuclei). Weathering from continents provides a supply of cations, as does the hydrothermal circulation. Carbon is removed from the hydrosphere/atmosphere system by precipitation, or by removal of organic carbon. The photochemistry of the troposphere and stratosphere may have been very different from today.

Belingwe belt, both the upper and lower stromatolite horizons (Martin *et al.* 1980) are in close stratigraphic proximity to mafic volcanics; similarly, in the *c.* 3.0 Ga old rocks of the Steep Rock Group, stromatolites are overlain by volcanics (Wilks & Nisbet 1988); in the 3.5 Ga old Barberton succession stromatolites and lavas are closely juxtaposed (Byerly *et al.* 1986). In all three cases, Walther's facies rule would suggest that the volcanics and the stromatolites were contemporaneous facies. Possibly, the hydrothermal and non-hydrothermal communities cross-fertilised each other, the photosynthetic community exporting oxidised seawater to enter at the base of the hydrothermal structure, while the hydrothermal community exported iron, sulphur and phosphorus, both directly (as, for instance, H_2S or iron in plumes of reduced water) and indirectly as bacterially captured iron and phosphorus, extracted by the bacteria around hydrothermal vents and then recycled into the overlying oceanic community of life living in more oxidized conditions.

The extent of the marine littoral biome is debateable. Archaean stromatolites are widespread in the record, but uncommon in that they constitute only a small part of the preserved sequence. Archaean shallow water sediments are generally clastic or ironstones, not carbonates. However, it can be argued that this simply reflects location and preservation: in modern areas such as California, limestone deposition is not widespread in comparison to clastics, but marine life is abundant nevertheless. The lack of limestone in the Californian geological record simply reflects contemporaneous tectonic conditions and chance of preservation. It is possible that in the Archaean a very widespread transient algal mat ecology existed, now generally lost because it was eroded or buried and metamorphosed to trace organic matter in clastic sediment. Some Archaean ironstones may be products of iron or magnetotactic bacterial blooms, in protected lagoonal settings (e.g. Nisbet *et al.* 1993). Thus, although the preserved record of the marine littoral life community is fragmentary, it is

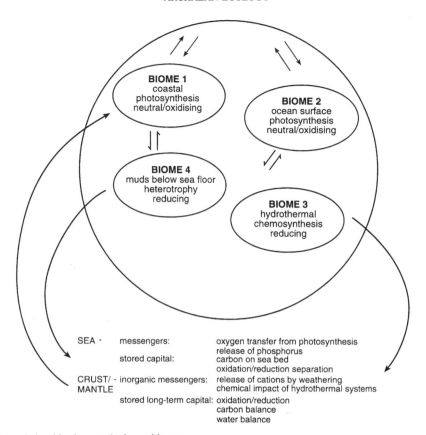

Fig. 5. Interrelationships betwen Archaean biomes.

nevertheless possible that the community at the time may have been very extensive.

Other biomes can also be imagined (Fig. 5). The most significant is the open ocean biome, now dominated by eukaryotic plankton, but in the Archaean probably with a picoplankton community based on floating cyanobacterial photosynthesis. Populations would have been closely linked to the positions of oceanic fronts: the 'weather' of the sea. Bacteria have varying degrees of barotolerance, depending on type and history (Jannasch 1984). Many bacteria are phototactic or magnetotactic (Rowbury *et al.* 1983), capable of migrating upward and downward diurnally. Cyanobacteria such as *Anabaena flos-aquae*, when growing in conditions of low light irradiance, produce enough gas vacuoles to make the cells buoyant. When moved to high light levels, the buoyancy is lost: in lakes populations of cyanobacteria have been observed to move upwards in the morning, then sink again as evening approaches (van Liere & Walsby, 1982). Nitrogen fixation in cyanobacteria is very sensitive to O_2, and movement may have

been a response to this problem. On the modern earth, nitrogen-fixing by the cyanobacterium *Trichodesmium* is widespread in tropical oceans, although not in colder waters: thus it is possible that the Archaean oceans were relatively poor in cyanobacteria outside the tropics. *Trichodesmium* uses flotation upwards to gather nutrient, in large blooms, though this strategy has the consequence that blooms of buoyant bacteria are eventually caught in calm weather on the surface and killed by sunlight (Fogg 1982). More generally, the distribution of cyanobacterial populations in the open ocean would have been very heavily dependent on phosphorus supply, and blooms would have been located by the currents, over major upwellings bringing nutrient from deep water, from volcanic sources and from run-off from land. Stratification of light gathering organisms would have occurred, with microzones of differing ecology developing, and both mutualistic and symbiotic associations, on the one hand, and, on the other, antagonistic relationships, including toxin warfare.

Outside the tropics, an extensive community of archaea lives in the present-day picoplankton of the cold surface waters of Antarctica (DeLong *et al.* 1994) where they form up to 34% of the prokaryotic biomass. Some of these archaea are related to hydrothermal hyperthermophiles such as *Sulfolobus*, others are closer to moderate thermophiles such as *Thermoplasma* and, more distantly, the methanogens and halobacteria. This present-day evidence should be seen as powerful support for the hypothesis that a widely varied community of archaea was present as Archaean picoplankton in colder waters.

In the oceanic sediments, an extensive bacterial biome must have been present. On the modern earth, in the Pacific, viable bacterial populations are present in muds in sediment as deep as 500 m below the sea floor (Parkes *et al.* 1994).

Terrestrial bacterial communities would also have developed in the Archaean. These could have been spread from bacteria in wind blown spray. Wetter environments, such as fresh-water rivers, lakes and swamps, would have been colonized first, but modern cyanobacteria inhabit environments as extreme as deserts (Stanier *et al.* 1987), where they grow in microfissures just below the rock surface. In the absence of eukaryotic grazers, it is possible that a cyanobacterially-based ecology occupied the chemically weathered soil profile in the more moist parts of the land surface, and may have played a significant role in mediating weathering processes.

The inflationary biosphere

Thus a series of Archaean biomes can be recognized: hydrothermal; marine littoral; open ocean; and terrestrial. What can also be recognised is the potential for developing distinct zones of oxic and anoxic habitat. By this separation, sequestering oxidative and reductive power in disequilibrium, life effectively inflates the potential productivity of the environment, and is no longer limited simply by the oxidation power from the inorganic world's contrast between the planet's interior and exterior; the new limits in the inflationary biosphere are the availability of essentials such as phosphorus and trace metals. In all biomes, distinct oxic and anoxic zones could develop, even in the open ocean where the near-surface zone of photosynthesis may rapidly have become oxic, but may have remained underlain by anoxia. The state of the atmosphere is ruled by the balance of disequilibrium, the oxygen, nitrogen, carbon dioxide and trace gases in the air simply being the standing crops, residing over times ranging from seconds (some trace gases) to millions of years (nitrogen, oxygen after it built up), in a set of cycles where the total flux over the aeons is much greater than the standing crop at any one time. Chemicals such as iron are biologically moved away from their primary sources (e.g. in iron bacteria, or as magnetite in bacteria, where the iron may have been derived ultimately from hydrothermal or continental sources, but moved biologically) and can thus be transported independently of the oxidation state or pH of the setting. Nevertheless, the atmosphere and the chemical state of the sea remain as links between the component biomes, shaping their general environment and determining their broad population.

Schidlowski (1984, 1991), in a penetrating insight into the puzzle of the Archaean biosphere, pointed out the implications of the evidence that the terrestrial carbon cycle has likely been under biological control since prior to 3.5 Ga, if not 3.8 Ga ago. To an order of magnitude, the oxygen production by photosynthesis has remained constant over the entire geological record over this period. The size of the oxygen reservoir actually present in the atmosphere is, however, determined by the kinetics of the responses of production and consumption to changes in the oxidation state of the environment. In the face of photosynthesis, to maintain a neutral atmosphere, without significant oxygen, extremely rapid processes of oxygen consumption must have occurred. Like the inflationary universe, the inflationary biosphere is able to make something out of very little: by using light to develop segregated zones of anoxic and oxic conditions, it can increase the potential scope of oxidation/reduction processes. The sum of a large positive number and a large negative number can remain zero. The total net chemical oxidation power of the environment remains limited by weathering and hydrothermal processes, but in practice by segregating oxidation power, the available habitat for life is vastly increased.

Evolution and interconnection of Archaean biomes

What connections could there have been between the Archaean biomes? Could they collectively have modified the planet's atmosphere and climate?

It is possible to imagine an interconnected post-photosynthetic biosphere (Figs 4 & 5) in which various Archaean biomes cooperate to create a global biosphere. In the marine littoral, cyano-bacterial stromatolitic and algal mat communities existed, oxic in their upper levels, but overlying anaerobic decay. The output of oxygen from photosynthesis would have provided oxidised seawater globally. In the open tropical ocean, planktonic cyanobacteria provided oxygen to the air, and organic debris from them would create anoxic bottom muds in which other guilds of bacteria thrived. Complex communites of archaea and eubacteria would have recycled organic matter in the colder waters The oxidation state of the various water levels in the ocean (and, by exchange, of the air) would have depended critically on the productivity of the planktonic community, and the balance between oxygen output and reduction elsewhere. Both the marine littoral and open ocean biomes would have liberated nitrogen by accident, and this would have built up in the air, increasing pressure with associated impact on the greenhouse gases (although nitrogen itself is not radiatively active). Bacteria such as *Hyphomicrobium S,* a prosthecate (Stanier *et al.* 1987), may have released dimethyl sulphide (see Large 1983) to air. Dimethyl sulphide has cloud nucleating effects, and also helps in returning sulphur to the land. A methane smog may have built up, protective against UV (Lovelock 1988).

The reductive power of the environment at any one moment depends on the availability of reductants exposed to fluids and air interacting with the surface settings of the Earth. Available reductive power includes some of the sequestered reduced carbon in muds, but not necessarily all, and rock surfaces created by the supply of new basalt lava at the ridges and volcanoes that are exposed to hydrothermal fluids and by erosion on continents. The hydrothermal biome is thus a counter to the oceanic and littoral biomes. At the hydrothermal systems, the input of oxidised seawater is used to produce H_2S, CH_4, H_2, etc., that help sustain the hydrothermal biome and also supply trace components, such as H_2S (possibly in marked disequilibrium with the air), to other parts of the biosphere. Walker (1990) has argued that there is an imbalance today between oxidants and reductants, and that over the aeons, by oxidising the atmosphere, life has also affected the oxidation state of the mantle. Thus the unmixing activity of the biosphere not only affects the planet's surface, but by its influence on temperature gradients, on hydration and on oxidation of subduction zones, it has shaped the evolution of the planet's interior (Nisbet 1987).

It is thus possible to imagine an interactive biosphere, each biome playing a role in a global exchange.

Memory and messengers

A characteristic of a modern biome is that it has memory. Biomes evolve first because there is variety of climate and topography on the planet. This physical variety is largely (though not wholly) gradational, yet the biological ecotones between biomes are typically sharp, not blurred. Part of the reason for this may be that as biomes become established, they develop memory. On the post-glacial plains of the modern northern hemisphere, scattered conifers may have seeded: once grown, they defined shaded area beneath them, created acid soils, competed with other species by their willingness to burn, changed the albedo of vast areas, to create a uniformity of habitat populated by conifers and conifer-tolerant or competitive species. Further south, where conditions were slightly more arid, grass predominated, with its specific fire ecology, different soils, soil moisture and light conditions. The boundary between the two biomes became sharply defined. Soil, albedo, bogs and ponds, in the two habitats are sharply different, and constitute the 'memory' or capital resources of the systems. Just as an immigrant to a country needs to learn the language and culture to survive, and, if not deported, once integrated, tends to repel newer immigrants, so biomes defend themselves against unaccultured immigrants.

Biomes also have messengers that convey information within biomes, connecting components, and between biomes globally. Today these may be as varied as insects or migratory birds, or dust or atmospheric trace gases that help in cloud nucleation. Messengers may be intra-biome, or between biomes, such as long distance migratory species. Prairie dust in North America and Asia sustains the fertility of the boreal forest. It can be argued that the Saharan biome and the Amazonian forest are symbionts, the Sahara needing the energy output of Amazonia, as latent heat in water vapour, which influences the location of the jet streams in the global climate system to maintain dry weather conditions over the desert, and Amazonia needing a constant supply of Saharan dust to replace trace elements lost via rivers to the sea. This argument is unverifiable at present, but illustrative. The atmosphere, in its major components (N_2, O_2), is the longest memory of the surface biosphere; it also, in its trace gases, contains many of the most active messengers.

Figures 4 and 5 illustrate possible interrelationships between Archaean biomes. Pattern and scale in marine ecosystems depend on a variety of factors (Levin 1992), but it is clear that a bacterial ecosystem capable of rudimentary photosynthesis would very rapidly spread across the planet, simultaneously rapidly evolving more sophisticated chemical techniques, changing a solo cello suite into a symphony of life. Secondary foci would appear, to occupy new biomes, and within a few million generations or so the planet would be occupied with all readily accessible metabolic pathways in action.

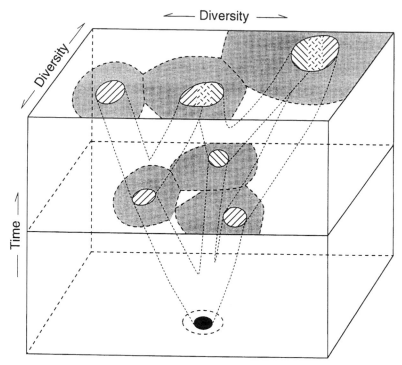

Fig. 6. Evolution of a diverse bacterial community in a set of biomes. (**1**) Bottom level. Original population of variants of a single progenote species in a hydrothermal setting, very limited in its ability to occupy habitat even within the confines of its hydrothermal environment. (**2**) Middle level. Divergence from an ancestral population, with pioneering types filling habitat niches. The original population, of a single progenotic species, evolves to the varied phyla of eubacteria and archaea . These occupy new living space, with diversity in various dimensions (e.g. temperature habitat, salinity, acidity). However, the total biosphere remains very restricted both in species numbers and in ability to occupy habitat,.and is still confined to the hydrothermal habitat, though it spreads globally along volcanic chains and mid-ocean ridges. (**3**) Top level. After the evolution of photosynthesis, development of wider communities, each with representatives of the major guilds of bacteria, to occupy the planet's surface widely. In each community a varied population of organisms exists and evolves. Some cross-over between biomes occurs, with immigrants seeding new tribes of organisms in the community, but the biomes are broadly distinct over long periods of time.

Change and collapse: biosphere evolution in the Archaean

Evolving communities develop organizational complexity (Fig. 6). By 3.5 Ga ago, the seas would have been occupied by a patchy but globally distributed and habitat filling network of communities. In each biome, guilds of micro-organisms would perform specific services to the system as a whole, with functional redundancy within each guild, and a broad temporal (but not locally spatial) stability would be achieved. The distribution of organisms, determined by environmental heterogeneity, would be patchy, allowing the coexistence spatially of competitors that could not exist locally together in a homogenous setting (Levin 1992). All this would make for stability, and

the biological world, based on eubacteria and archaea, would have been secure in its tenure of the planet once the modern carbon cycle had been established prior to 3.5 Ga ago.

Species occupying new habitat typically pass through a variety of stages (Fig. 7) as they create an ecosystem (Holling 1992). During the initial phase of *exploitation,* there is rapid colonization of new habitat. Immediately after the first water-splitting bacterium evolved, the seas would have been explosively occupied by its descendants, with rapid evolution of new species. At this stage, nutrient would be widely available, and opportunistic occupation of every habitat where even a bare minimum of nutrient was available would occur. Very rapidly, this phase would end, as the previous standing crop of nutrient would be depleted, and a

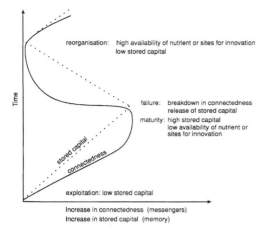

reorganisation: high availability of nutrient or sites for innovation
low stored capital

failure: breakdown in connectedness
release of stored capital

maturity: high stored capital
low availability of nutrient or
sites for innovation

exploitation: low stored capital

Increase in connectedness (messengers)
Increase in stored capital (memory)

Fig. 7. Stages in the evolution of a biome (after Holling 1992). In the early Archaean, catastrophes would have included natural events such as volcanic eruptions and meteorite impacts, and also the evolution of new metabolic pathways that suddenly opened up new habitats, or increased predation. Many of the major divergences in the evolutionary tree of Fig. 2 may reflect such events.

crisis would occur, with widespread death and surviving populations only in those areas most nutrient-rich. At this stage, other non-photosynthetic bacteria would have the chance of clearing up the wreckage by fermentation. The nitrogen and suphur cycles are crucial: bacteria in an initial hydrothermal community that participated in these cycles would rapidly spread to colonise the new habitat opened up by pioneers.

The next step, *conservation* (Holling 1992) or maturity, is the increase in connectedness and stored capital in the system. In a simple single layer system without predators, but with various guilds of photosynthesizers and respirers, connectedness may be more limited than in modern multi-layer food webs, but can nevertheless develop. The keystone cyanobacteria would exist in local regions of preferred habitat, where nutrient was most available. In each patch, other bacteria would flourish on the cyanobacterial basis, with locally specific representatives of guilds of bacteria and hence evolution of redundancy in species roles. Those products of these communities that were released to atmosphere, such as oxygen and nitrogen, would affect all other communities, globally. Rapidly, specific oxidising and reducing habitats would appear, separated by biological activity and with specific populations in each. As noted by Schidlowski (1984), it is not obvious what the global oxygen level should have been, because it was presumably set by the competing yet sym-

biotic demands of photosynthesizers and respirers, with probable bacterial transport of iron (and perhaps also uranium?). The flux of oxygen and the isolation of oxidizing from reducing environments were presumably sufficient that within a few million years, oxygen and nitrogen levels would have been biologically, not inorganically controlled.

Within perhaps a few tens of millions of years, a stable biosphere would have evolved: this, in essence, is the modern biosphere, as recorded by the antiquity of the carbon cycle, but in a system with only one composite level of photosynthesizers and respirers. The phase of maturity in a one-level system would have been marked by increasingly little tolerance of deviance and resilience to sharp change, either internally or externally driven, as all available nutrient was used, and the stored capital increased in the shape of the standing atmosphere and segregated zones of oxidation and reduction. Any new organism intolerant of these global conditions would not survive, except perhaps in the still weakly-connected hydrothermal biome.

Although in its broadest sense, this community has been stable since 3.5 Ga, catastrophes of varying intensity must have occurred, the failure and *release* stage of biome evolution (Holling 1992). Apart from natural disasters, such as eruptions, the appearance of predation would have been one of the first 'catastrophes', or events of rapid change. In diagrams like Fig. 2, the splitting points between evolutionary lines may reflect such catastrophes and consequent subsequent opportunity. The initial predators on cyanobacteria may have had a splendid time, perhaps very rapidly reducing the earlier populations to tiny numbers and threatening the stability of the entire whole. However, isolation between population patches would have slowed the spread of the catastrophe. Destruction is also opportunity (the time of reorganization or *exploitation* phase of Holling), and it is possible that the appearance of a new system utilising the released resources of the old would have been very rapid, leading to a major reorganization of the balance of life. The present defensive warfare by cyanobacteria, emitting toxins (Kirk & Gilbert 1992) to keep off predators may have begun rapidly.

The eukaryotes

The evolution of the eukaryotes is poorly understood (Knoll 1992). It is generally accepted that the modern eukaryotic cell results from symbiotic incorporation (e.g. see Margulis & Sagan 1986) of eubacterial organelles into an archaebacterial-like root stock, though the details of the process remain controversial. The chloroplast appears to be derived

from a cyanobacterium-like ancestor, while mitochondria appear to be derived from an ancestor
similar to purple bacteria (for discussion, see
Chapman 1992, including comment on the
question of the chloroplast origin). Mitochondrial
acquisition may have predated chloroplast incorporation (Runnegar 1992). The nuclear genome
today contains genes (e.g. for cytochrome c) that
are of specific eubacterial origin; it also contains
genes that are of archaebacterial character, and
some genes that appear to be specific to eukaryotes
(Woese 1983). The ancestral stock may have been
in the archaea, but it is also possible that the
eukaryotic root stock is of the very greatest
antiquity, diverging from the point of divergence
of the eubacteria and the archaea (e.g. see Achenbach-
Richter *et al.* 1988). Precisely *when* endosymbiont
eukaryotes first appeared remains obscure.
Arguments have been made in favour of a late
Archaean origin, possibly after 2.8 Ga ago
(Vossbrinck *et al.* 1987 and see Chapman 1992),
but the answer is yet unknown.

Multi-celled eukaryotes came later. The
massive radiation leading to the photosynthetic
protists, fungi and animals may have occurred
around 1–1.5 Ga ago (Chapman 1992), but with
remarkable conservation of common genetic
elements such as actins, tubulins and the
homeobox.

The appearance of abundant eukaryotes,
especially of zooplankton, would have radically
changed the global ecosystem (see Pace & Funke
1991), though it is possible that in many locations
the population remained nutrient rather than grazer
limited (e.g. Hansson 1992; see also Power 1992).
Highly productive two-level systems tend
to 'grow away' from grazing pressure, and to
become nutrient limited. With the radiation of the
eukaryotes, three-level systems would have
occurred, with further catastrophes.

Table 1. *Characteristics of Gaia*

Cybernetically self-regulating system that optimises the
environment for its living communities as they evolve
within it.

Properties
   Self-regulation of climate and chemical environment,
   in response to external input and internal evolution.
   Global in scope.

Memory
   Development of long-term chemical inventory, that
   helps to stabilize environment.

Messengers
   Components that can carry signals from one part of the
   system to another.

Affecting each later development is the memory
of the previous system, especially in the composition of the atmosphere and sequestration of
oxidation and reduction into specific zones: the
new world system must fit into the debris of the old.
An oxygen rich atmosphere must have existed
from at least prior to about 1.8 Ga (the oldest
unambiguous red-beds) but may have long
pre-dated that time. At the basis of the biosphere,
however, there is conservatism of great antiquity:
the fundamental biochemical reactions remain
the same; new plant is but old cyanobacterium
writ large; new mitochondria contain antique heat
shock proteins. From the point of view of the
eubacteria, the eukaryotes are simply space suits
for the inhabitants of the eubacterial world, means
by which they can create forests or travel to the
moon.

Was a Gaian system possible in the Archaean?

The Gaia hypothesis is that collectively the Earth,
in its living and non-living components, represents
a cybernetically self-regulating system (Table 1),
(Lovelock 1979, 1989; Lovelock & Margulis 1974;
Kirchner 1989). The objections to this hypothesis
that have been raised are very varied, but one of the
most fundamental is that there is no way in which it
could have evolved. On a local to biome scale, the
mechanisms for the evolution of functionally
organized communities have been discussed by
various authors (see Wilson 1992). However,
fundamental to the notion of Gaia is the demand
that it operate on a planetary scale. Watson &
Lovelock (1983), for instance, have constructed a
model in which global temperature is controlled
over time by biota, the 'daisy-planet' model.

Gaia is seen in cybernetic terms by Lovelock,
with the Earth's surface environment being a
self-regulating system. The word 'system' is
stressed. The criticisms of the hypothesis are based
on the implication that there will be interactions
within the system in which altruistic behaviour is
demanded of specific organisms. This, it is held (as
in the general problem of altruism), goes directly
against the Darwinian imperative that organisms
can only survive if they are individually competitive. The systems approach of Gaia, in contrast,
appears to imply widespread altruism. Is the
biosphere an accidentally linked set of cause and
effect chains, or is it a true network?

Can one imagine the evolution of a Gaian system
that is strictly Darwinian? Or perhaps Tennysonian
(in his rigorous pre-Darwin vision of 'nature red in
tooth and claw', eliminating uncompetitive species:
In Memoriam, LV). In answering this, the starting

point must be selection in the local setting. Life reproduces, reproduction creates variety. In the variety of offspring there will be some organisms that *accidentally* create biological by-products of their existence that are either toxic or beneficial to the environment as far as the organism and all other organisms in the setting are concerned. Initially, these by-products will be accidental, at no cost, free preadaptations – waste: the complexity of the biosphere is built on excrement. They may include trace gases or other chemical emissions, or even the to-be dead bodies of the organisms. Those organisms that 'enhance' the environment accidentally live in more favourable conditions, reproduce better, and are thus selected for. Those organisms that produce 'toxins' cause disadvantage to other organisms, which adapt, compete, or become extinct, until the community is only made of organisms that can co-exist. In higher animals, especially in species that are highly successful, apparently 'disadvantageous' or altruistic characteristics may perhaps evolve despite their cost, but in bacteria reproduction is so rapid and competition so intense that, initially at least, altruistic behaviour must surely be accidental preadaptation, at no cost.

The result of this process over time in any place is a local community, in which organisms adapt to the accidental altruistic behaviour of others. Inevitably, the altruistic species will adapt to the presence of the species benefitting from the altruism. Competition will occur, especially within guilds of bacteria, and predation will be matched by defence, but overall an interdependent, 'altruistic' community will evolve on a local scale (e.g. a hydrothermal pool). The word 'local' is stressed here, applying to a pool or to a local bacterial habitat.

The second mechanism is that these *local* populations of organisms, which locally change their environment, will also accidentally change their *regional* environment. The development of regional concentrations of oxygen-rich water is an example that can be imagined. Local populations that are adversely affected by this regional change will be disadvantaged and will tend to die out. In contrast, local populations that can exploit the regional change will flourish. Consequently, a system will evolve in which a set of local populations lives interdependently in the regional environment they have created, the messengers between the components of the system being, for instance, exchange of trace gases or chemicals, or even of bacterial populations, and the memory of the system being the long term genetic heritage and the regional environment created (a pool of water enriched in oxygen that lasts a few days is many generations of bacterial life).

Now let us generalize further, and imagine a set of regional systems. Some regional systems will prosper, and accidentally change the global environment. They may change climate by emitting dimethyl sulphide, for instance, or more radically by causing the build-up of oxygen. In the modern world, a tropical rain forest region, for instance, can so influence the global air circulation, accidentally initially, that rainfall is enhanced over the forest. The impact of successful forest growth in the first region may be such that rainfall is decreased elsewhere. Any forest that attempts to grow in the second area will fail. Competitive natural selection prevails. The planet becomes populated by ecosystems which each operate to enhance not only their local and regional environment, but also by proxy act to favour the coexisting pattern of climax vegetation in all other ecosystems too. Eventually, a global system evolves in which only the compatible, mutually-enhancing regional systems co-exist.

Obviously, cuckoos (spoilers that enjoy the conditions but do not contribute) will evolve too, but will be limited by the maximum number that can be accommodated without reducing the net benefit to zero. In nature, cuckoos do not wipe out their hosts. Technological advance comes too, producing super-competitive organisms that did not necessarily need the advantage of the mutualistic bacterial society. With each new bacterial 'invention' would have come a crisis of adaptation until a new mutualism was restored, and the spiral of evolutionarily stable but dynamic strategies continued.

Can 'optimum' occur?

What does the dynamic stability tend towards? What does it maximize? Indeed, does 'optimum' ever occur? Most stably efficient homeostatic systems, like a domestic heating system, are sub-optimal, oscillating around the optimum, and biology, in its patchiness, appears to act similarly.

Of the many possible answers — total biomass, total productivity, total use of energy, total entropy effect, and so on — perhaps the best answer is that natural selection will act to choose the dynamic state with the highest chance of survival in each temporal interval (until struck by either an external event, such as a meteorite, or an internal 'technological' advance). Emissions to the atmosphere will occur in such a way that is consistent with the survival of the constantly adapting bacterial communities, and those communities that are adversely affected will die.

Over the aeons, the Sun has brightened, yet the Earth's surface temperature, adequate in the early

Archaean to sustain marine littoral life, has not shown any sign of boiling the oceans.The global greenhouse has clearly changed over time, reducing the warming effect of the air as solar radiance has increased and CO_2 has been degassed from the volcanoes. Walker *et al.* (1981) suggested that an inorganic negative feedback mechanism operates. This invokes increased surface weathering as surface temperature and rainfall increased; the weathering results in capture and precipitation of CO_2 from the air/ocean, and hence causes decrease in temperature, returning to stability. Similarly, the reverse operates if temperature falls, and the feedback loop stabilises conditions over the aeons.

The Gaian response to this is that the atmosphere is a biological construct and is biologically 'fine-tuned' (Watson & Lovelock 1983). Although ultimately four-fifths of CO_2 is precipitated as $CaCO_3$ (essentially inorganic, though nowadays biologically mediated), one fifth of carbon is precipitated biologically, and this proportion can rapidly be altered as the biota encounter favourable or unfavourable conditions. Moreover, biological processes produce CH_4 as well as CO_2, and have impact on the most important greenhouse gas, water, by influencing evaporation, changing albedo of the surface or by helping cloud nucleation. Collectively, these processes can act much more quickly than inorganic processes, and over time, the total flux of biological operations is very large. An externally forced change to increase temperature (e.g. a brightening Sun) enhances biological activity, drawing down atmospheric carbon, changing albedo, increasing cloudiness. Reduction of temperature operates in reverse. Alternatively, 'technological change' can occur, adapting the biosphere so that it is optimised to the new external environment, as with the evolution of the C_4 plants (which thrive in a low CO_2 glacial atmosphere) in the Tertiary. Time scale is critical: biological and inorganic controls act on different time scales for the most part, as systems of nested cycles.

Management of the atmosphere?

Whether or not it is accepted that a Gaian control is possible, it is nevertheless clear that *the bulk of the modern atmosphere is a biological construct,* created and maintained by the biosphere. Nitrogen, Oxygen, Carbon dioxide, and the trace gases are all biologically cycled, and only the noble gases are essentially abiological. Water vapour, the other major component of the air, is intensively managed by photosynthesis and respiration, and also by production of cloud nucleation chemicals.

The parallel between the atmosphere and the recent example of 'Biosphere 2' is fascinating. In this experiment, a closed experimental ecosystem was set up in southern Arizona. Biosphere 2 is a 1.3 ha airtight structure roofed in glass and underlain by an impermeable liner, housing an artificial ecosystem. The system was closed, and during the first 16 months of closure the O_2 in the air decreased from 21% to 14% (Severinghaus *et al.* 1994), but with an increase in CO_2 of only about 2%, even after allowing for the CO_2 taken up by a scrubber. The obvious explanation of the O_2 decline is that photosynthesis had not kept up with respiration of soil organic matter, but this explanation does not account for the small change in CO_2. Severinghaus *et al.* (1994) demonstrated that the 'missing' respired CO_2 was actually taken up by reaction with the structure's concrete, to form calcium carbonate. The analogy with the planet as a whole is interesting; the Earth's crust is analogous to the concrete, and weathering and reaction with Ca from the crust are the planetary equivalent of the uptake. Why does the O_2 content of the atmosphere not fall also? What sets it? Is the control essentially inorganic (Walker *et al.* 1981) or biologically managed (Lovelock 1988).

Nevertheless, whatever the control on oxygen, the carbon isotope evidence is clear that the modern carbon cycle has been in operation through most (or all) of the geological record (Schidlowski, 1991). During this time, the biosphere has exerted a profound influence on the atmosphere. This idea is not new (Macgregor 1949), but the implication is that the thermal control exerted by the greenhouse gases in the atmosphere has set the surface temperatures and by stabilizing the ocean has hence controlled much of the physical and tectonic evolution of the planet.

This work originally began as part of the Archaean Crustal Project of the University of Zimbabwe, which was strongly supported by John Sutton (who was especially interested in the palaeontology!). I thank my 'Belingwe' colleagues, Mike Bickle, Tony Martin, John Orpen and Jim Wilson, for many years of generous cooperation and discussion. In the early stages of the work, Geoff Bond's support was crucial. The biological side of our research began during discussions with Mike Bickle on a visit to the ruins of the Huntsman stromatolites, described by A.M. Macgregor. Much of this present manuscript is based on work carried out during research sponsored by NSERC Canada at the University of Saskatchewan, and I thank W.G.E. Caldwell, W. Braun, the late H. E. Hendry and W.A.S.Sarjeant for their interest and support. I also thank J. Cann, the late P. Cloud, F. Dyson, J.E. Lovelock, L. Margulis, J. Walker, C. van Dover and M. and R. Viljoen for much discussion, corespondence and, with many, debate in the field, and O. Kandler for sending me his fascinating recent work. I particularly thank Andy Knoll and Jim Kasting for penetrating critical comment.

References

ACHENBACH-RICHTER, L., GUPTA, R., STETTER, K. O., & WOESE, C. R. 1987. Were the original eubacteria thermophiles? *Systematics and applied microbiology,* **9**, 34–39.

——, ——, ZILLIG, W. & WOESE, C. R. 1988. Rooting the archaebacterial tree: the pivotal role of *Thermococcus celer* in Archaebacterial evolution. *Systematics and applied microbiology,* **10**, 231–240.

BARBER, J. & ANDERSSON, B. 1994. Revealing the blueprint of photosynthesis. *Nature,* **370**, 31–34.

BOCK, A., JARSCH, M., HUMMEL, H. & Schmid, G. 1985. Evolution of translation. *In:* SCHLIEFER, K. H. & STACKEBRANDT, E. (eds) *Evolution of Prokaryotes.* Academic Press, London, 73–91.

BYERLY, G. R., LOWE, D. R. & WALSH, M. M. 1986. Stromatolites from the 3300–3500 Myr Swaziland Supergroup, Barberton Mountain Land, South Africa. *Nature,* **319**, 489–491.

CHAPMAN, D. J. 1992. Origin and divergence of protists. *In*: SCHOPF, J. W. & KLEIN, C. (eds) *The Proterozoic biosphere,* Cambridge University Press, 477–483

COLINVAUX, P. 1993. *Ecology 2.* J. Wiley and Sons, New York.

COMPSTON, W. & Pidgeon, R. T. 1986. Jack Hills, evidence of more very old detrital zircons in Western Australia. *Nature,* **321**, 766–9.

DARWIN, C. 1959. Some unpublished letters (1871). Ed. Sir Gavin de Beer. *Notes and Records of the Royal Society. London,* **14**, 1.

DELONG, E. F., WU, K. Y., PRÉZELIN, B. B. & JOVINE, R. V. M. 1994. High abundance of Archaea in Antarctic marine picoplankton. *Nature,* **371**, 695–697.

DONOHUE, T. M., HOFFMAN, J. H., HODGES Jr., R. R. & Watson, A.J. 1982. Venus was wet: a measurement of the ratio of deuterium to hydrogen. *Science,* **216**, 630–633.

ELLIS, R. J. 1994 The general concept of molecular chaperones. *Philosophical Transactions of the Royal Society, London,* **B339**, 257–261

FOGG, G. E. 1982. Marine plankton. *In:* CARR, N. G. & WHITTON, B. A. (eds) *The biology of cyanobacteria.* Blackwell, Oxford, 491–513.

GETHING, M.-J. & Sambrook, J. 1992. Protein folding in the cell. *Nature,* **355**, 33–45.

GREEN, D. H. 1975. Genesis of Archaean peridotitic magmas and constraints on Archaean geothermal gradients and tectonics *Geology,* **3**, 15–18.

HANSSON, L.-A. 1992. The role of food chain composition and nutrient availability in shaping algal biomass development. *Ecology,* **73**, 241–247.

HARTMAN, F. C. & HARPEL, M. R. 1994. Structure, function, regulation and assembly of D-Ribulose-1,5-Bisphosphate Carboxylase/Oxygenase. *Annual Review of Biochemistry,* **63**, 197–234.

HEDENQUIST, J. W. & LOWENSTERN, J. B. 1994. The role of magmas in the formation of hydrothermal ore deposits. *Nature,* **370**, 519–527.

HEMMINGSEN, S. M., WOOLFORD, C., VAN DER VIES, S. M., TILLY, K., DENNIS, D. T., GEORGOPOULOS, C. P., HENDRIX, R. W. & ELLIS, R.J. 1988. Homologous

plant and bacterial proteins chaperone oligomeric protein assembly. *Nature,* **333**, 330–334.

HENTSCHEL, U. & Felbeck, H. 1993. Nitrate respiration in the hydrothermal vent tubeworm *Riftia pachyptila. Nature,* **366**, 338–340.

HOLLING, C. S. 1992. Cross-scale morphology, geometry, and dynamics of ecosystems. *Ecological monographs,* **62**, 447–502.

HUNTEN, D. M. 1993. Atmospheric evolution of the terrestrial planets. *Science,* **259**, 915–919.

JANNASCH, H. W. 1984. Microbes in the oceanic environment. *In:* KELLY, D. P. & CARR, N. G. (eds) *The Microbe, 1984: Prokaryotes and Eukaryotes.* Cambridge University Press, 97–122.

JONES, C. W. 1985. The evolution of bacterial respiration. *In:* SCHLIEFER, K. H. & STACKEBRANDT, E. (eds) *Evolution of Prokaryotes.* Academic Press, London, 175–204.

KANDLER, O. 1992. Where next with the archaebacteria? *Biochemical Society Symposium,* **58**, 195–207.

—— 1994. The early diversification of life. *In:* BENGTSON, S. (ed.) *Early life on Earth.* Nobel symposium, **84**. Columbia University Press, New York, 152–161.

KASTING, J. F. 1992. Models relating to Proterozoic atmospheric and ocean chemistry. *In:* SCHOPF, J. W. & KLEIN, C. (eds) The Proterozoic biosphere, Cambridge University Press, 1185–1187.

—— 1993. Earth's early atmosphere. *Science,* **259**, 920–925.

——, EGGLER, D. H. & RAEBURN, S. P. 1993*a*. Mantle redox evolution and the oxidation state of the Archaean atmosphere. *Journal of Geology,* **101**, 245–257.

——, WHITMORE, D. P. & REYNOLDS, R. T. 1993*b*. Habitable zones around main sequence stars. *Icarus,* **101**, 108–128.

KIRCHNER, J. W. 1989. The Gaia hypothesis: can it be tested? *Reviews of Geophysics,* **27**, 223–236.

KIRK, K. L. & GILBERT, J. J. 1992. Variation in herbivore response to chemical defenses: zooplankton foraging on toxic cyanobacteria. *Ecology,* **7**, 2208–2217.

KNOLL, A. H. 1992. The early evolution of Eukaryotes: a geological perspective. *Science,* **256**, 622–627.

—— & BAULD, J. 1989. The evolution of ecological tolerance in prokaryotes. *Transactions of the Royal Society of Edinburgh, Earth Sciences,* **80**, 209–223.

LARGE, P. J. 1983. *Methylotrophy and methanogenesis.* van Nostrand, Wokingham.

LAWLOR, D. W. 1993. *Photosynthesis.* 2nd edition. Longmans, Harlow.

LEVIN, S. A. 1992. The problem of pattern and scale in ecology. *Ecology,* **73**, 1943–1967.

LOVELOCK, J. E. 1979. *Gaia – a new look at life on Earth.* Oxford University Press.

—— 1988. *The ages of Gaia.* W.W. Norton, New York.

—— 1989. Geophysiology: the science of Gaia. *Reviews of Geophysics,* **27**, 215–222.

—— & MARGULIS, L. 1974. Homeostatic tendencies of the earth's atmosphere. *Origins of Life,* **5**, 93–103.

MACGREGOR, A. M. 1949. The influence of life on the face of the Earth. *Rhodesia Scientific Association: Proceedings and Transactions,* **42,** 5–10.

MARGULIS, L. & SAGAN, D. 1986. *Microcosmos.* Simon and Schuster, New York.

MARTIN, A., NISBET, E. G. & BICKLE, M. J. 1980. Archaean stromatolites of the Belingwe greenstone belt, Zimbabwe (Rhodesia). *Precambrian Research,* **13,** 337–362.

MARTIN, J., HORWICH, A. L. & HARTL, F. U. 1992. Prevention of protein denaturation under heat stress by the chaperonin Hsp60. *Science,* **258,** 995–998.

NISBET, E. G. 1986. RNA and hot water springs. *Nature,* **322,** 206.

—— 1987. *The Young Earth.* G. Allen and Unwin, London.

—— 1991. *Living Earth.* Chapman and Hall, London.

——, CANN, J. R. & VAN DOVER, C. L. 1995. Origins of photosynthesis. *Nature,* **373,** 479–480.

——, MATTEY, D. P. & LOWRY, D. 1994. Are some diamonds dead bacteria? *Nature,* **367,** 694.

——, MARTIN, A., BICKLE, M. J. & Orpen, J. L. 1993. The Ngezi Group: komatiites, basalts and stromatolites on continental crust. *In:* BICKLE,M. J. & NISBET, E. G. (eds) *The geology of the Belingwe greenstone belt.* A. A. Balkema, Rotterdam, 121–166.

PACE, M. L. & FUNKE, E. 1991. Regulation of planktonic microbial communities by nutrients and herbivores. *Ecology,* **72,** 904–914.

PARKES, R. J., CRAGG, B. A., BALE, S. J., GETLIFF, J. M., GOODMAN, K., ROCHELLE, P. A., FRY, J. C., WEIGHTMAN, A. J. & HARVEY, S. M. 1994. Deep bacterial biosphere in Pacific Ocean sediments. *Nature,* **371,** 410–412.

POWER, M. E. 1992. Top-down and bottom-up forces in food webs: do plants have primacy? *Ecology,* **73,** 733–746

RAO, K. K., CAMMACK, R. & HALL, D.O. 1985. Evolution of light energy conversion. In: SCHLIEFER, K. H. & STACKEBRANDT, E. (eds) *Evolution of Prokaryotes.* Academic Press, London, 143–173.

ROWBURY, R. J., ARMITAGE, J. P. & KING, C. 1983. Movement, taxes and cellular interactions in the response of microorganisms to the environment. *In:* SLATER, J. H., WHITTENBURY, R. & WIMPENNY, J. W. T. (eds) *Microbes in their natural environment.* Cambridge University Press, 299–350.

RUNNEGAR, B. N. 1992. The tree of life. *In:* SCHOPF, J. W. & KLEIN, C. (eds) The Proterozoic biosphere. Cambridge University Press, 471–475.

SCHIDLOWSKI, M. 1984. Early atmospheric oxygen levels: constraints from Archaean photoautotrophy. *Journal of the Geological Society, London,* **141,** 243–250.

——.1989. Evolution of the sulphur cycle in the Precambrian. *In:* BRIMBLECOMBE, P. & LEIN, A. Y. (eds) *Evolution of the global biogeochemical sulphur cycle,* SCOPE 39, J. Wiley, Chichester, 3–20.

—— 1991. Organic carbon isotope record: index line of autotrophic carbon fixation over 3.8 Gyr of Earth history. *Journal of South-east Asian Earth Sciences,* **5,** 333–337.

SEVERINGHAUS, J. P., BROECKER, W. S., DEMPSTER, W. F., MacCALLUM, T. & WAHLEN, M. 1994. Oxygen loss in Biosphere 2. *EOS,* **75,** 33–37.

STACKEBRANDT, E. 1985. Phylogeny and classification of prokaryotes. In: SCHLIEFER, K. H. & STACKEBRANDT, E. (eds) *Evolution of Prokaryotes.* Academic Press, London, 309–334 .

STANIER, R. Y., INGRAHAM, J. L., WHEELIS, M. L. & PAINTER, P. R. 1987. *General microbiology,* 5th UK edition. Macmillan, London.

STETTER, K. O. 1986. Divesity of extremely thermophilic archaebacteria. *In:* BROCK, T. D. (ed.) *Thermophiles: general, molecular, and applied microbiology.* J. Wiley, New York, 39–99

SUMMERS, D. P. & CHANG, S. 1993. Prebiotic ammonia from reduction of nitrite by Iron(II) on the early Earth. *Nature,* **365,** 630–637.

SUMMONS, R. E. & HAYES, J. M. 1992. Principles of molecular and isotopic biogeochemistry. *In:* SCHOPF, J. W. & KLEIN, C. (eds) *The Proterozoic biosphere.* Cambridge University Press, 83–93

SUNDARM. T. K. 1986. Physiology and growth of thermophilic bacteria. *In:* BROCK, T. D. (ed.) *Thermophiles: general, molecular, and applied microbiology.* J. Wiley, New York, 75–106.

THAUER, R. K. & MORRIS, J. G. 1984. Metabolism of chemotrophic anaerobes. *In:* KELLY, D. P. & CARR, N. G. (eds) *The Microbe, 1984: Prokaryotes and Eukaryotes.* Cambridge University Press, 123–168

TUNNICLIFFE, V. 1991. The biology of hydrothermal vents: ecology and evolution. *Oceanographic marine biology Annual Review,* **29,** 319–407.

VAN DOVER, C. L., DELANEY, J., SMITH, M. & J. R. Cann, J.R. 1988. Light emission at deep-sea hydrothermal vents. *EOS,* **69,** 1498.

VAN LIERE, L. & WALSBY, A. E. 1982 Interactions of cyanobacteria with light. *In:* CARR, N. G. & WHITTON, B. A. (eds) *The biology of cyanobacteria,* Blackwell, Oxford, 9–45.

VOLBEDA, A., CHARON, M-H., PIRAS, C., HATCHIKIAN, E. C., FREY, M. & FONTCILLA-CAMPS, J. C. 1995. Crystal structure of the nickel-iron hydrogenase from *Desulfovibrio gigas. Nature,* **373,** 580–587.

VOSSBRINCK, C. R., MADDOX, J. V., FRIEDMAN, S., DEBRUNNER-VOSSBRINCK, B. A. & WOESE, C. R. 1987. Ribosomal RNA sequence suggests microsporidia are extremely ancient eukaryotes. *Nature,* **326,**411–414.

WACHTERSHAUSER, G. 1990. Evolution of the first metabolic cycles. *Proceedings of the National Academy of Sciences, USA,* **87,** 200–204.

WALKER, J. C. G. 1977. *Evolution of the atmosphere.* Macmillan, New York.

—— 1990 Origin of an inhabited planet. *In:* NEWSOM, H. E. & JONES, J. H. (eds) *Origin of the earth,* Oxford University Press, 371–376.

——, HAYS, P. B. & KASTING, J. F. 1981 A negative feedback mechanism for the long-term stabilisation of earth's surface temperature. *Journal of Geophysical Research,* **86,** 9776–9782.

WATSON, A. J. & LOVELOCK, J. E. 1983. Biological homeostasis of the global environment: the parable of the daisy world. *Tellus,* **35B,** 284–289.

——, DONOHUE, T. M. & KUHN, W. R. 1984 Temperatures in a runaway greenhouse on the evolving Venus. *Earth and Planetary Science Letters,* **68,** 1–6.

WHITTON, B. A. & POTTS, M. 1982 Marine littoral. *In:* CARR, N. G. & WHITTON, B. A. (eds) *The biology of cyanobacteria,* Blackwell, Oxford, 515–542.

WILKS, M. & NISBET, E. G. 1988 Stratigraphy of the Steep Rock Group, northwest Ontario: A major Archaean unconformity and Archaean stromatolites. *Canadian Journal of Earth Sciences,* **25,** 370–391.

WILSON, D. S. 1992. Complex interactions in meta-communities, with implications for biodiversity and higher levels of selection. *Ecology, 73,* 1984–2000.

WEN, J-S, PINTO, J. P. & YUNG, Y. L. 1989. Photochemistry of CO and H_2O: analysis of laboratory experiments and applications to the prebiotic Earth's atmosphere. *Journal of Geophysical Research,* **94,** 14957–14970.

WOESE, C. R. 1983. The primary lines of descent and the universal ancestor. In: BENDALL, D. S. (ed.) *Evolution from molecules to men.* Cambridge University Press, 209–234.

—— 1985. Why study evolutionary relationships among bacteria? *In:* SCHLIEFER, K. H. & STACKEBRANDT, E. (eds) *Evolution of Prokaryotes,* Academic Press, London, 1–30 .

——., KANDLER, O. & WHEELIS, M. L. 1990. Towards a natural system of organisms: proposal for the domains Archaea, Bacteria and Eucarya. *Proceedings of the National Academy of Sciences, USA,* **87,** 4576–4579

YUNG, Y. L., WEN, J-S., MOSES, J. I., LANDRY, B. M. & ALLEN, M. 1989. Hydrogen and deuterium loss from the terrestrial atmosphere: a quantitative assessment of nonthermal escape fluxes. *Journal of Geophysical Research,* **94,** 14971–14989.

ZILLIG, W., SCHNABEL, R., STETTER, K., THOMM, M., GROPP, F. & REITER, W. D. 1985. The evolution of the transcription apparatus. *In:* SCHLIEFER, K. H. & STACKEBRANDT, E. (eds) *Evolution of Prokaryotes,* Academic Press, London, 73–90.

Tectonic evolution of greenstone belts

R. M. SHACKLETON

The Croft Barn, Church Street, East Hendred, Wantage, Oxon OX12 8LA, UK

Abstract: Granite–greenstone terrains represent the main process by which continental crust was formed in the Archaean; the same process has continued since, at a diminishing rate. Granite plutonism added a layer about ten kilometres thick under the greenstones. From Early Archaean to Late Proterozoic, there was a progressive change in the volcanics; komatiites decreased from the Early Archaean, while the proportion of andesites and felsic volcanics increased. Continent-derived sediments, rare or absent from the Early Archaean, increased through time.

Tectonic deformation in the greenstone belts is separable into pre-diapiric, diapiric and post-diapiric. The first deformations were translational, mainly flat thrusts, the second compressional without crustal shortening, the third involved crustal shortening. Only the diapiric structures are an essential element of greenstone belt evolution but the others are generally important. The pre-diapiric and diapiric structures are related to subduction; the post-diapiric structures may be accretionary or collisional. Granite–greenstone terrains show only minor crustal thickening. Some adjacent high-grade terrains were intensely deformed, and the crust greatly thickened, by continent-continent collision.

John Sutton and Janet Watson, through their work on the Lewisian complex (Sutton & Watson 1951), demonstrated that there must have been an immense time interval between the Scourian and Laxfordian events. As Read, a great beer drinker, characteristically commented after their paper, they introduced great draughts of time into the Archaean. Subsequent geochronological work wholly vindicated their conclusions. They demonstrated the immensity of Archaean time.

The term Archaean is often used as if that is enough to specify age. The Archaean represents over a third of known geological time and, during that time, radiogenic heat production dropped by more than a half, greatly reducing the energy that drives tectonism. So one may expect there to be more differences in tectonic processes between the Early and the Late Archaean than between the Archaean and today. With that in view, I shall attempt to compare Early with Late Archaean tectonism, and that with its more recent styles, especially as seen in Africa since that is the continent where I have mostly worked. Because granite–greenstone terrains are often regarded as characteristic of the Archaean, make up more than three quarters of the areas of the Archaean cratons, and must represent the way the sialic crust grew, I shall look particularly at them.

The best known Early Archaean greenstone belt in Africa is Barberton (c. 3.5 Ga) in the Kaapvaal craton, South Africa (Fig. 1). This belt is very different from the Late Archaean Migori (SW Kenya) greenstone belt (c. 2.8 Ga), where I worked many years ago, whereas the Migori belt seemed not unlike Ordovician areas in North Wales that I knew. The proportions of ultramafic, mafic, intermediate and acid volcanic rocks, in Barberton and in the Migori belt are quite different. They are compared in Table 1, taking a rather larger area than the Migori belt, along with estimates for other greenstone belts. These estimates suggest an increase with time in the proportion of acid, and often, but not always, of intermediate (andesitic) volcanics relative to mafic (tholeiitic basalts) and a drastic decrease in ultramafic (komatiitic) volcanics. However, the variations between different belts are so large that the figures must be treated with caution.

The most striking and significant feature of the Archaean granite–greenstone terrains is the preponderance of the granitoids (Table 2). The proportions must partly depend on depth of erosion, which is often not established. The form and structure of the granitoid bodies is also dependent on depth of exposure: those at high levels, as in Arabia, are characteristically discordant, those at deeper levels often diapiric, diapirism requiring high viscosity of wall rocks. Still deeper crustal granitoids tend to have a regionally flat foliation and intense deformation. Despite these complications, it is clear that the volume of sialic magma, mostly of direct or indirect mantle origin, as shown by geochemistry, was far greater in the earlier stages of geological history. The actual rate of granitoid plutonism was also much higher earlier in the earth's history. The introduction of these huge

From COWARD, M. P. & RIES, A. C. (eds), 1995, *Early Precambrian Processes*, Geological Society Special Publication No. 95, pp. 53–65.

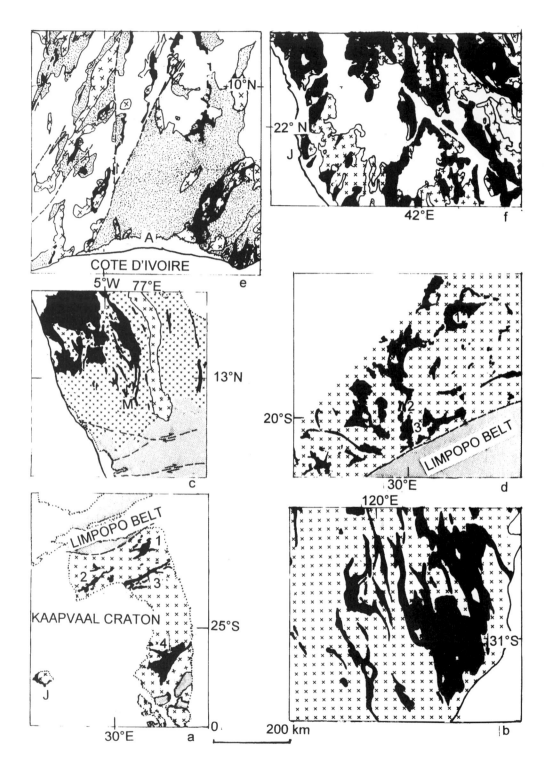

Table 1. *Proportions (by area) of different volcanics in greenstone belts*

Approx. age (Ga)	Barberton 3.5	Pilbara, W. Australia 3.5	U. Bulawayan, Zimbabwe 2.7	Migori-, Kendu, SW Kenya 2.9–2.7	Sekukumaland, Tanzania 2.8	Superior Province, Canada 2.75	Birrimian, W. Africa 2.2	Arabia 0.8
Acid	<5	c. 40	30	50	20	15	c. 50	c. 40?
Intermediate	<5		20	30	30	30	c. 50	c. 40?
Mafic	>80	c. 60	50	20	40	55	c. 50	c. 40?
Ultramafic	<10	c. 60	(minor)	—	—	—	—	c—

Table 2 *Proportion of granitoids to greenstone belts (by area)*

	Kaapvaal Craton, S. Africa	Pilbara, W. Australia	Yilgarn, SW. Australia	Slave Province, Canada	Tanzania Craton (Central & North)	Birrimian W. Africa	Arabia
Approx. age (Ga)	3.5–3.0	3.5–3.0	3.0–2.7	2.7	2.8	2.2	0.8
% Granitoids	90	60	70	75	75	60	40

volumes of granitoids, often amounting to about 10 km of crust, must have had major structural effects.

Structural effects of granite plutonism

The shapes of greenstone belts differ greatly, in different terrains and regions. At one extreme, there are the arrays so typical of greenstone belts, their cuspate outlines formed by a series of curved contacts with adjacent granitoid plutons. Typical examples are those in the Zimbabwe craton. At the other extreme are belts which are strongly linear, although flattened cuspate outlines can usually still be made out.

The structure of the cuspate-outlined greenstone belts in Zimbabwe (Fig. 1d) was already recognized by Maufe in 1919 and interpreted in the classic work of MacGregor (1951), who attributed their structures to the diapiric uprise of the granitic plutons and consequent down-drag of the synforms between them. This interpretation was rejected in the case of the prime example, the Chinamora batholith (Fig. 1d) and the surrounding greenstone belts in favour of a refolding hypothesis (Snowden & Bickle 1976; Snowden & Snowden 1979; Snowden 1984), although it was accepted that the structures were influenced by granite intrusions. Their argument was that the Chinamora pluton is a complex of multiple intrusions, including an extensive late sheet-like intrusion, and that the deformation of this sheet and the overlying greenstones occurred during a series of phases, F1, F2 and F3, over a long time. However, the diapiric interpretation was convincingly confirmed, first by detailed strain measurements using xenoliths in the

Fig. 1. Comparison of the geometry of Archaean and younger granite-greenstone terrains. (a) Kaapvaal craton, South Africa. Black, Early Archaean greenstone belts; crosses, Archaean granitoids; grey, granulite-facies gneisses etc. of Limpopo Belt (in N) and Swaziland (in S). Greenstone belts: 1, Sutherland; 2, Pietersburg; 3, Murchison; 4, Barberton. J, Johannesburg. (b) Yilgarn Block (part), Western Australia. Black, Late Archaean greenstone belts; crosses, Late Archaean granitoids and gneisses. (c) South India craton (part). Black, Late Archaean greenstone belts; spaced crosses, Closepet granite; close crosses, gneisses (basement of greenstone belts) and Late Archaean granites. Shaded, Late Archaean granulite-facies domain. M, Mysore. (d) Zimbabwe craton. Black, Late Archaean and minor Early Archaean greenstone belts. Crosses, granitoids and gneisses. 1, Chinamora; 2, Shurugwe; 3, Belingwe. (e) Part of Early Proterozoic (Birrimian) granite-greenstone terrain, West Africa. Black, volcanics (with some sediments); mottled; flysch, minor volcanics. Crosses, granitoids. A, Abidjan. (f) Late Proterozoic Arabian Shield (central part). Black, volcanics, minor ophiolites and sediments. Crosses, granites. J, Jeddah. All blank areas: younger cover (or sea).

pluton, which also showed that the pluton must have expanded by 'ballooning' which imposed flattening strains in the pluton, most intense towards its margin, and parallel to the contacts (Ramsay 1975, 1989). More recently, very detailed strain measurements in the surrounding greenstone belts also showed that the strains, although triaxial rather than pure flattening, were incompatible with the cross-folding interpretation and only explicable by diapirism and 'ballooning' (Jelsma 1993; Jelsma *et al.* 1993). Strain data have similarly demonstrated diapiric deformation in the Archaean Holenarsipur area, southern India (Bouhallier *et al.* 1993).

Other granitoid domes in greenstone terrains elsewhere have been more convincingly attributed to cross-folding, for example, an oval dome near Yalgoo, in the Yilgarn block, W. Australia (Fig. 1b) (Myers & Watkins 1985). An early subhorizontal gneissosity (D1 deformation) was imposed on pegmatitic gneiss, and a foliation on the overlying greenstones; a thick (*c.* 10 km) subhorizontal sheet of monzogranite was injected, mostly along or near the gneiss/greenstone interface, and this assemblage was then compressed into two interfering sets of folds with E–W and N–S vertical axial surfaces and foliations (D2 and D3 deformations). The result was a series of typical interference folds (Ramsay 1967, type 1). Most lineations (intersection?) are parallel to D2 or D3 fold axes; no radial extension lineations were seen in the dome. The authors conclude that their evidence supports reinterpretation of granite–greenstone structures elsewhere in the Yilgarn block, in Zimbabwe (Chinamora) and the Superior Provinces. Very detailed structural work in the Wabigoon subprovince of the western Superior Province (Schwerdtner 1990) led to the conclusion that the domes within granitoid complexes there, which include large volumes of gneissic tonalite–granodiorite, were tectonic, not diapiric. However, the Yilgarn and Wabigoon structures described are clearly different from, and probably represent deeper erosion than, many diapiric intrusions (e.g. Chinamora).

It may be questioned whether the interference folds are due to regionally successive deformation phases with mutually perpendicular principal compressive stresses σ_1. Nearly always, strong deformation results in oblate strain ellipsoids, implying extension in two directions (X & Y) at right angles. These deformations may be simultaneous (as in sheath folds), alternating, or successive, while regionally, σ_1 remains constant. The whole deformation belt does not usually extend. A change in regional σ_1 from normal to parallel to an arc or a collision zone or any result of plate convergence, is unlikely to happen often, yet 'cross

folds' are common, and usually attributed to successive deformation phases. More probably, many reflect alternating or successive responses to consistently oriented stress.

In the Canadian Cordillera and in the Late Proterozoic accreted arc complex of Egypt, yet another type of granite dome is recognized: the core complex (Crittenden *et al.* 1980; Sturchio *et al.* 1983). In these, a granitic gneiss core protrudes into lower-grade rocks as a result of gently-dipping extensional shearing and faulting sub-parallel to the gneissosity and foliation. Such structures have not yet been recognized in Archaean granite–greenstone terrains.

The principal criteria which can be used to distinguish these various types of granitoid domes are shown in Table 3. The most important criteria are stretching lineation patterns, strain characteristics and shear-sense indicators, none of which are usually recorded so it is difficult to determine which plutons are diapiric or ballooning and therefore difficult to estimate the structural effects of the voluminous plutonism. In the particular case of the tight synclinal greenstone belts round the Chinamora pluton in Zimbabwe, it appears that its ballooning pressure was responsible for most of the compressive deformation. Before trying to estimate the pluton-driven deformation elsewhere, it is necessary to summarize the evidence for pre-pluton structures, especially early gravitational structures, thrusts, bedding-plane slip and early (D1) foliations parallel to layering.

Early (pre-diapir) structures: gravity slides, thrusts, bedding-plane slip, extensional structures

Regional early (D1a) pre-diapir extensional slides have been described from the Barberton area, South Africa (Fig. 1a) (de Wit 1986; Lowe & Byerly 1986). They are entirely confined to the basal Onverwacht Jamestown ophiolite complex and are marked by horizons of anastomosing veins separated by schistose and protomylonitic folia of fuchsite, chlorite, sericite and serpentine. They are cut by Onverwacht simatic intrusions and deformed by later folds and thrusts. Gravity slides, recumbent folds and olistostromes in the upper Onverwacht and Fig Tree Groups have also been described (de Wit 1982). Similar early structural repetitions ascribed to gravity gliding from uplifts have been described from Western Australia (Martyn 1986).

However, from a tectonic viewpoint, the important structures are the many early (pre-diapir) thrusts which have been mapped in many greenstone belts, notably Barberton. They are associated

Table 3. *Some criteria which can be used to distinguish types of granitoid domes*

Diapirs and 'balloon s'	Core complexes	Interference-fold domes
Gneissose foliation intensifies outward to contact; continuous into greenstones, diminishing again away from contact; concentric, parallel to contact; flat on top of pluton.	Foliation in granitoid core and in overlying schists parallel, gently inclined. Intense shearing, sometimes mylonites, above core.	Early foliation often parallel to granite-greenstone contacts but two main foliations steep and approximately at right angles.
Stretching lineation, radial from diapir apex, where weak or absent, steepens outwards, plunging downwards parallel to contact.	Stretching lineation unidirectional across core; gentle plunge, in foliation.	Stretching lineations steep in tight domes, may be parallel to fold axes in more open domes. Intersection lineations steep.
No low-strain pressure shadows on opposite side of pluton.	Perhaps 'pressure-shadows' at both ends of core on line of lineation.	Low-strain 'pressure-shadows' at ends of elongated domes.
Synkinematic metamorphic T decreases away from contact.	Synkinematic metamorphism decreases rapidly, discontinuously upwards.	Metamorphism related to plutons pre-dates interfering structures.
Oldest members of greenstone sequence in contact with pluton, face away.	No systematic relation of cover sequence to contact with core.	No systematic relation between pluton contacts and envelope sequence.
Folds often concentric round pluton.	Folds immediately above core complex usually tight or rootless recumbent isoclines.	Two perpendicular sets of folds regionally consistent in orientation, transgress, or deflected around, plutons.
Shear sense indicators near pluton contact show pluton up, greenstones down.	Shear sense indicators near core show unidirectional movement.	Shear sense indicators show movement consistent with bedding-plane slip, top away from inter-pluton synforms.
At centres of cuspate triangular inter-dome synforms, steep biaxial constrictional strain.		Inter-dome synforms show fold set crossing the synforms at a high angle.
'Ballooning' plutons show near-oblate (flattening) strain ellipsoids near margins.		

with an early, originally flat-lying, schistosity (an almost universal feature of regional metamorphism) and often by recumbent folds (Platt 1980).

In the southeastern part of the Barberton belt there are many thrusts which dip gently SE towards the adjacent *c.* 3.0 Ga Piggs Peak (Lochiel) batholith; they are associated with tight NW-verging folds and are regarded as D2 structures (Lamb 1986). Recognition of the resulting repetition drastically reduces the estimated original thickness of the Onverwacht Group from *c.* 15 km (Viljoen & Viljoen 1970) to *c.* 3.5 km (Lamb 1986). The implied tectonic transport direction, towards the NW, driving this major deformation, is consistent with the regional trend over much of the

Kaapvaal craton (Fig. 1a). In the narrow Jamestown synform, a northwest spur of the Barberton Greenstone Belt, these and earlier D1 thrusts appear (evidence of transport direction on both limbs is needed to be sure) to be tightly folded by the main synform of the Jamestown Belt (de Wit, in press), causing repetitions of the stratigraphic units (Anhaeusser 1972). The synform and associated folds appear to be controlled by the *c.* 3.2 Ga Kaap Valley diapir on the south side and the *c.* 3.2 Ga Nelspruit granite on the north side.

The Barberton evidence of intense pre-diapiric greenstone belt deformation, as well as similar evidence elsewhere, clearly shows that regional tectonic stresses rather than diapirism were generally responsible for intense deformation.

However the evidence of the major role of ballooning diapirism around the Chinamora pluton in Zimbabwe and the non-linear pattern and cuspate outlines of most of the greenstone belts in Zimbabwe shows that such intense tectonism is not a necessary part of greenstone belt evolution. Only in the south of the Zimbabwe granite–greenstone terrain, where it is strongly affected by Limpopo Belt deformation (Coward *et al.* 1976) is the non-linear pattern changed to a strongly linear one (Fig. 1d). This deformation is post-diapiric.

The basal contacts of greenstone belts: tectonic or unconformities?

The basal contacts of the Early Archaean (3.5–3.4 Ga) greenstone belts in Africa, where not obscured by granitoid plutons, are tectonic, for example Barberton, in the Kaapvaal craton and Shurugwe (Selukwe) in Zimbabwe. The basal Barberton greenstones are thought to be ocean-floor ophiolites (de Wit 1982; but cf. Kröner & Tegtmeyer 1994); but it is not clear on to what they were obducted if they are allochthonous ophiolites. They may have been obducted from an oceanic area adjacent to the Swaziland gneisses (Hunter *et al.* 1992), but any such interpretation is speculative.

In the Shurugwe area, southern Zimbabwe (Stowe 1984; Fig. 1d), an early Archaean (*c.* 3.4 Ga) greenstone sequence of komatiitic lavas, with interlayered jaspilite and some siliceous tuffs, is associated with large shear-bounded lenses of metaperidotite and metaharzburgite, mostly altered to serpentinites or talc-carbonates, and containing large podiform masses of chromite — an assemblage strongly suggesting an obducted ophiolite of ocean-floor origin. These earliest volcanics were subaerially eroded and covered unconformably by boulder conglomerates, coarse sandstones, phyllites and banded ironstones of the Wanderer Formation, which is in turn unconformably overlain by more greenstones. The lower greenstones were eroded subaerially before being covered unconformably by the Wanderer Formation, which itself contains not only chromite from the Selukwe ultramafics, but also many pebbles of granite: if the lower greenstones are oceanic they must have been obducted before being covered by the Wanderer Formation. The region south and southeast of Shurugwe is underlain by gneisses (Tokwe Gneiss) which are dated at *c.* 3.5 Ga. The ophiolitic greenstones of the Shurugwe area may first have been obducted, perhaps onto older continental crust represented by the Tokwe Gneisses, then eroded and covered by the Wanderer Formation. The whole Selukwe succession was overturned, either

by northward thrusting before the intrusion of the *c.* 3.3 Ga Mont d'Or tonalite (Stowe 1974), or after it as a consequence of overturning of an earlier parautochthonous synclinal fold by compression from the Late Achaean (*c.* 2.7 Ga) Limpopo collision belt (Tasmondo *et al.* 1992).

In contrast to the Early Archaean greenstone belts, some at least of the Late Archaean greenstones are clearly unconformable on older sialic basement gneisses. In the Belingwe area (Fig. 1d) of Zimbabwe, two greenstone assemblages *c.* 2.9 Ga and 2.7 Ga overlie older (*c.* 3.5–2.9 Ga) gneisses, and well-exposed unconformities at the base of both these assemblages have been described (Bickle *et al.* 1975). Deformation and grade of metamorphism are generally low in the Belingwe belt, and the overall synclinal structure appears simple; primary sedimentary and igneous textures remain preserved. However, there are many major shear zones within the sequence, notably at a sulphidic banded ironstone horizon immediately below the *c.* 6 km thick komatiitic–tholeiitic volcanic assemblage. This particular shear zone has been interpreted (Kusky & Kidd 1992) as a major detachment, on which the whole of the volcanics within the Belingwe syncline have been transported from elsewhere. Since critical evidence of shear sense and of the transport direction on either side of the syncline was not obtained, this can only be viewed as speculation, especially in view of the detailed evidence of the simplicity of structure, and absence, aside from areas in the south which are within the influence of the Limpopo belt (Coward *et al.* 1976), of significant tectonic transport in other greenstone belts in Zimbabwe, such as those adjacent to the Chinamora pluton (Jelsma *et al.* 1993). The allochthon hypothesis implies supposedly younger over older; the maps show the same thin sedimentary sequence below the volcanics around most of the 65 km × 20 km syncline. It seems more likely that the deformation at the base of the volcanics is simply bedding plane slip, the magnitude of which depends on the thickness of the competent unit and the rotation due to the folding. In this case the competent unit could even be the whole 6 km volcanic sequence and the displacement could be more than 10 km, certainly far more than enough to explain the shear zone. The allochthon hypothesis is therefore rejected.

In southern India (Fig. 1c), a basal unconformity of the younger (*c.* 2.7 Ga) calcalkaline greenstones on older (*c.* 3.4–3.0 Ga) gneisses is also very well exposed (Ramakrishnan *et al.* 1976; Viswanatha *et al.* 1982; author's observations).

From these two examples it is evident that greenstones of the type seen in many Late Archaean terrains can be, and perhaps often were, erupted on older sialic basement.

Adjacent high-grade terrains: deeper levels of the greenstone belts, or different tectonic facies?

In many parts of the world, granite–greenstone terrains adjoin high-grade, often granulite-facies terrains, usually with a metamorphic transition. Their relationships have often been discussed: whether the high-grade terrains represent lateral, more metamorphosed, equivalents, or deeper levels

of the greenstone belts, or domains differing in tectonic evolution from the greenstone belts. There are examples of all three relationships, but often they are combined and are not always distinct.

Some areas of high-grade rocks associated with low-grade granite–greenstone terrains represent their lower crust, exposed by tilting, upthrusting or updoming. A classic example is the Vredefort dome in South Africa (de Wit *et al.* 1992; Figs 1 & 2a). Here, unconformably under the 2.91 Ga

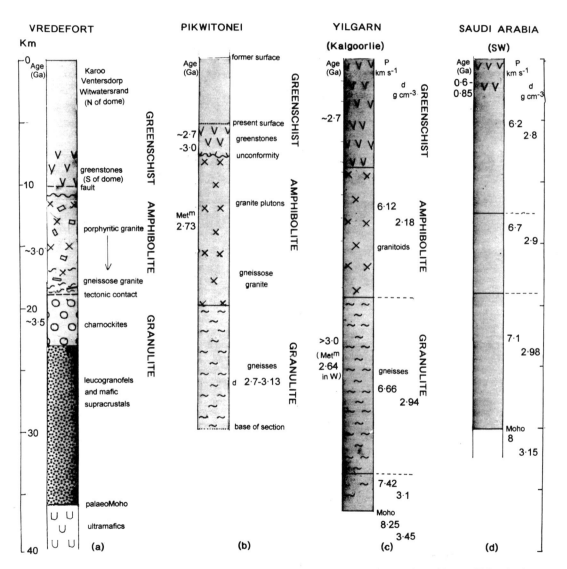

Fig. 2. Comparison of crustal sections of Archaean and younger granite–greenstone terrains, with ages (Ga), seismic velocities (*P*) and estimated densities (*d*). (**a**) Vredefort dome, Kaapvaal craton, South Africa. Based on de Wit *et al.* (1992). (**b**) Pikwitonei, NW Superior Province, Canada. Based on Ermanovics & Davison (1976). (**c**) Kalgoorlie, Yilgarn craton, SW Australia. Based on Archibald *et al.* (1981). (**d**) Saudi Arabia (Hizaz–Asir province). Based on Mechie *et al.* (1986).

Witwatersrand sediments, in the 20 km diameter core of this remarkable structure, almost the whole of the crust of the Kaapvaal craton is exposed in near-vertical section. The middle and lower crust are exposed in the dome proper; down-faulted upper crustal greenstones are seen on its southeast flank, and a borehole in the core of the dome revealed upper mantle harzburgites below the palaeo-Moho. Below the basal Witwatersrand unconformity is porphyritic granite, dated *c.* 3.0 Ga, which becomes increasingly gneissose downwards, the foliation dipping gently as so often occurs towards the base of the upper crust. This upper crustal section is underlain by granulite-facies charnockites and enderbites, and these by felsic granulites and supercrustal rocks. The contact between the upper and lower crust, formerly thought to be transitional, is marked by shearing and pseudotachylite and is recognized as tectonic (Hart *et al.* 1991), supposedly a thrust, but since it puts younger on older and causes omission not repetition, it is more probably extensional. A date of *c.* 3.5 Ga from the charnockites suggests that the granulite-facies supracrustals below may be still older.

This Vredefort section of central Kaapvaal crust, and also seismic evidence, shows that despite the reported absence of granulite xenoliths in kimberlites in southern parts of the craton (Griffin *et al.* 1979), the lower crust of at least a substantial part of the craton consists of granulite-facies rocks.

In the Yilgarn province in SW Australia (Fig. 2b), the greenstone belts become smaller and more scattered towards the southwest, the grade of melamorphism rises, eventually to granulite-facies, dated 2640 Ma, and patches of older (>3.0 Ga) gneisses appear (Myers 1993). The inference, that the whole province has been tilted, appears to be confirmed by seismic data which show that the base of the upper crust rises, from 20 km deep in the east, E of Coolgardie, to less than 6 km immediately east of the Darling fault, near Perth, before being dropped back to *c.* 15 km west of the fault (Archibald *et al.* 1981). Such regional tilt is similar in scale to many elsewhere. The Moho however deepens westwards, from 35 km to 45 km. This suggests lower crustal underplating as a cause of the tilting but the seismic section is necessarily very simplified and it is certainly possible that both lower and upper crusts were thickened tectonically in the west.

The Pikwitonei area in the northwest of the Superior Province in Canada exposes lower-crustal granulites, mid-crustal granites and gneisses, overlain unconformably by greenschist-facies metavolcanics (Fig. 2b). The lower crustal granites are brought up by thrusting (Ermanovics & Davison 1976).

Thus the Vredefort dome, the Yilgarn province and the Pikwitonei area exemplify different ways by which the lower crust of granite–greenstone terrains can be tectonically exposed. It appears generally to be in granulite-facies; in some cases at least, some components of the lower crust are older than those forming the upper crust (Fig. 2).

Very many high-grade terrains adjacent to granite–greenstone terrains represent younger collisional belts not directly related tectonically to the granite–greenstone terrains. Examples are the Late Proterozoic Mozambique belt adjacent to the Late Archaean Tanzanian craton, the eastern margin of which was reworked by the Mozambique belt tectonism. The collisional tectonism and metamorphism, between 2.7 and 2.6 Ga, in the Limpopo belt (Fig. 1a) is likewise quite distinct from that in the main part of the Kaapvaal craton, which was already stable before 3.0 Ga. The Early Archaean Pietersburg and Murchison belts, and others in the northern part of the Kaapvaal craton, were deformed and metamorphosed by the Limpopo collision. These and many other such high-grade belts next to granite–greenstone belts, superimposed on them by later tectonism, are irrelevant to the present discussion.

The most interesting relationships are those where deformation and metamorphism in adjacent high-grade and low-grade terrains are essentially contemporaneous but imply different tectonic settings. One such pairing in South Africa is that of the Early Archaean Swaziland Ancient Gneiss complex and the Barberton greenstone belt; they are everywhere separated by younger plutons or are in tectonic contact. The oldest dated member of the Ancient Gneiss complex, the Ngwane gneiss (3521 ± 23 Ma) itself yielded a 3683 ± 10 Ma zircon xenocryst, derived from a still older granitic source (Kröner & Tegtmeyer 1994). Although now mostly in upper amphibolite facies, parts of the Ngwane gneiss have been in granulite facies, with estimated T of 700–900°C and P of 6–7.5 kbar (Milisenda *et al.* 1987). Structurally overlying the Ngwane gneiss are small areas of serpentinites, komatiitic to andesitic metavolcanics, and metasediments, the Dwalile suite: their peak metamorphic conditions are estimated at 550–600°C, *c.* 4 kbar (Tegtmeyer 1989). Evidence that the *c.* 3450 Ma Dwalile metasediments were derived from an older continental source, possibly the Ngwane gnneiss, suggests that the Dwalile greenstones may originally have been unconformable on, or marginal to, continental gneisses.

Geochronological evidence (Kröner & Tegtmeyer 1994) indicates that the oldest of the Barberton volcanics are about the same age, *c.* 3.45 Ga, as, and may be correlated with, the Dwalile metavolcanics some 40 km south of them in Swaziland. The oldest member of the Barberton sequence, consisting of serpentinites, komatiites, oceanic-type tholeiites and minor acid volcanics, was named the Jamestown Ophiolitic Complex (de Wit *et al.* 1987). These ophiolites are thought to have been obducted (D1 deformation) north-northeastwards, between 3440 and 3450 Ma, soon after their formation (Armstrong *et al.* 1990). The major deformations, D2 and D3, dominantly thrusting towards the NNW, occurred much later, between 3229 and 3227 Ma (D2) in the terrane south of the Inyoka fault and at < 3164 Ma (D3) in the terrane north of it (de Wit in press). These major deformations are attributed to collision between the Swaziland Gneiss terrane to the south and another such terrane to the north, the Jamestown ophiolitic complex of the Barberton belt representing the intervening ocean (Myers & Kröner 1994). If these interpretations are right, the earlier granitoid plutons surrounding the oceanic arc Barberton greenstone belt are supra-subductional, the later ones post-collisional crustal melt and within-plate plutons. The Kaapvaal craton is thought to be made up of many separate terranes (de Wit in press).

A second example is the relation between the low-grade greenstones and granitoids of the Zimbabwe craton and the high-grade Limpopo belt to the south. The tectonics of the Limpopo belt, with intense and complex deformations, and evidence of crustal thickening, are quite different from those of the main part of the Zimbabwe and Kaapvaal cratons; however the marginal parts of both cratons were strongly deformed and metamorphosed by the Limpopo collision, the Kaapvaal craton for at least 200 km from the 'margin' of the Limpopo belt, the Zimbabwe craton only in a much narrower belt.

The Limpopo belt collision, between 2700 and 2670 Ma (Barton *et al.* 1991), produced major northward thrusting in its northern zone and a less intense system of southward thrusts in the southern zone. Northerly-directed thrusts separating the northern and central zones have been traced down towards the Moho by seismic reflections (de Beer & Stettler 1992). Because of these dominant north-vergent thrusts in the north of the belt, it has been inferred that subduction was southwards, and the Kanye Volcanics and Gaborone granite, about 100 km south of the Limpopo belt, in eastern Botswana, have been proposed as Andean arc magmas above a south-dipping subduction zone (Burke *et al.* 1986). However the Gaborone granite is dated at 2685 ± 70 Ma and the Kanye Volcanics at 2590 ± 100 Ma (Cahen & Snelling 1984): these dates look too young for subduction before the collision at 2700–2670 Ma. Eighteen chemical analyses of the Gaborone granite and Kanye

Volcanics (Harding *et al.* 1974) were plotted using discrimination diagrams for major elements (Maniar & Piccoli 1989): the plots showed them to be rift-related, not subductional. Southward subduction seems to be unsupported by evidence.

Two alternative interpretations may be suggested: either there was no oceanic crust between the Kaapvaal and Zimbabwe cratons, and so no subduction either way; the deformation and crustal thickening might be compared to that produced by the India-Asia collision in the Tien Shan far to the north of it, although the Tertiary deformation and metamorphism there are far weaker than those in the Limpopo belt. More significantly, this scenario leaves the huge extent of the calcalkaline granitoid magmatism in the Zimbabwe craton unaccounted for. A mantle plume origin has been suggested (Jelsma 1993): this does not explain the spatial and temporal relations between the Limpopo belt and the Zimbabwe granite-greenstone assemblages. This interpretation is rejected.

The other interpretation, preferred here, is that subduction, before the Limpopo collision, was northwards under Zimbabwe. The north-vergent thrusting could be analagous to the flake tectonics (Oxburgh 1972) in the Eastern Alps. The voluminous calc-alkaline granitoids in the Zimbabwe craton, and the calc-alkaline greenstone volcanics (after initial rift-related volcanism) would be supra-subductional.

The granitoids are grouped into the Chingesi suite of granodioritic gneisses, 2.9–2.8 Ga, the syn-tectonic Sesombi suite, *c.* 2.65–2.6 Ga, and the Chilimanzi suite of post-tectonic tonalites, granodiorites and granites, *c.* 2.6 Ga (Jelsma 1993). The Bulawayan greenstones cover almost the same period, between *c.* 2.9 and 2.6 Ga. All of this magmatism, north of the Limpopo belt, thus ended at, or soon after, collision there. Analyses (from Jelsma 1993) of syn-tectonic and post-tectonic granitoids from the Chinamora area were tested by me using major element plots (Maniar & Piccoli 1989) and trace elements (Pearce *et al.* 1984). From the major element plots, they were classified in the combined field of IAG (island arc granites), CAG (continental arc granites) and CCG (continental collision granites) and most as CAG, fewer as IAG and none as CCG on Shand's index. On the trace element diagrams, again all plotted in the field of VAG (volcanic arc granites), with seven of the eleven also within the COLG (collisional granite) field. The two quite different methods together strongly suggest that these granitoids are supra-subductional: whether island arc or continental (Andean) arc is uncertain but there is increasing evidence that much of the crust of the Zimbabwe craton was already sialic (continental) by the Late Archaean, so an Andean arc is more likely.

The Chinamora area is *c.* 300 km from the northern margin of the Limpopo belt; the northern margin of the Zimbabwe craton is *c.* 400 km from the Limpopo belt — a distance similar to the width of the Andes. If the Late Archaean Zimbabwe granite–greenstone assemblages are supra-subductional, they could well relate to one subduction zone rather than representing a series of accreted terranes.

Tectonic interpretation

The granite-greenstone terrains represent the main growth process of the continental crust. Their predominantly Archaean age reflects the much faster growth rate in earlier times. There was no sudden change in their character at the end of the Archaean, but rather a gradual change which involved more continent-derived sediments, a reduction in the proportion of granitoids (plutons and underlying gneisses) relative to greenstone volcanics and a gradual change in the nature of the volcanics: the proportion of komatiites decreases from Early Archaean to Late Archaean and subsequently, and the proportion of andesitic and felsic volcanics increases.

The magmas represented by the greenstone volcanics and the far more voluminous granitoid plutons have been variously attributed to rifting, mantle plumes or subduction. Initial rifting is often identified from the geochemistry of the earliest tholeiites but that setting cannot explain the main magmatism, the enormous volumes of granitoids, the overall calc-alkaline geochemical signature, nor the sometimes weak but always significant compressional tectonic environment. Plumes seem incapable of explaining the huge volume of the predominant tonalitic–granodioritic granitoids, and especially their systematic association with compressional tectonics. There is no sign of the uni-directional overlap of successive plutons even though plate movements were probably faster than in later times. An island or Andean arc origin or alternatively a marginal basin origin now seem the most probable (Windley 1977). Cogent arguments in favour of the island arc origin have recently been advanced (Taira *et al.* 1992). The many greenstones which are autochthonous on older basements may be on arcs which, like the Hida Belt in Japan, were detached from an adjacent continent, or from a trans-oceanic continent as in the Tibet plateau, or they may have been built on microcontinents. An arc origin for Archaean granite-greenstone terrains is supported by the similarity of their crustal sections to that of the Late Proterozoic arc-origin Saudi-Arabian crust (Fig. 2).

Lateral accretion of exotic terranes bounded by strike-slip faults rather than by accretion on ophiolite-decorated sutures was probably important in the Early Proterozoic Birrimian greenstone terrains (Mortimer 1992) and in the Late Proterozoic Tuareg (W Africa) and Arabian–Nubian Shields (Black *et al.* 1994; Stern 1994); it has not yet been demonstrated in the Archaean in Africa but it has been inferred in SW Greenland (Friend *et al.* 1988).

The deformation in the granite–greenstone terrains may be separated into pre-diapiric, diapiric and post-diapiric stages. The earliest of the pre-diapiric structures recognized in the Barberton area (de Wit 1986), confined to the Onverwacht simatic rocks, are attributed to hydraulic fracture gliding on near-horizontal surfaces during the formation of the ophiolites.

The main pre-diapiric structures, in the Barberton and many other areas, are a complex array of thrusts, originally near horizontal, and associated tight folds which are attributed to obduction.

Diapiric and ballooning deformations, dominant in some terrains, notably in the Zimbabwe and Dharwar cratons, caused strong compression and shortening of the inter-diapir greenstone belts, steep dips into the inter-diapir synclines, and the characteristic cuspate outlines of most greenstone belts, but no crustal shortening. Because the diapiric plutons are geochemically identified as subduction-related, these structures too can be regarded as sub-duction-related.

The post-diapiric structures, responsible for the linearity of many of the terrains, such as the Yilgarn block in west Australia, the Birrimian in west Africa and the eastern part of the Dharwar craton in southern India, are the most problematic. They may still be subduction-driven, or collisional, or within-plate as a result of distant collision, like the Tien-Shan deformation in response to the India–Asia collision. The intensity of these deformations, ranging from negligible in Zimbabwe (apart from the Limpopo-imposed deformation in the south) and in much of the Dharwar craton, to quite strong in the Yilgarn block in SW Australia and in the Birrimian in west Africa, is still far less than that of typical continent–continent collisional belts; but it is comparable to that in arc accretion complexes such as the Arabian–Nubian Shield, or the accreted microcontinents in Tibet, north of the Indus–Tsangpo suture.

Interpretation of much of the deformation in the greenstone belts as subduction-related, as is most of that in the Andes now, suggests that tectonic interpretations generally underestimate the importance of subduction as compared to collision.

Most greenstone belts yield metamorphic pressures of 2.5–4.5 kbar; estimates of the average thickness eroded (Condie 1981) are 5–8 km. Since their crustal thicknesses now are mostly about 35 km or less, the crustal thickness before any erosion, assuming no later underplating, was less than 45 km. This is less than the Andes but more than most island arcs, although not more than their probable thickness after compression during accretion or collision. The Late Proterozoic Arabian arc complex was about that thickness before erosion. The relatively moderate crustal thickness implies only moderate crustal shortening, although without knowing the pre-deformation crustal thickness it is not possible to make a numerical estimate. It must however be very much less than the intense local shortening (>80%) seen in parts of the Barberton belt (de Wit *et al.* 1992). That deformation is probably a reflection of thin-skinned translation, for example during obduction.

High-grade terrains adjacent to greenstone terrains are of several different kinds. Some represent deeper levels, exposed by tilting or upthrusting, although possibly not all greenstone belt terrains are underlain by granulite-facies lower crust.

Some of the bordering high-grade terrains, for example the Limpopo Belt adjacent to the Kaapvaal granite–greenstones terrain in southern Africa, show a high-grade, deep-crustal metamorphism, contemporaneous with lower-grade metamorphism in the adjacent greenstone belts. The deformation in the low-grade greenstone belts is much less intense than in the high-grade terrains, in which the crust must have been greatly thickened. Such high-grade terrains are interpreted as Himalayan-type continent–continent collisions. They are not just lower crustal basement of the adjacent upper crust greenstone terrains but are tectonically distinct. They are however related, in that the collision implies previous subduction, which may have given rise to the adjacent granite-greenstone magmatism and part of their deformation.

Conclusions

(1) The granite–greenstone terrains represent the main growth process of the sialic crust. Their granitoids (mostly tonalite and granodiorite) represent addition of a thickness of at least 10 km to the crust below the much less voluminous greenstones.

(2) The Early Proterozoic Birrimian and the Late Proterozoic Arabian–Nubian Shield are interpreted as direct analogues of the Archaean greenstone belts.

(3) The Archaean granite–greenstone terrains and their younger analogues represent Andean or island arcs produced by subduction and amalgamated by accretion.

(4) The components of the granite–greenstone terrains changed gradually from Early to Late Archaean and subsequently; the proportion of ultra-mafic (komatiitic) volcanics decreased and that of andesitic volcanics increased. Continent-derived sediments, rare or absent from the Early Archaean greenstones, increased through time. Some, perhaps most, of the Late Archaean greenstones and younger equivalents accumulated auto-chthonously on sialic basement; these are not oceanic, so not ophiolites.

(5) The structures in the greenstone terrains are separable into pre-diapiric gravity slides, thrusts (obductional?) and related compressional structures; structures resulting from the diapiric rise and ballooning of plutons; and post-diapiric compressional structures. Of these, only the diapir-related structures are essential elements in the growth of the granite–greenstone terrains. The others vary in intensity from negligible to very strong.

(6) Apart from areas adjacent to collision belts, the crust of some granite–greenstone terrains was not much thickened tectonically.

(7) Some high-grade terrains adjacent to granite–greenstone terrains represent tectonically exposed lower levels of the crust below them while some represent tectonically distinct collisional zones.

References

ANHAEUSSER, C. R. 1972. The geology of the Jamestown Hills area of the Barberton Mountain Land, South Africa. *Transactions of the Geological Society of South Africa*, **75**, 225–263.

ARCHIBALD, N. J., BETTANY, L. F., BICKLE, M. J. & GROSS, D. I. 1981. Evolution of Archaean crust in the eastern goldfields province of the Yilgarn Block, western Australia. *In*: GLOVER, J. E. & GROVES, D. I. (eds) *Archaean Geology*. Geological Society of Australian Special Publications, **3**, 491–504.

ARMSTRONG, R. A., COMPSTON, W., DE WIT, M. J. & WILLIAMS, I. S. 1990. The stratigraphy of the 3.5–3.2 Ga Barberton Greenstone Belt revisited: a single zircon ion microprobe study. *Earth and Planetary Science Letters*, **101**, 90–106.

BARTON, J. M., BRANDL, G., VAN REENEN, D. D., VAN SCHALKWYK, J. F. & STEVINS, G. 1991. Northern Kaapvaal Craton and Limpopo Belt. *In::* ASHWAL, L. D. (ed.) *Traverse through two cratons and an orogen: excursion guidebook and review articles for a field workshop through selected Archaean terranes of Swaziland, South Africa, Zimbabwe,* Rand Africaans. Univ Johannesburg, South Africa, 209–218.

BICKLE, M. J., MARTIN, A. & NISBET, E. G. 1975. Basaltic and peridotitic komatiites and stromatolites above a basal unconformity in the Belingwe Greenstone Belt, Rhodesia. *Earth and Planetary Science Letters,* **27**, 155–162.

BLACK, R., LATOUCHE, L., LIEGEOIS, J. P., CABY, R. & BERTRAND, J. M. 1994. Pan-African displaced terranes in the Tuareg shield (Central Sahara) *Geology,* **22**, 641–644.

BOUHALLIER, H., CHOUKROUNE, P. & BALLÈVRE, M. 1993. Diapirism, bulk homogeneous shortening and trans-current shearing in the Archaean Dharwar craton: the Holenarsipur area, southern India. *Precambrian Research,* **63**, 43–58.

BURKE, K., KIDD, W. S. F. & KUSKY, T. M. 1986. Archaean foreland basin tectonics in the Witwatersrand, South Africa. *Tectonophysics,* **5**, 439–456.

CAHEN, L. & SNELLING, N. J. 1984. *The geochronology and evolution of Africa.* Oxford University Press.

CONDIE, K. C. 1981. *Developments in Precambrian Geology: v. 3 Archean Greenstone Belts.* Elsevier N. Holland Inc. New York.

CRITTENDEN, M. D., CONEY, P. J. & DAVIS, G. H. (eds) 1980. *Cordilleran Metamorphic core complexes.* Geological Society of America Memoirs, **153**.

COWARD, M. P., LINTERN, B. C. & WRIGHT, L. I. 1976. The pre-cleavage deformation of the sediments and gneisses of the northern part of the Limpopo Belt. *In:* WINDLEY, B. F. (ed.) *The Early History of the Earth.* Wiley-Interscience, London, 323–330.

DE BEER, J. H. & STETTLER, E. H. 1992. The deep structure of the Limpopo belt from geophysical studies. *Precambrian Research,* **55**, 173–186.

DE WIT, M. J. 1982. Gliding and overthrust nappe tectonics in the Barberton Greenstone Belt. *Journal of Structural Geology,* **4**, 117–136.

—— 1986. Extensional tectonics during the igneous emplacement of the mafic-ultramafic rocks of the Barberton Greenstone Belt. *In::* Workshop on *Tectonic Evolution of Greenstone Belts.* LPI Technical Report. **86-10**. Lunar & Planetary Institute, Houston, 84–85.

—— in press *In:* DE WIT, M. J. & ASHWAL, L. D. (eds) *Tectonic Evolution of Greenstone Belts.*

——, HART, R. A. & HART, R. J. 1987. The Jamestown ophiolite complex, Barberton Mountain belt: a section through 3.5 Ga oceanic crust. *Journal of African Earth Sciences,* **6**, 681–730.

——, ROERING, C., HART, R. J., ARMSTRONG, R. A., DE RONDE, C. E. J., GREEN, R. W. E., TREDOUX, M., PEBERDY, E. & HART, R. A. 1992. Formation of an Archaean continent. *Nature,* 553–562.

ERMANOVICS, I. F. & DAVISON, W. L. 1976. The Pikwitonei granulites in relation to the northwestern Superior Province of the Canadian Shield. *In:* WINDLEY, B. F. (ed.) *The Early History of the Earth.* Wiley-Interscience, London, 331–350.

FRIEND, C. R. L., NUTMAN, A. P. & McGREGOR, V. P. 1988. Late Archaean terrane accretion in the Godthåb region, southern west Greenland. *Nature,* **335**, 535–538.

GRIFFIN, W. L., CARSWELL, D. A. & NIXON, P. W. 1979. Lower crustal granulites and eclogites from Lesotho, Southern Africa. *In:* BOYD, F. H. (ed.) *The Mantle: The Sampled Inclusions in Kimberlites and Other Volcanics.* Proceedings of the 2nd International Kimberlite Conference, 1977, **11**, 59–80.

HARDING, R. R., CROCKETT, R. N. & SNELLING, N. J. 1974. *The Gaborone Granite, Kanye Volcanics and Ventersdorp Plantation Porphyry: geochronology and review.* Report of the Institute of Geological Sciences, London, **74/5**.

HART, R. J., ANDREOLI, M. A. G., REIMOLD, W. V. & TREDOUX, M. 1991. Aspects of the dynamic and thermal metamorphic history of the Vredefort cryptoexplosion structure: implications for its origin. *Tectonophysics,* **192**, 313–331.

HUNTER, D. R., SMITH, R. G. & SLEIGH, D. W. W. 1992. Geochemical studies of Archaean granitoid rocks in the Southeastern Kaapvaal Province: implications for crustal development. *Journal of African Earth Sciences,* **15**, 147–151.

JELSMA, H. A. 1993. *Granites and greenstones in northern Zimbabwe: Tectono-thermal evolution and source regions.* Akademisch Proefschrift, Vrie Universiteit te Amsterdam, Netherlands.

——, VAN DER BECK, P. A. & VILNYU, M. L. 1993. Tectonic Evolution of the Bindura-Shamva greenstone belt (northern Zimbabwe): progressive deformation around diapiric batholiths. *Journal of Structural Geology,* **15**, 163–176.

KRÖNER, A. & TEGTMEYER, A. 1994. Gneiss-greenstone relationships in the Ancient Gneiss Complex of southwestern Swaziland, southern Africa, and implications for early crustal evolution. *Precambrian Research,* **67**, 109–139.

KUSKY, T. M. & KIDD, W. S. F. 1992. Remnants of Archean oceanic plateau, Belingwe greenstone belt, Zimbabwe. *Geology,* **20**, 43–46.

LAMB, S. H. 1986. Synsedimentary deformation and thrusting on the eastern margin of the Barberton Greenstone Belt, Swaziland. *In: Workshop on Tectonic Evolution of Greenstone Belts.* LPI Technical Report. **86-10**. Lunar & Planetary Institute, Houston, 140–141.

LOWE, D. R. & BYERLY, G. R. 1986. The rock components and structures of Archean greenstone belts: an overview. *In: Workshop on Tectonic Evolution of Greenstone Belts.* LPI Technical Report. **86-10**. Lunar & Planetary Institute, Houston, 142–146.

MACGREGOR, A. M. 1951. Some milestones in the Precambrian of Southern Rhodesia. *Proceedings of the Geological Society of South Africa,* **54**, 27–71.

MANIAR, P. D. & PICCOLI, P. M. 1989. Tectonic discrimination of granitoids. *Geological Society of American Bulletin*, **101**, 635–643.

MARTYN, J. E. 1986. Evidence for structural stacking and repetition. *In*: *Workshop on Tectonic Evolution of Greenstone Belts*. LPI Technical Report. **86-10**. Lunar & Planetary Institute, Houston, 152–153.

MECHIE, J., PRODEHL, C. & KOPTSCHALITSCH, G. 1986. Ray path interpretation of the crustal structure beneath Saudi Arabia. *Tectonophysics*, **131**, 331–352,

MILISENDA, C., KRÖNER, A. & LIEW, T. C. 1987. Granulites in the Ancient Gneiss Complex of Swaziland: evidence for thick continental crust in the early Archaean. *Terra Cognita*, **7**, 420.

MORTIMER, J. 1992. Lithostratigraphy of the early Proterozoic Toumodi Volcanic Group in central Côte d'Ivoire: implications for Birrimian stratigraphic models. *Journal of African Earth Sciences*, **14**, 81–91.

MYERS, J. S. 1993. Precambrian history of the west Australian craton and adjacent orogens, *Annual Reviews of Earth and Planetary Sciences*, **21**, 453–487.

—— & KRÖNER, A. 1994. Archaean Tectonics. *In*: HANCOCK, P. L. (ed.) *Continental Deformation*. Pergamon Press, 355–369.

—— & WATKINS, K. P. 1985. Origin of granite-greenstone patterns, Yilgarn Block, Western Australia. *Geology*, **13**, 778–780.

OXBURGH, E. R. 1972. Flake Tectonics and Continental Collision. *Nature*, **239**, 202–204.

PEARCE, J. A., HARRIS, N. B. W. & TINDLE, A. G. 1984. Trace Element Discrimination Diagrams for the Tectonic Interpretation of Granitic Rocks. *Journal of Petrology*, **25**, 956–983.

PLATT, J. P. 1980. Archaean greenstone belts: a structural test of tectonic hypotheses. *Tectonophysics*, **65**, 127–150.

RAMAKRISHNAN, M., VISWANATHA, M. M. & NATH, J. S. 1976. Basement-cover relationships of Peninsular Gneiss with high grade schists and greenstone belts of southern Karnataka. *Journal of the Geological Society of India*, **17**, 97–111.

RAMSAY, J. G. 1967. *Folding and Fracturing of Rocks*. McGraw Hill, New York.

—— 1975. The structure of the Chindamora Batholith. *19th Annual Report of the Institute of African Geology, University of Leeds*, 81.

—— 1989. Emplacement kinematics of a granite diapir: the Chindamora batholith, Zimbabwe. *Journal of Structural Geology*, **11**, 191–209.

SCHWERDTNER, W. H. 1990. Structural tests of diapir hypotheses in Archaean crust of Ontario. *Canadian Earth Sciences*, **27**, 387–402.

SNOWDEN, P. A. 1984. Non-diapiric batholiths in the north of the Zimbabwe Shield. *In*: KRÖNER, A. & GREILING, R. (eds) *Precambrian Tectonics Illustrated*. Nägele und Obermiller, Stuttgart, 135–145.

—— & BICKLE, M. J. 1976. The Chindamora Batholith: diapiric intrusion or interference fold? *Journal of the Geological Society, London*, **132**, 131–137.

—— & SNOWDEN, D. V. 1979. Geology of an Archaean batholith, the Chinamora batholith, Rhodesia. *Transactions of the Geological Society of South Africa*, **82**, 7–22.

STERN, R. J. 1994. Arc assembly and continental collision in the neoproterozoic East African orogen: implications for the consolidation of Gondwanaland. *Annual Reviews of Earth and Planetary Science*, **22**, 319–351.

STOWE, C. W. 1974. Alpine type structures in the Rhodesian basement at Selukwe. *Journal of the Geological Society, London*, **130**, 411–425.

—— 1984. The early Archaean Selukwe nappe, Zimbabwe. *In*: KRÖNER, A. & GREILING, R. (eds) *Precambrian Tectonics Illustrated*. Nägele und Obermiller, Stuttgart, 41–56.

STURCHIO, N. C., SULTAN, M. & BATIZA, R. 1983. Geology and origin of Meatiq Dome, Egypt: a Precambrian metamorphic core complex? *Geology*, **11**, 72–76.

SUTTON, J. & WATSON, J. V. 1951. The pre-Torridonian metamorphic history of the Loch Torridon and Scourie areas in the north-west Highlands and its bearing on the chronological classification of the Lewisian. *Quarterly Journal of the Geological Society of London*, **106**, 241–307.

TAIRA, A., PICKERING, R. T., WINDLEY, B. F. & SOH, W. 1992. Accretion of Japanese island arcs and implications for the origin of Archean Greenstone Belts. *Tectonics*, **11**, 1224–1244.

TEGTMEYER, A. R. 1989. *Geochronologie und Geochemie im Präkambrium des Südlichen Afrika*. Doctoral dissertation, Univ. Mainz.

TSOMONDO, J. M., WILSON, J. F. & BLENKINSOP, T. G. 1992. Reassessment of the structure and stratigraphy of the early Archaean Selukwe Nappe, Zimbabwe. *In*: GLOVER, J. E. & HO, S. E. (eds) *The Archaean: terrains, processes and metallogeny*. Geology Department (Key Centre) & University Extension, University W. Australia Publications **22**, 123–135.

VILJOEN, M. J. & VILJOEN, R. P. 1970. Archaean volcanicity and continental evolution in the Barberton region, Transvaal. *In*: CLIFFORD, T. N. & GASS, I. G. (eds) *African Magmatism and Tectonics*. Oliver & Boyd, Edinburgh, 27–49.

VISWANATHA, M. M., RAMAKRISHNAN, M. & NATH, J. S. 1982. Angular unconformity between Sargur and Dharwar supracrustals in Sigegudda, Karnataka craton, South India. *Journal of the Geological Society of India*, **23**, 85–89.

WINDLEY, B. F. 1977. *The Evolving Continents* (2nd Edn.) John Wiley & Sons, Chichester.

Soft lithosphere during periods of Archaean crustal growth or crustal reworking

P. CHOUKROUNE, H. BOUHALLIER & N. T. ARNDT

Géosciences, UPR CNRS 4661, Université de Rennes 1,
Campus de Beaulieu, 35042 Rennes Cedex, France

Abstract: Field observations, and an analysis of the strain field, have been carried out in Archaean terranes of the Hebei province in the Dharwar craton of India, the Sino-Korean craton in China, and the Man shield in the Ivory Coast.

Two broadly different situations can be recognized. In the Dharwar and Man Shields the deformation shows a range of characteristic features. The foliation trajectories outline dome-and-basin structures in which supracrustal rocks of greenstone belts invariably occupy the basins. The foliation trajectories are perturbed only by bands of transcurrent shearing. Variations in deformation directions depend essentially on the geometry of plutonic or migmatitic bodies within which the foliation invariably has a domal form. Changes in the intensity of deformation are essentially limited to boundaries between granitoids and supracrustal series. These changes are largely horizontal. Characteristics of the finite strain ellipsoid (the parameter k) vary extremely rapidly.

The second situation is manifested in the Hebei province where the Archaean crust has deformed without the preservation of domes and basins. Here the deformation is characterized by the presence of a foliation and a very strong vertical flattening which extends into catazonal domains. This deformation is homogeneous. To explain these observations it is proposed that this homogeneous deformation was superimposed on pre-existing dome-and-basin structure, obliterating the original geometry. The Hebei example represents a stage of deformation greater than that of the Ivory Coast, which in turn is greater than that in the India craton.

It is apparent that the Archaean crust has specific characteristics that controlled the manner in which it deformed. During accretion, or reworking, it was dominated by body forces related to differences in the densities of the two major Archaean lithological units: granitoid and greenstones. Deformation resulting from boundary forces then led to a second type of behaviour controlled by a vertical planar anisotropy induced during the first stage.

These characteristics are never reported in modern orogenic belts, which generally are marked by a uniform structural trend and vergence. It is proposed that Archean cratons completely lacked a rigid element during the accretion events responsible for their formation, that the thermal regime in these periods was quite unlike that of today, and that the continental crust formed or was reworked during episodes of enhanced mantle plume activity.

The concept that Archaean cratons have a structure fundamentally different from that of modern orogenic belts arose when it was noticed that these regions contain a great abundance of granitoids and felsic gneisses, and that metavolcanics and metasediments are commonly confined to discontinuous, often cusp-shaped belts between felsic bodies. McGregor (1951) popularized the concept in his description of the 'gregarious' granites of the Rhodesian (now Zimbabwe) Craton, and the satellite view of the domes and basins of the Pilbara craton has become a geological classic. Nonetheless, the precise manner in which these structures formed and, more specifically, whether they resulted from processes like those on the modern Earth, have long been subjects of debate (see for instance Fyfe 1974; Platt *et al.* 1980). In the past five years a wave of papers has appeared in which it is argued that the model of plate tectonics can be adapted to Archaean times, and that thrusting on a scale comparable to that seen in the Alpine chain and other modern mountain belts has contributed to the thickening of an Archean protocrust that formed part of a rigid lithosphere.

These ideas have to reckon with certain remarkable lithological, metamorphic and structural characteristics of Archaean cratons which, although sometimes ignored, are of immediate relevance to the question of how the Archaean Earth operated.

(a) High-pressure, low-temperature metamorphism has not been convincingly documented in regions older than 2.5 Ga.

(b) The volume of magmatic material in Archaean terrains is exceptionally large: the

From COWARD, M. P. & RIES, A. C. (eds), 1995, *Early Precambrian Processes*,
Geological Society Special Publication No. 95, pp. 67–86.

Archaean was without doubt a major period of continental crust formation and most of this growth took place in short, sharp episodes (e.g. Taylor & McLennan 1985; Nelson & DePaolo 1985).

(c) Neither oceanic crust, nor sutures, have been identified in Archaean regions. In certain papers, typical greenstone belt volcanics are compared with mid-ocean ridge basalts (e.g. Helmstaedt *et al.* 1986; de Wit *et al.* 1987), but in our opinion, none of these reports is convincing (see also Bickle *et al.* 1994). On the other hand, the characteristic lithological duality of Archaean granite–greenstone belts persists in certain Proterozoic regions (e.g. the Birimian of West Africa, Abouchami *et al.* 1990), but is absent from more modern regions. Archaean sequences commonly comprise two assemblages of contrasting composition, a granitoid assemblage (TTG, or tonalite–trondhjemite–granodiorite) and volcano-sedimentary material (greenstone belts).

(d) Many Archaean terranes lack an obvious linear trend. This is especially noticeable in regions older than 3.0 Ga such as the Pilbara of Australia (Collins 1989) where the strain field is highly unusual, being marked by abrupt changes in strike and dip of the strain axes, and large variations in the axial planes of structures: a dome-and-basin geometry is characteristic and widespread. In younger regions, such as the Superior Province of Canada or the Yilgarn of Australia, linear belts are developed (e.g. Card 1990; Clowes *et al.* 1992) but the dome-and-basin geometry remains.

The debate about Archaean dynamics can be phrased in terms of two extreme models. In the first, boundary forces are assigned a dominant role. The existence of rigid microcontinents is assumed, even at the earliest stages of accretion, and the margins of these continents are thought to deform during collisions comparable to those on the modern Earth. In the second model, the role of body forces is seen as more important. These forces, which act in the interiors of the proto-continents, result in vertical displacements related to gravitational instabilities that stem from crustal fusion and the density differences between the two major Archaean lithological assemblages, granitoids and greenstone belts.

In the following sections, three examples that may illustrate different stages of the deformation that has affected Archaean continental crust are described, e.g. the Dharwar craton in India, the Hebei province of the Sino-Korean craton in China, and the Man shield in West Africa. These examples provide evidence that the strain field had certain characteristics that were peculiar to this period of Earth history. We recognize that certain features that might be regarded as characteristic of Archaean cratons are also found in modern orogenic belts, but is emphasized that the scale of

these structures is very different. Thrusts are present in Archaean terranes but it is argued that, where convincingly documented, they are relatively minor features and not to be compared with those that characterize modern orogenic belts. Thus this paper defends the idea that the tectonic processes active during the events that formed or reworked Archaean continental crust were fundamentally different from those which dominate modern orogenic domains.

Examples

Dharwar craton

Lithologies. The Archaean terrains of southern India mainly consist of linear and arcuate, low- to high-grade volcano-sedimentary belts ('greenstone', 'supracrustal' or 'schist' belts) surrounded by larger regions of high-grade infracrustal rocks (Naqvi & Rodgers 1983; Swami Nath & Ramakrishnan 1981; Rhadakrishna & Naqvi 1986; Rodgers 1986). These are associated with low-K tonalitic, trondhjemitic and granodioritic gneisses (TTG series) with ages between 3.35 Ga and 2.5 Ga (Crawford 1969; Venkatasubramanian 1975; Beckinsale *et al.* 1980, 1982; Monrad 1983; Stroh *et al.* 1983; Taylor *et al.* 1984; Drury *et al.* 1986; Naha *et al.* 1990; Meen *et al.* 1992). The greenstone belts have been divided into two types (Swami Nath *et al.* 1976) on the basis of differences in metamorphic grade and structural evolution: the highly deformed Sargur type, and the less deformed Dharwar type (see Ramakrishnan *et al.* 1988; Radhakrishna & Ramakrishnan 1990). Although contacts between Sargur and Dharwar rocks are very scarce, there is some evidence that the Dharwar type rests uncomformably on rocks of the Sargur type and on TTG series gneisses (Chadwick *et al.* 1979, 1981; Viswanatha *et al.* 1982).

Metamorphism. A characteristic feature of the Dharwar craton is the transition from a low- to medium-grade terrain in the north to a high-grade granulitic terrain in the south. Palaeopressures in gneissic and mafic rocks increase from *c.* 3 kbar in central Karnataka to *c.* 8–9 kbar in the Sargur area (Harris & Jayaram 1981; Janardhan *et al.* 1982; Raith *et al.* 1982; Hansen *et al.* 1984; Raase *et al.* 1986). The southward *P–T* gradient is perpendicular to the general N–S structural trend of the craton (Drury & Holt 1980).

The Closepet granite, a Late Archaean batholith dated *c.* 2.5 Ga (Crawford 1969; Friend & Nutman 1991; Jayananda *et al.* 1992), extends for almost 400 km from north to south of the craton, and is clearly linked to a period of major migmatization that affects the surrounding gneisses and supracrustal rocks (Friend 1983). The southernmost part of the Closepet batholith is overprinted by granulites (Janardhan *et al.* 1979*a*, 1982; Friend 1981, 1983; Hansen *et al.* 1987; Stähle *et al.* 1987).

Structure. The strong N–S-trending fabric of the Dharwar craton is partly the result of a Late Archaean transcurrent ductile shearing episode (Fig. 1) (Drury & Holt 1980; Drury 1983; Drury *et al.* 1984; Chadwick *et al.* 1989) that was

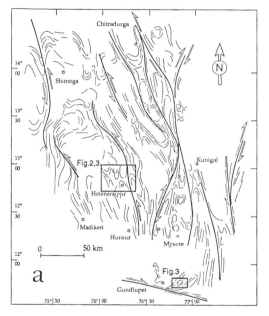

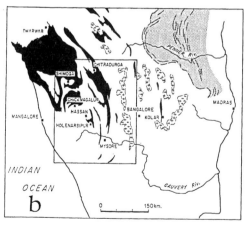

Fig. 1. (a) Structural map of the central Dharwar craton showing domal structures affected by later transcurrent shearing. **(b)** Geological map of the Dharwar craton showing the location of (a). Greenstone belts are black, late granitoids with crosses, Proterozoic belts with dotted pattern.

contemporaneous with the emplacement of the Closepet granite (Jayananda & Mahabaleswar 1991). Here the Holenarsipur and the Gunlupet areas, two complementary areas that typify the structural evolution of the craton, are described.

(a) Holenarsipur belt. Holenarsipur is a cusp-shaped greenstone belt surrounded by domes of felsic gneisses, in the central medium-grade part of the Dharwar craton (Hussain & Naqvi 1983; Ramakrishnan *et al.* 1981). Several generations of intermediate to felsic plutonic rocks comprise a TTG infracrustal sequence. These rocks, which have now been transformed into highly strained gneisses, are grouped together in the Holenarsipur region under the term 'Gorur gneisses'. Mafic to ultramafic volcano-sedimentary supracrustal units make up the second lithology (Naqvi *et al.* 1983*b*).

The gneisses have yielded Rb–Sr (Beckinsale *et al.* 1980, 1982) and Pb–Pb ages (Taylor *et al.* 1988) between 3.35 Ga and 3.305 Ga. They are intruded by trondhjemitic plutons dated using various methods between 3.2 and 3.0 Ga (Beckinsale *et al.* 1982; Bhaskar Rao *et al.* 1983; Monrad 1983; Stroh *et al.* 1983; Taylor *et al.* 1984; Meen *et al.* 1992; Peucat *et al.* 1989). The entire area was metamorphosed at amphibolite facies.

The main structural feature of the Holenarsipur area is a regional foliation that affects both supracrustal and infracrustal rocks. This foliation (S1) is characterized by a wide variability in dip and strike (Figs 2 & 3). Foliation trajectories (Fig. 2) outline a dome-and-basin structure within which supracrustal rocks occupy the synforms and infracrustal rocks the elliptical domal antiforms. At contacts between supracrustal and infracrustal rocks, which are invariably vertical, the foliation is strongly developed and steeply dipping: in the central part of the domal structures, it is less pronounced and has very shallow dips. At these places the intrusive rocks have a relatively isotropic fabric.

Between elliptical domes, a triangular arrangement of the vertical S1 foliation defines triple junctions (Brun 1983*a*). These are present in both supracrustal and infracrustal rocks. The triple junctions are preferred sites for trondhjemitic intrusions and quartz-tourmaline veins which indicate major channelized fluid circulation.

The S1 trajectory map also shows a highly foliated linear zone along the eastern boundary of the supracrustal belt. The foliation here is always near vertical and the strike is 170°. This linear part of the belt differs significantly from the dome-and-basin pattern of the surrounding areas.

The attitudes of lineations, like those of the foliations, gradually change from gently plunging in the cores of domes to near vertical at contacts of supra- and infracrustal rocks. In supracrustal rocks,

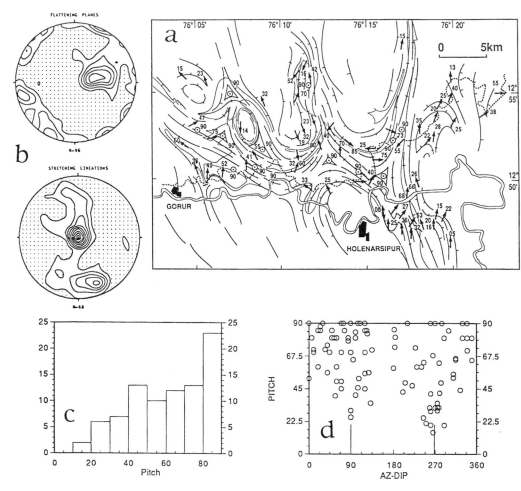

Fig. 2. (a) Map of strain trajectories in the Holenasipur area. Greenstones shown in grey, TTG in white. (b) Stereograms of foliation planes and stretching lineations. (c) Histograms showing the predominance of high pitches of lineation. (d) Pitch v. orientations of foliation planes (AZ-dip) showing that the lowest pitches of lineations are confined to N–S foliation planes associated with strike-slip motions.

the stretching lineations are generally downdip, except at the linear, easternmost boundary of the supracrustal belt where they are mainly horizontal or plunge gently northward. Within the triple junctions, stretching lineations are well developed and invariably vertical (Fig. 2).

The shape of the finite strain ellipsoid (Flinn 1965) was estimated using four different types of rock fabric (Schwerdtner *et al.* 1977) (Fig. 3), each easily distinguishable in the field. A high variability in the distribution of fabric types throughout the area is demonstrated by the following observations.

(i) In the central parts of the domes, fabrics are poorly developed.

(ii) Planar fabrics characterize two-thirds of the study area. This fabric is present in areas where the S1 foliation plane exhibits major variations in dip. It is dominatant in sectors where the foliation is vertical, and is ubiquitous throughout the supracrustal belt, except near its eastern boundary.

(iii) Planar–linear fabric zones are present in two different settings. At margins of triple junctions, planar elements define the triangular arrangement, and lineations are vertical. In the linear zone at the eastern N–S boundary of the belt, lineations dip more gently.

(iv) In the centres of triple junctions the fabric is entirely linear and vertical.

Using kinematic indicators of the type defined by

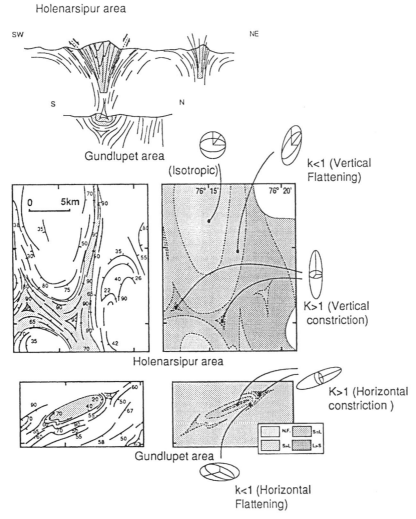

Fig. 3. Cross-sections and maps of foliations and fabrics in the Holenasipur and Gundlupet areas, Dharwar craton. For further explanation see text. The cross-section shows the present interpretation of the structural relationship between the two areas. The Gunlupet area exposes a deeper level than the Holenasipur area.

Choukroune *et al.* (1987), Bouhallier *et al.* (1993) analysed the sense and direction of shearing and outlined two areas of distinctly contrasting data. The first, which includes the dome-and-basin structures, displays kinematic criteria indicating vertically downward displacements of supracrustal belts relative to the TTG basement (Fig. 2). The second, which is restricted to the N–S linear zone at the eastern margin of the belt, is dominated by horizontal sinistral displacements. The structural data, which are discussed in detail by Bouhallier *et al.* (1993), are compatible with those observed by Drury (1983), Chadwick *et al.* (1989) and

Jayananda & Mahabaleswar (1991) in shear zones in surrounding parts of the Dharwar craton (Fig. 1).

(b) Gundlupet area. In the Gunlupet area, *c.* 100 km south of Holenasipur, supracrustal rocks are highly metamorphosed and limited to small, discontinuous lenses within the gneisses (Figs 1 & 3). As in the Holenasipur belt, a strong and penetrative foliation affects both supracrustal and infracrustal rocks and shows a wide variation of dip and strike. At map scale, foliation trajectories outline domes of infracrustal rocks and basins of supracrustal rocks. However, in the Gundlupet area the basins are elliptical and the domes linear, in

contrast to the situation in the Holenarsipur area where both elements are more circular. Another important feature is the local presence of a strong horizontal foliation in the central parts of basins.

At contacts between supra- and infracrustal rocks, the foliation is well developed and the strain is very high. At each end of the synformal basins, a triangular arrangement of gently dipping S1 foliations defines a triple junction (Fig. 3). The main axes of triple junctions plunge gently and are oriented parallel to: (a) the long axes of the elliptical basins, (b) contacts between supra- and infracrustal units, and (c) the trajectories of stretching lineations, which at a regional scale, plunge gently or are almost horizontal.

The distribution of fabrics throughout the area varies as follows.

(i) Isotropic zones, and zones of poorly developed fabric, appear absent.

(ii) Planar fabric zones are common in the infracrustal rocks, particularly in regions far from supracrustal rocks.

(iii) Planar-linear fabric zones are developed in the central parts of supracrustal basins, around triple junctions, and at contacts between supra- and infracrustal rocks.

(iv) Linear horizontal fabrics are restricted to the centres of basins, and to triple junctions outside the basins (Fig. 3).

(v) Linear vertical fabrics are absent.

(vi) Apparently superimposed structures occur wherever linear fabrics are mapped. Although some of these were described by Janardhan *et al.* (1979*b*) in terms of polyphase deformation, they are simply related to the constrictive nature of the finite strain ellipsoid.

(vii) Evidence of non-coaxial deformation during the main tectonic event is lacking.

Discussion. The dome-and-basin structures affected both granitoids and greenstones, and the domes appear geometrically unrelated to the intrusion of any of the dominant TTG lithologies. The centres of the domes are not occupied by granitic intrusions. Throughout the region, in the amphibolite-facies rocks of Holenasipur as in the granulites of Gundlupet, evidence of partial melting in the form of veins and migmatitic patches is observed. Some of the migmatites are strongly deformed by the doming event, others retain their magmatic textures. These observations suggest that the domes formed during the deformation of a pre-existing granite-greenstone sequence during a major period of metamorphism and migmatization.

The style of deformation associated with the dome-and-basin structures is characterized by a number of highly significant features: (a) the locations of superimposed structures coincides with domains in which the deformation ellipsoid is constrictive; (b) the spatial variations of strain intensities and strain ellipsoid type are highly variable, and (c) regions of non-coaxial deformation are restricted to the contacts between supra- and infracrustal series. This combination of features allows only one interpretation, that the deformation resulted from relative vertical displacements between ascending masses of

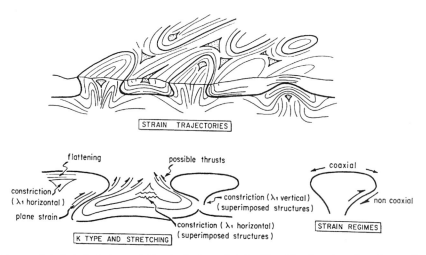

Fig. 4. Diagrams of strain trajectories, variations in the type of strain ellipsoid and strain regimes in a medium affected by diapirism. The pattern is inferred from the experimental data of Dixon (1975) and Dixon & Summers (1983), augmented by field observations made by the authors.

migmatitic gneisses and descending regions of supracrustal rocks (Dixon 1975; Gorman *et al.* 1978; Schwerdtner *et al.* 1978; Bouhallier *et al.* 1993).

Key observations that support a diapiric origin include the parallelism of foliation trajectories with contacts between supra- and infracrustal rocks, the increase of strain at these contacts, the locations of triple junctions and the specific organization and distribution of the strain regimes, as illustrated in Figs 4 & 5. Alternative explanations can be ruled out. Fold interference cannot explain the restriction of greenstones to synformal structures (in a folded sequence the greenstones would also occupy anticlines). During folding strain variations should be linked directly to the competence of the different layers, and regions of maximum strain would not be restricted to gneiss-greenstone contacts. In a folded sequence the sense of shearing is always directed towards the dome axes, not away from them, as is observed in the field area.

In the Holenarsipur area, supracrustal rocks are confined to linear synformal basins and infracrustal rocks form elliptical domes. In the Gundlupet area, supracrustal rocks are confined to elliptical synformal basins whereas infracrustal rocks form linear domes. From the southward increase of palaeopressures (Harris & Jayaram 1981; Janardhan *et al.* 1982; Raith *et al.* 1982), it might be expected that the contrasting structural features of the two areas might be related to different levels in the crust. The same structures probably are exposed in each area, and differences between them can be attributed to differences in crustal level – deeper for Gundlupet, where syntectonic assemblages are granulitic, than for Holenarsipur

where assemblages are amphibolitic (Bouhallier & Guiraud, work in progress). For these reasons we propose that the roots of the infracrustal and elliptical domes that outcrop in the Holenarsipur area may be found in the long and linear domes of the Gundlupet area, and the supracrustal elliptical basins of the Gundlupet area may represent deeper levels of inverted diapirs whose upper reaches are exposed as linear connected synforms in the Holenarsipur area (Fig. 3).

The principal conclusion is that crust-scale dome-and-basin structures can explain the geometry and kinematics of all the features revealed by the structural analysis of Holenarsipur and Gundlupet areas. These diapiric structures, which are locally affected by zones of horizontal shearing, are the only recognizable structures in the Archaean continental crust of Dharwar craton.

It is useful to recall the numerous experimental studies that serve as a basis for the interpretation of deformation fields related to gravitational instabilities. These analogue or numerical models allow discussion of the geometry and mechanics of these instabilities (Biot & Ode 1965; Anketell *et al.* 1970; Berner *et al.* 1972; Talbot 1977; West & Mareschal 1979; Mareschal & West 1980; Marsh 1982; Schmeling *et al.* 1988), as well as the deformation fields within them (Dixon 1975; Dixon & Summers 1983), and the patterns of interference between them (Brun 1983*a, b*). Numerous field studies have demonstrated the utility of applying these concepts to natural structures, and the majority of these come from Archaean terrains (Schwerdtner 1984; Schwerdtner *et al.* 1977, 1978, 1980, 1983, 1985; Brun *et al.* 1981; Drury 1977; Collins 1989; Ramsay 1989; Delor *et al.* 1991).

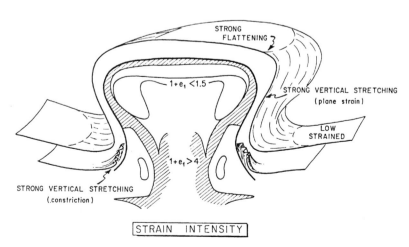

Fig. 5. Variations of strain intensity within a diapiric dome (from Dixon 1975), and in the surrounding area (from Brun 1983).

Recently Schwertner (1984) reported that certain aspects of the strain field associated with dome-and-basin structures in the Superior Province of Canada, do not conform to that predicted by the diapirism model. Although this author was not able to present a convincing explanation for certain features, such as high K-values in triple junctions, a model of superimposed folding was preferred essentially because of a lack of tensile bending in the domes, the presence of a synformal granitic sheet and a coherent vergency of folds. Such an interpretation cannot be applied to the examples studied by the present authors in India where strain intensity, *K* values and strain regimes are consistent from one dome to another and from one synformal greenstone unit to another.

It is also important to note that, in the Dharwar craton, absolutely no record of crustal thickening by stacking of thrust slices, as may have operated before the initiation of gravitational instabilities, was found by the present authors. Either such thrusting never happened, or all trace of it has been lost, which seems highly improbable.

Finally, the deformation field is disturbed by transcurrent shearing which was contemporaneous with the emplacement of the Closepet granite (Jayananda & Mahabaleswar 1990) and which affected the entire thickness of the Archaean crust from greenschist- to granulite-facies regions. It is stressed that this heterogeneous deformation was related to a thermal episode that affected a vast area of the proto-continent, and from near the surface to close to its base.

Hebei province, China

The Hebei province is situated in the northern part of the Sino-Korean craton, an Archaean terrane that makes up the entire substratum of northern China (Fig. 6). The region has been the object of various geochemical and geochronological investigations, but structural work in the Archaean terrains has been limited to several studies of a local and applied nature. In this paper a 140 km long structural section is presented which demonstrates that the entire region has been intensely deformed and transformed under catazonal conditions.

Lithologies. In the study area, the distinction between various lithological units and the establishment of a sequence of evolution is not straightforward. Liu *et al.* (1990) have proposed that, in the west, a single term, the Qianxi complex, should be applied to a complex group of gneissic and supracrustal rocks. However, the same type of material in the eastern Suichang–Coazhuang section has been divided by Sills *et al.* (1987) and Wang *et al.* (1990) into two units on the basis of geochemical and geochronological data (Pidgeon 1980; Jahn & Zhang 1984; Huang *et al.* 1986; Jahn *et al.* 1987; Liu *et al.* 1985; Liu *et al.* 1992).

The first unit, called the Coazhuang gneissic complex, is older than 3.5 Ga and is made up of supracrustal rocks intruded by felsic, TTG-type gneisses. The supracrustal sequences include mafic to felsic metavolcanics and metasediments such as fuchsite or sillimanite quartzites, marbles,

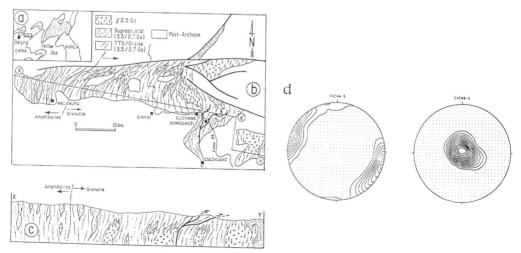

Fig. 6. (a) Location in China of the part of the Hebei province treated in this investigation. (b) Map adapted from an unpublished map of Y. Gang. (c) Section showing the geology and structure of the region. (d) Stereograms of foliations and lineations (112 measurements). Note the vertical foliation along the entire section, except in the thrust zone which is Proterozoic.

kinzigites and banded ironstones. The second unit, the Suichang gneissic complex, contains equal portions of supracrustal formations and granitoid intrusions with intermediate (monzogranitic and granodioritic) compositions. The depositional age of the supracrustal units is quite well constrained: they are intruded by 2.7 granitoids, and were probably deposited c. 3.0 Ga, the age of rhyolites intercalated with the sedimentary rocks (Liu et al. 1990). As in the Coazhuang complex, the supracrustal units contain mafic to felsic meta-volcanics and sedimentary rocks with compositions ranging from pelite to greywacke to BIF. The Coazhuang gneissic complex is observed only as disrupted fragments within the Suichang gneisses.

Both sequences, which appear to correspond to two distinct Archaean cycles, were intruded 2.7–2.45 Ga ago by felsic plutonic rocks with widely variable compositions (gabbro–diorite, monzo-diorite, granodiorite, monzogranite). All were affected by a major tectometamorphic event dated c. 2.5 Ga (Pidgeon 1980; Compston et al. 1983; Jahn & Zhang 1984; Liu et al. 1985, 1990). The youngest magmatic event was the emplacement of pegmatites at about 2.2–2.3 Ga (Liu et al. 1986).

Metamorphism. The entire Qianxi complex is metamorphosed and, in most of the region studied, the grade was granulite facies. Zhang et al. (1981), Zhang & Cong (1982), Sills et al. (1987) and Chen (1990) estimated conditions in the Coating region, in the SE part of our section, as $P = 8$ kbar and $T = 820$–$840°$. In contrast, in the western part of the section, the metamorphism was in the lower amphibolite facies with $P = 4.5$–6 kbar and T c. 600–$650°$ (Sills et al. 1987). Widespread, but irregular, retrograde metamorphism affected the granulites resulting in the replacement of ortho-pyroxene, clinopyroxene, and in some areas garnet, by hornblende and biotite. This event probably was post-Archaean (Jahn et al. 1984).

Structure (Fig. 6). Along the entire 140 km length of our section, all the lithologies described above display a regional vertical foliation, the nature of which attests to a pronounced homogeneous vertical flattening (Choukroune et al. 1993). The intensity of finite strain is documented by markers such as pillows in metabasalts from the Shangshuan region, in which the ratio of long to short axes is now 10:1. The only units to have escaped this deformation are the youngest granodioritic plutons, which have essentially undeformed cores surrounded by extremely foliated margins.

The vertical foliation within the ortho- and paragneisses provides evidence of very strong flattening. The direction of major extension is vertical wherever displayed. Furthermore, the deformation seems entirely coaxial: micro-structures such as pressure shadows and shear bands are almost always symmetrical, and features such as rotated mineral grains and sygmoidal structures in synkinematic minerals were not found.

On a map scale, the foliation is oriented roughly north–south (020°) in the southern part of the study area, and changes progressively to NE or ENE in the northern part. Local refraction at the margins of competent objects are occasionally observed. The change in regional trend in the north is consistent with a zone of dextral E–W shearing that forms a northern limit to the Archaean domain. This phase is clearly younger than the major deformation and is related to an episode of transcurrent movement that affected potassic pegmatites dated at 2.2–2.3 Ga (Liu et al. 1986).

The deformation in mafic rocks is readily corre-lated with granulite-facies metamorphism in the west and amphibolite-facies metamorphism in the east. In the gneisses, the same phenomena are generally observed, even though certain charnockites display more static textures which probably result from post-deformation re-crystallization. The intimate relationship between deformation and metamorphism receives further convincing support in thin sections which reveal deformation textures of a type that only form at high temperatures, such as the deformation of quartz by a prismatic 'C-type' slip mechanism.

To conclude, the structure of the Archaean part of the Hebei province is the result of deformation that led to a strong, homogeneous horizontal shortening which resulted in vertical foliations. This deformation, which removed all trace of earlier tectonic episodes, was contemporaneous with major metamorphism at 2.5 Ga (Pidgeon 1980).

The trajectories of the regional foliation appear to be well ordered along most of the section, and are only perturbed in the north by Proterozoic dextral shearing. In the small southeast segment of the cross-section, which follows the valley of the Luan river, the vertical foliation is affected by multiple horizontal shears, along which metre-wide cataclastites are developed. These cataclastites display E–W lineations and asymmetric micro-structures that indicate movement towards the east. The shearing also affects the 2.3 Ga pegmatites and therefore has the same age as the dextral movement evident at the northern margin of the Archaean massif. In both cases the Archaean material under-went greenschist-facies retrograde metamorphism. The Proterozoic deformation was clearly the last tectonic event to have affected the Hebei province, because the entire sequence is overlain dis-cordantly by a thick tabular sedimentary series of

conglomerates, quartzites and carbonates with ages
between 1.7 and 0.6 Ga.

Discussion. The Archaean of the Hebei province
is characterized by a style of deformation that is
highly unusual and has never been described in the
deep zones of 'modern' mountain ranges. In the
latter, granulite episodes are related to a late stage
in the evolution of the belts, and are manifested by
the synmetamorphic development of a horizontal
foliation that results from strong vertical non-
coaxial shortening. Our understanding of the fabric
of the present deep continental crust comes mainly
from vertical seismic reflection studies. From such
studies we know that the entire sub-basement of
Western Europe between 20 and 30 km depth is
made up of a highly reflective domain that is
occasionally exposed in Tertiary orogenic zones
(Bois *et al.* 1991). In the Alps and Pyrenees, Brodie
et al. (1989) and Bouhallier *et al.* (1991) have
demonstrated that such domains are composed of
horizontally foliated Hercynian granulites whose
metamorphic evolution is related to tectonism that
was locally non-coaxial and corresponded to
global vertical shortening. Wherever exposed, such
rocks have a generally horizontal foliation, as in the
granulitic arc of Finland (Barbey 1986). In fact we
know of no example in the deep zones of recent
mountain ranges of granulites subjected to strong
horizontal shortening in a coaxial regime. The great
intensity, and the peculiar style, of the syngranulitic
deformation that accompanied the last Archaean
magmatic event in the deep crust of the Hebei
province, constitutes important evidence of unusual and specific behaviour. In contrast with the
situation in the Dharwar craton, where body forces
apparently were dominant, boundary forces appear
to have played an important role. During this event,
the Archaean crust behaved differently from that of
crust involved in modern orogenic events.

Man shield

The Man shield is an Archaean core preserved in
largely Eburnean or Birrimian (2 Ga) terranes of
the southern part of the West African craton
(Fig. 7). Most of the shield is in Liberia, but parts
extend into Sierra Leone and western Ivory Coast.
In the Man region, the subject of the present study,
the Archaean region is limited to the east by the
major Proterozoic Sassandra shear zone (Bessolles
1977; Caen-Vachette 1988; Camil *et al.* 1984).

Lithologies. The oldest rocks are *c.* 3.2 Ga TTG-
type gneisses (Camil 1984). A supracrustal series
containing abundant quartzite (with or without
magnetite) and aluminous paragneisses, and lesser

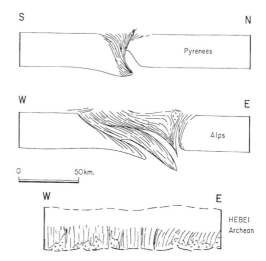

Fig. 7. Strain pattern observed in the Hebei region
compared with that of two segments of the Alpine chain.
The diagrams are drawn at the same scale in order to
highlight the great distance over which foliations are
consistently vertical in the Hebei region.

mafic metavolcanics, constitutes a greenstone belt
that extends 150 km in a NE–SW direction into
Liberia. In the study region near the town of Man,
the belt is confined between two granitoid domes.

The two classic Archaean assemblages are
largely migmatized and intruded by charnockites
dated at 2.7 Ga and by late pegmatites dated at
2.5 Ga (Rb–Sr ages of Camil *et al.*, 1984) This
history is closely comparable to that proposed by
Hedge *et al.* (1975) for similar lithologies in
Liberia. There, a formation made up essentially of
TTG has been assigned an age around 3.2 Ga on
the basis of Rb–Sr dating, and a period of major
granitoid intrusion and migmatization has been
related to a tectonothermal event during the period
2.7–2.5 Ga. In both regions the age of the supracrustal series is known only to be between 3.2 and
2.7 Ga.

Metamorphism. Almost all the study area has been
subjected to high grade, granulite facies metamorphism, and only a small portion of the
supracrustal material, in the centre of the greenstone belt, is in amphibolite facies. Conditions
during the major metamorphism associated with
the intrusion of charnockites and the widespread
migmatization of supracrustal rocks, are estimated
as $P = 5–7$ kbar and $T = 750–850°C$. This metamorphism is part of the tectonothermal event dated
at 2.7 Ga. A final stage of retrograde amphibolite-
facies metamorphism is recognized in zones of
late intense deformation (Camil 1984).

Structures. In a cross section of the Man shield (Fig. 8), the supracrustal series can be seen to occupy a central region confined between two gneissic domes (Man and Mt Pekou domes) with vertical to slightly reversed margins. Within the domes, the style of deformation that affects the intrusive charnockites is closely comparable to that of the Dharwar craton, in that vertical lineations are very strongly developed on the flanks of the domes. There are, however, two important differences: the first is that the domes are more linear, and the domains at their summits, where foliations are flat, are elongated. In other words, the ellipticity of the domes is very pronounced. The second difference is in the style of deformation of the supracrustal series where both foliations and lineations are vertical, alternating repeatedly with numerous zones of deformation in which the lineation is horizontal. These zones are oriented preferentially in two directions; at 010°, the direction of sinistral ductile shearing and at 060°, the direction of dextral ductile shearing. In the relatively sparse regions between the shear zones, the structural characteristics of the

migmatized supracrustals are well preserved. A strong syn-migmatitic vertical flattening, with a vertical principal stretching direction, is typical of these regions. Within the deformation zones, in contrast, the migmatites are retromorphosed and intensely sheared. At the margins of the gneissic domes the late deformation phase is responsible for the development of mylonitic zones, particularly at the borders of charnockitic bodies. Structures indicative of deformation in a viscous state, as well as C/S structures, provide evidence that the late 2.5 Ga pegmatites were sheared during their emplacement.

Discussion. In the Man region it is difficult to constrain the kinematics of the processes responsible for the formation of the gneissic domes and supracrustal basins using field criteria. This is because most early structures were obliterated by the late shearing marked by a dense pattern of retromorphosed ductile-deformed vertical deformation zones which post-date the doming event. However, what remains of the early strain field is compatible with doming.

Archean tectonics: a hypothesis

As a working hypothesis, it is proposed that the regions described above represent three stages of Archaean crustal deformation and that the Archaean continents were affected by tectono-metamorphic events on a scale approaching or exceeding that of the exposed cratons. In India, a dome-and-basin geometry was developed throughout a region that encompasses all of the presently exposed Dharwar craton. The original extent of these terranes is not known, but they may once have been parts of a much larger continent. Certainly the regions studied provide only a minimum estimate of the area that developed a dome-and-basin structure related to a tectonothermal episode. It is important to add that these processes affected the entire thickness of the Archaean crust. Contrary to the idea of Park (1982) who envisaged a zone of decoupling beneath a diapiric crust, we found evidence of domes at depths from 25–30 km (corresponding to the 8–9 kbar palaeopressures in the southern part of the craton) to a few kilometres below the surface.

Such structures are never produced in the same volumes in more recent orogenic belts. Gneissic or migmatitic domes certainly exist, but they are restricted to specific domains in which crustal thickening by thrust stacking is important. During the Archaean, crustal thickening by this mechanism is neither demonstrable nor necessary; the enormous quantity of juvenile material that accreted to the growing continents was alone

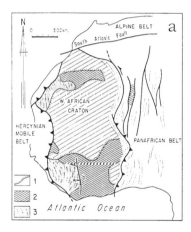

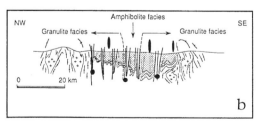

Fig. 8. (a) Location of the Man shield within the West African craton. 1, Archean domains; 2, Birrimian domains; 3, Upper Proterozoic to Hercynian basins. (b) Structural cross-section near the city of Man: the supracrustal belt between two granitoid domes is strongly affected by vertical shears, implying horizontal shortening.

sufficient to account for the thickening (Condie 1990). In the Dharwar craton, the episodes of dome formation are invariably related to the addition of material and to a general migmatization of pre-existing crustal rocks. Thermal phenomena thus accompanied the tectonic events, and this all happened on a scale very much larger than that represented by migmatitic or high-grade metamorphic domains in modern belts.

Consider again the examples of India and the Man shield. Although body forces and diapirism were responsible for the major structures of these cratons, the regions were also subjected to an additional episode of regional deformation contemporaneous with a large-scale thermal event, which was the result of application of boundary forces. The entire Dharwar craton was affected by transcurrent shearing, the organization of which points to a shortening direction consistent with the ellipticity of the domes. This heterogeneous, globally coaxial, deformation is also contemporaneous with a major thermal event that is manifested in widespread migmatization and the emplacement of the enormous Closepet batholith, as discussed by Drury *et al.* (1984), Jayananda & Mahabaleshwar (1990) & Bouhallier *et al.* (1993).

The Late Archaean deformation is better developed in the Man shield where zones of deformation related to the doming are cut by ductile shearing that modified the original structure. Here as well, a late magmatic event accompanied the heterogeneous deformation, the causes of which must be sought at the non-observable limits of the Archaean craton.

Finally, the Hebei province of the Sino-Korean craton could be considered as the region where the post-dome phase of deformation is the most intense. This vertical, homogeneous and coaxial deformation affected even deep granulite-facies parts of the Archaean crust regions during the Late Archaean thermal event. The deformation also involved a large volume of the crust. We assume that the dome-and-basin geometry is no longer observable in the region because these structures were completely obliterated by the intense deformation at 2.5 Ga.

There are strong analogies between the three regions. Each contains similar rock types, each was subjected to intense and widespread tectonic and metamorphic events, and each was affected by a specific and unusual type of deformation. The differences in structural style illustrated in Fig. 9, are explained by variations in the intensity of post-diapiric deformation. The explanation is entirely in accord with Ermanovics & Davison (1976), Anhaeusser (1984), Hickman (1984), and Padgham (1992) for other Archaean regions (although these authors are in the minority and

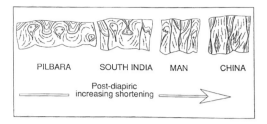

Fig. 9. Evolution of Archean crust, interpreted on the basis of the authors' investigations in China, India and Ivory Coast. The intensity of post-diapiric horizontal shortening is minimal in the Pilbara (Delor *et al.* 1991) and reaches a maximum in the Chinese case.

that for each region there are many others who advocate alternative interpretations). The differences between the three regions discussed earlier could be due solely to variations in the action of boundary forces: the regions are distinguished only by the intensity of the horizontal shortening that followed the development of domes and basins (Fig. 9).

Distinctive character of Archaean deformation

Archaean thrusting

Archaean thrusting has been described in all cratons and, particularly in the past few years, sections through these cratons have been drawn in a style which suggests that the deformation was much the same as in modern orogenic belts (e.g. Drury *et al.* 1984; Swager *et al.* 1992; Hammond & Nisbet 1992; Treloar *et al.* 1992, Van Kranendonk & Helmstaedt 1992). These interpretations are here disputed and it is argued that (i) the prevalence of thrusting has been strongly over-emphasized in certain regions, and that certain examples postulated on the basis of field studies are suspect, (ii) that a certain type of thrusting can develop during entirely vertical movements, and (iii) even if certain thrusts signify truly tangential tectonics, their presence is not a sufficient criterion to refute a specific, non-modern behaviour of the Archaean crust. Of the various examples of thrusting that have been attributed to tangential movements, several categories can be distinguished.

The first case is concerned with inferred thrusting that is shown to be restricted to contacts between granite and greenstone, as in the examples described by Stowe (1984), Nutman *et al.* (1989), Ralser and Park (1992) and Van Kranendonk & Helmstaedt (1992). In these examples, the geometry of the domes and the vertical orientation of the supracrustal series are acknowledged, but these characteristics are considered have developed late, following a regime of tangential tectonics which produced the large isoclinal folds responsible for multiple repetition of the series (Bickle *et al.* 1980; Kusky 1992). These examples are inconclusive because they are not supported by data of the type needed to document the style of deformation. If the situation is indeed as described, it should be readily demonstrated by the super-imposition (or interference) of two strain fields: the first, a result of tangential shearing, should be disturbed by a second phase related to the doming. It has been shown that such superimposition cannot be demonstrated in India, where the deformation style is entirely compatible with the diapirism model. It is useful to recall that deformation accompanying doming introduces a foliation and may cause isoclinal folding, even at the earliest stage of doming. During continued development of the domes, the axial planes of the folds may be refolded, and segments of the gneissic and supracrustal series may be superimposed (Talbot 1974, 1977). This is the case in the Gunlupet region of the Dharwar craton, where such structures have developed in the complete absence of tangential tectonics. It is argued here that many of the thrusts described or drawn at gneiss-greenstone contacts are neither necessary nor proven.

The second case concerns thrusting within, and confined to, greenstone belts (Collins 1989; De Wit 1991; Corfu *et al.* 1989; Kidd *et al.* 1988). This thrusting, which never penetrates the boundaries with gneissic rocks, can be readily explained by progressive deformation within portions of a greenstone belt descending between rising gneissic domes (Fig. 4). Gorman *et al.* (1978) noted that the trajectories of such thrusting follow contours within the gneissic domes, and they developed the 'sagduction' model to explain structures in the Abitibi belt, Canada. Collins (1989) applied similar reasoning to examples from the Pilbara in Australia. Neither the character, nor the signifi-cance of such thrusting can be compared with that associated with the regional-scale tangential tectonics that causes crustal thickening in the interiors of modern, linear orogenic belts.

The third case is that of thrusts at craton margins, such as those recognized between the Kaapvaal & Zimbabwe cratons (Light 1982; McCourt & Wilson 1992; Treloar *et al.* 1992) or in the Pontiac Group in the southern Superior Province Canada (Camiré & Burg 1993). The Limpopo Belt is commonly interpreted to have developed during Late Archaean (2.6–2.46 Ga) collision between pre-existing cratons. The extrusion of material from the belt, which had been granulitized in the course of the collision, was achieved by thrusting towards the north onto the Zimbabwe craton, and towards the south onto the Kaapvaal craton. The well-argued account of McCourt & Wilson (1992) contains several interesting points. The first is the age of the thrusting, which is confined to the latest Archaean and thus to the latest period of crustal evolution, and not to an earlier stage, as commonly envisaged in other regions (Stowe 1984, Ralser & Park, 1992, Van Kranendonck & Helmstaed, 1992). Indeed, recent dating by Kamber *et al.* (1993) has shown that much of the thrusting took place in the Proterozoic. Under these circumstances it can readily be accepted that the cratons had a certain rigidity at the time of their collision. The second is the location of the thrusts, which are restricted to a zone between the two cratons. The third is the fact that the collision failed to obliterate earlier deformation related to doming events in the interiors of the cratons. A parallel can thus be drawn with the examples above discussed above, in particular within the Dharwar craton, where the latest Archaean deformation might be related to collision between cratons (Drury *et al.* 1984; Newton 1990) even if no clear collisional zones can be directly observed.

Important differences between Archaean and modern thermotectonic events

Certain features of the three Archaean regions are crucial to the discussion of the characteristics of Archaean crust.

(i) In each region, repeated major thermal and tectonic events involved immense volumes of crust.

(ii) Thermal events in the Indian shield, and probably also the Man shield, were responsible for triggering large-scale gravitational instabilities that led to the restructuring of the entire protocrust. The planar anisotropy resulting from this restructuring was on average vertical, at least in the root zones of the diapirs.

(iii) Late tectonometamorphic events had the effect of accentuating the vertical structure through heterogeneous shortening. This ranged from only moderate in the Dharwar and Man shields to intense in the Sino-Korean craton.

(iv) If major thrusting had ever taken place in the interiors of these cratons, all evidence of this process, be it geometric or kinematic, was completely erased by the later events.

These features: either resulted from the intrinsic properties of the Archaean crust or the thermal or kinematic boundary conditions that reigned during the Archaean events were fundamentally different from those during the formation of modern belts. With the first explanation, the development of

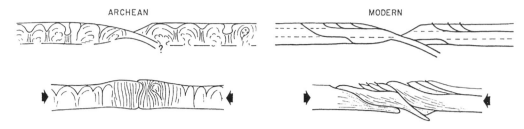

Fig. 10. Differences between the internal structures of Archaean and modern crust, and the control these features exert when the crust is shortened. The principal difference is the presence in modern crust of roughly horizontal layers.

dome-and-basin structures during the early stages of the continental crust, implies the absence of a zone of horizontal decoupling within this young crust (Fig. 10). Furthermore, the Archaean crust has a distinct fabric defined by vertical contacts between domes and supracrustal belts, a predominantly vertical layering within these two units, and vertical orientations of the basic dykes that fed the basalts of the greenstone belts. These features imposed a strong planar and vertical anisotropy that is absent from modern crust, which is more heterogeneous and in which the levels of resistance to deformation are horizontally disposed.

Less can be said about the second explanation, because any discussion of boundary conditions and kinematics in the Archaean is inevitably poorly constrained. By contrast, it is relatively easy to discuss the thermal structure. It can be stated with some certainty that, during Archaean tectonic events, the entire crust was intensely and totally reheated. This had two major effects. First, the reheating led to the migmatization of enormous volumes of sialic material, and the resultant decrease in density was the trigger for gravitational instabilities. Second the protocrust was softened, a process that is evident from the absence of contrast in the deformability of the different lithologies and the development of the exceptionally homogeneous character of the deformation as for example, in the Hebei province.

It is here emphasized that the presence of a soft crust and absence of rigid lithosphere was temporary and not the normal situation during the Archaean. The evidence for the existence of thick lithosphere beneath the continents for much of the Archaean (the existence of Archaean diamonds, Richardson *et al.* 1984; the relatively subdued thermal gradients in Archaean metamorphic terrains, Bickle 1986, Richter 1985). Nonetheless it is proposed that the rigid substratum of the continents was temporally destroyed during periodic events that either created or reworked continental crust.

The Source of heat and the nature of Archaean tectonism

From the above descriptions and interpretations, it is evident that Archaean continental crust was subjected to tectonothermal events during which new material was added (basalt and TTG) and the entire crust was deformed in a manner unlike that of modern crust. In developing a model to explain these processes we stress certain key points.

(1) During the main events, diapirism on a major scale affected the entire crust and was followed by a later period of large-scale homogeneous shortening: these observations indicate that body forces exceeded the yield strength of the crustal rocks.

(2) To change the strength of material with essentially the same composition as modern crust (granitoids with minor supracrustals), heat must have been added, again on a major scale.

(3) The nature and the scale of homogeneous shortening in the Man and Hebei shields indicates the absence of a rigid substratum during the deformation, i.e. during these events, the crust apparently was not underlain by resistant lithospheric mantle.

(4) The scale of such events, the volumes of crust involved and the quantity of heat added are so large that only mantle sources can be considered. The eruption of thick sequences of basalt, and the inclusion of komatiite in many greenstone belts, is commonly explained in terms of melting in mantle plumes (Campbell *et al.* 1990; Abouchami *et al.* 1990).

(5) Compilation of the ages of crustal rocks from all continents, and studies of ages of detrital zircons in modern river sediments by Goldstein & Arndt (1993) indicate that tectonothermal events were episodic throughout the Precambrian. Major, sharp events at e.g. 3.1, 2.7 or 2.5 Ga were separated by longer periods of inactivity. This pattern contrasts with the quasi-continuous nature of modern crustal evolution (Arndt 1992).

(6) The events in India, China and West Africa, whose ages correspond to those of the global events, mainly involved reworking of pre-existing continental crust. In other regions, such as the southern Superior craton at 2.7 Ga, the events started with komatiitic to basaltic volcanism, and climaxed with the production of large volumes of juvenile continental crust.

(7) During the Archaean the mantle was hotter. Oceanic crust probably was thick (40–50 km, Sleep & Windley 1982; Vlaar 1986), and the lithosphere relatively thin (Bickle 1986). The thickness of Archaean continents is difficult to judge, but cannot have been much more than 40–50 km: a marked contrast between the thicknesses of continental and oceanic crust, as now exists, was not likely in the Archaean.

These features are explained by postulating that the tectono-thermal events that create or modify continental crust are related to mantle plumes of the type that yielded continental or oceanic flood basalts in more recent earth history. Three stages are considered, each involving interaction between plume and crust.

(a) The first stage involves the arrival of a plume in a purely oceanic environment. High degree melting within the plume, as it approaches the surface, leads to flood volcanism that spreads a thick layer of basaltic to komatiitic rocks on the 40 km thick oceanic crust (Desrochers *et al.* 1993). The thick and relatively buoyant plateaus resist subduction, and new subduction zones form at their margins, leading to calc-alkaline volcanism and plutonism, and the emplacement of granitic plutons. The consequence is the formation of juvenile continental crust, as in the 2.7 Ga southern Superior province in Canada (Card 1990), or the 3.5 Ga Pilbara craton (Bickle *et al.* 1983).

(b) The second stage involves the arrival of a plume beneath pre-existing continental crust of the type that might have formed in the first stage. Here the crust and lithosphere were initially thick and the plume stalled at greater depths. The degree and the amount of mantle melting is therefore less. Subduction at the margins again led to calc-alkaline magmatism and, at a later stage, as heat from the plume head migrated to higher levels in the crust, migmatization, metamorphism and diapirism of the type recognized in the Man shield and Indian craton.

(c) The third stage is that of a plume arriving beneath thick, mature lithosphere. The plume stalls at a greater depth and melting is minor. Only low-degree melts form and these penetrate only to the lower crust. Advection or conduction of heat from the plume provokes crustal melting and the emplacement of intrusions like the Closepet granite of the Dharwar craton. The softened crust is then deformed, probably because of boundary forces whose origin is in the surrounding thick rigid oceanic crust. At this stage, the thick but soft continent was compressed and deformed homo-geneously between thick, subduction-resistant oceanic plates, a tectonic set-up far removed from the modern situation.

This work was supported by the CNRS-INSU program 'Dynamique et bilan de la terre'. A. S. Janardhan from Mysore University, B. Mahabaleshwar from Bangalore University and M. Jayananda were involved in numerous discussions about the geology of India. Lao Liu and Y. Geng introduced us to the Archaean of Hebei. J. Camil helped us in Ivory Coast. We are greatly indebted to our collegues from Rennes, B. Auvray, D. Chardon, G. Gruau, S. Fourcade, H. Martin and J. J. Peucat for stimulating discussions that greatly improved the paper.

References

ABOUCHAMI, W., BOHER, M., MICHARD, A. & ALBARÈDE, F. 1990. A major 2.1 Ga old event of mafic magmatism in West Africa. *Journal of Geophysical Research*, **95**, 17 605–17 629.

ANHAEUSSER, C. R. 1984. Structural elements of Archaean granite-greenstone terranes as exemplified by the Barberton Moutain Land, southern Africa *In*: KRÖNER, A. & GREILING, R. (eds), *Precambrian Tectonics Illustrated*. IUGS, 57–78.

ANKETELL, J. M., CEGLA, J. & DZULINSKI, S. 1970. On the deformational structures in systems with reversed density gradients. *Annales de la Société Géologique de Pologne,* 40, 3–29.

ARNDT, N. T. 1992. Rate and mechanism of continent growth in the Precambrian. *Abstract, IGC Kyoto*, 48.

BARBEY, P. 1986. *Signification géodynamique des domaines granulitiques: la ceinture des granulites de Laponie (Fennoscandie). Memoires et Documents du CAESS*, Rennes,7.

BECKINSALE, R. D., DRURY, S. A. & HOLT, R. W. 1980. 3.360 Myr old gneisses from the south Indian craton. *Nature*, **283**, 469–470.

——, REEVES-SMITH, G., GALE, N. H., HOLT, R. W. & THOMPSON, B. 1982. Rb-Sr and Pb-Pb isochron ages and REE data for Archaran gneisses and granites, Karnataka state, South India. *In*: ASHWAL, L. D. (eds), *Indo-U.S. Workshop on the Precambrian of south India* (abs). National Geophysical Research Institute Hyderabad, 35–36.

BERNER, H., RAMBERG, H. & STEPHANSSON, O. 1972. Diapirism theory and experiment. *Tectonophysics*, **15**, 197–218.

BESSOLES, B. 1977. *Géologie de l'Afrique: le craton Ouest Africain*. Memoires du BRGM, Orléans.

BHASKAR RAO, Y. J., BECK, W., RAMA MURTHY, V.,

NIRMAL CHARAN, S. & NAQVI, S. M. 1983. Geology, geochemistry, and age of metamorphism of Archaean gray gneisses around Channarayapatna, Hassan district, Karnataka, South India. *In*: NAQVI, S. M. & ROGERS, J. J. W. (eds), Precambrian of South India. *Geological Society of India Memoirs*, **4**, 309–328.

BICKLE, M. J. 1986. Global thermal histories. *Nature*, **319**, 13–14.

——, BETTENEY, L. F., BOULTER, C. A., BLAKE, T. S. & GROVES, D. I. 1980. Horizontal tectonic interaction of an Archean gneiss belt and greenstones, Pilbara block, Western Australia. *Geology*, **8**, 525–529.

——, ——, BARLEY, M. E., CHAPMAN, H. J., GROVES, D. I., CAMPBELL, I. H. & DE LAETER, J. R. 1983. A 3500 Ma plutonic and volcanic calc-alkaline province in the Archean East Pilbara block. *Contributions to Mineralogy and Petrology*, **84**, 25–35.

——, NISBET, E. G. & MARTIN, A. 1994. Archaean greenstone belts are not oceanic crust. *Journal of Geology*, **102**, 121–138.

BIOT, M. A. & ODE, H. 1965. Theory of gravity instability with variable overburden and compaction. *Geophysics*, **30**, 213–227.

BOIS, C. & ECORS SCIENTIFIC PARTY 1991. Late and Post-Orogenic Evolution of the Crust studied from ECORS Deep Seismic Profiles. *In*: MEISSNER *ET AL.* (eds), *Continental Lithosphere Deep Seismic Reflexions*. AGU Geodynamic Series, **22**, 53–68.

BOUHALLIER, H., CHOUKROUNE, P. & BALLEVRE, M. 1991. Evolution structurale de la croute Hercynienne profonde: exemple du massif de l'Agly (Pyrénées orientales France). *Comptes Rendus Acadamie de'l Sciences Paris* , **312**, 647–654.

——, —— & —— 1993. Diapirism, bulk homogeneous shortening and transcurrent shearing in the Archean Dharwar craton: the Holenarsipur area, South India. *Precambrian Research*, **63**, 43–58.

BRODIE, K. H., REX, D. & RUTTER, E. H. 1989. On the age of deep extentional faulting in the Ivrea zone, Northern Italy. *In*: Park, R. G., Coward, M. P. & Dietrich, D. (eds), Alpine Tectonics. *Geological Society, Special Publications*, **45**, 203–210.

BRUN, J. P. 1983*a*. Isotropic points and lines in strain fields, *Journal of Structural Geology*, **5**, 321–327.

—— 1983*b*. L'origine des domes gneissiques: modèles et tests. *Bulletin de la Société Geologique de France*, **25**, 219–228.

——, GAPAIS, D. & LE THEOFF, B. The mantle gneiss domes of Kuopio (Finland): interfering diapirs. *Tectonophysics*, **74**, 283–304.

CAEN-VACHETTE, M. 1988. Le craton Ouest africain et le Bouclier Guyannais: un seul craton au Proterozoique inférieur? *Journal of African Earth Sciences*,.**7**, 479–488.

CAMIL, J. 1984. *Pétrographie, chronologie des ensembles granulitiques archéens et formations associées de la région de Man (Cote d'Ivoire)*. Abidjan Univ. Thesis.

——, TEMPIER, P. & CAEN-VACHETTE, M. 1984. Schéma chronologique et structural des formations de la région de Man (Cote-d'Ivoire). *In*: *Géologie*

Africaine. Musée Royale d'Afrique Centrale, Tervuren, 1–10.

CAMPBELL, I. H., HILL, R. E. T. & GRIFFITHS, R. W. 1990. Melting in an Archaean mantle plume: heads its basalts, tails its komatiites. *Nature*, **339**, 697–699.

CARD, K. D. 1990. A review of the Superior Province of the Canadian Shield, a product of Archaean accretion, *Precambrian Research*, **48**, 99–156.

CAMIRÉ, G. E. & BURG, J. P. 1993. Late Archean thrusting in the Northwestern Pontiac Subprovince, Canadian shield. *Precambrian Research*, **61**, 51–66.

CHADWICK, B., RAMAKRISHNAN, M. M., VISWANATHA, M. N. & SRINIVASA MURTHY, V. 1978. Structural studies in the Archaean Sargur and Dharwar supracrustal rocks of the Karnataka craton, *Journal of the Geological Society of India*, **19**, 531–549.

——, —— & —— 1981. Structural and metamorphic relations between Sargur and Dharwar supracrustal rocks and Peninsular gneisses in central Karnataka. *Journal of the Geological Society of India*, **22**, 557–569.

——, ——, VASUDEV, V. N. & VISWANATHA, M. N. 1989. Facies distributions and structures of a Dharwar volcanosedimentary basin: evidence for late Archaean transpression in Southern India? *Journal of the Geological Society of London*, **146**, 825–834.

CHEN, M. 1990. Metabasic dyke swarms in a high-grade metamorphic terrane: a case study in the Taipingzhai-Jingchangyu area, Eastern Hebei Province. *Acta Geologica Sinica*, **64**, 169–183.

CHOUKROUNE, P., GAPAIS, D. & MERLE, O. Shear criteria and structural symmetry. *Journal of Structural Geology*, **9**, 525–530.

——, AUVRAY, B., JAHN, B. M., CHEN, T., GENG, Y. & LIU, D. 1993. Coupe structurale de la croute archéenne en Hebei (Craton sino-coréen, Chine du nord). *Comptes Rendus de l'Acadamie des Sciences, Paris*, **316**, 669–675.

CLOWES, R. M., COOK, F. A., GREEN, A. G., KEEN, C. E., LUDDEN, J. N., PERCIVAL, J. A., QUINLAN, & WEST, G. F. 1992. Lithoprobe: new perspectives on crustal evolution. *Canadian Journal of Earth Sciences*, **29**, 1813–1864.

COLLINS, W. J. 1989. Polydiapirism of the Archean Mont Edgar Batholith, Pilbara Block, Western Australia. *Precambrian Research*, **43**, 41–62.

CONDIE, K.C. 1990. Growth and accretion of continental crust: inferences based on Laurentia. *Chemical Geology*, **83**, 183–194.

COMPSTON, W., ZHANG, F. P., FOSTER, J. J., COLLERSON, K. D., BAI, J. & SUN, D. Z. 1983. Rubidium–strontium geochronology of Precambrian rocks from the Yenshan Region, North China. *Precambrian Research*, **22**, 175–202.

CORFU, F., KROGH, T. E., KWOK, Y. Y. & JENSEN, L. S. 1989. U-Pb zircon geochronology in the southwestern Abitibi greenstone belt, Superior Province. *Canadian Journal of Earth Sciences*, **26**, 1747–1763.

CRAWFORD, A. R. 1969. Reconnaissance Rb-Sr dating of the Precambrian rocks of Southern Peninsula India. *Journal of the Geological Society of India*, **10**, 117–166.

DELOR, C., BURG, J. P & CLARKE, G. 1991. Relations diapirisme-métamorphisme dans la province du Pilbara (Australie occidentale): implications pour les régimes thermiques et tectoniques à l'Archéen. *Comptes Rendus Acadamie des Sciences, Paris,* **312,** 257–263.

DESROCHERS, J.-P., HUBERT, C., LUDDEN, J. N. & PILOTE, P. 1993. Accretion of Archaean oceanic plateau fragments in the Malartic Composite Block, Abitibi Greenstone Belt, Canada. *Geology,* **21,** 451–454.

DE WIT, M. J. 1991. Archean greenstone belt tectonism and basin development:some insights from the Barberton and Pieterburg greenstone belts, Kaapvaal craton,South Africa. *Journal of African Earth Sciences,* **13,** 45–63.

——, HART, R. A. & HART, R. J. (1987). The Jamestown ophiolite complex, Barberton mountain belt: a section through 3.5 Ga oceanic crust. *Journal of African Earth Sciences,* **6,** 681–730.

DIXON, J. M. 1975. Finite strain and progressive deformation models of diapiric structures, *Tectonophysics,* **28,** 89–104.

—— & SUMMERS, J. M. 1983. Patterns of total and incremental strain in subsiding troughs: experimental centrifuged models of inter-diapiric synclines. *Canadian Journal of Earth Sciences,* **20,** 1843–1861.

DRURY, S. A. 1977. Structures induced by granite diapirs in the Archaean greenstone belt at Yellowknife, Canada: implications for Archaean geotectonics. *Journal of Geology,* **85,** 345–358.

—— 1983. A regional tectonic study of the Archaean Chitradurga greenstone belt, Karnataka based on Landsat interpretation, *Journal of the Geological Society of India,* **24,** 167–184.

—— & HOLT,R. W. 1980. The tectonic framework of the south India craton: a reconnaissance involving LANDSAT imagery. *Tectonophysics,* **65,** 111–115.

——, HARRIS, N. B. W., HOLT, R. W., REEVES-SMITH, G. J. & Wightman, R. T. 1984. Precambrian tectonics and crustal evolution in south India. *Journal of Geology,* **92,** 3–20.

——, VAN CALSTEREN, P. C. & REEVES-SMITH, G. J. 1986. Sm-Nd isotopic data from Archean metavolcanic rocks at Holenarsipur, south India. *Journal of Geology,* **95,** 837–843.

ENGLAND, P. & BICKLE, M. J. 1984. Continental thermal and tectonic regimes during the Archean. *Journal of Geology,* **92,** 353–367.

ERMANOVICS, I. F. & DAVISON, W. L. 1976. The Pikwitonei Granulites in relation to the Northwestern Superior province of the Canadian shield. *In*: Windley, B. F. (ed.), *The early history of the Earth.* John Wiley & Sons, London, 331–347.

FLINN, D. 1965. On the symmetry principle and the deformation ellipsoid. *Geological Magazine,* **102,** 36–45.

FRIEND, C. R. L. 1981. Charnockite and granite formation and influx of CO2 at Kabbaldurga. *Nature,* **294,** 550–552.

—— 1983. The link between charnockite formation and granite production: Evidence from Kabbaldurga,

Karnataka, South India. *In*: ATHERTON, M. P. & CRIBBLE, C. D. (eds), *Migmatites, melting, metamorphism.* Shiva, Nantwich, 264–276.

—— & NUTMAN, A. P. 1991. Shrimp U-Pb geochronology of the Closepet granite and Peninsular gneisses, Karnataka, South of India. *Journal of the Geological Society of India,* **38,** 357–368.

FYFE,W. S. 1974. Archaean tectonics. *Nature,* **24,** 338.

GOLDSTEIN, S. L. & ARNDT, N. T. & STALLARD, R. F. 1989. Use of Nd isotopes and U-Pb zircon ages in sediments to delineate crust formation histories: the Orinoco River basin. *Terra abstracts,* **1,** 335–336.

GORMAN, B. E., PEARCE, T. H. & BIRKETT, T. C. 1978. On the structure of Archean greenstone belts. *Precambrian Research,* 6, 23–41.

HAMMOND, E. C. & NISBET, B. W. 1992. Towards a structural and tectonic framework for the central Norseman-Wiluna greenstone belt, Western Australia. *In*: GLOVER, J. E. & HO, S. E. (eds), *The Archaean: Terrains, Processes and Metallogeny.* The Geology Key Centre & University extension, the University of Western Australia, 39–50.

HANSEN, E. C., NEWTON, R. C. & JANARDHAN, A. S. 1984. Pressures, temperatures, and metamorphic fluids across an unbroken amphibolite-facies to granulite-facies transition in Southern Karnataka, India. *In*: KRÖNER, A., GOODWIN, A. M. & HANSON, G. N. (eds), *Archaean geochemistry.* Springer-Verlag, Berlin, 161–181.

——,JANARDHAN, A. S., NEWTON, R. C., PRAME, K. B. M. & RAVINDRA KUMAR, G. R. 1987. Arrested charnockite formation in Southern India and Sri Lanka. *Contributions to Mineralogy and Petrology,* **96,** 225–244.

HARRIS, N. B. W. & JAYARAM, S. 1981. Metamorphism of cordierite gneisses from the Bangalore region of the Indian Archaean. *Lithos,* **15,** 89–98.

HEDGE, C. E., MARVIN, R. F. & NAESSER, C. W. 1975. Age provinces in the basement rocks of Liberia. *Journal of Research, US Geological Survey,* **3,** 425–429.

HICKMAN, A. H., Archaean diapirism in the Pilbara Block, Western Australia, *In*: KRÖNER, A. & GREILING, R. (eds), *Precambrian Tectonics Illustrated.* IUGS, 113–127.

HELMSTAEDT, H., PADGHAM, W. A. & BROPHY, J. A. 1986. Multiple dykes in the lower Kam Group, Yellowknife Greenstone Belt: evidence for Archaean sea floor spreading? *Geology,* **14,** 562–566.

HUANG, W., BI, Z. & DE PAOLO, D. J. 1986. Sm-Nd isotope study of early Archean rocks, Qianan, Hebei province, China. *Geochimica et Cosmochimica Acta,* **50,** 625–631.

HUSSAIN, S. M. & NAQVI, S. M. 1983. Geological, geophysical and geochemical studies over the Holenarsipur schist belt, Dharwar craton, India. *In*: NAQVI, S. M. & ROGERS, J. J. W. (eds), Precambrian of South India. *Geological Society of India, Memoirs,* **4,** 73–95.

JAHN, B. M. & ZHANG, Z. Q. 1984. Archean granulite gneisses from Eastern Hebei Province,China: rare earth geochemistry and tectonic implications.

Contributions to Mineralogy and Petrology, **85**, .224–243.

——, AUVRAY, B., CORNICHET, J., BAI, Y. L., SHEN, Q. H. & LIU, D. Y. 1987. 3.5 Ga old amphibolites from Eastern Hebei Province, China: field occurrence petrography, Sm-Nd isochron age and REE geochemistry. *Precambrian Research*, **34**, 311–346.

.NARDHAN, A. S., NEWTON, R. C. & SMITH, J. V. 1979a. Ancient crustal metamorphism at low pH2O: charnockite formation at Kabbaldurga, south India. *Nature*, **278**, 511–514.

——, RAMACHANDRA, H. M. & RAVINDRA KUMAR, G. R. 1979b. Structural history of Sargur supra crustals and associated gneisses, Southwest of Mysore, Karnataka. *Geological Society of India*, **20**, 61–72.

——, NEWTON, R. C. & HANSEN, E. C. 1982. The transformation of amphibolite facies gneiss to charnockite in Southern Karnataka and northern Tamil Nadu, India. *Contributions to Mineralogy and Petrology*, **79**, 130–149.

JAYANANDA, M. & MAHABALESWAR, B. 1991. Relationship between shear zones and igneous activity: the Closepet Granite of southern India. *Proceedings of the Indian Academy of Sciences (Earth and Planetary Science)*, **100**, 31–36.

——, MARTIN, H. & MAHABALESHWAR, B. 1992. The mechanism of recycling of the Archaean continental crust:example from the Closepet granite, southern India. *In*: GLOVER, J. E. & HO, S. E. (eds), *The Archaean: Terrains,Processes and Metallogeny*. The Geology Key Centre & University extension & the University of Western Australia,Perth, 213–222.

KAMBER, B. S., KRAMER, J. D. & ROLLINSON, H. R. 1993. The Triangle Shear Zone, a Proterozoic kill-joy in the tectonic models for the Archean Limpopo belt. *Terra abstracts*, **1**, **5**, 316.

KIDD, W. S. F., KUSKY, T. & BRADLEY, D. C. 1988. Late Archean greenstone tectonics: evidence for thermal and thrust loading, lithospheric susidence from stratigraphic sections in the Slave Province, Canada. In: Ashwald, L. D. (ed.), *Workshop on the deep crust of Southern India*. Lunar and Planetary.Institute, Houston, 79-80.

KRANCK, E. H. 1957. On folding-movements in the zone of the basement. *Geologische Rundschau*, **47**, 261–282.

KUSKY, T. M. 1992. Relative timing of deformation and metamorphism at mid- to upper crustal levels in the Point Lake Orogen, Slave Province,Canada. *In*: GLOVER, J. E. & HO, S. E. (eds), *The Archaean: Terrains, Processes and Metallogeny*. The Geology Key Centre & University extension, University of Western Australia, 59–72.

LIGHT, M. P. R. 1982. The Limpopo mobile belt: a result of continental collision. *Tectonics*, **4**, 325–342.

LIU, D. Y., PAGE, R. W., COMPSTON, & WU, J. S. 1985. U-Pb Zircon geochronology of late Archean metamorphic rocks in the Taihangshan-Wutaishan area, North China. *Precambrian Research*, **27**, 85–109.

——, SHEN, Q. H., JAHN, B. M., ZHANG, Z. Q., AUVRAY, B., ZHANG, Q. Z. & YE, X. J. 1986. U-Pb geochronology of granitoids from the archaean metamorphic terranes of eastern Hebei Province, China (extended abstract). ICOG VI, Cambridge.

——, ——, ZHANG, Z. Q., JAHN, B. M. & AUVRAY, B. 1990. Archean crustal evulution in China:U-Pb geochronology of the Qianxi complex. *Precambrian Research*, **48**, 223–244.

——, NUTMAN, A. P., COMPSTON, W., WU, J. S. & SHEN, Q. H. 1992. Remnants of > 3800 Ma crust in the Chinese part of the Sino-Korean Craton. *Geology*, **20**, 339–342.

McCOURT, S. & VEARNCOMBE, J. R. 1992. Shear zones of the Limpopo Belt and adjacent granitoid-greenstone terranes: implications for late Archean collision tectonics in southern Africa. *Precambrian Research*, **55**, 539–552.

—— & WILSON, J. F. 1992. Late Archaean and early Proterozoic tectonics of the Limpopo and Zimbawe Provinces,Southern Africa. *In*: GLOVER, J. O. & HO, S. E. (eds), *The Archaean Terranes, Processes and Metallogeny*. The Geology Key Centre & University extension, University of Western Australia, Perth, 237–245.

McGREGOR, A.M., 1951. Some milestones in the Precambrian of Southern Rhodesia. *Transactions of the Geological Society of South Africa*, **54**, 27–70.

MARESCHAL, J. C. & WEST, G. F. 1980. A model for tectonism. Part 2. Numerical models of vertical tectonism in greenstones belts. *Canadian Journal of Earth Sciences*, **17**, 60–71.

MARSH, B. D. 1982. On the mechanism of igneous diapirism, stoping and zone melting. *American Journal of Science*, **282**, 808–855.

MEEN, J. K., ROGERS, J. J. & FULLAGAR, P. D. 1992. Lead isotopic composition of the Western Dharwar craton, southern India: Evidence for distinct Middle Archaean terranes in a Late Archean. *Craton.Geochimica et Cosmochimica Acta*, **56**, 2455–2470.

MONRAD, J. R. 1983. Evolution of sialic terrains in the vicinity of the Holenarsipur belt, Hassan District, Karnataka, India. *In*: NAQVI, S. M. & ROGERS, J. J. W. (eds), *Precambrian of South India*. Geological Society of India Memoirs, **4**, 343–364.

NAHA, K., SRINIVASAN, R. & JAYARAM, S. 1990. Structural evolution of the Peninsular Gneiss: an early Precambrian migmatitic complex from South India. *Geologische Rundschau*, **79**, 99–109.

NAQVI, S. M. & RODGERS, J. J. W. (eds) 1983. *Precambrian of South India*. Geological Society of India Memoirs, **4**.

——, ALLEN, P. & 12 others. 1983b. Geochemistry of gneisses from Hassan district and adjoining areas, Karnataka, India. *In*: NAQVI, S. M. & ROGERS, J. J. W. (eds), *Precambrian of South India*. Geological Society of India Memoirs, **4**, 401–416.

NELSON, B. K. & DE PAOLO, D. G. 1985. Rapid production of continental crust 1.7–1.9 b.y. ago: Nd and Sr isotopic evidence from the basement of the North American midcontinent. *Geological Society of America Bulletin*, **96**, 746–754.

NEWTON, R. C. 1990. The late Archean high-grade terrain of South India and the deep structure of the Dharwar craton. *In*: SALISBURY, M. H. & FOUNTAIN, D. M. (eds), *Exposed Cross-Sections of the Continental Crust*. Kluwer Academic Publishers, Amsterdam, 305–326.

NUTMAN, A. P., FRIEND, C. R. L., BAADSGAARD, H. & MCGREGOR, V. R. 1989. Evolution and assembly of Archean gneiss terranes. *Tectonics*, **3**, 573–589.

PADGHAM, W. A. 1992. Mineral deposits in the Archean Slave Structural Province; lithological and tectonic setting. *Precambrian Research*, **58**, 1–24.

PARK, R. G. 1982. Archaean Tectonics. *Geologisches Rundschau*, **71**, 22–37.

PEUCAT, J. J., VIDAL,P., BERNARD-GRIFFITHS, J. & CONDIE, K. C. 1989. Sr, Nd and Pb isotopic systematics in the archaean low- to high- grade transition zone of southern India: syn-accretion vs. post-accretion granulites. *Journal of Geology*, **97**, 537-550.

PIDGEON, R. T. 1980. 2480 Ma old zircons from granulites facies rocks from East Hebei Province, North China. *Geological Reviews*, **26**, 198–207.

PLATT, J. P. 1980. Archean greenstone belts: a structural test of tectonic hypotheses. *Tectonophysics*, **65**, 127–150.

RAASE, P., RAITH, M., ACKERMAND, D. & LAL, R. K. 1986. Progressive metamorphism of mafic rocks from greenschist to granulite facies in the Dharwar craton of south India. *Journal Geology*, **94**, 261–282.

RADHAKRISHNA, B. P. & NAQVI, S. M. 1986. Precambrian continental crust of India and its evolution. *Journal of Geology*, **94**, 145–166.

—— & RAMAKRISHNAN, M. (eds) 1990. *Archaean greenstone belts of South India*. Geological Society of India, Memoirs, **19**.

RAITH, M., RAASE, P. & ACKERMAND, D. 1982. The Archaean craton of Southern India: metamorphic evolution and P-T conditions. *Geologische Rundschau*, **71**, 280–290.

RALSER, S. & PARK, A. F. 1992. Tectonic evolution of the Archaean rocks of the Tavani Area, Keewatin, N.W.T., Canada. *In*: GLOVER, J. E. & HO, S. E. (eds), *The Archaean: Terrains, Processes and Metallogeny*. The Geology Key Centre & University extension, University of Western Australia, 99–106.

RAMAKRISHNAN, M. & VISWANATHAL, M. N. 1981. Holenarsipur belt. *In*: SWAMI NATH, J. & RAMAKRISHNAN, M. (eds), *Early Precambrian supracrustals of Southern Karnataka*. Geological Survey of India Memoirs, **112**, 115–141.

RAMSAY, J. G. 1989. Emplacement kinematics of granite diapir: the Chindamora batholith, Zimbabwe. *Journal of Structural Geology*, **11**, 191–209.

RICHARDSON, S. H., GURNEY, J. J., ERLANK, A. J. & HARRIS, J. W. 1984. Origin of diamonds in old enriched mantle. *Nature*, **310**, 198–203.

RICHTER, F. M. 1985. Models for the Archean thermal regime. *Earth and Planetary Science Letters*, **73**, 350–360.

RODGERS, J. J. W. 1986. The Dharwar craton and the assembly of Peninsular India. *Journal of Geology*, **94**, 129–143.

SCHMELING, H., CRUDEN, A. R. & MARQUART, G. 1988. Finite deformation in and around a fluid sphere moving through a viscous medium: implications for diapiric ascent. *Tectonophysics*, **149**, 17–34.

SCHWERDTNER, W. M. 1984. Archaean gneiss domes in the Wabigoon Subprovince of the Canadian Shield,

northwestern Ontario. *In*: KRÖNER, A. & GREILING, R. *Precambrian Tectonics Illustrated*, Nägele und Obermiller, Stuttgart, 129–134.

—— & LUMBERS, S. B. 1980. Major diapiric structures in the Superior and Greenville provinces of the Canadian shield. *Geological Association of Canada, Special Paper*, **20**, 149–180.

——, BENETT, P. J. & JANES, T. W. 1977. Application of L-S fabric scheme to structural mapping and paleostrain analysis. *Canadian Journal of Earth Sciences*, **14**, 1021–1032.

——, SUTCLIFFE, R. H. & TROENG, B. 1978. Patterns of total strain within the crestal region of immature diapirs. *Canadian Journal of Earth Sciences*, **15**, 1437–1447.

——, STONE, D., OSADETZ, K., MORGAN, J. & STOTT, G. M. 1979. Granitoid complexes and the Archean tectonic record in the southern part of northwestern Ontario. *Canadian Journal of Earth Sciences*, **16**, 1965–1977.

——, STOTT, G. M. & SUTCLIFFE, R. H. 1983. Strain patterns of crescentic granitoid plutons in the Archean greenstone terrain of Ontario. *Journal of Structural Geology*, **5**, 419–430.

——, MORGAN, J.& STOTT, G. M. 1985. Contacts between greenstone belts and gneiss complexes within the Wabigon subprovince,Ontario. *In*: AYRES, L. D., THURSTON, P. C., CARD, K. D. & WEBER, W. (eds), Evolution of Archaean supracrustal sequences. *Geological Association of Canada*, 117–124.

SHACKELTON, R M. 1976. Shallow and deep level exposures of Archaean crust India and Africa. *In*: WINDLEY, B. F. (ed.), *The Early History of the Earth*, Wiley, London and New York, 317–321.

SILLS, J. D., WANG, K. Y., YAN, Y. H. & WINDLEY, B. F. 1987. The Archaean granulite-gneiss terrain in E. Hebei Province, N.E. China: geological framework and metamorphic conditions. *In*: PARK, R. G. & TARNEY, J. (eds), *Evolution of the Lewisian and Comparable Precambrian High Grade Terrains*. *Geological Society of London, Special Publication*, **27**, 297–305.

SLEEP, N. H. & WINDLEY, B. F. 1982. Archean plate tectonics: constraints and inferences. *Journal of Geology*, **90**, 363–379.

STÄHLE, H. J., RAITH, M., HOERNES, S. & DELFS, A. 1987. Element mobility during incipient granulite formation at Kabbaldurga, southern India. *Journal of Petrology*, **28**, 803–834.

STOWE, C. W. 1984. The early Archean Selukwe nappe, Zimbabwe. *In*: KRONER, A. & GREILING, R. (eds), *Precambrian Tectonics Illustrated*. Schweitzerbart: Stuttgart, 41–56.

STROH, P. T., MONRAD, J. R., FULLAGAR, P. D., NAQVI, S. M., HUSSEIN, S. M. & ROGERS, J. J. W. 1983. 3000-m.y.-old Halecote trondjhemite: a record of stabilisation of the Dharwar craton. *In*: NAQVI, S. M. & ROGERS, J. J. W. (eds), *Precambrian of South India*. Geological Society of India, Memoirs, **4**, 365–376.

SWAGER, C. P., WITT, W. K., GRIFFIN, T. J., AHMAT, A. L., HUNTER, W. M., MCGOLDRICK, P. J. & WYCHE, S. 1992. Late-Archaean granite-greenstones of the

Kalgoorlie Terrane, Yilgarn Craton, Western Australia. *In*: GLOVER, J. E. & HO, S. E. (eds), *The Archaean: Terrains, Processes and Metallogeny.* The Geology Key Centre & University extension, University of Western Australia, 107–122.

SWAMI NATH, J. & RAMAKRISHNAN, M. (eds) 1981. Early Precambrian supracrustals of Southern Karnataka. *Geological Survey of India, Memoirs*, 112.

——, —— & VISWANATHA, M. N. 1976. Dharwar stratigraphic model and Karnataka craton evolution. *Records of the Geological Survey of India*, **107**, 149–175.

TALBOT, C. J. 1974. Fold nappes as asymmetric mantled gneiss domes and ensialic orogeny, *Tectonophysics*, **24**, 259–276.

—— 1977. Inclined and asymmetric upward-moving gravity structures. *Tectonophysics*, **42**, 159–181.

——, RONNLUND, P., SCHMELING, H., KOYI, H. & JACKSON, M. P. A. 1991. Diapiric spoke patterns. *Tectonophysics*, **188**, 187–201.

TAYLOR, P. N., CHADWICK, B., MOORBATH, S., RAMAKRISHNAN, M. & VISWANATHAL, M. N. 1984. Petrography, chemistry and isotopic ages of Peninsular gneiss, Dharwar acid volcanic rocks and the Chitradurga granite with special reference to the late Archaean evolution of the Karnataka craton, Southern India. *Precambrian Research*, **23**, 349–375.

——, CHADWICK, B., FRIEND, C. R. L., RAMAKRISHNAn, M. & VISWANATHA, M. N. 1988. New age data on the geological evolution of Southern India. *In*: ASHWAL, L. D. (ed.), *Indo-US Workshop on the deep continental crust of south India*. National Geophysical Research Institute, Hyderabad, 181–183.

—— & MCLENNAN, S. M. 1985. *The Continental Crust: its Composition and Evolution.* Blackwell, Oxford.

TRELOAR, P. J., COWARD, M. P. & HARRIS, N. B. W. 1992. Himalayan–Tibetan analogies for the evolution of the Zimbabwe craton and Limpopo Belt. *Precambrian Research*, **55**, 571–587.

VAN KRANENDONK, M. J. & HELMSTAEDT, H. 1992. Late Archaean sructural history of allochtonous Upernarvik supracrustal rocks in the high-grade Nain Province, Labrador: evidence of a link between the tectonic evolution of gneiss terrains and greenstone belts. *In*: GLOVER, J. E. & HO, S. E. (eds), *The Archaean: Terrains, Processes and Metallogeny.* The Geology Key Centre & University extension, University of Western Australia, 137–150.

VENKATASUBRAMANIAN, V. S. 1975. Studies in the geochronology of the Mysore craton. *Geophysical Research Bulletin*, NGRI, **13**, 239–246.

VISWANATHA, M. N., RAMAKRISHNAN, M. & SWAMI NATH, J. 1982. Angular uncomformity between Sargur and Dharwar supracrustals in Shigegudda, Karnataka craton, South India. *Journal of the Geological Society of India*, **23**, 85–89.

VLAAR, N. J. 1986. Archean global dynamics. *Geologieen Mijnboum*, **65**, 91–101.

WANG, K., WINDLEY, B. F., SILLS, J. D. & YAN, Y. 1990. The Archean gneiss complex in E. Hebei Province, North China. *Precambrian Research*, **48**, 245–265.

WEST, G. F. & MARESCHAL, J. C. 1979. A model for Archean tectonism. Part 1: The thermal conditions. *Canadian Journal of Earth Sciences*, **16**, 1942–1950.

ZHANG, R. Y & CONG, B. L. 1982. Mineralogy and P-T conditions of cristallisation of early Archean granulites from Qianxi County, NE China. *Scientia Sinica*, **25**, 96–112.

——, ——, YING, Y. P. & LI, J. L. 1981. Ferrifayalite-bearing eulysite from Archaean granulites in Qianan county, Hebei, North China. *Tschermaks, Mineralogische Petrologisches Mitteilungen*, **28**, 167–187.

Archaean deformation patterns in Zimbabwe: true indicators of Tibetan-style crustal extrusion or not?

PETER J. TRELOAR[1] & TOM G. BLENKINSOP[2]

[1]School of Geological Sciences, Kingston University, Kingston-upon-Thames, Surrey KT1 2EE, UK

[2]Department of Geology, University of Zimbabwe, PO Box MP167, Mount Pleasant, Harare, Zimbabwe

Abstract: Recent models of the structural evolution of the Zimbabwe Archaean craton have stressed the role of WSW-trending sinistral strike-slip shears in accommodating the WSW-directed extrusion of crust thickened in response to NNW–SSE compression and shortening. This relationship between crustal thickening and crustal extrusion is analogous to the structural response of Tibetan crust to the Tertiary collision of India and Asia. We have re-examined many of the structures of the Zimbabwe craton and find the extrusion model largely untenable. Instead, a conjugate set of ESE-striking dextral shears and NNE-striking sinistral shears have been identified, on both regional and outcrop scale, which apparently reflect part of the structural response of the Zimbabwe crust to NNW–SSE shortening resultant from collision of the craton with the Central zone of the Limpopo belt to the south. This collision, at about 2.58 Ga, post-dated the 2.68 Ga aged collision between the Central zone and the Kaapvaal craton which represents the Limpopo orogeny sensu stricto. The suture between the Zimbabwe craton and the Central Zone was re-activated at *c.* 2.0 a as a major intracontinental dextral strike-slip shear zone. Partial melting of the lower crust at *c.* 2.58 Ga accompanied thickening of the Zimbabwe Archaean crust following collision with the Central zone. The conjugate shears deform fabrics that date from that collision, are synchronous with, or post-date, granite emplacement and may have acted as conduits for transport of the Chilimanzi monzogranitic magmas from the lower to the upper crust. Contacts between greenstone belts and granites are commonly sheared. These marginal shear zones are often parallel to the regional conjugate shears, although local variations in stress trajectories imposed by pluton emplacement may result in them being non-parallel to the major shears.

One of the problems that has increasingly exercised structural geologists studying Archaean terrains is the extent to which working models for the tectonic and magmatic evolution of Archaean crust may be derived from processes operative at the present. Due to long term changes in global thermal budgets, direct analogues of many modern processes may not be relevant to the Archaean, and consequently, there has been an involved debate as to whether plate tectonic processes of the kind recognized today, were operative during the Archaean. As most plate tectonic processes are essentially thermally driven, part of this debate has resolved around the extent to which the thermal flux from mantle to crust may be lower now than in the Archaean (Bickle 1978). In addition, as both the composition and temperature of the mantle reservoir from which crustal rocks are ultimately derived may be substantially different now from that during the Archaean, the chemical evolution of the early crust may differ from that of modern crust at subduction zones (Ellam & Hawkesworth 1988; Martin 1986, 1987, 1993) and mid-ocean ridges

and the volcanic products may differ, a difference most noticeably indicated by the widespread occurrence of komatiites during the Archaean (Arndt 1983; Nisbet *et al.* 1993). Overall uncertainties are still such that vivid debates continue concerning the role of mantle plumes in controlling Archaean magmatism (Arndt 1992) and the mechanisms by which trondhjemite–tonalite–granodiorite (TTG) gneiss suites were extracted from the mantle reservoir (compare Kramers 1988 with Drummond & Defant 1990), or by which early crustal nuclei developed and subsequently grew (compare Kramers 1988 with de Wit *et al.* 1992a).

There is, thus, an inherent inexactitude as to the extent to which deep crustal processes may have differed in the Archaean from today. Partly this inexactitude arises from differences in thermal structure and crustal rheology. However, even though rates of processes may have been different (marginally or significantly) during the Archaean from now, and although continental geotherms may have been somewhat higher then than now (see discussion in Ridley 1991) such that the brittle/

From COWARD, P. P. & RIES, A. C. (eds), 1995, *Early Precambrian Processes,*
Geological Society Special Publication No. 95, pp. 87–108.

ductile transition may have been located at shallow crustal levels than today, rocks will still have responded to stress as they do today, continental crust would have thickened in response to overall shortening and compression and deep crustal rocks would have melted if the thermal environment was suitable. That nuclei of old continental crust surrounded by increasingly young crust can be recognized, as at Barberton in the Kaapvaal craton or in the Sebakwian inliers of the Zimbabwe craton, suggests that plate collisions

should have occurred even in the Archaean. Even though it is arguable as to whether continental growth necessarily involved collision, or even cordilleran-style accretion, it is likely that the growing platelets would have deformed internally in response to collision or to strains resulting from their outward growth by whatever process that involved.

It is not the aim here to comment directly on the validity of various models of Archaean crustal growth and accretion, nor to evaluate whether or

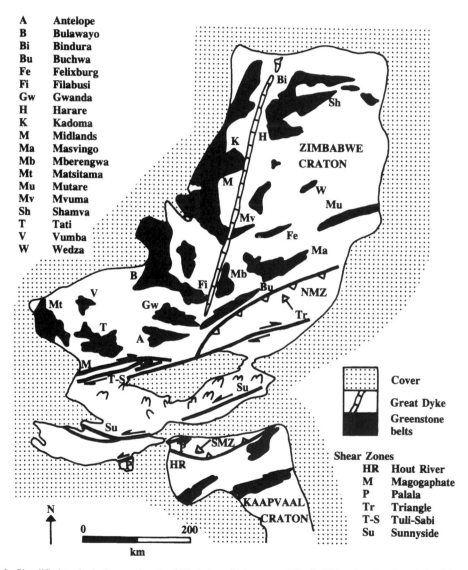

A	Antelope
B	Bulawayo
Bi	Bindura
Bu	Buchwa
Fe	Felixburg
Fi	Filabusi
Gw	Gwanda
H	Harare
K	Kadoma
M	Midlands
Ma	Masvingo
Mb	Mberengwa
Mt	Matsitama
Mu	Mutare
Mv	Mvuma
Sh	Shamva
T	Tati
V	Vumba
W	Wedza

Cover

Great Dyke

Greenstone belts

Shear Zones
HR	Hout River
M	Magogaphate
P	Palala
Tr	Triangle
T-S	Tuli-Sabi
Su	Sunnyside

Fig. 1. Simplified geological map of parts of Zimbabwe, Botswana and South Africa showing the relationships between the Limpopo Belt and the adjoining granite–greenstone terrains of the Zimbabwe and Kaapvaal cratons. NMZ: North Marginal zone. SMZ: South Marginal zone. CZ: Central zone.

not the contention that modern style plate tectonics, or something like them, may have been operative during the Archaean. Rather, the purpose of this paper is to address the problem of craton wide deformation during the Archaean, in particular the extent to which present day styles of crustal deformation may be analogues for Archaean crustal deformation patterns, and the relationships between crustal thickening, granite emplacement and cratonization. To do this the primary example used is the Zimbabwe Archaean craton, although with some reference to the Kaapvaal craton and the intervening Limpopo belt.

Zimbabwe Archaean craton

The geology and stratigraphy of the Zimbabwe craton (Fig. 1) have been described elsewhere, most notably and coherently by Wilson (1979). Four periods of greenstone belt generation are documented: the Sebakwian at *c.* 3.5 Ga, the Lower Bulawayan at *c.* 3.0–2.8 Ga, the Upper Bulawayan

at *c.* 2.85–2.66 Ga, and the younger Shamvaian (Wilson 1979; Taylor *et al.* 1991). New, ion probe derived, U–Pb age data on single zircons broadly confirm these major divisions (Wilson *et al.* this volume). The Sebakwian rocks outcrop in the southern central part of the craton, the Tokwe segment. Within the craton, age and isotope data (Taylor *et al.* 1991) are consistent with westerly and northerly crustal accretion and growth through time. Although the data presented by Taylor *et al.* (1991) are restricted to the Zimbabwe part of the craton, they are supported by young lead ages from the westernmost part of the craton in Botswana (Coomer *et al.* 1977; Aldiss 1991). Tonalitic rocks within the Tokwe segment are dated at *c.* 3.5 Ga, ages decreasing to both west and north. Westward, ages decrease to *c.* 2.7 Ga in the Sesombi suite of plutons which intrude the westernmost parts of the *c.* 2.9–2.8 a Shangani and Rhodesdale plutons (Fig. 2). To the north plutons dated at *c.* 2.67 Ga (Jelsma *et al.* 1993) are intrusive into bimodal volcanics of the 2.66–2.74 Ga aged Upper

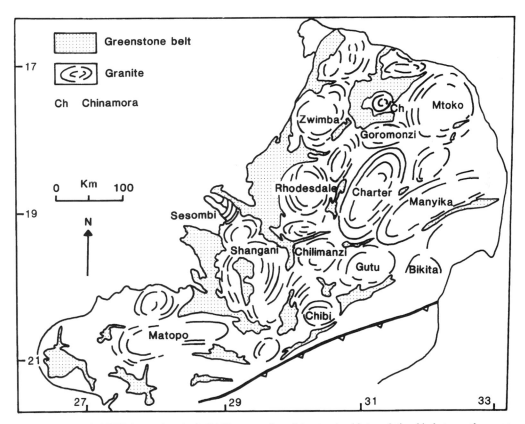

Fig. 2. McGregor's (1951) 'gregarious batholith' interpretation of the structural inter-relationship between the greenstone belts and granitic batholiths of the Zimbabwe craton. This map has been redrawn to accommodate the shapes of greenstone belts as established by mapping subsequent to McGregor's original interpretation. Ch, Chinamora.

Bulawayan Group. The processes by which this cratonic growth occurred are uncertain. It could have been by subduction-related magmatic processes analogous to those of today (Martin 1986; Drummond & Defant 1990), by accretion of small crustal fragments involving deformation of intervening sialic crust (the MARCY model of Kramers 1988), or by a process analogous to present day cordilleran accretion (de Wit *et al.* 1992*a*). Rollinson (1993), for instance, argues that greenstone belt sequences in the Buchwa and Matsitama belts on the southern and southwestern edges of the craton, respectively, are sufficiently anomalous to suggest that they represent distinct terranes accreted to the craton in the Late Archaean.

Similar outward accretion from an old cratonic nucleus occurred on the Kaapvaal craton (de Wit *et al.* 1992*a* and references therein). Formation of the central nucleus of the Kaapvaal craton, which includes the Barberton belt, occurred during the period 3.7–3.1 Ga when the initial shield area was stabilized by widespread melting and granite emplacement. Subsequent northward growth of the craton continued until about 2.7 Ga. De Wit (1991) and de Wit *et al.* (1992*a, b*) argued that this was through a cordilleran-type accretion of exotic terranes onto the northern and western margins of the Kaapvaal craton, accretion being on the hanging wall of a south-dipping subduction zone. This period of accretion ended with collision between what is now defined as the Central Zone of the Limpopo belt and the northern margin of the Kaapvaal craton and the consequent metamorphism and deformation associated with the Limpopo orogeny. This deformation is dated at between 2.7 and 2.65 Ga (Barton & van Reenen 1992).

The last major period of plutonism which affected the Zimbabwe Archaean craton was the craton-wide intrusion of the Chilimanzi monzogranite suite at 2.570 ± 0.025 Ga (Hickman 1978). Although this major phase of intrusion marks the cratonization of the Archaean complex, it is stressed that it is only the last of a series of granitoid intrusive events that mark the evolution of the craton from 3.5 Ga onwards.

Previous models for Archaean tectonics on the Zimbabwean craton

The first attempt to explain the tectonics of the Zimbabwe Archaean craton was that of McGregor (1951). He noted that a number of granitic bodies within the craton had an ovoid shape, with margins sheared against adjacent greenstone belts which were characteristically folded and flattened with an arcuate shape (Fig. 2). He explained these observations in terms of some form of diapirism

or mantled gneiss dome formation, the space constrictions imposed by the upward movement of the granitic bodies resulting in downward sagging of the intervening greenstone belts accommodated by shortening across, and deformation within, them. On the basis of our present knowledge of the Archaean craton, partly derived from more detailed mapping and the ensuing division of the granitic parts of the terrain into older granite gneisses and younger granites of a variety of ages, McGregor's 'gregarious batholith' model appears to be a grossly oversimplified one of limited applicability to the whole craton. Although ovoidal to spheroidal batholiths may be recognizable in the Rhodesdale and Chinamora regions (Fig. 2), they are not a feature of the whole province. In addition diapiric models of granite emplacement have been severely criticized on thermal, mechanical and rheological grounds (Clemens & Mawer 1992). Increasingly models of granite emplacement have invoked space creation in the upper crust through faulting (i.e. Hutton 1988; Hutton *et al.* 1990; D'Lemos *et al.* 1992; Hutton & Reavy 1992; Grocott *et al.* 1994) combined with models for dyke-like emplacement of internally sheeted granite plutons (Petford *et al.* 1993). However, despite this, the role of diapirism in accommodating emplacement of granite bodies on the Archaean craton has recently been persuasively argued by Ramsay (1989) and Jelsma *et al.* (1993) for the plutons of the Bindura–Shamva region of northern Zimbabwe.

The second generation of tectonic models involved the recognition of major strike-slip shear zones of large strike length with the implication that intracratonic tectonics was dominated by horizontal displacements. In a series of papers Coward and his co-workers (Coward 1976, 1980; Coward *et al.* 1973, 1976; Coward & James 1974; Coward & Daly 1984) demonstrated that the southern part of the craton was marked by strongly arcuate foliation trends (Fig. 3). They mapped regionally developed ENE-striking, ductile strike-slip shear zones with sub-horizontal stretching lineations which curve westward into NW-trending shears with down-dip stretching lineations. On the basis of regional fabric curvatures, displacement along the ENE-trending shears was described as being dominantly sinistral, although the presence of several second order dextral shears was noted. Hence Coward and his co-workers proposed a kinematic model for the southern part of the Zimbabwe craton in which the granite-gneiss terrain of the East Midlands had been displaced towards the WSW along ENE-trending sinistral shears. This block of crust is bounded to the southwest by a zone of thickened crust with a NNW-trending grain, which essentially marks the frontal ramp of a thrust system, the lateral ramp of which

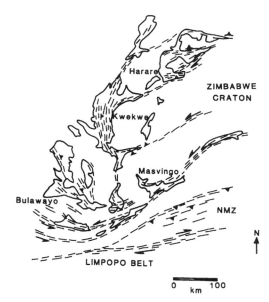

Fig. 3. Structural map of the Zimbabwe Archaean craton (after Treloar et al, 1992) showing the major fabric trends identified by them, together with zones of thrusting and faulting.

terms of a single plate collision dominated by non-plane strain conditions such that an overall NNW–SSE shortening was accommodated, initially, by crustal thickening but with blocks of thickened crust subsequently extruded laterally along ENE-directed shears (Fig. 4). This model is analogous to that proposed by Tapponnier *et al.* (1986) for Tibet where, in response to an approximately north-south collision with India, Tibetan crust initially thickened to a critical value before discrete blocks of thickened crust were extruded eastward along crustal-scale strike-slip faults oriented at a high angle to the shortening direction.

Any model which successfully explains the tectonic evolution of the Zimbabwe craton must resolve the combination of NNW–SSE shortening with associated strike-slip displacements along shear zones developed at a high angle to the shortening direction. In addition, as an intimate relationship between deformation in the southern part of the Zimbabwe craton and that within the Limpopo belt has frequently been inferred, an understanding of the tectonics of the Limpopo belt is an important co-requisite to understanding the tectonics of the Zimbabwe craton. This paper attempts to generate a unified model which explains both the structural evolution of the Zimbabwe craton and the intimate relationship of that structural evolution to that of the Limpopo belt.

is marked by the ENE-trending shears. That the timing of this deformation was approximately synchronous with the emplacement of the Chilimanzi granite suite is indicated both by the variable deformation of these granites, and by the ductile nature of the deformation which implies that the granites were not cold and brittle.

Coward & Treloar (1990) and Treloar *et al.* (1992) took the Coward model and extended it in a new, and somewhat speculative, direction. They suggested that the structural grain throughout the eastern part of the craton has an overall ENE to E trend with shear fabrics characterized by sub-horizontal lineations; that to the west these master shears curve into high-strain zones with N to NW trends and down-dip lineations such that two major crustal blocks can be recognized, each bounded by ENE-trending shears: one in southern Zimbabwe to the south of the Mvuma and Felixsburg greenstone belts, and one in northern Zimbabwe to the north of the Mvuma and Felixsburg belts (Fig. 3). Although deformation on a regional scale is most clearly defined by ENE-trending shear zones there is plenty of evidence for NNW–SSE-directed shortening, together with a northward migration in the timing of deformation from the Kaapvaal craton, across the Limpopo belt and into the Zimbabwe craton. The model outlined by Treloar *et al.* (1992), dubbed the 'Tibetan analogy', accounted for the two movement directions in

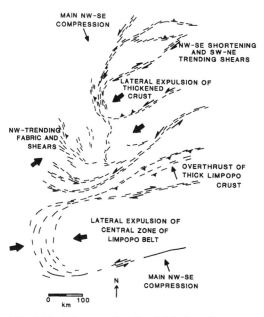

Fig. 4. The 'Tibetan analogy' model for lateral extrusion of blocks of thickened crust (after Treloar *et al.* 1992). The large arrows mark inferred displacement trajectories.

Tectonic evolution of the Limpopo belt

The Zimbabwe and Kaapvaal cratons are separated by the Limpopo belt (Fig. 1), main phase orogenic deformation and metamorphism within which is cited, largely on the basis of age data from the Alldays gneiss and the Bulai and Matok plutons, as being at between 2.70 and 2.65 Ga (Barton & van Reenen 1992). However, significant quantities of new geochronological data published during the preparation of this paper question some fundamental perceptions of the tectonothermal evolution of the belt. Not all of these data are consistent with previously aquired data, and the data sets for the Central and South Marginal zones of the Limpopo belt are becoming increasingly ambiguous. Caution needs to be taken in analysing these data and their implications, especially where they may appear to be inconsistent with well documented structural features and intrusive and metamorphic relationships.

The Limpopo belt is divided into three zones (Fig. 1), a Central zone (CZ) flanked by the North (NMZ) and South Marginal zones (SMZ). The two marginal zones have generally been considered as the deformed equivalents of the Zimbabwe and Kaapvaal cratons to which they are adjacent, and to be symmetric equivalents of each other. However, Ridley (1992) and Rollinson & Blenkinsop (1995) have shown the NMZ to be dominated by igneous rocks of a granitoid plutonic assemblage, probably emplaced after 2.8 Ga and mainly between 2.7 and 2.6 Ga (Berger 1994), with only minor amounts of supracrustal rocks, and argue that it is not simply the deformed equivalent of the Zimbabwe craton. By contrast, within the SMZ supracrustals are common and Stevens & van Reenen (1992) argue that rocks of the SMZ can clearly be identified as deformed equivalents of the Kaapvaal craton. The Central zone is clearly exotic and has been displaced to the SSW, with respect to the North and South Marginal zones, along the dextral Triangle and Tuli-Sabi shear zones on its northern side, and the sinistral Sunnyside and Palala Shear Zones on its southern side (Fig. 1; McCourt and Vearncombe 1987, 1992). The timing of this displacement is uncertain, with evidence for both Archaean and Proterozoic displacements (Kamber et al. 1995). Many misconceptions, however, remain about the tectonic relationships between the two marginal zones and the cratons that flank them. As these are relevant to the tectonic evolution of the Zimbabwe craton they will be addressed below.

There has long been a tendency to consider the Limpopo orogeny as the result of a north-south collision between the Zimbabwe and Kaapvaal cratons (Light 1982; Winter 1987; Roering et al. 1992a; de Wit & Hart 1993). As an alternative to this simple view, McCourt and Vearncombe (1987)

interpreted the Central zone as an exotic, allochthonous west-directed thrust sheet made up of crust already thickened during an earlier shortening event, thrusting emplacing the sheet onto the already collided Zimbabwe and Kaapvaal cratons (Fig. 5a). Treloar et al. (1992) saw such a sense of displacement as being consistent with the Tibetan analogue. Van Reenen et al. (1987) used McCourt & Vearncombe's model, together with the apparent symmetry of the two marginal zones, to explain outward thrusting within the marginal zones, northward in the NMZ and southward in the SMZ, onto the adjacent cratons. Essentially, this represented a tectonic collapse of the rocks onto which the CZ had been emplaced (Fig. 5b), a suggestion adopted by McCourt & Vearncombe (1992). As an extension of the model, McCourt & Wilson (1991) argued that the NNE-trending fabrics in the Bulawayan region were the result of a collision between the main body of the Zimbabwe craton and the Motloutse fragment to its southwest (Fig. 5c), and that subsequent collision between the Kaapvaal and Zimbabwe cratons, rather than involving one of N–S shortening, was sinistrally oblique and predated the westward thrusting of the Central zone of the Limpopo Belt onto the collided cratons (Fig. 5c & d). This explained the sinistral fabrics described by Coward (1976) and Coward et al. (1977) in the southern part of the Zimbabwe craton.

The models described by McCourt and van Reenen and their co-workers, imply that the Kaapvaal and Zimbabwe cratons formed a structural entity prior to the west-directed collision that emplaced the Central zone onto the supercraton and that deformation within the marginal zones was an effect of the SSW-directed emplacement of the Central zone allochthon. However, there are a number of problems with these models. Firstly, Treloar et al. (1992) pointed out that although lineations within the NMZ are consistently down dip and indicate NNW-vergent thrusting (James 1975; Ridley 1992), those within the SMZ are consistent with SW- or even WSW-directed oblique-slip displacement (Roering et al. 1992b; Smit et al. 1992). Secondly, timing of deformation and metamorphism is not synchronous across the belt. On the basis of Pb–Pb data, Barton et al. (1994) document a high-grade metamorphism within the SMZ as early as 3.165 ± 0.05 Ga, although timing of the main phase of southward thrusting within the SMZ is constrained at between 2.70 and 2.65 Ga (Barton & van Reenen 1992). In part, this latter age is constrained by the emplacement into the SMZ, probably under granulite facies conditions, of the syntectonic, charnokitic, Matok pluton at c. 2.67 Ga (Barton et al. 1992). It is unclear as to whether rocks of the SMZ were at granulite facies conditions throughout the period 3.16–2.65 Ga, or whether they experienced more

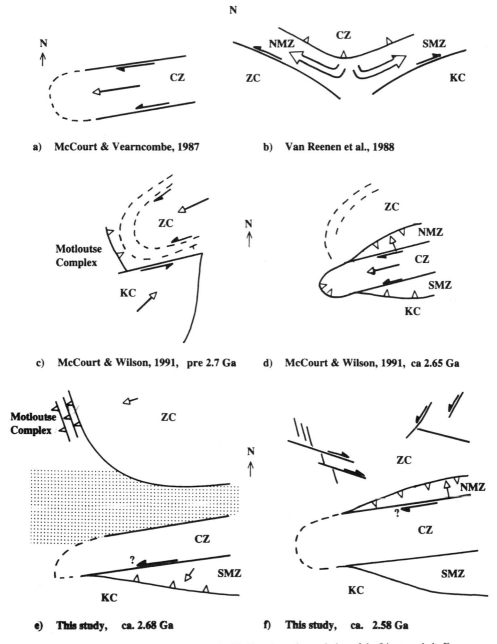

Fig. 5. Synoptic diagrams showing recent changes in thinking about the evolution of the Limpopo belt. For discussion see text. (e) and (f) summarize the model proposed in this paper. The arrows adjacent to the shears on both the northern and southern margins of the Central Zone refer to the uncertainty of timing of that shearing, most of which is probably at about 2.0 Ga. CZ: Central zone. NMZ: North Marginal zone. SMZ: South Marginal zone. KC: Kaapvaal craton. ZC: Zimbabwe craton. The shaded area indicates crust of unknown type which disappeared during convergence at about 2.58 Ga.

than one granulite-facies event within that period. Timing of deformation and metamorphism within the NMZ is younger than this and can be dated by the syn-deformational emplacement, under granulite facies conditions (Rollinson & Blenkinsop 1995) of the Chilimanzi-type Razi granite suite. These ages range from a U–Pb zircon age of 2.627 ± 0.007 Ga, derived from a granite sheet that cuts the Umlali Shear (a NNW-vergent thrust that marks the western end of the contact between the NMZ and the Zimbabwe craton), to a Rb–Sr whole rock age of 2.583 ± 0.05 Ga (Berger et al. 1993; Mkweli et al. 1995). This Rb–Sr whole rock errorchron age is essentially identical to that of 2.574 ± 0.014 Ga derived by Hickman (1978) for the emplacement of Chilimanzi type granites into the southern part of the Zimbabwe craton. Thirdly, cratonization of the two cratons by major crustal anatexis and granite emplacement was also not synchronous. On the Kaapvaal craton cratonization was complete by 2.70 Ga (Barton et al. 1992, de Wit et al. 1992a, b) and thus pre-dated the generally accepted age for orogenesis in the Limpopo belt, whereas on the Zimbabwe craton, where it may be dated by the Chilimanzi granite suite, it post-dated the Limpopo orogeny. Therefore, although the common impression of the Limpopo belt is one of lithological symmetry across the two cratons and the two marginal zones, the structural and geochronological picture is, in fact, one of gross asymmetry. Fourthly, there is growing evidence that dextral displacements along the Triangle–Tuli-Sabi shear zone, rather than being Late Archaean as suggested by McCourt & Vearncombe (1987), were in fact active at about 2.0 Ga (van Breemen & Hawkesworth 1980; Kamber et al. 1993, 1995) as part of a Proterozoic crustal thickening event, although this data cannot exclude a possible re-activation of an earlier, dextral Archaean shear zone (see discussion in Roering et al. 1992a).

These differences are resolved as follows. It is that the Limpopo orogeny did not involve the westward thrusting of the Central zone onto a crustal block made up of the already collided Kaapvaal and Zimbabwe cratons. Instead a model is preferred in which the structural concordance across the CZ and SMZ, together with the similarities in timing of metamorphism and in P–T–t paths of metamorphism and deformation (van Reenen et al. 1990; Roering et al. 1992a), suggest that the Limpopo orogeny sensu stricto involved the convergence, or accretion, at between 2.70 and 2.65 Ga, of the exotic Central zone to the northern margin of the Kaapvaal craton (Fig. 5e). The accretion of the Central Zone to the northern margin of the Kaapvaal craton may, thus, have been the last stage of the northward growth of that craton by cordilleran-style terrane accretion as described by de Wit et al. (1992).

Convergence of the Zimbabwe craton with the Central zone–Kaapvaal craton pair, dated by the syn-tectonic Chilimanzi-type Razi granites at about 2.58 Ga (Berger et al. 1993; Mkweli et al. 1995), took place later and thus post-dates rather than pre-dates the Limpopo orogeny (Fig. 5f). The down-dip nature of the stretching lineations within the NMZ suggests that this convergence may have been truly orthogonal. Although conventional U–Pb and Rb–Sr age data from the Bulai gneiss in the Central zone demonstrate emplacement, into gneisses older than 3.0 Ga, at about 2.69 (Barton et al. 1979) or 2.63 Ga (van Breemen & Dodson 1972), conventional U–Pb analyses of zircons from the granodiorites of the Bulai complex yield ages of 2.572 ± 0.004 Ga (Barton et al. 1994), which may reflect decompressive melting following on tectonic uplift associated with NW-vergent thrusting along the contact between the NMZ and the Zimbabwe craton. Prior to this, the southern margin of the Zimbabwe craton has it own history of magmatism, metamorphism and terrane accretion (Rollinson 1993; Rollinson & Lowry 1992; Rollinson & Blenkinsop 1995). The key relationship is that cratonization of the Zimbabwe craton, through widespread crustal anatexis and emplacement of the Chilimanzi-type granites, was approximately synchronous with convergence of that craton with the CZ. These relationships essentially mean that all of the structures within the Zimbabwe craton which pre-date emplacement of the Chilimanzi granite suite and the synchronous convergence with the CZ–Kaapvaal craton block at 2.58 Ga must relate to internal deformation of the craton, or to accretion related cratonic growth processes. They cannot, however, relate to an earlier convergence with the Kaapvaal craton as suggested by McCourt & Wilson (1991) and McCourt & Vearncombe (1992).

Although the available geochronological and structural data argue against symmetrical and synchronous deformation and metamorphism across the SMZ–CZ–NMZ triad, the fact that the majority of ages fall within the time interval 2.70–2.58 Ga suggests that that main phase evolution of the belt may be defined by some form of continuum of deformation and metamorphism. However, the Limpopo orogeny, as defined by Barton & van Reenen (1992), can be seen to be but part of a longer scale tectonic event that involved the northward migration of crustal deformation and accretion across the entire Zimbabwe craton-Limpopo belt–Kaapvaal craton region and which involved the processes of transpressive terrane accretion argued for by Rollinson (1993) and de Wit et al. (1992a). It is noted that both the timing and asymmetry of deformation excludes the McCourt & Vearncombe (1992) model where deformation in the NMZ and SMZ was a function of collapse following the overthrusting of the CZ onto those two zones at c. 2.68 Ga. The timing of sinistral strike slip tectonics along the Sunnyside and Palala shear zones, that separate the SMZ from the CZ, and of dextral strike slip displacement along the Triangle and Tuli-Sabi shear zones, that separates the CZ from the NMZ in the east and from the Zimbabwe craton in the west are uncertain, but probably polyphase. Although there may have been a period of Late Archaean movement along the latter shear zone, isotopic data strongly supports a Proterozoic age (c. 2.0 Ga) of displacement, associated with metamorphism within the CZ and possibly the NMZ (van Breemen & Hawkesworth 1980; Kamber et al. 1993, 1995; Barton et al. 1994) possibly associated with crustal thickening. A consideration of the tectonics of that event is, however, beyond the scope of this paper. Similarly, late movements along the Sunnyside and Palala shear zones post-date the emplacement of the Bushveld complex (McCourt & Vearncombe 1992).

Validity of the Tibetan analogue model

Although superficially attractive, it is shown, largely on the basis of our new field data, that the Tibetan-analogue model (Treloar et al. 1992) is invalid. In its nascent form it proposes a continuity of NNW–SSE-directed shortening resolvable, over a period of at least 300 Ma (2900–2550 Ma), into a northward migrating zone of crustal thickening and subsequent crustal extrusion that encompassed the final stages of terrane accretion on the north margin of the Kaapvaal craton, the Limpopo orogeny and subsequent internal deformation of the Zimbabwe craton. In the following sections the validity of the Tibetan-analogue model is re-assessed, specifically to ascertain whether the structures exposed at surface on the Zimbabwe craton are consistent with such a model.

That there is a relationship between orogenic shortening, crustal thickening and lateral extrusion of blocks of thickened continental crust has been clearly demonstrated both experimentally, by plasticine experiments (Peltzer et al. 1982; Tapponnier et al. 1982), and theoretically (Villotte et al. 1982; Houseman and England 1986; England & Houseman 1986, 1989). In all cases, the initial result of shortening is crustal thickening, with faulting being dominantly of a thrust-type and with a plateau of thickened crust forming in front of, and spreading away from, the indentor. With time there is a transition from dominantly thrust faulting in the brittle upper crust to dominantly strike-slip faulting that accommodates crustal extrusion. In describing what controlled extrusion processes in the ductile lower crust, Bird (1991) demonstrated that a planar channel flow will develop within the lower crust, the flux of which is a function of the topographic gradient of the crustal surface, the temperature of the lower crust and the geothermal gradient. Essentially, Bird demonstrated that flow within the

lower crust will remove material from beneath regions of high topography and hence smooth and level the topography, and that the time scale over which this happens is dependant largely on the thermal structure of the lower crust and upper mantle and the thickness of the crustal column. High Moho temperatures and geothermal gradients, normally the result of high heat fluxes across the Moho, are as conducive to flow as are significant increases in crustal thickness. Thus, not only will channelized, deep-crustal flow be expected beneath regions of high topography and great crustal thickness, as in Tibet, but also in regions of more normal crustal thickness but with abnormally high heat flux. There is a rheological constraint as well in that quartzo-feldspathic rocks, characterized by minerals with low brittle–ductile transition temperatures, flow more readily than mafic to ultramafic rocks characterized by minerals with higher brittle–ductile transition temperatures. What is evident is that, given the correct rheology, either unusually great crustal thickness or unusually high lower crustal temperatures or both will favour the development of crustal extrusion tectonics which may affect not only the entire crustal column, but also at least the upper part of the mantle lithosphere.

Although there is no uniformity of opinion, the majority of workers who have considered the problem of Archaean thermal fluxes in recent years concur that surface heat flows in the Archaean were higher than today (Schubert et al. 1980; Ridley 1991). Heat flux from the mantle across the Moho and internal heat generation within the crust were both probably higher than today (Ridley 1991). Given the time scales over which crustal layering may be expected to develop, internal heat generation may have been more evenly distributed over the entire crustal thickness rather than being concentrated within the upper part of the crust. High heat fluxing during the Archaean has been invoked to explain the development of large magma lakes within the mantle or lower crust (for instance the LLLAMA and MARCY models of Nisbet & Walker 1982, and Kramers 1988, respectively) and the generation of craton-wide granitization events in the Late Archaean (Ridley 1991). Geothermobarometric evidence (Grambling 1981) suggests that geothermal gradients may have been somewhat higher than today. However, England & Bickle (1986) have argued that if mountains existed during the Archaean, as suggested by the sedimentary record, they could only have been supported for significant periods if Moho temperatures were not significantly higher than those of the present. We would prefer to follow Ridley (1991) in inferring that, although the Archaean cratonic crust may have been of normal, present day thickness, the geothermal gradient would have been somewhat, although not necessarily unusually, high by present standards. This could yield Moho and lower crustal temperatures which, although somewhat hotter than at present, were still low enough to support mountain ranges but which were also physically suitable for the kind of deep-crustal flow described by Bird (1991). The latter could control the regional development of crustal-scale strike-slip faults capable of accommodating crustal extrusion. Thus there is no theoretical reason why Tibetan-style crustal extrusion should not have been operative during the Archaean in response to crustal shortening.

Structures of the Zimbabwe craton

Treloar *et al.* (1992) presented a map, here reproduced as Fig. 3, of the major shear zones present within the Zimbabwe craton. The map indicated shear sense criteria for some of these shears, and showed that over the eastern half of the craton fabrics had a dominantly ENE trend with sub-

horizontal lineations, curving into steeply dipping NW- or N-trending fabrics. However, subsequent fieldwork, partly by the authors and partly by S. D. G. Campbell and P. E. J. Pitfield (summarized in Campbell *et al.* 1991 and Pitfield *et al.* 1991) allows the map to be redrawn (Fig 6). Essentially, the new map shows that not all of the shear zones are of the obvious continuity as depicted in Treloar *et al.* (1992). Neither are the shear senses always of the sense inferred by Treloar *et al.* In addition we are able to consider in more detail the problems of shear zone reactivation, of second order shearing, of conjugate shearing and the relationships between shearing and granite emplacement.

Mutare to Masvingo

Coward (1976), Coward *et al.* (1976), McCourt & Wilson (1991) and Treloar *et al.* (1992) argued that the southern part of the craton is characterized by sinistral, strike-slip shear zones which extend from Mutare in the northeast through Masvingo and Mberengwa before splaying into a series of parallel

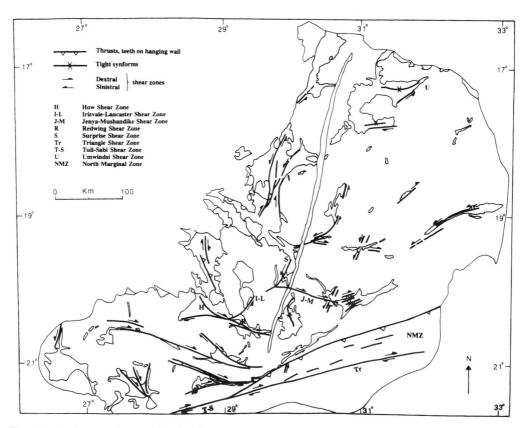

Fig. 6. Revised structural map of the Zimbabwe craton showing, in particular, the conjugate ESE-striking dextral and NNE-striking sinistral conjugate shear zones.

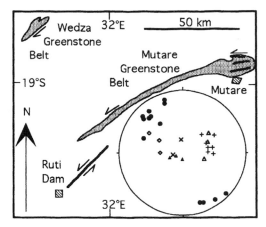

Fig. 7. Outline of structures within the Mutare and Wedza greenstone belts (shaded). Thick lines: shear zones, with shear sense indicated where known. Lower hemisphere equal area projection of structural data from the Mutare belt: solid circles, poles to foliation; solid triangles and crosses, fold axes and mineral stretching lineations from the central section of the southwestern part of the belt; open triangle and crosses, sinistral kink bands, fold axes and mineral stretching lineations from the northern margin of the southwestern part of the belt; diamonds, axes of sinistral kink bands from shear zones in the northeast of the belt.

sinistral shears through the Gwanda and Bulawayan greenstone belts. However, none of these studies provided any detailed evidence for the existence of these shear zones. Our own fieldwork indicates that the region between Mutare and Masvingo, where a ductile, sinistral shear zone has been previously mapped, is marked instead by an irregular compositional banding in a granitic gneiss trending generally northeast. The gneiss contains angular unstrained xenoliths and the banding is folded disharmonically. These features are consistent with a magmatic fabric which has possibly been mistaken for a major shear zone. We have also examined the Mutare and Masvingo greenstone belts for evidence of major sinistral shear. In the southwestern part of the Mutare belt, the southern contact between the greenstone belt and the adjacent gneiss is steeply dipping with a down-dip stretching lineation (Fig. 7). Within the belt itself, all foliations dip steeply, and occasionally carry a steep lineation (Fig. 7). Along the northern margin, the foliation carries a lineation that plunges moderately towards the northeast, parallel to the axes of a set of sinistral kink-bands that deform the foliation (Fig. 7). In the northeastern part of the belt, an L–S fabric is developed within ultramafic schists on the northern margin. Here asymmetric S-kink bands that plunge moderately towards the

west also indicate sinistral movement. There is also evidence for two other parallel zones of intense strain in the centre and to the south of the belt with sub-horizontal and steeply-plunging extension respectively (Chenjarai *et al.* in press). The northern part of the greenstone belt is intruded by an unsheared granite dated at 2.522 ± 0.036 Ga (Rb–Sr whole rock; Schmidt Mumm *et al.* 1993). Therefore the evidence for sinistral shear within, and at the edges of, the Mutare belt is limited to narrow zones of sinistral kink-bands along the northern margin, the age of which must predate 2.50 Ga. No other indicators of non-coaxial strain are seen. A prominent lineament can be traced on a Landsat thematic mapper image parallel to, and about 10 km south of, the Mutare belt trending southwest towards Ruti Dam (Fig. 7). In the field, this is a zone of mylonites and protomylonites up to a few hundred metres wide with s-porphyroclasts, S–C' fabrics and rolling structures that clearly indicate sinistral shear. Strain is very heterogeneous in this ill-defined zone and, at Ruti Dam itself, the fabrics are weak, suggesting that there is negligible strain farther to the southwest. This is confirmed by the termination of the lineament, mappable on the Landsat image, at Ruti Dam.

Most of the Masvingo belt is characterized by tight folds with axial planes dipping steeply towards the southeast, interpreted (Coward & James 1974; James 1975) as dating from the NNW–SSE shortening related to thrusting along the Northern Marginal zone of the Limpopo belt. Late, east-trending, dextral shear zones, one of which extends into the Jenya–Mushandike dislocation zone, deform the main fabrics (Coward & James 1974). Within the northern part of the belt, however, lineations on steeply-dipping cleavage surfaces and rodded pebbles and clasts, are sub-horizontal to gently plunging. Lineation plunges are towards both east and west (James 1977), although the majority of our data have an ENE-plunge at angles of between 15 and 36°. James (1977) interpreted these S–L tectonites, together with weak shear zones developed in granites immediately to the north of the belt, as being due to a low-strain, dextral shear. Thus, even in the zones of sub-horizontal stretching near the northern margin of the Masvingo belt, we find no evidence for a sinistral shear, and conclude that there is no consistent pattern of persistent major sinistral shear zones between the Mutare and Masvingo greenstone belts. To the contrary, within the Mutare belt deformation varies from dip-slip along the southern margin and within the southwestwern part of the belt, to minor sinistral reverse-oblique-slip with NE-plunging kink-band axes along the northern margin, to sinistral-slip with southwest-plunging kink-band axes in the northern part of the belt. The lack of persistent major structures and the variable orientations of the minor structures may point to a local, rather than regional, deformation pattern.

West of Masvingo

Having indicated the lack of persistent sinistral structures between Mutare and Masvingo, the area to the west of Masvingo is now considered. The only clear sinistral shear zone in this area is that on the south margin of the Buchwa greenstone belt (Fedo 1994), although the age of this shearing is uncertain. Although some of the ESE-trending shear zones in the Bulawayo and Gwanda region show occasional ductile indicators of sinistral shear, kinematic indicators including s-porphyroclasts and S-C′ fabrics are dominantly dextral in nature. Campbell et al. (1991) identify this entire suite of shear zones as being essentially dextral. Although the Jenya Shear Zone has previously been described as a sinistral shear zone (Coward et al. 1976), the Mushandike–Jenya dislocation zone is clearly dextral (Blenkinsop & Treloar 1995). The Redwing shear zone in the Filabusi greenstone belt is also dextral (Campbell et al. 1990), as is the How shear zone in the Bulawayo greenstone belt. These two ESE-trending shears curve into the conjugate NNE-trending, sinistral Irisvale–Lancaster shear zone (Fig. 6). The southern margin of the Gwanda greenstone belt is marked by a zone of ductile, dextral strike-slip shearing, deformation within which was synchronous with granite emplacement. (Coward et al. 1977). Campbell et al. (1991) indicate that shear zones in the Antelope greenstone belt, near Fort Nicholson and northwest of the Gwanda greenstone belt are all dextral. On the basis of occasionally preserved sinistral kinematic indicators, S. D. G. Campbell (pers. comm.) indicates that, in the Antelope region, the dominant ductile, dextral displacement may have been preceded by an earlier sinistral, ductile deformation. However, rather than that part of the craton exposed in southwestern Zimbabwe being dominated by sinistral, strike-slip displacements as suggested by earlier workers, it is clear that it is dominated by ductile, dextral strike-slip displacements, although possibly with an earlier sense of sinistral strike-slip movement.

In the southwestern part of the craton, the ESE-trending shear zones cut a series of NNW- to N-trending fabrics with down dip lineations. Within the zone of N-trending fabrics around, and to the north of, the Bulawayo region the rocks are the right way up, although in Botswana the Tati (Coward & James 1974) and Vumba (Key et al. 1976) greenstone belts, which may be correlated with the Upper Bulawayan belts in Zimbabwe, are overturned with dips dominantly towards the WSW

and with an ENE-directed sense of thrusting. This broad zone of NNW-trending fabric probably reflects a WSW–ENE shortening event, attributed by McCourt & Wilson (1991) to collision between the Zimbabwe craton to the east and the Matsitama fragment of the Motloutse complex to the west (Fig. 5c & e). Aldiss (1991) clearly demonstrates that this deformation predated both granite plutonism which, although undated, may be correlated with emplacement of the Chilimanzi suite within Zimbabwe, and NW–SE shortening which involved both folding of the overturned units and deformation of the granites. For instance, granites within the Tati belt were displaced dextrally with respect to each other along NNW-trending shears, movement on which was probably syn- to post- granite emplacement (Coward & James 1974). All of these structures were postdated by a major phase of ductile, dextral shearing along the WSW-trending Magogaphate shear zone which Aldiss (1991) interprets as the along-strike continuation of the 20 to 25 km wide zone of shearing defined by McCourt & Vearncombe (1992) as the Tuli-Sabi shear zone.

The central part of the craton

In the central part of the craton Coward & Treloar (1990), Treloar et al. (1992) and McCourt & Wilson (1991) identified a zone of sinistral shear passing through the Mvuma greenstone belt and extending towards the ENE. Fabiani (1989) mapped the Mvuma belt and surrounding areas and a map of the area combining his data with that of the present authors is shown in Fig. 8. Within the greenstone belt, folds plunge moderately towards the west with an axial-planar cleavage trending parallel to the belt, and with a strong vertical lineation associated with prolate strains for which $K = 5.4$ (Fabiani 1989). Along the northern margin of the belt, the gneisses of the Rhodesdale granitic terrain are sinistrally sheared with sub-horizontal to gently SW-plunging lineations. To the east this shear zone curves northeast and then northward toward the Mwanesi greenstone belt. S-C′ fabrics give a clear sense of sinistral shear in NNE-trending mylonites within the centre of this curving belt. An E-trending mylonitic shear zone, several tens of metres wide, is located along the southern margin of the belt in gneisses of the Chirumanzu granitic terrain (Fig. 8). Porphyroclasts here give an unambiguous dextral shear sense. Fabiani interpreted the deformation history of the area by a combination of SW movement and subsequent clockwise rotation of the Rhodesdale granite terrain to produce shortening across the belt together with the strike-slip shearing along the northern margin of the belt and in the mylonites to the northeast.

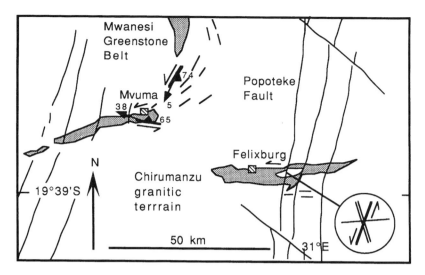

Fig. 8. Outline of structures in and around the Mvuma, Felixburg and Mwanesi greenstone belts (shaded) based on the author's fieldwork and that of Fabiani (1989). Thick lines and arrows: shear zone foliation and mineral stretching lineations with dip and plunge respectively. Short thin lines: foliation. Double short thin lines: tension fractures. Long thin lines: faults.

However, this model does not adequately explain the dextral shear along the southern margin of the belt.

The Felixsburg greenstone belt is an, extremely poorly exposed, E-trending belt located to the east of the Mvuma belt. Campbell *et al.* (1991) show dextral shears along both north and south margins of the belt. To the north of the belt, gneisses show a weak foliation parallel to the belt with some evidence for sinistral shear. Within the belt, an elongate granite body, probably of the Chilimanzi suite, contains an east-striking (080°) foliation offset by a set of sinistral ductile shears which trend at *c.* 020°. These are parallel to a prominent set of NNE-trending sinistral faults which offset both the Mvuma and Felixburg greenstone belts and which form part of the Popoteke fracture set described by Wilson (1990). Granites to the south of the Felixburg belt, are virtually undeformed and show only a weak east-trending foliation.

To the NE of the Felixburg belt, and about 50 km NW of the Mutare belt, is the small Wedza greenstone belt. Here, the southeastern contact between the greenstones and adjacent deformed granites trends NE (045°) with a sub-vertical mylonitic foliation and sub-horizontal lineations. Both sinistral and dextral shear senses can be identified within rotated porphyroclasts, but the majority show dextral shear (Tsomondo 1986: Jawi 1986). The age of the deformed granite is not known.

Within the central part of the Zimbabwe craton, therefore, there is no evidence for major ENE-striking sinistral shears. Rather, as in the Mutare

greenstone belt, there is evidence for horizontal shortening and vertical stretching within the Mvuma belt together with some sub-horizontal displacements along strike-slip shear zones. Characteristically, sinistral strike-slip shears north of the Mvuma belt and around the Wedza and Felixburg belts tend to be NNE- or NE-striking with lineations plunging gently towards the SSW or SW. In addition, the southern margin of the Mvuma belt is characterized by an ESE-trending dextral shear zone. This pattern of NNE-trending sinistral shears and ESE-trending dextral shears is similar to that described by the dextral How and Redwing shears and the sinistral Irisvale–Lancaster shear in the Bulawayo region. An identical pattern has been described to the east of the Bulawayo greenstone belt by Tsomondo & Blenkinsop (1991) for the sinistral Surprise and dextral Jenya–Mushandike shear zones. They demonstrated that the two shear zones form a conjugate pair of ductile shear zones which accommodated the ENE-directed expulsion of a small crustal block, and the orientation of which is consistent with overall NNW–SSE directed shortening (Fig. 9), although this analysis is only valid if both sets of shears were operative at the same time. Such conjugate shears are widely distributed, at all scales from the regional to the outcrop, across the central part of the craton, shear sets of similar orientations also occurring within both the southern part of the craton (James 1975) and in the NMZ (Blenkinsop & Mkweli 1992). We note that the orientation of the dextral shears in the central part of the craton is parallel to that of

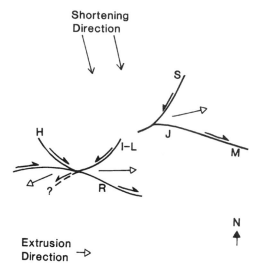

Fig. 9. Schematic map showing the relationship between regional compression and shortening, and crustal extrusion accommodated by the Surprise – Jenya–Mushandike and How – Redwing – Irisvale–Lancaster fault systems.

dextral shears in the southwestern part of the craton. Timing of this deformation is constrained only to be post-upper Bulawayan, the presumed age of the Mutare, Mvuma and Felixsburg greenstones, although the NNE-trending sinistral shears in the Felixburg belt clearly postdate emplacement of the Chilimanzi granite suite.

Finally, Coward & Treloar (1990), Treloar et al. (1992) and McCourt & Wilson (1991) indicate that major ENE-trending shears developed within the northern and central parts of the craton curve into a zone of N-trending fabrics developed within the Midlands greenstone belt. Theoretically, therefore, this should be a zone of flattening with the shears marked by dip-slip displacements. However, Pitfield & Campbell (1990) and Pitfield et al. (1991) show the N-trending shears within the greenstone belt to have a variety of slip senses from dip-slip thrusting, through oblique-slip to dominantly dextral, strike-slip shearing (Fig. 6). All these structures are associated with large-scale folds and it may be possible that much of the shearing post-dates initial shortening and thrusting. One of the ENE-trending shears cited as curving into the N-trending fabrics, the Umwindsi shear zone (Fig. 6) is not, as depicted by Treloar et al. (1992) a dextral shear zone of significant extent but a sinistral oblique-slip shear zone with lineations plunging at c. 40° towards the SW. At its most intense, the shear zone is ductile but is of limited along strike length and cannot be traced from its type outcrop in the Umwindsi River southwest into

the Harare greenstone Belt as shown by Baldock (1991) and, as such, cannot have any regional significance.

Summary

In summary the southern and central parts of the Zimbabwe craton are marked by three sets of fabric trends and shear zones: a NNW-trending fabric with down-dip lineations in the southwestern part of the craton; WNW-trending, dextral shear zones and conjugate NNE-trending sinistral shear zones, and in the Mvuma, Mutare and Masvingo greenstone belts, sub-vertical extension in the centre of the belt and sub-horizontal extension parallel to the trend of the belt along the margin. There is no clear evidence for ENE-trending, regional-scale sinistral shear zones. The conjugate shear zones appear to be regionally distributed across the southern and central parts of the craton although have not, as yet, been identified in the northern part of the craton.

Discussion

The points outlined above indicate to us that the structural framework depicted by Coward (1976) and adopted by Coward & Treloar (1990), Treloar et al. (1992) and McCourt & Wilson (1991) is not just an over simplification but is qualitatively incorrect. It was the combination of regional-scale strike-slip displacements along ENE-trending shears with clear evidence for NNW–SSE shortening that led Treloar et al. (1992) to propose the Tibetan analogue. The data presented here are supportive of approximately N–S shortening but we find no evidence for the regionally developed sinistral strike-slip shear zones mentioned by earlier workers. To the contrary, there is clear evidence for regionally distributed conjugate sets of ESE-trending dextral and NNE-trending sinistral shears. It is obvious, therefore, that those models, including the Tibetan-analogue, which invoke the SW-directed movement of Archaean crust (Coward et al. 1976; Coward & Treloar 1991; McCourt & Wilson 1991) need to be questioned. Essentially any model has to address at least the following points. What is the regional significance, if any, of the conjugate shear sets? What is the relationship between NNW–SSE shortening and regional-scale, lateral displacements along dextral, strike-slip shear zones such as the Jenya–Mushandike shear zone? What is the timing of deformation across the Zimbabwe craton, with respect to that in the Limpopo belt and Kaapvaal craton and along the Triangle–Tuli-Sabi shear zone? What is the relationship between deformation and granite generation and emplacement, and to what extent does the combination of local strike-slip shear zones and vertical extension within greenstone belts represent regionally distributed strains as opposed to local strains resulting from granite emplacement? To what extent do local structures reflect rotations of relatively small-scale crustal blocks within an overall regional compressive or transpressive environment? To answer these questions we need to address three points: firstly, the relationship between the Limpopo belt and the Kaapvaal and Zimbabwe cratons; secondly, the internal deformation of the Zimbabwe craton; and thirdly the relationship between granite emplacement and deformation.

The relationship between the Limpopo belt and the Kaapvaal and Zimbabwe cratons

As stated above the evolution of the Limpopo belt involved: firstly, deformation and metamorphism at *c.* 2.68 Ga associated with the convergence of the Central zone with the Kaapvaal craton; secondly, convergence dated by emplacement of the Razi granite suite, at *c.* 2.58 Ga between that block and the Zimbabwe craton to the north; and, finally, renewed deformation at *c.* 2.0 Ga which involved transposition of Late Archaean fabrics along the intracontinental Triangle–Tuli-Sabi–Magogaphate shear zone that separates the CZ from the NMZ, together with an amphibolite- to granulite-facies metamorphism. That emplacement of the craton-wide Late Archaean Chilimanzi monzogranites was synchronous with convergence of the Zimbabwe craton with the CZ, suggests that deformation and thickening of the craton was intimately linked with partial melting of the lower crust and emplacement of the monzogranitic suite. Whether initial convergence between the Zimbabwe craton and the Central zone/Kaapvaal craton block involved a strike-slip sense of displacement or was truly orthogonal is uncertain, although the down-dip nature of the stretching lineations throughout the NMZ (James 1975; Treloar *et al.* 1992) suggests the latter. If this is so, then the dextral strike-slip displacements along the Triangle–Tuli-Sabi shear zone may be entirely Proterozoic in age.

The internal deformation of the Zimbabwe craton

The first, regionally distributed, deformation which we can identify with any certainty on the craton is an ENE-WSW flattening that resulted in the zone of NNW-trending fabrics in the southwestern part of the craton and recumbent folding of the Botswanan greenstone belts. McCourt & Wilson (1991) attribute this to collision between the Zimbabwe craton and a putative 'Matsitama' fragment. The present authors interpret this as

part of the outward growth, perhaps by terrane accretion, of the Zimbabwe craton, accretion of the Matsitama fragment of the Motloutse complex (Aldiss 1991) being analogous to the cordilleran-type terrane accretion described by de Wit *et al.* (1992*a*, *b*) for the northern part of the Kaapvaal craton. Although an apparent curvature of these NNW-trending fabrics into ENE-trending sinistral strike-slip shear zones has been inferred in the past (Coward *et al.* 1976; Treloar *et al.* 1992), there is little evidence for such strike-slip displacements, and would argue that the ESE-striking dextral shears are later than, and cut, the NNW-trending fabrics. The possibility that some sinistral displacement was associated with accretion of new crust to the southern margin of the growing craton is not excluded. Rollinson (1993) describes the Buchwa greenstone belt as being possibly an accreted fragment and we see no reason why such accretion could not involve a transcurrent sense of movement, although would stress that there is as yet no direct evidence for this.

The most important phase of deformation that we identify within the Zimbabwe craton results in the combination of steeply dipping NNW-trending shortening fabrics with down-dip stretching lineations (as in the NMZ of the Limpopo belt, and in the Masvingo, Mvuma, Mutare and Arcturus greenstone belts) and the development of conjugate NNE-trending sinistral and ESE-trending dextral shears. The distribution of the conjugate shear sets is consistent throughout with an overall sense of NNW–SSE shortening. Although, there is no absolute certainty as to the age or synchroneity of all of these structures, it is inferred that they are related to a major phase of NNW–SSE shortening. This shortening, which was broadly synchronous with the emplacement of the Chilimanzi suite of monzogranites, was associated with thrusting of the NMZ of the Limpopo belt onto the southern margin of the Zimbabwe craton, and was thus most likely related to convergence between the Central zone/Kaapvaal craton block and the Zimbabwe craton. We infer that the conjugate shear sets have a regional significance. Tsomondo & Blenkinsop (1991) showed that the dextral Jenya–Mushandike and sinistral Surprise conjugate pair of shears could have accommodated the eastward extrusion of a significant part of the central region of the Zimbabwe craton, and it is likely that the How, Redwing and Irisvale–Lancaster shears could have had the same effect. If the shears that are seen at surface represent part of a set of conjugate shears, regionally developed in response to NNW–SSE shortening, then there is a structural mechanism for eastward and westward extrusion of discrete crustal blocks of varying size. Coward & James (1974) and Coward *et al.* (1976) used the term 'billiard ball tectonics' to describe localized rotations and displacements of small crustal blocks and granite bodies relative to each other in response to the overall regional displacements. Many of these inferred rotations, as in the Rhodesdale batholith (Fabiani 1989), could be accounted for by displacements along conjugate shear sets.

One of the key problems in the analysis of Archaean shear zone patterns is that posed by subsequent transposition of Archaean fabrics during subsequent deformations and the localization of later shearing along Archaean structural lineaments. This is clearly demonstrated within the Zimbabwe craton. Although the ESE-trending dextral shears in the southwestern part of the craton near Bulawayo and Gwanda are interpreted as part of the conjugate shear pattern, with deformation along them synchronous with emplacement of the Chilimanzi granite suite (Coward et al. 1976), they are located just north of the dextral Triangle–Tuli-Sabi–Magogaphate shear zone. Timing of deformation along this ductile shear, which marks the northern margin of the Central zone of the Limpopo belt, may be as late as c. 2.0 Ga and it is probable that the dextral shears on the southwestern part of the craton, although being fundamental Archaean structures, have a variable history of transposition and re-activation during Proterozoic-aged intracontinental shearing along the Triangle–Tuli-Sabi shear zone. Similarly the NNE-trending sinistral shears within the central part of the craton are paralleled by both the Great Dyke trend and the early Proterozoic Popoteke fault trend (Wilson 1990), and it is likely that these fundamental Archaean structures are also sites where significant post-Archaean strain accumulated.

The relationship between granite emplacement and deformation

The main phase of deformation identified within the Zimbabwe craton is characterized by conjugate shearing and homogeneous flattening controlled by an overall NNW–SSE compression that is probably related to thrusting of the granulites of the North Marginal zone onto the southern margin of the Zimbabwe craton. This deformation spans the period of Chilimanzi granite emplacement. Blenkinsop & Treloar (1995) describe the dextral Mushandike shear zone as probably predating Chilimanzi granite emplacement whereas other Chilimanzi granites have margins sheared by the conjugate shears. It is necessary to consider the extent to which, firstly, the deformation controlled granite generation and emplacement, and, secondly, the extent to which regionally developed structures are locally modified by granite emplacement.

Much work has been published recently on mechanisms of granite emplacement. A persistent theme of this work is that diapirism is not a sensible solution to problems of granite emplacement and that space needs to be created tectonically into which granite magmas can migrate. Ideally granites are emplaced into zones of tension, commonly dilational jogs in strike slip or normal fault systems (Hutton 1988; Hutton et al. 1990; D'Lemos et al. 1992). Despite this, a coherent model for the tectonic and geochemical evolution of the Harare–Bindura–Shamva greenstone belt, based on mantle plume activity and diapiric ascent of granitic magmas, has recently been proposed by Jelsma (1993). Structural evidence was used by Jelsma et al. (1993) in support of an argument that greenstone belt deformation is strongly related to the diapiric ascent of magma and ballooning of plutons during emplacement into the upper crust. Radial and oblique tangential lineation trajectories, variation of strain from vertical constriction at cleavage triple points between three plutons to triaxial flattening strain in inter-domal synclinoria between two plutons, increase of strain towards plutons. and shear zones with pluton-up senses of movement all support some form of pluton-related deformation. Similar strain trajectories have also been described from the Indian Archaean craton (Bouhallier et al. 1993). There is a similarity between the Harare–Shamva–Bindura belt and the Mvuma and Mutare belts in that all of them show a sub-vertical extension in their central parts, and there must be a possibility that, if the diapiric model works for the Harare–Bindura–Shamva belt, then it may be an appropriate one for the other belts as well. However, most greenstone belts are not arcuate to the extent of the Harare–Shamva–Bindura belt.

Although the strain patterns described by Jelsma et al. (1993) would be consistent with diapirism, growth of a pluton by emplacement of a series of sill-like bodies or a laccolith (Corry 1988) could result in strain patterns very similar to those observed and modelled for ballooning 'diapirs' (cf. Jackson & Pollard 1990). In addition, thermal modelling of granite diapirs (Mahon et al. 1988) predict that, unless continental crust is significantly less viscous than generally believed, diapirs should suffer thermal death and lock up solid in the middle crust, and to this extent there must be a possibility that there is an alternative explanation to the structures described by Jelsma et al. (1993). Consequently, rather than explaining all greenstone belt structures within the Zimbabwean craton within a framework of diapiric emplacement of the craton-wide Chilimanzi suite of granites, a non-diapiric model may be contemplated, ideally one based on a fault control of granite emplacement. It is obvious that structural features preserved around plutons may not uniquely define emplacement mechanisms, and that a clear distinction needs to be made between mechanisms of magma transport and of pluton emplacement.

What thermal impetus drove the generation of the Chilimanzi granites is uncertain, although as Chilimanzi granite emplacement is synchronous with collision between the Zimbabwe craton and the Central zone/Kaapvaal craton crustal block, it seems logical to infer that granite generation was a response to that collision. Given the volume of granite it appears unlikely that they could be derived by fractionation from new, mantle-derived basic magma, and thus they must represent lower crustal melts. It is possible that widespread partial melting of the lower crust and subsequent plutonism may have been due to an added mantle heat source, for instance through a hot spot, although this mechanism may, perhaps, be ruled out due to both the complete lack of synchronous basic magmatism and the temporal relationship with crustal collision. Thus granite generation is more likely to be part of the thermal response to crustal thickening.

Whether the Chilimanzi suite of Late Archaean monzogranites was derived from melting of the lower crust due to thermal relaxation once it had thickened, in response to convergence between the Zimbabwe craton and the Central zone/ Kaapvaal craton, or by fractional crystallization of a sub-crustal tonalitic–trondhjemitic–granodioritic magma ocean as suggested by Kramers & Ridley (1989), Ridley & Kramers (1990) and Ridley (1991) is beyond the scope of this paper. However, we would argue that monzogranite melts were developed by partial melting of the lower crust in direct response to crustal thickening resulting from collision of the Zimbabwe craton with a crustal block to the south. Rather than invoking diapirism to explain ascent of these granitoid magmas, we would argue that emplacement was, instead, by a series of dyke-like fissures through which magma was transferred from the lower crust to high crustal levels. The widespread presence of andalusite and of cordierite–orthoamphibole assemblages in greenstone belts argues that emplacement was regionally at shallow depths. Ballooning of plutons would have resulted from continual magma injection through the dyke-like conduits although the pluton shape need not have been spherical. Clemens & Mawer (1992) argue that the ascent of granitoid magmas through fractures could feed magma ponds located along sub-horizontal discontinuities, leading to the growth of lens-like laccolithic plutons made up of a series of individual sill-like bodies. Such sill-like granite bodies, with a sub-horizontal internal layering, have been clearly described from the Tingi Hills region of the Sierra Leone Archaean craton (Williams 1978; Rollinson & Cliff 1982). There, tabular, late-kinematic granites appear compositionally similar to granite dykes up to 1 km wide. Preliminary fieldwork has shown that parts of the Chilimanzi Murehwa batholith in the north central part of the Zimbabwe craton have horizontal foliations and contacts between intrusive phases, suggesting that it may comprise a number of sill-like bodies (T. Sweatman, pers. comm. 1990).

The obvious candidates for dyke-like conduits are the conjugate sets of dextral and sinistral shears. Tsomondo & Blenkinsop (1991) have shown that the conjugate Surprise and Jenya–Mushandike shears could have accommodated the eastward extrusion of a large block of the central Zimbabwe craton. If sufficient numbers of such shears were operative during the Late Archaean, then the Zimbabwe craton would have been constituted of a series of small crustal blocks the geometry of which was controlled by conjugate shears and which could have moved laterally with respect to each other with granite transport located along the shear zones. Ballooning of plutons could result in shortening across, and vertical extension within, greenstone belts, lateral displacements along shear zones resulting in sub-horizontal displacements along greenstone margins and in the adjacent granites. Indeed, Coward & James (1974) argued that much of the deformation along the contacts between the greenstone belts and adjacent granites was the result of some form of block sliding or 'billiard ball tectonics'. Here it is argued that the marginal deformation was in fact controlled by the conjugate shears that result from NNW–SSE shortening and which accommodated granite emplacement.

Although conjugate shears can be widely identified in the southern and central part of the craton, they have not, as yet, been identified in the northern part. This does not, however, mean that they are not present there. It should be borne in mind, as well, that the broad synchroneity between granite emplacement and regionally-developed conjugate shearing implies that some such shears may have been intruded and destroyed by late-stage Chilimanzi-granites. The model of pluton emplacement along conjugate shears, combined with ballooning plutonism, needs to be tested by examining strain patterns for the whole Zimbabwe craton. However, these data do not yet exist.

Although the conjugate shears described here represent structures operative at shallow crustal levels, their regional extent implies that they may be the near-surface representation of deep-seated structures which could have permitted lateral flow of deep-crustal material. Lateral flow (Bird 1991) is facilitated by high temperatures, increased crustal thicknesses and a reduction in rheological strength of the lower crust. Increased crustal thickness and elevated temperature are implied by the crustal thickening model. As melting will lead to a decrease in lower crustal strength the three components which facilitate lateral crustal flow are all present. In that melt generation weakens the lower crust and therefore permits more crustal thinning, so that thinning will drive yet more melting through the medium of biotite and/or hornblende vapour absent melting reactions.

Conclusions

(1) Structures within the Limpopo belt record at least three major tectonic events. The first of these, at *c*. 2.68 Ga, saw the collision with associated deformation and metamorphism of the Central zone with the Kaapvaal craton. This marked the end of the outward growth of the Kaapvaal craton through continued accretion. Subsequently the Central zone–Kaapvaal craton block converged with the Zimbabwe craton at about 2.58 Ga. The suture zone separating the Central zone from the North Marginal zone and Zimbabwe craton was re-deformed at about 2.0 Ga during a major transpressional crustal thickening event.

(2) Prior to convergence with the Central zone of the Limpopo belt, the Zimbabwe craton had grown outwards through one or more of a series of accretion and collisional processes. Basement formation ages clearly decrease west and north from the Tokwe block. Collision, after deposition of the Lower Bulawayan sequences, between the southwestern part of the craton and the Motloutse Complex of the Matsitama fragment resulted in shortening across the southwestern part of the craton represented by a NNW-trending steep fabric with down dip lineations.

(3) Convergence between the Zimbabwe craton and the Central zone, at about 2.58 Ga, resulted in thrusting of the NMZ onto the southern margin of the craton. As a result of this convergence, both the NMZ and the entire Zimbabwe craton to the north were affected by major NNW–SSE shortening. Within the craton this shortening was marked by flattening fabrics with down-dip stretching lineations developed in many of the greenstone belts, and the development of generally NNE-trending sinistral and ESE-trending conjugate shears.

(4) Emplacement of potassic monzogranites of the Chilimanzi suite was syn- to post- the NNW–SSE flattening and shortening phase. What the thermal causes of granite generation were is beyond the scope of this paper, except that it is inferred that they were related to shortening of the whole crust and sub-crustal lithosphere. Their emplacement may have been controlled by the array of conjugate shear zones.

(5) A closer study of cratonic structure suggests the broad structural framework depicted by earlier workers to be incorrect, with little evidence for wholescale southwest-directed extrusion of thickened continental crust. However, the conjugate shears may have accommodated some localised, dominantly eastward, extrusion of small crustal blocks rotations and displacements of which would have partially accommodated granite emplacement.

Field work has been supported by the Royal Society of London (P.J.T.) and University of Zimbabwe Research Council (T.J.B.). We acknowledge discussions over many years with a large number of colleagues including J. F. Wilson, M. P. Coward, J. Ridley, J. D. Kramers, H. R. Rollinson, S. McCourt, B. Kamber, M. Berger and J. D. Clemens.

References

ALDISS, D. T. 1991. The Motloutse Complex and the Zimbabwe craton/Limpopo belt transition in Botswana. *Precambrian Research*, **50**, 89–109.

ARNDT, N. T. 1983. Role of a thin, komatiite-rich oceanic crust in the Archaean plate tectonic process. *Geology*, **4**, 372–375.

—— 1992. Rate and mechanism of continental growth formation in the Archaean. *In*: MARUYAMA, S. (ed.) *Evolving Earth Symposium*. Okazaki, 38–41.

BALDOCK, J. W. 1991. *The geology of the Harare Greenstone Belt and surrounding granite gneiss terrain*. Geological Survey of Zimbabwe Bulletin **94**.

BARTON, J. M. & VAN REENEN, D. D. 1992. When was the Limpopo orogeny? *Precambrian Research*, **55**, 7–16.

——, DOIG, R., SMIT, C. A., BOHLENDER, F. & VAN

REENEN, D. D. 1992. Isotopic and REE characteristics of the intrusive charnoenderbite and enderbite geographically associated with the Matok pluton, Limpopo Belt, southern Africa. *Precambrian Research*, **55**, 429–450.

——, HOLZER, L., KAMBER, B., DOIG, R., KRAMERS, J. D. & NYFELER, D. 1994. Discrete metamorphic events in the Limpopo belt, southern Africa: implications for the application of P-T paths in complex metamorphic terrains. *Geology*, **22**, 1035–1038.

——, RYAN, B., FRIPP, R. E. P. & HORROCKS, P. 1979. Effects of metamorphism on the Rb-Sr and U-Pb systematics of the Singelele and Bulai gneisses, Limpopo Mobile Belt, southern Africa. *Transactions of the Geological Society of South Africa*, **82**, 259–269.

BERGER, M. 1994. U-Pb and Sm-Nd geochronology on

magmatic granulites of the northern marginal zone of the Limpopo Belt, Zimbabwe. *Geological Society of America, Abstracts with programs*, **26**, 49.

——, KAMBER, B., MKWELI, S., BLENKINSOP, T. G. & KRAMERS, J. D. 1993. New zircon U-Pb data on the North Marginal zone, Limpopo Belt, Zimbabwe. *Terra Abstracts, Supplement no. 1 to Terra Nova*, **5**, 313.

BICKLE, M. J. 1978. Heat loss from the earth: a constraint on Archaean tectonics from the relationship between geothermal gradients and the rate of plate production. *Earth and Planetary Science Letters*, **40**, 301–315.

BIRD, P. 1991. Lateral extrusion of lower crust from under high topography, in the isostatic limit. *Journal of Geophysical Research*, **96**, 10275–10286.

BLENKINSOP, T. G. & MKWELI, S. 1992. The relation between the Zimbabwe craton and the North Marginal Zone of the Limpopo belt. *Proceedings of the second symposium of Science and Technology, Research Council of Zimbabwe*, **IIIc**, 236–248.

—— & TRELOAR, P. J. 1995. Geology, classification and kinematics of S-C and S-C′ fabrics in the Mushandike area, Zimbabwe. *Journal of Structural Geology*, **17**, 397–408.

BOUHALLIER, H., CHOUKROUNE, P. & BALLEVRE, M. 1993. Diapirism, bulk homogeneous shortening and trans-current shearing in the Archaean Dharwa craton: the Holenarsipur area, southern India. *Precambrian Research*, **63**, 43–58.

CAMPBELL, S. D. G., BAGLOW, N. & PITFIELD, P. E. J. 1990. The structural framework of the Filabusi Greenstone Belt, and its relevance to gold mineralisation. *Annals of the Zimbabwe Geological Survey*, 15, 29–38.

——, OESTERLEN, P. M., BLENKINSOP, T. G., PITFIELD, P. E. J. & MUNYANYIWA, H. 1991. A provisional 1:2,500,000 scale tectonic map and the tectonic evolution of Zimbabwe. *Annals of the Zimbabwe Geological Survey*, **16**, 31–51.

CHENJERAI, K. G., SCHMIDT MUMM, A., BLENKINSOP, T. G. & CHATORA, D. R. In Press. Tectonic setting and regional exploration significance of the Mutare greenstone belt, Zimbabwe: the Redwing gold deposit. *In*: BLENKINSOP, T. G. & TROMP, P. L. (eds) *Sub-Saharan economic geology*. Special Publication of the Geological Society of Zimbabwe. **3**.

CLEMENS, J. C. & MAWER, C. K. 1992. Granitic magma transport by fracture propagation. *Tectonophysics*. **204**, 339–360.

COOMER, P. G., COWARD, M. P. & LINTERN, B. C. 1977. Stratigraphy, structure and geochronology of ore leads in the Matsitama schist belt of northern Botswana. *Precambrian Research*, **5**, 23–41.

CORRY, C. E. 1988. *Laccoliths: mechanisms of emplacement and growth*. Geological Society of America Special Paper **220**.

COWARD, M. P. 1976. Archaean deformation patterns in southern Africa. *Philosophical Transactions of the Royal Society of London*, **A283**, 313–331.

—— 1980. Shear zones in the Precambrian crust of southern Africa. *Journal of Structural Geology*, 2, 19–27.

——. & DALY, M. P. 1984. Crustal lineaments and shear zones in Africa: their relationship to plate movements. *Precambrian Research*, **24**, 27–45.

—— & JAMES, P. R. 1974. The deformation patterns of two Archaean greenstone belts in Rhodesia and Botswana. *Precambrian Research*, **1**, 235–258.

—— & TRELOAR, P. J. 1990. Tibetan models for Archaean tectonics in southern Africa. *In*: BARTON, J. M. (ed.) *The Limpopo Belt: a field workshop on Granulites and deep crustal tectonics*. Department of Geology, Rand Afrikaans University, 35–38.

——, GRAHAM, R. H., JAMES, P. R. & WAKEFIELD, J. 1973. A structural interpretation of the northern margin of the Limpopo orogenic belt, southern Africa. *Philosophical Transactions of the Royal Society of London*, **A273**, 487–491.

COWARD, M. P., JAMES, P. R. & WRIGHT, L. I. 1976. Northern margin of the Limpopo mobile belt, southern Africa. *Geological Society of America Bulletin*, **87**, 601–11.

DE WIT, M. J. 1991. Archaean greenstone belt tectonism and basin development: some insights from the Barberton and Pietersburg greenstone belts, Kaapvaal craton, South Africa. *Journal of African Earth Sciences*, **13**, 45–63.

—— & HART, R. A. 1993. Earth's earliest continental lithosphere, hydrothermal flux and crustal re-cycling. *Lithos*, **30**, 309–335.

——, JONES, M. G. & BUCHANAN, D. 1992b. The geology and tectonic evolution of the Pietersburg greenstone belt, South Africa. *Precambrian Research*, **55**, 123–153.

——, ROERING, C., HART, R .J., ARMSTRONG, R. A., DE RONDE, C. E. J., GREEN, R. W. E., TREDOUX, M., PERBEDY, E. & HART, R. A. 1992a. Formation of an Archaean continent, *Nature*, **357**, 553–562.

D'LEMOS, R. S., BROWN, M. & STRACHAN, R. A. 1992. Granite magma generation, ascent and emplacement within a compressional orogen. *Journal of the Geological Society, London*, **149**, 487–490.

DRUMMOND, M. S. & DEFANT, M. J. 1990. A model for trondhjemite-tonalite-dacite genesis and crustal growth via slab melting: Archaean to modern comparisons, *Journal of Geophysical Research*, **95**, 21 503–21 521.

ELLAM, R. M. & HAWKESWORTH, C. J. 1988. Is average continental crust generated at subduction zones? *Geology*, **16**, 314–317.

ENGLAND, P. C. & BICKLE, M. J. 1986. Continental thermal and tectonic regimes during the Archaean. *Journal of Geology*, **92**, 353–368.

—— & HOUSEMAN, G. 1986. Finite strain calculations of continental deformation. 2: Comparison with the India-Asia collision zone. *Journal of Geophysical Research*, **91**, 3664–3676.

—— & —— 1989. Extension during continental convergence,with application to the Tibetan Plateau. *Journal of Geophysical Research*, **94**, 17 561–17 579.

FABIANI, W. M. B. 1989. *The geology of the eastern portion of the Mvuma greenstone belt with special reference to the mineralisation and structure of Athens Mine*. PhD thesis. University of Zimbabwe.

FEDO, C. M. 1994. Structural geology of the 3.0 Ga Buchwa greenstone belt, Zimbabwe: any relationship to the "Limpopo" orogeny? *Geological Society of America, Abstracts with programs*, **26**, 407.

GRAMBLING, J. A. 1981. Pressures and temperatures in Precambrian metamorphic rocks. *Earth and Planetary Science Letters*, **73**, 63–68.

GROCOTT, J., BROWN, M., DALLMEYER, R. D., TAYLOR, G. K. & TRELOAR, P. J. 1994. Mechanisms of continental growth in extensional arcs: an example from the Andean plate boundary zone. *Geology*, **22**, 391-394.

HICKMAN, M. H. 1978. Isotopic evidence for crustal reworking in the Rhodesian Archaean craton, southern Africa. *Geology*, **6**, 214–216.

HOUSEMAN, G. & ENGLAND, P. C. 1986. Finite strain calculations of continental deformation. 1: Method and general results for convergent zones. *Journal of Geophysical Research*, **91**, 3651–3663.

HUTTON, D. H. W. 1988. Granite emplacement mechanisms and tectonic controls: inferences from deformation studies. *Transactions of the Royal Society of Edinburgh: Earth Sciences*, **79**, 245–255.

—— & REAVY, R. S. 1992. Strike slip tectonics and granite petrogenesis. *Tectonics*, **11**, 960–967.

——, DEMPSTER, T. J., BROWN, P. R. & BECKER, S. M. 1990. A new mechanism of granite emplacement: intrusion in active extensional shear zones. *Nature*, **343**, 452–455.

JACKSON, M. D. & POLLARD, D. D. 1990. Flexure and faulting of sedimentary host rocks during growth of igneous domes, Henry Mountains, Utah. *Journal of Structural Geology*, **12**, 185–206.

JAMES, P. R. 1975. *A deformation study across the northern margin of the Limpopo belt, Rhodesia*. PhD thesis, University of Leeds.

JAWI, J. 1986. *The geology of the greenstone-granite terrain between Adzwe mine and St Josephs School, Wedza*. BSc (Hons) thesis, University of Zimbabwe.

JELSMA, H. A. 1993. *Granites and greenstones in northern Zimbabwe: tectonothermal evolution and source regions*. PhD thesis, Free University, Amsterdam.

——, VAN DER KEEK, P. A. & VINYU, M. L. 1993. Tectonic evolution of the Bindura-Shamva greenstone belt (Zimbabwe): progressive deformation around ballooning diapirs. *Journal of Structural Geology*, **15**, 163–176.

KAMBER, B., BLENKINSOP, T. G., ROLLINSON, H. R., KRAMERS, J. D. & BERGER, M. 1992. *Dating of an important tectono-thermal event in the Northern Marginal Zone of the Limpopo Mobile Belt, Zimbabwe: first results*. North Limpopo Field Guide and Abstracts Volume. Geological Society of Zimbabwe, **39**.

——, KRAMERS, J. D. & ROLLINSON, H. R. 1993. The Triangle Shear Zone, a Proterozoic kill-joy in the tectonic models for the Archaean Limpopo belt. *Terra Abstracts, Supplement no. 1 to Terra Nova*, **5**, 316–317.

——, ——, NAPIER, R., CLIFF, R. A. & ROLLINSON, H. R. 1985. The Triangle Shear Zone, Zimbabwe, revisited: new data document an important event at 2.0 Ga in the Limpopo Belt. *Precambrian Research*, **70**, 191–213.

KEY, R. M., LITHERLAND, M. & HEPWORTH, J. V. 1976. The evolution of the Archaean crust of northeast Botswana. *Precambrian Research*, **3**, 275–413.

KRAMERS, J. D. 1988. An open system fractional crystallization model for very early continental crust formation. *Precambrian Research*, **38**, 281–295.

—— & RIDLEY, J. 1989. Can Archaean granulites be direct crystallisation products from a sialic magma layer. *Geology*, **17**, 281–295.

LIGHT, M. F. R. 1982. The Limpopo Mobile Belt: a result of continental collision. *Tectonics*, 1, 325–342.

MAHON, K. I., HARRISON, T. M. & DREW, D. A. 1988. Ascent of a granite diapir in a temperature varying medium. *Journal of Geophysical Research*, **93**, 1174–1188.

MARTIN, H. 1986. Effect of steeper Archaean geothermal gradient on geochemistry of subduction zone magmas. *Geology*, **14**, 753–756.

—— 1987. Petrogenesis of Archaean trondhjemites, tonalites and granodiorites from eastern Finland: major and trace element geochemistry. *Journal of Petrology*, **28**, 921–953.

—— 1993. The mechanisms of petrogenesis of the Archaean continental crust -comparison with modern processes. *Lithos*, **30**, 373–388.

McCOURT, S. & VEARNCOMBE, J. R. 1987. Shear zones bounding the central zone of the Limpopo Mobile Belt, southern Africa. *Journal of Structural Geology*, **9**, 127–137.

—— & —— 1992. Shear zones of the Limpopo belt and adjacent granite-greenstone terranes: implications for late Archaean collision tectonics in southern Africa. *Precambrian Research*, **55**, 553–570.

—— & WILSON, J. F. 1991. Late Archaean and early Proterozoic tectonics of the Limpopo and Zimbabwe provinces, southern Africa. *In: The University of Western Australia. Publication*, **22**, 237–245.

McGREGOR, A. M. 1951. Some milestones in the Precambrian of Southern Rhodesia. *Transactions of the Geological Society of South Africa*, **54**, 27–71.

MKWELI, S., KAMBER, B. & BERGER, M. 1995. Westward continuation of the craton-Limpopo Belt tectonic break in Zimbabwe and new age constraints on the timing of the thrusting. *Journal of the Geological Society, London*, **152**, 77–83.

NISBET, E. G. & WALKER, D. 1982. Komatiites and the structure of the Archaean mantle. *Earth and Planetary Science Letters*, **60**, 105–113.

——, CHEADLE, M. J., ARNDT, N. T. & BICKLE, M. J. 1993. Constraining the potential temperature of the Archaean mantle: a review of the evidence from komatiites. *Lithos*, **30**, 291–307.

PELTZER, G., TAPPONNIER, P. & COBBOLD, P. 1982. Les grands dechrochments de l'Est asiatique, evolutions dans le temps et comparaison avec un modele experimental. *Comptes Rendus de l'academy des Sciences, Paris*, **294**, 1341–1348.

PETFORD, N., KERR, R. C. & LISTER, J. R. 1993. Dyke transport model of granitoid magmas. *Geology*, **21**, 845–848.

PITFIELD, P. E. J. & CAMPBELL, S. D. G. 1990. Integrated exploration; Midlands Goldfield project. *Annals of the Zimbabwe Geological Survey*, **15**, 21–35.

——, —— & MUGUMBATE, F. 1991. Exploration geochemical orientation and reconnaissance studies in the Midlands Goldfield, Zimbabwe. *Annals of the Zimbabwe Geological Survey*, **16**, 52–69.

RAMSAY, J. G. 1989. Emplacement kinematics of a granite diapir: the Chindamora batholith, Zimbabwe. *Journal of Structural Geology*, **11**, 191–209.

RIDLEY, J. 1991. The thermal causes and effects of voluminous, late Archaean monzogranite plutonism. *In: The University of Western Australia. Publication*, **22**, 275–285.

—— 1992. On the origins and tectonic significance of the charnokite suite of the Archaean Limpopo Belt, Northern Marginal Zone, Zimbabwe. *Precambrian Research*, **55**, 407–427.

RIDLEY, J. & KRAMERS, J. D. 1990. The evolution and tectonic consequences of a tonalitic magma layer within Archaean continents. *Canadian Journal of Earth Sciences*, **27**, 219–228.

ROERING, C., VAN REENEN, D. D., SMIT, C. A., BARTON, J. M., DE BEER, J. H., DE WIT, M. J., STETTLER, E. H., VAN SCHALKWYK, J. F., STEVENS, G., & PRETORIUS, S. 1992a. Tectonic model for the evolution of the Limpopo Belt. *Precambrian Research*, **55**, 539–552.

——, —— & VAN SCHALKWYK, J. F. 1992b. Structural, geological and metamorphic significance of the Kaapvaal craton – Limpopo Belt contact. *Precambrian Research*, **55**, 69–80.

ROLLINSON, H. R. 1993. A terrane interpretation of the Archaean Limpopo belt. *Geological Magazine*, **130**, 755–765.

—— & BLENKINSOP, T. G. 1995. The magmatic, metamorphic and tectonic evolution of the Northern Marginal Zone of the Limpopo Belt in Zimbabwe. *Journal of the Geological Society, London*, **152**, 65–75.

—— & CLIFF, R. A. 1982. New Rb-Sr age dates on the Archaean basement of Eastern Sierra Leone. *Precambrian Research*, **17**, 63–72.

—— & LOWRY, D. L. 1992. Early basic magmatism in the evolution of the North marginal zone of the Archaean Limpopo belt. *Precambrian Research*, **55**, 33–45.

SCHUBERT, G., STEVENSON, D. & CASSEN, P. 1980. Whole planet cooling and the radiogenic heat source contents of the earth and moon. *Journal of Geophysical Research*, **85**, 2531–2538.

SCHMIDT MUMM, A., OBERTHUR, T., BLENKINSOP, T. G. & CHENJARAI, K. G. 1993. A fluid hydraulic model for shear zone related gold deposits: the Redwing gold mine, Zimbabwe. *In:* PARNELL, J., RUFFEL, A. H. & MOLES, N. R. (eds) *Geofluids '93*. Department of Geology, Queen's University, Belfast, 194–197.

SMIT, C. A., ROERING, C. VAN REENEN, D. D. 1992. The structural framework of the southern margin of the Limpopo belt, south Africa. *Precambrian Research*, **55**, 51–67.

STEVENS, G. D. & VAN REENEN, D. D. 1992. Constraints on the form of the P-T loop in the southern marginal zone of the Limpopo belt, South Africa. *Precambrian Research*, **55**, 279–296.

TAPPONNIER, P., PELTZER, G., LE DAIN, A. Y., ARMIJO, R. & COBBOLD, P. 1982. Propagating extrusion tectonics in Asia: new insights from plasticine experiments. *Geology*, **10**, 611–616.

——, —— & ARMIJO, R. 1986. On the mechanics of the collision between India and Asia. *In:* COWARD, M. P. & RIES, A. C. (eds) *Collision Tectonics*. Geological Society of London, Special Publications, **19**, 115–157.

TAYLOr, P. N., KRAMERS, J. F., MOORBATH, S., WILSON, J. F., ORPEN, J. L. & MARTIN, A. 1991. Pb/Pb, Sm-Nd and Rb-Sr geochronology in the Archaean craton of Zimbabwe. *Chemical Geology (Isotope Geosciences)*, **87**, 175–196

TSOMONDO, C. M. 1986. *The geology of the northeastern part of the Wedza Greenstone belt and the adjacent eastern granitoids*. BSc (Hons) thesis, University of Zimbabwe.

—— & BLENKINSOP, T. G. 1991. New perspectives on Archaean tectonics: an overview of the Shurugwi Greenstone Belt. *In: Traverse through two cratons and an orogen: excursion guidebook and review articles*. Geological Survey of Swaziland, 275–291.

TRELOAR, P. J., COWARD, M. P. & HARRIS, N. B. W. 1992. Himalayan-Tibetan analogies for the evolution of the Zimbabwe craton and Limpopo Belt. *Precambrian Research*, **55**, 571–587.

VAN BREEMEN, O. & DODSON, M. H. 1972. Metamorphic chronology of the Limpopo Belts, Southern Africa. *Geological Society of America Bulletin*, **83**, 2005–2013.

—— & HAWKESWORTH, C. J. 1980. Sm-Nd isotopic study of garnets and their metamorphic host rocks. *Transactions of the Royal Society of Edinburgh, Earth Sciences*, **71**, 97–102.

VAN REENEN, D. D., BARTON, J. M., ROERING, C., SMIT, C. A. & VAN SCHALKWYK, J. F. 1987. Deep crustal response to continental collision: the Limpopo belt of southern Africa. *Geology*, **15**, 11–14.

——, ROERING, C., BRANDL, G., SMIT, C. A., VAN SCHALKWYK, J. F. & BARTON, J. M. 1990. The granulite facies rocks of the Limpopo belt, southern Africa. *In:* VIELZEUF, D. & VIDAL, P. (eds) *Granulites and crustal evolution*. NATO ASI, **c311**. Kluwer, Dordrecht, 257–289.

VILLOTTE, J. P., DAIGNIERES, M. & MADARIAGA, R. 1982. Numerical modelling of intraplate deformation: simple mechanical models of continental collision. *Journal of Geophysical Research*, **87**, 10 709–10 728.

WILLIAMS, H. R. 1978. The Archaean of Sierra Leone. *Precambrian Research*, **6**, 251–268.

WILSON, J. F. 1979. A preliminary reappraisal of the Rhodesian basement complex. *In:* Anhaeusser, C. R., Foster, R. P. & Stratten, T. (eds) *A symposium on mineral deposits and the transportation and deposition of metals*. Geological Society of South Africa, Special Publications, **5**, 1–23.

—— 1990. A craton and its cracks: some of the behaviour of the Zimbabwe block from the late Archaean to the Mesozoic in response to horizontal movements and the significance of some of its mafic dyke patterns. *Journal of African Earth Sciences*, **10**, 483–501.

Zircon geochronology of Archaean felsic sequences in the Zimbabwe craton: a revision of greenstone stratigraphy and a model for crustal growth

JAMES F. WILSON[1], ROBERT W. NESBITT[2] & C. MARK FANNING[3]

[1]Department of Geology, University of Zimbabwe, PO Box MP 167, Mount Pleasant, Harare, Zimbabwe

[2]Department of Geology, The University, Southampton SO17 1BJ, UK

[3]Research School of Earth Sciences, ANU Canberra, ACT, Australia

Abstract: U–Pb ion-microprobe (SHRIMP) work on zircon populations from 13 Zimbabwean Archaean felsic rocks are presented and interpreted. Samples were extracted from felsic volcanic sequences from most of the major greenstone belts and represent the first zircon geochronological data from within the greenstone belts themselves. The data demonstrate a Late Archaean volcanicity spanning 250 Ma which began at least 2900 Ma ago and ended at 2650 Ma. The intrusion of extensive granitoid sills of the Chilimanzi suite at *c.* 2.6 Ga marks the widespread stabilization of the craton. Based on the new zircon data and a re-evaluation of published mapping, a new stratigraphic subdivision is presented for the Late (< 2.9 Ga) Archaean of Zimbabwe. A feature of the stratigraphic model is the cyclicity of magmatism which begins with ultramafic–mafic rocks, progresses through felsic volcanism and ends with a granitoid event. These cycles are repeated at least three or four times in the 250 Ma time span. An important characteristic of the felsic volcanic rocks is that the bulk of the material examined contains inherited, xenocrystic zircons whose ages range from 1000 Ma to 20 Ma older than the host rocks. The oldest xenocrystic zircons (*c.* 3.6 Ga) are restricted to volcanic rocks which erupted through the Tokwe segment; itself the only known > 3.3 Ga fragment of Archaean crust in Zimbabwe. These data suggest that the Early Archaean crust is restricted to the Tokwe segment in the south of the country. Since even the oldest of the felsic volcanics (2.90 Ga, Lower Belingwean) have zircons which are 50 Ma older, it is suggested that remnants of earlier cyclic greenstone–granitoid events must underlie the present craton and that all of the currently exposed greenstone belts of Zimbabwe were developed on sialic crust.

The Zimbabwe Archaean craton of southern Africa contains granite-greenstone terrains of different ages whose broad subdivisions, distinguished on field relationships and age determinations, have been recognized for some years (e.g. Wilson *et al.* 1978). Based on this previous work, the development of the craton can be viewed in terms of two major events each involving volcanism, crustal growth and orogeny: the first, in the early Archaean, spanning *c.* 3500–3300 Ma with hints of things as old as *c.* 3800 Ma and as young as *c.* 3200 Ma (Dodson *et al.* 1988); and the second, in the Late Archaean, at *c.* 3000–2600 Ma. All that has so far been recognized of the earlier event is the complex of tonalitic gneisses and tightly infolded greenstone-belt remnants that constitute the Tokwe crustal segment exposed in a roughly triangular area in the south of the craton. The greenstone remnants represent what is left of the presumed *c.* 3500 Ma Sebakwian Group: they are dominantly mafic and ultramafic with associated quartz-magnetite banded iron formation (BIF) and some diopside quartzites; some of the ultramafic rocks contain stratiform chromitites. No meta-volcanics of more silicic compositions have so far been described but these would be difficult to recognize within the orthogneisses. The original relationship of the infolded greenstones to the ancient gneisses is not clear but the latter possibly represent a combination of deformed intrusive granitoids and original granitic basement. Gneisses and greenstones have reached amphibolite facies and both show retrogression. The segment has been intruded by Late Archaean granitoids.

In contrast, the Late Archaean granite–greenstone terrains are more widespread; metamorphic grade in the greenstone belts away from intrusive granite contacts is low and original textures are frequently well preserved. A succession of more compositionally varied greenstone sequences, constituting the Belingwean, Bulawayan and Shamvaian Supergroups can be recognized

From COWARD, M. P. & RIES, A. C. (eds), 1995, *Early Precambrian Processes*, Geological Society Special Publication No. 95, pp. 109–126.

together with episodes of tonalitic–granodioritic plutonism. A flood of monzogranites heralded the final stabilization of the craton.

Attempts at rationalization of the greenstone belt stratigraphy of the Zimbabwe craton were begun by Macgregor (1947, 1951) and continued later by Wilson (Wilson 1979; Wilson et al. 1978, 1990). Using the two sequences of the Belingwe greenstone belt as a model, Wilson evaluated the lithostratigraphy of the greenstone belts of the southcentral region of the craton. Figure 1 summarizes some of the subdivision terminologies that have been used for this area and for ease of reference introduces a new numerical unit nomenclature, L1–L4 and U1–U6, based largely on the upper and lower sequences of formations found in the Belingwe belt. A detailed account of the Belingwe belt can be found in Bickle & Nisbet (1993). Our use of the terms Belingwean and Lower Bulawayan follows that of Wilson et al. (1990).

Previous geochronological work

Most age determinations on the Zimbabwe Archaean have involved Rb–Sr whole-rock isochron studies with some results from Pb–Pb and from Sm–Nd isochrons and Sm–Nd model ages. The latest and most comprehensive summary of geochronological data is in Taylor et al. (1991).

Most of the determinations are from granitoids and gneisses but there are some from the Upper Bulawayan volcanics and on these latter rocks errors on apparent ages by Rb–Sr are large. A more reliable age on the Upper Bulawayan is by Pb–Pb isochron on komatiitic basalts from the Reliance Formation on the east side of the Belingwe greenstone belt; this gives a figure of 2692 ± 9 Ma (Chauvel et al. 1993). There are no reliable results from rocks of the Belingwean or Lower Bulawayan and their presumed ages arise from their stratigraphic position based on the field evidence and from a consideration of the assumed synvolcanic granitoid plutonism. U–Pb dating on single zircons using SHRIMP (sensitive high resolution ion microprobe) has been done only on detrital grains in sediments from what are now considered to be the basal Upper Bulawayan at Shurugwi and probable Upper Belingwean in the Buchwa greenstone belt (Dodson et al. 1988; Tsomondo et al. 1992). These zircons have yielded abundance peaks at c. 3800, 3600, 3460, 3350 and 3200 Ma and emphasize the antiquity and complex history of the Tokwe segment.

In this paper the first SHRIMP-determined U–Pb ages on comagmatic zircons from a spectrum of late Archaean felsic (intermediate–silicic) volcanics that range geographically from near Bulawayo in the south to around Harare in the north

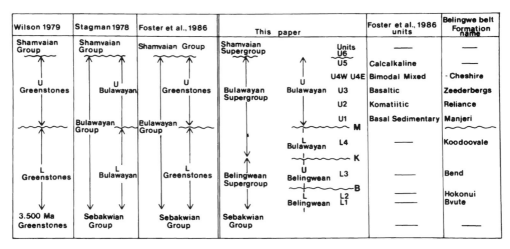

Fig 1. Stratigraphic nomenclature employed by various workers for the greenstone belts of Zimbabwe and applied in particular to the south-central part of the craton. In this paper the Supergroup-Group nomenclature follows that of Wilson et al. (1990); the Sebakwian Group is restricted to the Tokwe segment; the Belingwean, around and within the Shangani batholith, includes rocks previously assigned to the c. 3500 Ma greenstones/Sebakwian of Wilson (1979), Stagman (1978), Foster et al. (1986) and Bickle & Nisbet (1993). M, K and B designate Manjeri, Koodoovale and Bend unconformities, respectively. A new L–U numerical nomenclature is introduced for the Late Archaean greenstones. In the Belingwe belt the units L1–L4 and U1–U4E are respectively the Mtshingwe and Ngezi Groups of Bickle & Nisbet (1993) but, on the SE side of the Belingwe belt, the Brooklands Formation of Bickle and Nisbet's Mtshingwe Group includes rocks which are part of the Sebakwian Group.

and stratigraphically from the Lower Belingwean to the Upper Shamvaian are presented; Late Archaean greenstone stratigraphy and crustal growth in the light of these results and of recent ideas on the development of the craton are evaluated.

Late-Archaean granite–greenstone terrains

In the following, the main elements of the Late Archaean geology of the Zimbabwe craton and a new stratigraphy for the greenstone belts are presented.

Greenstones

In the south-central region, the Belingwean (L1–L3, Fig. 1) occurs west of the Tokwe segment on the western side of the Belingwe belt (Fig. 2). It is also found sporadically rimming, and within,

the polydomal granitoid–gneiss complex that constitutes Macgregor's (1951) Shangani batholith (Fig. 2). Major developments also occur at Filabusi and northwest of Gweru. On the eastern side of the Belingwe belt, where the belt abuts the western margin of the Tokwe segment, L1 and L3 occur with some lithofacies variations (L3) and the formations may also be present to the north in the Shurugwi area. The Bend (B) unconformity allows separation into an Upper and Lower Belingwean. The Lower Belingwean in its occurrence west of the Tokwe segment consists of the remains of a volcanic sequence dominantly mafic in its lower part (L1) passing upwards into a bimodal succession with mafic and felsic alternations on various scales (L2) and with minor BIF development. The Upper Belingwean is a mafic-ultramafic volcanic sequence (L3) characterized by persistent BIF interlayers.

A major break, the Koodoovale (K) unconformity, separates the overlying Bulawayan from

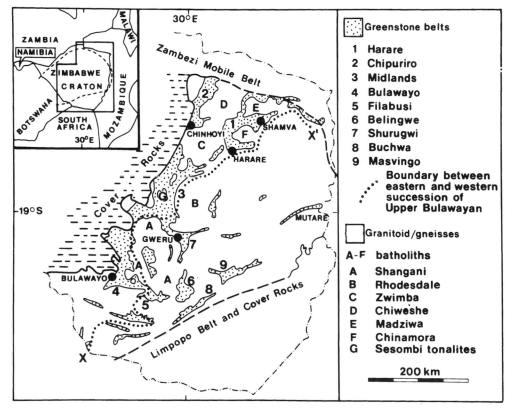

Fig. 2. Generalized map of the Zimbabwe greenstone belts. Numbers identify the different belts and letters give the names of the major batholiths. The boundary (X–X[1]) between the western and eastern successions of the Upper Bulawayan is after Wilson (1979) and Foster *et al.* (1986). Greenstone belt nomenclature is modified after Campbell *et al.* (1992) Inset shows the general position of Zimbabwe.

the Belingwean. The Lower Bulawayan (L4) is essentially a felsic unit with dacitic flows, some andesites and generally fine-grained volcaniclastic sediments. It occurs around most of the Shangani batholith and, where not in contact with Belingwean rocks, it is intruded by granitoids. In its development in the southwest part of the Belingwe belt, clastic sediments constitute a major component of L4; these include polymict conglomerates whose clasts contain a variety of the underlying Belingwean Supergroup lithologies and granitoids. A feature of L4 is the apparent lack of mafic volcanism and it represents one of the few events within the history of the craton where such volcanism does not occur.

The Upper Bulawayan (U1–5) is much more extensive. In the south-central region the lower-most unit (U1) is a laterally persistent development of clastic and chemical sediments with dominant BIF and minor, locally stromatolitic, limestones; it is generally thin, in places no more than a few tens of metres, but reaches up to 0.5–1.0 km in certain areas. Around the Shangani batholith it is at its thinnest and caps the Lower Bulawayan with no obvious structural break. On the western side of the Belingwe belt it has a unconformable relationship with the Belingwean and Lower Bulawayan whereas on the eastern side it overlies and transgresses Belingwean rocks to rest directly on the ancient gneisses and Sebakwian greenstones of the Tokwe segment. This Manjeri (M) unconformity is recognized at Shurugwi on older greenstones (Tsomondo et al. 1992) and at Masvingo on older granitoid and gneiss (Orpen & Wilson 1981). U1 is overlain by a largely BIF-free volcanic pile comprising a lower high-magnesium unit (U2), dominated by komatiitic basalts and komatiites, that passes upwards into a mafic unit (U3) of massive and pillowed flows with contemporaneous sills. The maximum thicknesses at Belingwe are about 1.5 km for U2 and 5 km for U3. This volcanic pile, with or without the development of U2, is present throughout the south-central region. Kusky & Kidd (1992) have suggested that this volcanic association in the Belingwe belt (the Reliance and Zeederbergs Formations, see Fig. 1) is allochthonous and is an obducted fragment of oceanic plateau. Following the views of Blenkinsop et al. (1993) and Bickle & Martin (1993), there is no field evidence for these suggestions.

Above U3 the eastern development of the Upper Bulawayan differs from that in the west. In the east, U3 is overlain by fine-grained clastic sediments with some intercalated BIF and locally developed stromatolitic-bearing limestones; these sediments are in turn overlain and in part interbedded with, further tholeiitic flows and the whole constitutes unit U4E. This appears to pass laterally into a western, bimodal mafic-felsic unit, U4W, which shows considerable shallow-water reworking of the felsic volcanic products as well as locally developed intercalations of fine-grained clastic and chemical sediments. A further major lithostratigraphic unit, which we designate U5, occurs in the western succession. It is dominated by andesites and related volcaniclastics but it also contains some dacites and, in some areas, basalts and rhyolites. Its exact positioning within the Upper Bulawayan stratigraphy is not altogether clear but, following Wilson (1979), it is placed above U4W in Fig. 1.

The final unit, U6, in the Late Archaean stratigraphy, is the locally occurring Shamvaian Supergroup. This is a predominantly sedimentary association with felsic volcanic intercalations in some areas. In the south-central region the major development is in the Masvingo greenstone belt.

Two further important igneous features of the south-central region are the several mafic dyke swarms that constitute the Mashaba-Chibi dykes; and the various dominantly ultramafic and composite sills that make up the Mashaba Ultramafic suite (Wilson 1979). These intrusions are here regarded as part of the feeder system of the Upper Bulawayan U2–U3 volcanic association with some of the sills perhaps representing the remains of open-system, high level magma chambers that fed the lavas. The dykes and sills help delineate the pre-Upper Bulawayan basement. For the most part, the dykes occur within the Tokwe segment but they also occur west of this segment where they cut the granitoids of the Chingezi suite in the type area (west of the Belingwe greenstone belt). To a large extent the sill occurrences are stratigraphically controlled with emplacement along the M unconformity and in the underlying L4. Within the Tokwe segment, faults related to the listric-rift system of the U1 basin may also have been a possible control.

Outside the south-central region, the recognition of Belingwean and Lower Bulawayan rocks is uncertain especially in the more highly deformed greenstone belts of the southwest of the craton where both might well be present. In the oldest gneisses of the Rhodesdale batholith, greenstone enclaves have been described (e.g. Harrison 1970; Robertson 1976; Bliss 1962) and these were originally assigned to the Sebakwian Group (i.e. pre-L1). These are now regarded by the present authors as Belingwean (see Wilson 1979).

The Upper Bulawayan, in contrast, is much more extensively developed and delineation of its western and eastern successions has been attempted across the craton (Fig. 2). In the Harare and Chipuriru greenstone belts, in the northern part of the craton in particular south of Shamva and south of Chinhoyi, sheared contacts occur between granitoids and greenstones (Stagman 1961; Stidolph 1977) which are interpreted as sheared unconformities between Upper Bulawayan green-stones and granitic basement. In the Shamva area the basement gneisses contain scattered remnants of older greenstones that may be Belingwean. On the basis of the stratigraphic sequence of the south-central region, the greenstone cover sequence, especially in the Harare belt, has many features that allow it to be equated with the U1–U2–U3–U4W of the Upper Bulawayan western succession and is unconformably overlain by a major develop-ment of the Shamvaian Supergroup U6. West of the Shamva area, this basal Upper Bulawayan un-conformity is underlain by a felsic unit (Iron Mask Formation) that rims much of the intrusive Chinamora batholith. This felsic unit has a marked development of interbedded BIF but otherwise is lithologically similar to the Lower Bulawayan, L4, subdivision of the south-central region. On the basis of the U-Pb zircon geochronology presented here a correlation with L4 now appears untenable (see below and also Jelsma 1993).

Granites

Wilson et al. (1978) and Wilson (1979, 1981) divided the Late Archaean granitoids into two loosely defined groupings which they called the Chilimanzi and Sesombi suites. The c. 2600 Ma Chilimanzi suite of monzogranitic sills marked the last major pre-Great Dyke granite event: the c. 2700 Ma, tonalitic–granodioritic Sesombi suite embraced a number of massive to slightly gneissic plutons intrusive into the Upper Bulawayan and occupying a NNE-trending zone, some 300 km long, on the west side of the south-central region of the craton. The parallelism of this zone to the trend of the Upper Bulawayan western succession, coupled with the rather limited chemical and Rb–Sr age data, led these authors to suggest that the Sesombi suite was a possible plutonic expression of the same major magmatism that produced the intermediate–silicic volcanism of the upper part of the western succession (our U4W and U5). An evaluation of subsequent field and laboratory studies allows the present authors to expand the Sesombi suite to include certain gneisses of the granite terrains and to extend the concept of synvolcanic plutonism to two further tonalitic–granodioritic groupings which are here termed the Chingezi and Wedza suites. The c. 2900–2800 Ma Chingezi suite, which includes the 'c. 2900 Ma terrain' of Wilson (1981) and Martin et al. (1993), also contains some development of discrete diorites and monzogranites: the c. 2700 Wedza Ma suite is even more complex.

The Chingezi and redefined Sesombi suites can be recognized in the south-central region from a study of the work of Baglow (1987), Bliss (1962), Cheshire et al. (1980), Garson (1991), Harrison (1969, 1970, 1981), Martin (1978), Martin et al. (1993), Orpen et al. (1986), Stowe (1968, 1979) and Taylor et al. (1991); and both suites can be recognized in the polymodal granitoid–gneiss complexes that constitute the Rhodesdale and Shangani batholiths. The type area of the Chingezi suite lies on the western and southern sides of the Belingwe greenstone belt as the Chingezi gneiss complex of Martin et al. (1993). Here, and to the west in the Filabusi area, folded, Belingwean greenstones and the Chingezi suite and granitoids intrusive into them, are cut by numerous Mashaba-Chibi dykes and a major sill of the Mashaba ultra-mafic suite.

The recognition of the Wedza suite arises from a study of the work of Anderson (1978), Baldock (1991), Fey (1976), Lauderdale (1988), Snowden (1975), Stidolph (1977) and Stocklmayer (1978, 1980). The suite is recognizable in a number of areas, in particular in the Chinamora and Madwiza batholiths, north of Harare, where it intrudes and deforms the western succession of the Upper Bulawayan of the Harare greenstone belt (Jelmsa et al. 1993). Its full extent is not known but it possibly underlies most of the northeast and eastern areas of the craton. Its relationship to the Sesombi suite is not clear but the two converge in the north of the exposed craton, west of Harare, in the Chiweshe and Zwimba batholiths.

The c.2600 Ma Chilimanzi suite is a major, craton-wide, multiphase, post-Shamvaian event. Its greatest development is in the eastern half of the craton as a number of areally extensive composite, monzogranitic sills that overlie, and intrude, a complex floor of Wedza suite.

In the context of the evolution of the Zimbabwean craton the Late Archaean granitoid suites are important crustal-forming events and mark specific stages in crustal development. With the exception of the Chilimanzi suite, the suites are here viewed as plutonic equivalents of, and thus broadly synchronous with, the intermediate–silicic volcanism which itself forms part of the cyclic nature of the greenstone magmatism. Thus the Chingezi suite is correlated with the felsic volcanism of the Belingwean and the Sesombi and Wedza suites with the felsic volcanism of the Upper Bulawayan. The relationship of the Lower Bulawayan (L4) felsic volcanism to the Chingezi suite is not clear. In tectonic terms the Chingezi, Sesombi and Wedza suites are syn- to post-tectonic relative to the doming-cum-diapiric induced deformation they have produced in their related greenstone cycles.

It is also pertinent to consider the various granite suites in relation to the granite-greenstone configuration of the craton. While earlier deformation affecting the Belingwean (the diapiric intrusion of the Chingezi suite) and Sebakwian greenstones contribute to this configuration, as does the framework of the pre-Upper Bulawayan basement, the gross greenstone pattern is that of the Upper Bulawayan with its western succession areally the most important. The structure and distribution pattern of these greenstone belts results from the Late Archaean 'main deformation'. At its simplest this deformation can be viewed in terms of two major stages: an early doming stage of vertical tectonics induced by the intrusion of the Sesombi and Wedza suites; and a later shear-zone stage of essentially horizontal tectonics involving zones of intense flattening and strike-slip movement that can be explained as resulting from a NNW–SSE-directed, regional compressive stress. The two stages overlapped in time and each was in itself diachronous. The Chilimanzi suite is post-tectonic to the doming stage and late- to post-tectonic to the shear-zone stage, especially in the south where the NNW–SSE-directed stress resulted in the thrusting of the North Marginal zone of the Limpopo belt, NNW onto the craton. The basins of the Shamvaian Supergroup were variously generated as syntectonic responses to both stages of the deformation. Wilson (1990) has proposed that the NNW–SSE stress field continued after the final cratonic stabilization to produce the fracture pattern that in part controlled the emplacement of the Great Dyke at *c.* 2460 Ma.

Rationale and methodology of zircon geochronology

The understanding of the Archaean geology of Zimbabwe has resulted largely from an on-going field-based study, particularly in the south-central part of the craton, of the stratigraphic relationships within, and between, each greenstone belt. Using this as a base, geochronological studies, particularly on the granitoid rocks has constructed an interpretable time-base. What has been missing is a detailed understanding of an internal time frame within the greenstone belts themselves. This arises from the difficulty of dating extensively altered rocks which make up the greenstone belts. Quoted errors in the published data from greenstone rocks (see the comprehensive review by Taylor *et al.* 1991), give ample proof of this difficulty.

The present study, although largely exploratory, demonstrates that the greenstone belts can be satisfactorily dated by using zircons from the extensive developments of felsic volcanic and related volcaniclastic rocks that are an integral part of greenstone stratigraphy.

About 2 kg of material was collected at each sample site and this was crushed, washed and sieved to < 360 μm

and the zircons separated by a combination of heavy liquids, magnetic separation and hand picking. Zircons were mounted in epoxy resin discs together with a zircon standard (SL13) This standard is concordant with an age of 572 Ma and has a radiogenic $^{206}Pb/^{238}U$ ratio of 0.0928. The techniques of SHRIMP analysis have been described elsewhere (e.g. Compston *et al.* 1984, 1986; Williams & Claesson 1987; Creaser & Fanning 1993). The SL13 standard zircon was analysed 79 times in the course of seven analytical sessions. The analyses consist of five scans through the mass range. The observed coefficient of variation in the Pb/U ratio measured for the SL13 standard over the seven sessions ranged from 3.18% to 1.73%. The observed value of 1.73% is below what is considered to be a reasonable uncertainty for the Pb/U ratio for this standard and an assumed value of 2.5% has been used for the relevant two sessions. Th/U has been calculated from ThO/UO using a conversion factor of 1.11. The 'unknowns' have been corrected for isobaric interferences at mass 204 and 207, the latter from $^{206}PbH^+$.

Results

Of the 19 felsic volcanic rocks crushed, 13 provided suitable zircon populations. A feature of the zircons was their small grain size (generally 50 μm or less) and their variability of morphology, colouring, style of zonation and degree of preservation. In general more than ten grains were analysed from each population (except for ZIM92/24 where only eight grains were recovered). Final results are given in Table 1, together with localities and stratigraphic position. The majority of the detailed descriptions of the analytical results are given elsewhere (Nesbitt *et al.* in prep.). Some of the more important data are presented in Fig. 3.

The range of data provided by the 13 samples varies fairly widely. However, in the majority of samples, some of the zircons were concordant and an acceptable age was obtainable by statistical analysis of discordia. More than half of the samples contained grains which were clearly older, and these inherited zircons provided an unexpected window into the development of the Zimbabwean Archaean crust.

Discussion

The results from the thirteen samples (Table 1) show that the Late Archaean volcanicity of the Zimbabwe craton took place during a time span of some 250 million years from about 2900 to 2650 Ma (Fig. 4). The ages fall into the following four groups; (i), 2900 Ma; (ii), 2830–2800 Ma; (iii) *c.* 2700 Ma, and (iv), *c.* 2650 Ma. These groupings largely confirm the stratigraphic sequence established for the south-central region but indicate that the extension of these correlations to the northern region is too simplistic.

Table 1. *Sample details and results.*

Sample	Map reference	Grid ref.	Formation name	Lithology	Unit BS	Unit AS	Crystallization age (Ma)	Inheritance age	Pb loss
ZIM92/2	Gatooma 1829 B4	QK931543	What Cheer[1]	Felsic clast in reworked volcanic breccia	U4W	U4W	2683 ± 8	2780	c.1750, c.600–500?
ZIM92/6	Sebakwe Poort 1829 D3	QK654154	Maliyami[2]	Flow banded siliceous andesite	U5	U5	2702 ± 6		
ZIM92/10	Torwood 1929 B1	QJ744739	Arizona[3]	Flow banded rhyodacite.	L4	L4	2805 ± 6		
ZIM92/12	Guinea Fowl 1929 D2	RJ114301	Surprise[11]	Deeply weathered felsic volcanic	L2	U6?	2698 ± 27	2820	
ZIM92/13	The World's View 2028 B3	PH588522	Avalon[4]	Rhyodacite from vent	U4W	U4W	2696 ± 7	2895	Present
ZIM92/16	Filabusi 2029 C2	QH547282	Eldorado[5]	Foliated felsic volcanic	L4	L4	2788 ± 10	2823	Present
ZIM92/17	Filabusi 2029 C2	QH534250	Eldorado[5]	Flaggy felsic gneiss	L4	L4	2799 ± 9	2850	Present
ZIM92/24	Mberengwa 2029 B4	QH991385	Hokonui[6]	Felsic clasts from volcanic breccia	L2	L2	2904 ± 9	2950	Present
ZIM92/27	Lake Kyle 2031 A1	TN974747	Upper Shamvaian[7]	Flow banded felsite in vent breccia	U6	U6	2661 ± 17	3620, 3530	c.1750, c. 530
ZIM92/29	Domboshawa 1731 C1	TR897481	Iron Mask[8]	Flow banded rhyolite	L4	H	2645 ± 4		
ZIM92/30	Bushu 1731 B1	UR432972	Maparu[9]	Water lain tuff	U4W	U4W	2679 ± 9	2720	900–800
ZIM92/31	Lake McIlwaine 1730 D4	TR878348	Passaford[8]	Felsic clasts in reworked tuff	U4W	H	2643 ± 8	2700	Present?
ZIM92/32	Belingwe Peak 2029 D2	RH070178	Koodoovale[10]	Dacite clasts in reworked volcanic breccia	L4	L4	2831 ± 6	2880	Present?

Map and grid references apply to the 1:50 000 topographical maps issued by the Department of the Surveyor General, Harare, Zimbabwe. Formation name references are: (1). Bliss 1970; (2). Harrison 1970; (3). Cheshire et al. 1980; (4.) Garson 1991; (5). Baglow 1987; (6). Martin 1978; (7). Wilson 1964; (8). Baldock 1991; (9). Stidolph 1977; (10). Orpen 1978 and Orpen et al. 1985; (11). Tyndale-Biscoe 1949. Columns labelled Unit BS and Unit AS refer to unit designation before (BS) and after (AS) SHRIMP dating.

Lower Belingwean and Lower Bulawayan

The first two of our groups (see above) delineate the Lower Belingwean and Lower Bulawayan, respectively. The oldest zircon date obtained is from one successful attempt at finding suitable zircons in the Lower Belingwean (L2) unit. Sample 92/24 is from the Hokonui Formation of the Belingwe belt and gives an age of 2904 ± 9 Ma.

Group 2 results from the Lower Bulawayan (L4) unit were much more satisfactory in that nearly all samples from this stratigraphic position provided zircon populations. All are significantly younger than L2. Sample 92/10 from the Arizona Formation, west of Gweru, and sample 92/17 from the Eldorado Formation at Filabusi, give 2805 ± 6 Ma and 2799 ± 9 Ma, respectively; statistically indistinguishable ages from two different greenstone belts. A second sample from the Eldorado Formation, (92/16), gives two populations at 2788 ± 10 and 2823 ± 11 Ma respectively. The younger age is here interpreted as the age of crystallization of the volcanic rock and the older, as the age of inherited xenocrysts. The final sample (92/32), from the Koodoovale Formation at

Belingwe, gives a significantly different age (2831 ± 6 Ma) which is the oldest recorded from this group. Taken overall, the results indicate two felsic events within L4, the first at *c.* 2830 and the second at *c.* 2800 Ma.

Upper Bulawayan

The Upper Bulawayan samples fall into the remaining two groups. The first at *c.* 2700 Ma is made up of 5 samples and the second (three samples) is about 2650 Ma.

Samples 92/13 (Avalon Formation, near Bulawayo), 92/2 (What Cheer Formation, in the Midlands greenstone belt) and 92/30 (Maparu Formation north of Shamva) are all from the U4W stratigraphic level. With ages of 2696 ± 7 Ma, 2683 ± 8 Ma and 2697 ± 9 Ma , they are statistically indistinguishable and confirm the wide geographical range of the Upper Bulawayan volcanism in the west and north of the craton (Fig. 5). In the same grouping is sample 92/6 from the Maliyami Formation (U5) in the Midlands area. This sample has an age of 2702 ± 6 Ma and is also indis-

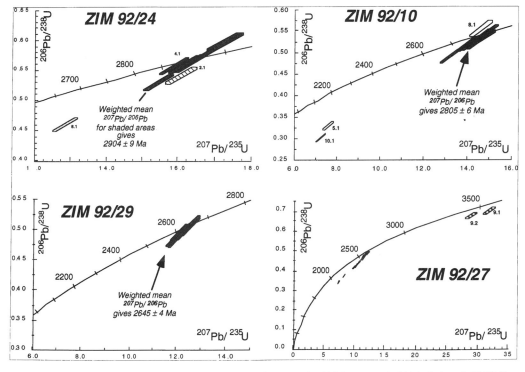

Fig. 3. Concordia plots for four zircon populations taken from samples 92/24 (Hokonui), 92/10 (Arizona), 92/29 (Iron Mask Formation) and 92/27 (Upper Shamvaian). Note the presence of inherited zircons in samples from the Upper Shamvaian (analysis spots 9.1 and 9.2) and Hokonui (analysis spot 2.1).

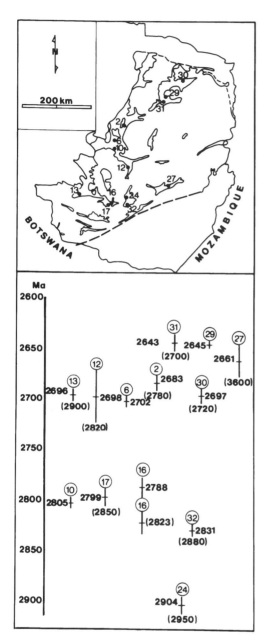

Fig. 4. Upper part shows the location of sampled felsic volcanics and lower part gives details of SHRIMP U–Pb zircon ages. Circled figures indicate sample location; the crystallization age of the sample, in Ma, is given immediately below. Lengths of the vertical lines are proportional to error (95% confidence level) and number in brackets indicate approximate age of xenocrystic zircons. Ages for inherited zircons are approximate since they are based on very few grains. See Table 1 for more detail.

tinguishable in age from the others in the group. This sample again highlights the problem of the position of U5 in the Upper Bulawayan stratigraphy (cf. Wilson 1979), because the present results group the Maliyami and U5 with the rest of Upper Bulawayan. In the Midlands area, the Maliyami Formation has been regarded as both older (Harrison 1970; Robertson 1976) and younger (Bliss 1962) than the U4W unit; whereas in the Bulawayo area and northwards its presumed extension lies near the centre of the greenstone belt suggesting it is younger (cf. Garson 1991; Macgregor 1928). The problem remains and the one result presented here does not solve it. Perhaps U4W unit passes laterally and upwards into U5 with the lower part of U5 interbedded with U4W.

Sample 92/30 from the Maparu Formation gives a discordant result with a lower intercept at 800 to 900 Ma. This may reflect Pan-African events in the Zambezi mobile belt to the north. Sample 92/2 (2683 ± 8 Ma) is not easy to interpret as it appears to have suffered lead loss twice, one at c. 1750 Ma and one possibly at 600–500 Ma. These are meaningful numbers in the context of the geological history of Zimbabwe and could be reflecting the c. 1800 Ma Magondi orogeny and Pan-African events, respectively.

Sample 92/12, from the Surprise Formation, west of Shurugwi, appears to belong to group 3. The only material available from this formation is heavily weathered and is extremely friable and it is therefore not surprising to find that the bulk of the zircon grains from this sample have lost significant lead. A discordia regression line fitted through eight grains gives an age of 2698 ± 27 Ma with two of the grains lying on concordia. These give an independent $^{207}Pb/^{206}Pb$ age of 2702 Ma. Interpretation of field data had previously led us to interpret this Formation as pre-Upper Bulawayan, probably equatable with L2 or at youngest, with L3. The zircon result giving an Upper Bulawayan age was quite unexpected and the Formation therefore lives up to its name. The area needs further field and geochronological investigation and at this stage the result remains unexplained. It can only be speculated that perhaps these volcanics are related to some kind of strike-slip, Shamvaian-type basin related to the nearby major shear zone known as the Surprise Fault.

The fourth grouping involves the two samples from near Harare: 92/29 from the Iron Mask Formation on the south side of the Chinamora batholith and 92/31 from the Passaford Formation from within the Harare city limits. These give statistically indistinguishable ages of 2645 ± 4 Ma and 2643 ± 8 Ma with 92/31 showing evidence of c. 2700 Ma inherited zircons. These are somewhat surprising results because for some years the Iron

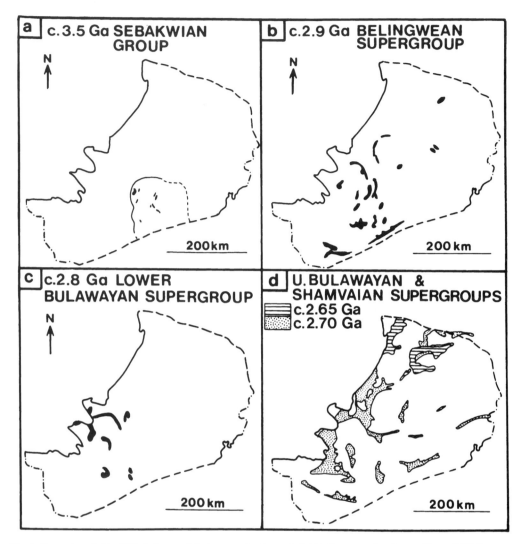

Fig. 5. Greenstone belts of Zimbabwe and their approximate areal distribution according to age. The *c*. 2.65 Ga Upper Bulawayan (d) represents the Harare sequence discussed in the text.

Mask Formation and the Upper Bulawayan, which unconformably overlies it, have been regarded as the northern time equivalents of the Lower Bulawayan (L4) and the Upper Bulawayan (U1–U4W) of the south central region (see Foster *et al*. 1986). The results presented here for samples 92/32, 92/29, 92/16 & 17 place L4 at about 2800 Ma and negates such a correlation for the Iron Mask Formation.

A further problem is the disparity between our results and those of Jelsma (1993). Using multigrain TIMS techniques, this author obtained ages of 2713 ± 20 Ma from the Iron Mask Formation on the northwest side of the Chinamora batholith;

2672 ± 12 Ma from a syn-tectonic porphyry (the Black Cat Porphyry) intruding rocks regarded as the lateral equivalents of the Passaford Formation, east of Shamva; and 2667 ± 4 Ma from a Wedza suite gneiss from the outer NE margin of the Chinamora batholith. Whilst these results support the idea that the Iron Mask is not L4 in age, they are significantly older than our *c*. 2650 Ma results from the Iron Mask Formation on the south side of the Chinamora batholith and from the Passaford Formation within Harare.

One possible explanation for the discrepancy is that samples analysed by Jelsma (1993) contained a significant percentage of older inherited zircons which would raise the overall age of the zircon population. If, however, this is not the case, then a geological explanation must be sought. The present authors see no reason to look upon the Passaford result as a metamorphic age, as suggested by Jelsma (1993), and, while an intrusive origin for 92/23 cannot be ruled out at this stage, there is also the possibility that, in spite of its apparent continuity around the outer margin of the western half of the Chinamora batholith, the Iron Mask Formation on the south side of the batholith is younger than lithologically similar material to the north. It is considered that the apparent young ages of samples 92/29 and 92/31, taken in conjunction with the age obtained from the Maparu Formation (92/30) and our assessment of the overall picture, indicate the presence of two greenstone sequences within the Upper Bulawayan of the Harare greenstone belt: one at $c.$ 2700 Ma and equating with the south-central region of the craton and one at $c.$ 2650. This interpretation would also imply that the unconformity south of Shamva and by inference the presumed unconformity west of Chinhoyi, is not the M unconformity of the south-central region but marks the likely base of this younger sequence. It is accepted that the age data are tantalizingly meagre and more work is needed. However, the field evidence in conjunction with the new SHRIMP data, suggest that a considerable area of younger sequence may exist in the northern part of the craton (Fig. 5). For the purposes of this paper this younger sequence is referred to as the Harare sequence and the underlying unconformity as the Harare, H, unconformity (Fig. 8). As well as Upper Bulawayan rocks, the Harare sequence includes rocks designated Shamvaian by previous workers (cf. Stidolph 1977; Stagman 1978).

Sample 92/27 from the Upper Shamvaian of the Masvingo area, with a discordant age of 2661 ± 17 Ma, is also here included in this fourth grouping. The result puts an older limit on the shear-zone stage of the main deformation in the Masvingo greenstone belt and on the NNW-directed thrusting of the North Marginal Zone of the Limpopo Belt onto the Zimbabwe craton. The Masvingo Shamvaian rocks were possibly deposited in a strike-slip basin related to early deformation during the shear zone stage. The lower intercept of 530 Ma is probably a meaningless age but the interpreted multi-stage Pb loss is in keeping the 2000–1800 Ma mineral ages from the Central Zone of the Limpopo belt and possible major movement on the Triangle shear zone at this time (cf. Wilson 1990).

Implications for crustal growth

Wilson *et al.* (1990) have suggested that the age and isotope data presented by Taylor *et al.* (1991),

especially their results involving Rb–Sr and Pb/Pb isochrons and Sm–Nd model ages on the same samples, are consistent with westward crustal growth with time in the south-central region. The data base is very limited but tonalite plutonism, (dated by Rb–Sr and/or Pb/Pb) youngs westwards from the $c.$ 3500 Ma gneisses of the Tokwe segment through the c. 2900–2800 Ma Chingezi suite in the southeastern area of the Shangani batholith, to the $c.$ 2700 Ma Sesombi suite. A similar trend is shown by the Sm–Nd depleted mantle model ages (T_{DM}) from a range of 3600–3400 Ma in the Tokwe segment to $c.$ 3000 Ma and $c.$ 2700 a in the Chingezi and Sesombi suites, respectively. In the Rhodesdale batholith to the north, the gneisses mapped as the oldest unit by Stowe (1979), and the Sesombi tonalite intruding the Midlands greenstone belt to the west, also show T_{DM} model ages which similarly conform to the Chingezi and Sesombi suites. If these model ages indicate the approximate mantle derivation ages of the granitoids, or their precursors, then the Chingezi and Sesombi suites can be regarded at least in part as successive new additions to the crust. A westward shift with time is also apparent in the geographical location of the presumed complementary volcanism with the remains of the Lower Belingwean, L2, occurring sporadically around the Shangani batholith but not extending on to the Tokwe segment; and the Upper Bulawayan western succession, U4W and U5, making its first appearance on the western side of the Shangani and Rhodesdale batholiths.

This apparent westward growth is accompanied by, and is perhaps also a reflection of, the diachronous stabilization of the craton. The Tokwe segment stabilized much earlier than the rest of the craton and over much of its extent it retains the NNE trend of the ancient Tokwe gneisses and the complexly infolded Sebakwian greenstone remnants. Progressive stabilization westwards is implied by the Mashaba–Chibi dykes which, by Upper Bulawayan times, appear undeformed, not only in the Tokwe segment but to the west where they are found cutting the Chingezi suite granitoids within the Shangani batholith. The lineament of the Tokwe segment's western margin seems also to have influenced the development of some of the Late Archaean supracrustal sequences. The Belingwean L3 unit shows a marked increase in clastic sediment input where it abuts the western margin of the segment on the eastern side of the Belingwe belt compared with its development elsewhere around the Shangani batholith. In addition L4 is absent from the east side of the Belingwe belt and has not been recognized in the Tokwe segment; and the basal sedimentary unit U1 of the Upper Bulawayan transgresses Belingwean and

Sebakwian rocks to rest directly on gneisses of the segment. Figure 6 attempts to illustrate these various features by means of a composite cartoon cross-section from the Tokwe segment across the Shangani batholith.

Inherited zircons

The model thus far developed for the Archaean greenstone belts of Zimbabwe suggests a series of repetitive cycles. In general a cycle consists of a series of ultramafic-mafic to mafic–felsic units which are associated with a series of granitoid plutons which themselves may be syn-volcanic to the felsic extrusions. The new zircon U–Pb data support such a model. The data also support the concept of westward crustal growth in the south-central region, (whatever might have been the methods of such growth) and shed some light on the question of the original westward extent of the Tokwe segment and the Chingezi–Belingwean granite–greenstone terrain by virtue of their content of inherited zircons. The data base is small but it is perhaps significant that the only sample yielding inheritance ages in the c. 3600 Ma range is 92/27 from the Upper Shamvaian of the Masvingo belt, a belt which is known to lie unconformably on the Tokwe segment (Orpen & Wilson 1981).

The lack of Tokwe segment related zircons in the felsic volcanic rocks found outside the Tokwe segment, raises the question of its original areal extent. Its eastern extent remains unknown. The lineament of the western margin of the Tokwe segment has influenced Archaean geological events from the lower Belingwean onwards, and even beyond to the early Proterozoic Great Dyke (Wilson 1990). If the Tokwe segment rocks extended westwards as a floor to the lowest Belingwean (L1), then it seems to have been as a thinner less stable mass that may have been obliterated by later plutonism.

Of the other samples from the south-central region that show evidence of inherited zircons, the oldest inherited age is c. 2950 Ma from sample 92/24 from the Hokonui Formation of the Belingwe belt. This is a discordant minimum result and however it is to be explained, it does not reflect the ancient rocks of the Tokwe segment. Sample 92/32 from the L4 unit at Belingwe, shows an inheritance age of c. 2880 Ma which is interpreted as reflecting the Lower Belingwean L2 volcanism and the Chingezi suite plutonism. Sample 92/2 from the U4W unit in the Midland areas shows a c. 2800 Ma age inheritance as does the enigmatic Surprise Formation of the Shurugwi belt. These all serve to illustrate the extent of the L4 igneous activity outside the now preserved outcrop areas. The zircon population extracted from sample 92/31 from the Passaford Formation of the Harare sequence shows a c. 2700 Ma inheritance age, some 50 Ma older than the estimated crystallization; these zircons presumably come from the

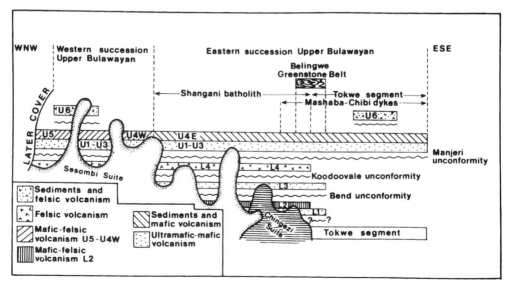

Fig. 6. Cartoon (not to scale) depicting an ESE–WSW schematic cross-section from the Tokwe segment across the Shangani batholith and beyond. Information from the northern areas of the Shangani batholith has been incorporated into the diagram. For the sake of clarity, the Chilimanzi suite has been omitted. The Manjeri, Koodoovale and Bend unconformities (M, K and B in Fig. 8) are discussed in the text, (see also Fig. 1).

underlying Upper Bulawayan cycle. The most westerly sample from our collection is 92/13 from the Avalon Formation (2696 Ma). This zircon population shows inherited zircons with an age of c. 2900 Ma, suggesting that L2 crustal material (Chingezi?) was sampled by the felsic magmatism.

While westward growth seems applicable to the south-central region, the northern and eastern parts of the craton present a different picture. The preservation of pre-Upper Bulawayan basement is minimal and nearly all the granite terrain is late or post-Bulawayan (Figs 5 & 7). Much of the north-eastern region south and east of Harare is devoid of

greenstone belts, which may be reflecting a lower crustal level, and consists of Chilimanzi suite sills overlying, and in many places obscuring, a complex floor of Wedza suite (Fig. 7). The felsic volcanics of the Upper Shamvaian of the Masvingo belt, 92/27 with an age of 2661 ± 17 Ma, are probably also related to the Wedza suite plutonism. Both suites occur in the four northern batholiths of Chinamora, Madziwa, Zwimba and Chiweshe. To what extent the Sesombi suite is present is not known but diapiric-induced deformation of the Harare greenstone belt involved the Wedza suite of the Chinamora and Wedza batholiths as a major

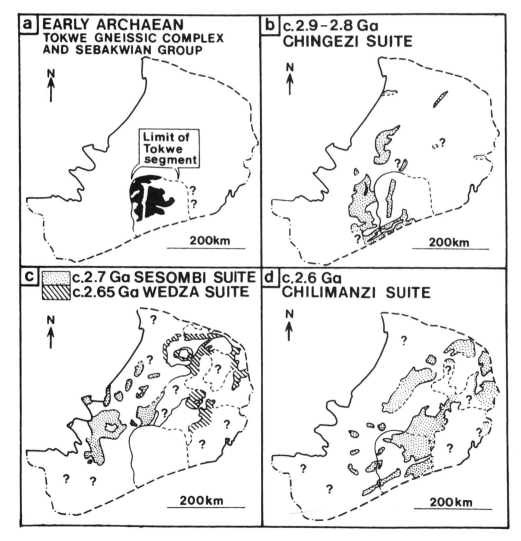

Fig. 7. Archaean granitoid suites and crustal growth in the Zimbabwe craton. (**a–d**) Approximate areal extent of each of the major granitoid suites. Note that the extreme SE and SW of the craton are largely unmapped and both regions probably contain extensive areas of Chilimanzi suite intrusions.

driving force (Jelsma *et al.* 1993). This deformation affects the *c.* 2650 Ma Harare sequence and suggests that the felsic volcanism of this sequence equates with the plutonism of the Wedza suite, and that the Wedza suite is thus younger than the Sesombi suite. The areal distribution of the Wedza suite in conjunction with the evidence from the south-central area suggests that outward growth from the Tokwe segment occurred not only west-wards but also northwards and eastwards (Fig. 7) with the intensity of intrusion of the Wedza suite largely obliterating and assimilating any earlier granitic terrain(s) in the eastern area.

Greenstone tectonic setting and origin of the granitoid suites

Numerous models for the tectonic setting of green-stone belts have been proposed and need not be elaborated on here. Currently popular models favour modern plate-tectonic analogues and, emphasizing oceanic affinities, invoke sea-floor spreading followed by subduction. More recently, some of the ultramafic–mafic greenstone belt sequences have been regarded as obducted oceanic plateaux (Windley 1993). In the Superior Province of the Canadian shield subduction-driven terrane accretion has been invoked to explain the juxta-positioning of granite–greenstone terrains and mega-sedimentary belts with the latter interpreted as accretionary prisms (Hoffman 1989; Card 1990).

The Upper Bulawayan greenstone belts of Zimbabwe are not readily explained by such models. Earlier suggestions that the Upper Bulawayan western succession represented rift-controlled magmatism that evolved into a mag-matic arc (Nisbet *et al.* 1981; Wilson 1981), were based on the apparent linear distribution of the magmatism and of the Sesombi suite plutonism. The recognition of two greenstone volcanic cycles in the Upper Bulawayan, the widespread Wedza suite and apparent crustal growth outward from the Tokwe segment during the Late Archaean, challenges this simple linear picture. The Upper Bulawayan greenstones are ensialic. They were deposited on an extensive basement of earlier granite–greenstone terrains that included the Tokwe segment, the Belingwean and Lower Bulawayan, and the Chingezi granitoid suite (Figs 5 & 7). There is no evidence of oceanic crust forming any part of either the *c.* 2.7 or *c.* 2.65 Ga greenstone cycles. It is considered here they are best explained as having been deposited on continental crust undergoing extension and limited rifting above a mantle plume. It is argued that partial melting of mafic lower crust would produce the felsic

volcanism and related Sesombi and Wedza suite plutonism, with this lower crust formed in part by underplating by the basic magmas of these and earlier greenstone cycles. The presence of more evolved monzogranitic intrusions in the Wedza suite implies partial melting of a more sialic crust.

The Belingwean rocks, although less well-preserved, have lithological volcanic associations similar to the Upper Bulawayan. While an ensialic origin is evident for the Bulawayan, such a con-clusion is not so evident for the Belingwean. The Lower Belingwean consists of the remains of a greenstone cycle similar to that of the Upper Bulawayan whilst Upper Belingwean rocks are represented by the remains of the lower ultra-mafic–mafic part of a second cycle. In the case of this cycle, the mafic–felsic part was either not developed or has been lost to erosion and the present authors therefore have no zircon data from it. The related Chingezi suite plutonism is little understood and its age is not well constrained but an origin similar to that of the Sesombi suite is possible, although more evolved granites also occur. However, while field evidence indicates that some of the Belingwean rocks were deposited on the Tokwe segment to the west of the segment in the Belingwe and Filabusi belts, the base of the mafic lower part (L1) of the first cycle is not seen and the unit is intruded by Chingezi suite gneisses. This combined with the lack of Tokwe segment zircon inheritance leaves open the possibility, at this stage, of an oceanic crustal origin for unit L1 outside the Tokwe segment.

The Lower Bulawayan L4 unit and its relation-ship, if any, to the Chingezi suite still presents a problem. The zircon ages presented here indicate two events, at *c.* 2.83 and *c.* 2.8 Ga, respectively. These two ages could explain the discrepancy between the occurrence of an obvious M-uncon-formity between L4 and U1 in the SW of the Belingwe belt and the lack of such an obvious tectonic break at this contact around the Shangani batholith to the west. It can only be speculated on at this stage. Perhaps the Belingwe belt southern development of L4, with its polymict con-glomerates and felsic volcanism, represents a late Belingwean, Shamvaian-like response to the Chingezi plutonism; and the younger *c.* 2.8 Ga event a volcanic expression of later, post-tectonic intrusions of the Chingezi suite. Whatever the explanation, a near 100 Ma of volcanic and tectonic inactivity appears to separate the L4 unit from the U1 basal sediments of the first greenstone cycle in the overlying Upper Bulawayan.

The final granite event in the Late Archaean history is the craton-wide emplacement of the *c.* 2.6 Ga Chilimanzi suite monzogranites; no surface manifestations of these intrusions have been recognized. Their magmas represent large scale melting of lower sialic crust but how this occurred, and for what reasons, is still unknown; perhaps it is a function of crustal thickening of the craton during the regional shear-zone stage of the main deformation. Their occurrence as sills indicates widespread stabilization of the craton by c. 2.6 Ga. Figure 8 summarises the major events in the Zimbabwe craton from the Late Archaean up to the Early Proterozoic Great Dyke.

Summary and conclusions

The Late Archaean greenstone terrains of the Zimbabwe craton constitute the Belingwean, the Bulawayan and Shamvaian Supergroups. SHRIMP U–Pb ages on co-magmatic zircons from their felsic volcanics show a time span of more than 250 Ma. The Belingwean and Bulawayan, each contains a number of greenstone volcanic cycles. These are ultramafic–mafic in their lower parts and mafic–felsic in their upper parts and can be coupled to episodes of syn-volcanic to late-tectonic, tonalite–granodioritic plutonism. Taken together these emphasize the role of the granite–greenstone pair in the development of Archaean granite–greenstone terrains.

The Belingwean Supergroup contains the remains of a c. 2.9 Ma cycle (L1, L2) and the ultramafic–mafic lower part of a second cycle (L3, undated). The two cycles relate to the Chingezi suite granitoid intrusions dated around 2.9–2.8 Ga (Taylor et al. 1991). The Lower Bulawayan (L4) contains a major development of felsic volcanics, apparently devoid of mafic rocks, and shows evidence of two events, at c. 2.83 and c. 2.8 Ga respectively. L4 rests with marked unconformity on deformed Belingwean rocks and may represent

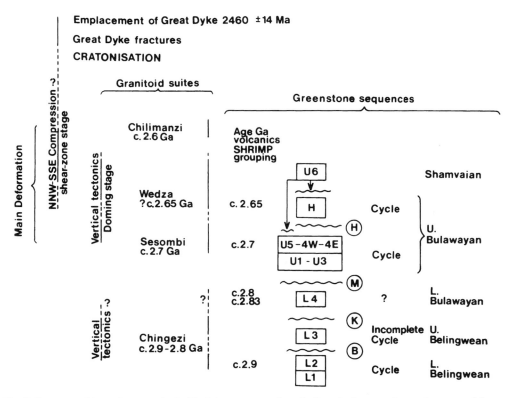

Fig. 8. Summary of the main events in the Zimbabwean craton from the Late Archaean to the emplacement of the Early Proterozoic Great Dyke. Ringed letters B, K, M and H represent Bend, Koodoovale, Manjeri, and Harare unconformities, respectively. H in the stratigraphic column represents the Harare sequence. A cycle indicates the full development of a lower ultramafic-mafic sequence which passes upwards into a mafic–felsic sequence. The felsic volcanics are dominantly dacites but may also include andesites and rhyolites. Unit U5 is andesite dominant. The Shamvaian Supergroup occurs unconformably on the Harare sequence and also on the eastern and western successions of the c. 2.7 Ga Upper Bulawayan cycle. In the Masvingo greenstone belt, felsic volcanics from the Upper Shamvaian give an age of 2661 ± 17 Ma.

the surface expression of late- and post-tectonic Chingezi suite plutonism. The Upper Bulawayan shows two further cycles: the older and most widespread at *c.* 2.7 Ga and the younger at *c.* 2.65 Ga developed in the northern part of the craton. These relate respectively to the *c.* 2.7 Ma Sesombi and presumed *c.* 2.65 Ma Wedza granitoid suites.

Field evidence (Blenkinsop *et al.* 1993) has shown that Upper Bulawayan greenstone sequences were deposited in a limited rifting environment, on an extensive sialic basement that existed by *c.* 2.8 Ga. Xenocrystic zircons found in these and almost all other felsic volcanic sequences support an ensialic model for the majority of the greenstone belts. Of the 13 samples examined, 9 contain inherited (xenocrystic) zircons ranging in age from 20 Ma to 1 Ga older than the host felsic volcanic. Even the oldest volcanic sampled (L2, Upper Belingwean) contains an inherited zircon some 50 Ma years older than the rock itself. These data indicate the presence of a pre-existing sialic basement beneath nearly all of the felsic rocks sampled with the oldest inheritance age (almost 1 Ga) being

found in a felsic volcanic which erupted through the Tokwe segment. Since such old inheritance ages have not been found in any other felsic rock sampled, we suggested that the >3.6 Ga Tokwe segment (which itself represents a cratonized ancient greenstone cycle or cycles) was confined to a small region in the south of Zimbabwe. Its absence elsewhere leads to the conclusion that at that time, oceanic crust must have underlain the bulk of what is now the Zimbabwe craton. On this oceanic crust, granite–greenstone cycles developed which although   not at present exposed were certainly sampled by later felsic magmatic activity.

This research was supported by a grant to R.W.N. from the Natural Environmental Research Council (GR9/609). Zircon separations were carried out at the ANU by R. Rudowski. A. Dunkley drafted the diagrams. K. Walsh is thanked for his help in sample collecting. J.F.W. acknowledges research support from the University of Zimbabwe. This work forms part of the ZAPIT (Zimbabwe Archaean Proterozoic Investigation Team) programme of the Geology Department of the University of Zimbabwe.

References

ANDERSON, C. B. 1978. *The Geology of the East Charter Area.* Rhodesia Geological Survey Bulletin, 84.

BAGLOW, N. 1987. *Filabusi geological map.* To accompany Zimbabwe Geological Survey Bulletin, **81**, in preparation.

BALDOCK, J. W. 1991. *The Geology of the Harare Greenstone Belt and Surrounding Granitic Terrain.* Zimbabwe Geological Survey Bulletin, 94.

BICKLE, M. J. & NISBET, E. G. 1993. (eds) *The Geology of the Belingwe Greenstone Belt, Zimbabwe.* Geological Society of Zimbabwe Special Publications, 2.

——, —— & MARTIN, A. 1993. Archaean greenstone belts are not oceanic crust. *Journal of Geology,* **102**, 121-138.

BLENKINSOP, T. G., FEDO, C. M., BICKLE, M. J., ERIKSSON, K. A., MARTIN, A., NISBET, E. G. & WILSON, J. F. 1993. Ensialic origin for the Ngezi Group, Belingwe Greenstone belt. *Geology,* **21**, 1135-1138

BLISS, N. W. 1962. *The Geology of the Country around Umvuma and Felixburg.* Southern Rhodesia Geological Survey Bulletin, 56.

—— 1970. *The Geology of the Country around Gatooma.* Rhodesia Geological Survey Bulletin, 64.

CAMPBELL, S. D. G., OSTERLEN, P. M., BLENKINSOP, T. G., PITFIELD, P. E. J. & MUNYANYIWA, H. 1992. A provisional 1 : 250000 Tectonic Map and the Tectonic Evolution of Zimbabwe. *In:* NCUBE, S. M. N., (ed.) Zimbabwe Geological Survey Annals, **16** (for 1991), 52–69.

CARD, K. D. 1990. A Review of the Superior Province of the Canadian Shield: a product of Archaean accretion. *Precambrian Research,* **48**, 99–156.

CHAUVEL, G., DUPRE, B., & ARNDT, N. T. 1993. Pb and

Nd isotopic correlation in Belingwe komatiites and basalts. *In:* BICKLE, M. J. & NISBET, E. G. (eds) *The Geology of the Belingwe greenstone belt, Zimbabwe.* Geological Society of Zimbabwe Special Publications, **2**, 167-174.

CHESHIRE, P. E., LEACH, A. & MILNER, S.A. 1980. *The Geology of the Country South and East of Redcliff.* Zimbabwe Geological Survey Bulletin, 86.

COMPSTON, W., WILLIAMS, I. S. & MEYER, C. 1984 U-Pb geochronology of zircons from lunar breccia 73217 using a high sensitive high mass-resolution ion microprobe. *Journal of Geophysical Research,* **89**, supplement, B525–534

——, KINNY, P. D., WILLIAMS, I. S. & FOSTER, J. 1986. The age and Pb loss behaviour of zircons from the Isua supracrustal belt as determined by ion-microprobe. *Earth and Planetary. Science Letters,* **80**, 71–81.

CREASER, R. A. & FANNING, C. M., 1993. A U-Pb zircon study of the Mesoproterozoic Charleston Granite, Gawler Craton, South Australia. *Australian Journal of Earth Sciences,* **40**, 519–526

DODSON, M. H., COMPSTON, W., WILLIAMS, I. S. & WILSON, J. F. 1988. A search for ancient detrital zircons in Zimbabwean sediments. *Journal of the Geological Society, London,* **145**, 977–983.

FEY, P. 1976. *The Geology of the Country South and East of Wedza.* Rhodesia Geological Survey Bulletin, 77.

FOSTER, R. P., MANN, A. G., STOWE, G. W. & WILSON, J. F. 1986. Archaean Gold Mineralisation in Zimbabwe. *In:* ANHAEUSSER, G. A. & MASKE, S. (eds) *Mineral Deposits of Southern Africa,* **1**. Geological Society of South Africa, 43-112.

GARSON, M. S. 1991. *Bulawayo geological map.* To

accompany Zimbabwe Geological Survey Bulletin, **93**, in press.

HARRISON, N. M. 1970. *The Geology of the Country around Que.* Rhodesia Geological Survey Bulletin, **67**.

HOFFMAN, P. F. 1989. Precambrian geology and tectonic history of North America. *In*: BALLY, A. W. & PALMER, P. (eds) *The Geology of North America*, **A**, *The geology of North America – an overview*, Geological Society of America, 447–512.

JELSMA, H. A. 1993. *Granites and greenstones in northern Zimbabwe tectono-thermal evolution and source regions.* PhD thesis Free University of Amsterdam, The Netherlands.

——, VAN DEN SEEK, P. A. & VINYU, M. L. 1993. Tectonic evolution of the Bindura-Shamva greenstone belt (northern Zimbabwe): progressive deformation around diapiric batholiths. *Journal of Structural Geology*, **15**, 163–176.

KUSKY, T. M. & KIDD, W. S. F. 1992. Remnants of an Archean oceanic plateau, Belingwe greenstone belt, Zimbabwe. *Geology*, **20**, 43–46

LAUDERDALE, J. N. 1988. *Dorowa-Shawa geological map.* To accompany Zimbabwe Geological Survey Bulletin, **75**, in press.

MACGREGOR, A. M. 1928. *The Geology of the Country around the Lonely Mine, Bubi district.* Southern Rhodesia Geological Survey Bulletin, **11**.

—— 1947. *An Outline of the Geological History of Southern Rhodesia.* Southern Rhodesia Geological Survey Bulletin, 38.

—— 1951. Some Milestones in the Precambrian of Southern Rhodesia. *Proceedings of the Geological Society of South Africa*, **54**, 27–71.

MARTIN, A. 1978. *The Geology of the Belingwe–Shabani Schist Belt.* Rhodesia Geological Survey Bulletin, 83.

——, NISBET, E. G., BICKLE, M. J. & ORPEN, J. L. 1993. Rock units and stratigraphy of the Belingwe Greenstone Belt: The complexity of the tectonic setting. *In*: BICKLE, M. J. & NISBET, E. G. (eds) *The Geology of the Belingwe greenstone belt, Zimbabwe.* Geological Society of Zimbabwe Special Publications, **2**, 13–37.

NISBET, E. G., WILSON, J. F. & BICKLE, M. J., 1981 The evolution of the Rhodesian Craton and adjacent Archaean terrain: tectonic models. *In*: KRÖNER, A. (ed.) *Precambrian Plate Tectonics.* Elsevier, Amsterdam, 161–183.

ORPEN, J. L. 1978. *The Geology of the southwestern part of the Belingwe greenstone belt – the Belingwe Peak area.* Dphil thesis, University of Rhodesia, Harare, Zimbabwe.

—— & WILSON, J. F. 1981. Stromatolites at c. 3500 Myr and a greenstone-granite unconformity in the Zimbabwean Archaean. *Nature*, **291**, 218-220.

—— BICKLE, M. J., NISBET, E. G. & MARTIN, A. 1986. *Belingwe Peak geological map.* To accompany Zimbabwe Geological Survey Short Report, **51**, in press.

ROBERTSON, I. D. M. 1976. *The Geology of the Country around Battlefields.* Rhodesia Geological Survey Bulletin, **76**.

SNOWDEN, P. A. 1976. *The geology of the granitic terrain north and east of Salisbury with particular reference to the Chinamora batholith.* Dphil thesis, University of Rhodesia, Harare, Zimbabwe.

STAGMAN, J. G. 1961. *The Geology of the Country around Sinoia and Banket, Lomagundi District.* Southern Rhodesia Geological Survey Bulletin, **49**.

—— 1978. *An Outline of the Geological History of Rhodesia. Rhodesia Geological Survey Bulletin*, **80**.

STIDOLPH, P. A. 1977. *The Geology of the Country around Shamva.* Southern Rhodesia Geological Survey Bulletin, **78**.

STOCKLMAYER, V. R. 1978. *The Geology of the Country around Inyanga.* Rhodesia Geological Survey Bulletin, **79**.

—— 1980. *The Geology of the Inyanga–North-Makaha Area.* Zimbabwe Geological Survey Bulletin, **89**.

STOWE, C. W. 1979. A sequence of plutons in the central portion of the Rhodesdale granitic terrane. *Transactions of the Geological Society of South Africa*, **82**, 277–285.

TAYLOR, P. N., KRAMERS, J. D., MOORBATH, S., WILSON, J. F., ORPEN, J. L. & MARTIN, A. 1991. Pb/Pb, Sm–Nd and Rb–Sr geochronology in the Archaean Craton of Zimbabwe. *Chemical Geology (Isotope Geoscience Section)*, **87**, 175–196.

TSOMONDO, J. M., WILSON, J. F. & BLENKINSOP, T. G. 1992. Reassessment of the structure and stratigraphy of the early Archaean Selukwe Nappe, Zimbabwe. *In*: GLOVER, J. E. & HO, S. E. (eds) *The Archaean: Terrains, Processes and Metallogeny.* Geology Department (Key Centre) & University Extension, The University of Western Australia, **22**, 123–135.

TYNDALE-BISCOE, R. 1949. *The Geology of the Country around Gwelo.* Southern Rhodesia Geological Survey Bulletin, **87**.

WILLIAMS, I. S. & CLAESSON, S. 1987. Isotopic evidence for the Precambrian provenance and Caledonian metamorphism of high grade paragneisses from the Seve Nappes Scandinavian Caledonides. II Ion microprobe zircon U-Th-Pb. *Contributions to Mineralogy and Petrology*, **97**, 205–217.

WILSON, J. F. 1964. *The Geology of the Country around Fort Victoria.* Southern Rhodesia Geological Survey Bulletin, **58**.

—— 1979. A Preliminary Reappraisal of the Rhodesian Basement Complex. *In*: ANHAEUSSER, C. R., FOSTER, R. P. & STRETTON, T. (eds) *A symposium on Mineral Deposits and Transportation and Deposition of Metals.* Geological Society of South Africa, Special Publications, **5**, 1–23.

—— 1981. The granite-gneiss greenstone shield, Zimbabwe. *In*: HUNTER, D. H. (ed.) *Precambrian of the Southern Hemisphere.* Elsevier, Amsterdam, 454–488

—— 1990. A craton and its cracks: some of the behaviour of the Zimbabwe block from the late Archaean to the Mesozoic in response to horizontal movements and the significance of some of its mafic dyke fracture patterns. *Journal of African Earth Sciences,* **10**, 483–501.

——, BICKLE, M. J., HAWKESWORTH, C. J., MARTIN, A., NISBET, E. G. ORPEN, J. L. 1978. The granite-greenstone terrains of the Rhodesian Archaean craton. *Nature,* **271**, 23–27.

——, BAGLOW, N., ORPEN, J. L. & TSOMONDO, J. M. 1990. A reassessment of some regional correlations of greenstone-belt rocks in Zimbabwe and their significance in the development of the Archaean craton. *Extended Abstracts: Third International Archaean Symposium, Perth 1990.* Geoconferences, Perth, 43–44.

WINDLEY, B. F. 1993. Uniformitarianism today: plate tectonics is the key to the past. *Journal of the Geological Society, London,* **150**, 7–19.

Paradigms for the Pilbara

ALEC F. TRENDALL

Geological Survey of Western Australia, Mineral House, 100 Plain Street,
East Perth, Western Australia, 6004

Abstract: The Pilbara Craton is a discrete and coherent body of well preserved and exposed early continental crust in the northwestern part of the Australian continent, with a present surface area of about 180 000 km². It has two tectonostratigraphic components: an older 'basement' of granite–greenstone terrane, formed between about 3.5 and 2.9 Ga, unconformably overlain by a supracrustal sequence, the Mount Bruce Supergroup, laid down between about 2.8 and 2.4 Ga in the Hamersley Basin. The main body of the craton has retained its integrity since that time, and has undergone only regional uplift, mild metamorphism, and gentle folding. A tendency to interpret the geology of the craton in terms of paradigms, involving a simplistic comparison with other areas thought to be better understood, has impeded the development of independent hypotheses of crustal evolution based on objective integration of observational evidence from within the Pilbara craton itself.

Geoscientific documentation and discussion of the Pilbara Craton, one of the best exposed areas of ancient continental crust on Earth, is largely restricted to the Australian literature. The primary purpose of this contribution is to attract wider attention to this still poorly understood tectonic unit by reviewing present scientific knowledge of it. A brief outline of the geology is followed by a summary of the published work on which that account is based, evaluated in terms of the various sub-disciplines of geoscience. Finally, the main paradigms which have been used in the interpretation of the Pilbara Craton are noted, and selected examples are given of their adverse effects. The paper closes with some suggestions for future studies.

Definitions: Pilbara, craton, paradigm

The name 'Pilbarra', derived from a local Aboriginal language, was applied to an early gold-mining centre about 100 km SSW of Port Hedland. Although a thriving town in the 1880s, all that can be seen today among the spinifex grass are traces of the town well and rubbish tip, as it was abandoned when prospectors' interest moved south after gold strikes at Coolgardie (1892) and Kalgoorlie (1893). The name 'the Pilbara' later came to be used as a general name for the far northwestern part of Western Australia. 'Pilbara Craton' was first used as a tectonic term by Gee (1979), and later became the recommended usage of the Geological Survey of Western Australia (GSWA 1990), replacing and extending the term 'Pilbara Block' (GSWA 1975).

Bates & Jackson (1987) define a craton loosely as a part of the Earth's crust that has attained stability, and has been little deformed for a long period. Current usage in Western Australia (GSWA 1990, p. 4) requires that stability was attained before 2.4 Ga. The Pilbara Craton thus includes both a 'basement' of granite–greenstone terrane as well as relatively mildly deformed supracrustal rocks, the Mount Bruce Supergroup, which unconformably overlie it. This is consistent with African usage in respect to the Kaapvaal Craton, which is generally used to include the supracrustal successions of the Witwatersrand Triad and Transvaal Supergroup.

The term 'paradigm' (Ancient Greek παραδειγμα — pattern) is used here to mean a set of beliefs that enables geologists to convince themselves that their areas of study so closely resemble some other area, or areas, generally accepted to be well (or better, or even completely) understood that they need not seek independent solutions to interpretational problems that arise within their own data; the 'pattern' etymologically implicit in the word paradigm thus refers here primarily to a pattern of thought. The application of paradigmatic thinking to particular sets of geological phenomena is often linked with the use of simple labels, and it is convenient to refer also to the resultant interpretations as paradigms. Thus the 'X' paradigm refers to the interpretation and labelling (as 'X') of some locally studied set of geological phenomena on the basis that they are elsewhere reliably identified and associated as 'X', usually with the implication that the geological interpretation of 'X' is unequivocally established.

Outline of geology

The Pilbara Craton has an area of some 180 000 km², and is the smaller of the two main

From COWARD, M. P. & RIES, A. C. (eds), 1995, *Early Precambrian Processes*,
Geological Society Special Publication No. 95, pp. 127–142.

areas of Early Precambrian crust in the Australian continent (Fig. 1). Figure 2 shows the simplified geology of the craton. Although a conceptual craton edge is shown on that map the structure, precise positions, and evolution of its margins are not well known. Geophysical evidence, noted later, provides some evidence, particularly along the southern edge.

The craton has two components: an older 'basement' of granite–greenstone terrane (GGT) and an unconformably overlying supracrustal sequence formally named the Mount Bruce Supergroup. The largest outcrop area of GGT extends across the northern part of the craton and is called the north Pilbara GGT (GSWA 1990), but scattered inliers also occur in the south. The main outcrop of the Mount Bruce Supergroup covers most of the southern two-thirds of the craton, but elongate outcrops extend northwards at both the eastern and western ends; there are important outliers on the north Pilbara GGT, the largest of which lies immediately east of Marble Bar (Fig. 2). While GGT may reasonably be assumed to be present at depth over the entire area of the craton, it is more doubtful whether rocks of the Mount Bruce Supergroup covered the whole craton before subsequent erosion reduced their outcrop area to its present extent of about 100 000 km^2.

Within the GGT, sinuous, crudely synformal, volcanosedimentary greenstone belts enclose granitoid domes of generally rounded form. In the formally defined stratigraphic succession of the greenstone belts, for which a total thickness of

Fig. 1. The location of the Pilbara craton within the Australian continent, and the comparative size of the British Isles. Frame shows outline of Fig. 2a.

some 30 km is claimed, the basal Warrawoona Group, consisting mainly of mafic volcanic rocks with subordinate felsic units, is successively overlain by the younger Gorge Creek, De Grey, and Whim Creek Groups; these are characterized by a greater proportion of sandstones and other epiclastic rocks than the Warrawoona Group, although mafic and felsic volcanic rocks are also present. Correlation of the various units between adjacent belts, and more particularly between the eastern and western parts of the north Pilbara GGT, is controversial, as noted later. Metamorphism of the greenstone belts rarely exceeds greenstone facies, and primary textures are often well-preserved in both volcanic and sedimentary rocks. The intensity of folding is variable; steep dips and tight folds with associated strong foliation generally occur where greenstone belts are narrow, with lower dips and open folding of undeformed rocks in broader belts.

The intervening granitoid domes, for which the name 'granitoid complex' is now preferred to 'batholith', generally have tectonic margins against the greenstones, and there is little evidence of contact metamorphism. They contain a wide range of compositional types, but are mainly granodioritic or monzogranitic. The more closely studied complexes have all been shown to consist of numerous component plutons, often interleaved with swirled mafic xenolith sheets. A substantial body of geochronological results, from Rb–Sr, Pb–Pb, Sm–Nd and zircon U–Pb methods (both multi- and single-grain) indicates that the volcanosedimentary rocks of the greenstone belts range in age from c. 3.5+ to c. 2.9 Ga, and that ages in the intervening granitoid complexes extend over the same time span.

The northern boundary of the main southern outcrop area of the Mount Bruce Supergroup is an erosional unconformity over the north Pilbara GGT. In the northern part of the outcrop area strata of the supergroup have gentle dips, but towards the southern outcrop margin folding increases in intensity, to become tight and complex against the normally tectonic contact with the adjacent Capricorn Orogen; greater tectonic complexity also exists along the eastern margin. Metamorphism rarely rises above greenschist facies. The supergroup has three major subdivisions, which are, in order of age, the Fortescue, Hamersley, and Turee Creek Groups. The Fortescue Group has a maximum thickness of c. 7 km, and consists predominantly of superimposed basalt flows, mainly subaerial but usually subaqueous in the south, with subordinate local silicic volcanics as well as units of tuffaceous, epiclastic, and carbonate rocks, the latter often stromatolitic. The uppermost Jeerinah Formation includes a persistent black shale. The

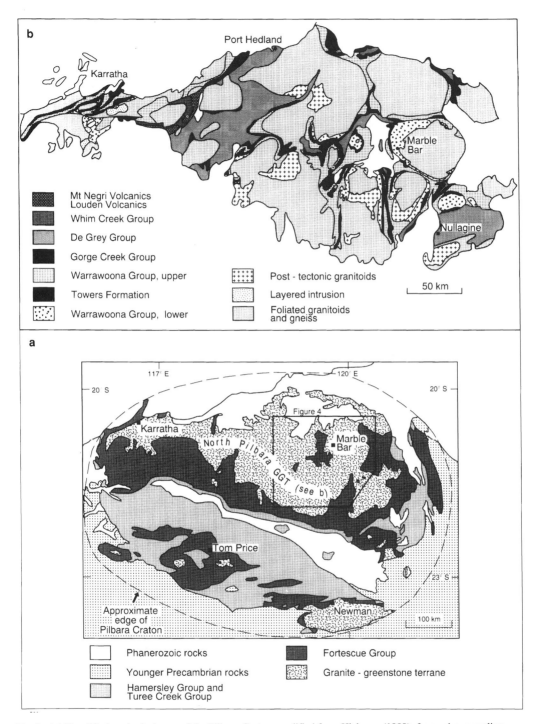

Fig. 2. (**a**) Simplified geological map of the Pilbara Craton, modified from Hickman (1980); frame shows outline of Fig. 3. The line marking the approximate edge of the craton is an ovoid envelope enclosing the present outcrop of craton components; although its course is consistent with regional geophysical patterns the craton boundary is nowhere either exposed or precisely defined. (**b**) Simplified geological map of the north Pilbara granite–greenstone terrane, modified from the same source.

2.5 km thick Hamersley Group, which overlies the Fortescue Group with almost perfect conformity, is characterized by thick units of banded iron formation (BIF), in which lateral persistence of fine-scale stratigraphy has been demonstrated. As in the underlying Fortescue Group, mafic sills are present at some levels, with local variations in abundance. A rhyolite unit in the upper part of the Hamersley Group is exceptionally voluminous, and is interpretationally challenging (Trendall 1992). The uppermost Turee Creek Group, which consists mainly of epiclastic rocks, shows extreme lateral thickness variations, and its discontinuous outcrop areas are confined to the southwest. Fortescue Group volcanism began at c. 2.8 Ga, and had virtually ceased at c. 2.7 Ga. Sedimentation in the depository of the Mount Bruce Supergroup (the Hamersley basin) probably ended at about 2.4 Ga.

The Pilbara Craton is transected by a number of mafic dyke swarms, both older and younger than the Mount Bruce Supergroup. Their ages are not well understood. North-northeast dykes of the Black Range suite, in the eastern part of the north Pilbara GGT, may be feeders for some volcanic rocks of the Fortescue Group.

This brief introduction to Pilbara geology is grossly simplified from a substantial body of published literature which is evaluated below. In addition to that evaluation it will be useful to note here some previous overviews of Pilbara geology. Systematic descriptions are included in Memoirs 2 and 3 of the Geological Survey of Western Australia (GSWA 1975, 1990). Shorter summaries appear in Knight (1975), Hallberg & Glikson (1981) and Hughes (1990). A recent special issue of *Precambrian Research* (Blake & Meakins 1993) brings together eleven papers on aspects of Pilbara Craton geology, but does not provide systematic coverage or a general account. This omission is partly filled by a brief interpretive overview by Myers (1993); the guidebook for excursion no. 5 of the Third International Archaean Symposium (Hickman 1990) also goes some way towards filling this gap, but will not be widely available internationally. The major bulletin of Hickman (1983), although now eleven years old, stands as a definitive landmark publication, systematically summarizing the geology and mineralization of the north Pilbara GGT, as well as of the immediately contiguous Mount Bruce Supergroup. Trendall's (1983) review of the Hamersley Basin was published in the same year, but still provides a useful summary account.

History of investigation

Hickman (1983) gave a detailed account of geoscientific work on the craton to that date,

especially on its granite–greenstone component. A complete bibliography of the Precambrian geology of the Pilbara Craton to 1994, with a commentary placing it in a historical context, will shortly be available (Trendall in press).

The scope of present geoscientific knowledge

This survey is necessarily brief, and to some extent selective; it focuses on what are seen as the major components of present knowledge and understanding, and evaluates these by reference to comparable terranes elsewhere on the Earth. Note that, although many general papers carry useful information on specialized topics, only papers focused specifically on specialized subdisciplines are counted as contributions to them.

Granite–greenstone terrane

Mapping and stratigraphic definition. These two topics are taken together because of their close linkage: the earlier Australian Code of Stratigraphic Nomenclature (ACSN 1964) defined lithostratigraphic *formations* as the basic units required for representation of stratigraphy *on a map*, and this association continues under the current code (ACSN 1973).

The best available geological maps of the craton are the fifteen contiguous 1 : 250 000 scale sheets published by the Geological Survey of Western Australia (GSWA) as the result of a systematic mapping program started in the early 1960s. These maps are lithological rather than lithostratigraphic, but Hickman's (1980) reduced-scale (1 : 1 000 000) maps enable lithostratigraphic units defined during mapping of the various greenstone belts (e.g. Lipple 1975; Hickman 1977a, b) to be identified. Hickman (1983) consolidated the early nomenclature of Maitland (1909), as well as later local schemes (e.g. Ryan 1964; Fitton et al. 1975; Ingram 1977) into a single Pilbara Supergroup with three constituent groups and a total of 20 formations; a fourth group was later added (Fig. 2b). Hickman (1983) describes the supergroup as having a maximum thickness of 30 km, of which about 15 km is usually present in any single area. In spite of local unconformities, and substantial stratigraphic differences between different greenstone belts, Hickman (1983) clearly affirmed his beliefs not only that the Pilbara Supergroup forms a coherent and recognizable lithostratigraphic entity across the entire northern GGT, but also that the depositories of most units within it (and especially the older ones) were craton-wide, and bore no relationship to the present synclinoria in which they are exposed; these were seen as due solely to later tectonism associated with the diapiric rise of the granitoid domes between the greenstone belts. In respect to some of the younger components of the supergroup Hickman (1990) later agreed, on the basis of sedimentological work referred to further below, that their present outcrop areas roughly coincide with local depositories.

Other authors (e.g. Di Marco & Lowe 1989*a*) have challenged Hickman's stratigraphic correlations between different belts of the northern GGT, and have also suggested (e.g. Krapez 1993) that some of his defined units may be tectonic repetitions. There are now radical differences of published opinion concerning, in particular, the correlation between the western and eastern belts. Thus Horwitz (1979, 1986, 1987, 1990*a*, *b*, *c*; and earlier in Fitton *et al.* 1975) sees the outcrop of the Whim Creek Group as widely extensive outside the belt from which it is named; Horwitz & Pidgeon (1993) give a historical summary of that debate. And Krapez (1993) has now put forward an alternative nomenclatural scheme for the greenstone belts of the northern GGT based on the sequence stratigraphic terminology of Haq *et al.* (1988), involving six separate sets of names for differing orders of rock units, time–rock units, and time units; he also divides the northern GGT into five tectonostratigraphic domains separated by northeast lineaments with a long history of development and reactivation.

The GSWA is planning a 1 : 100 000 mapping program on parts of the GGT (Clarke 1992), and until mapping at this scale has been completed geologists without personal knowledge of the craton will not have access to the comfort of a clear consensus view of greenstone belt stratigraphy, structure, and correlation.

Sedimentology. The sedimentological work initiated by the University of Western Australia (UWA) in the late 1970s (e.g. Dunlop 1978; Barley et al. 1979; Dunlop & Buick 1981) was greatly boosted by the work of Eriksson (1981, 1982, 1983), largely on the epiclastic sedimentology of units in the higher part of the Pilbara Supergroup. Continuing UWA work (Fleming 1982; Krapez 1984, 1989; Wilhelmij & Dunlop 1984; Wilhelmij 1986; Buick 1985; Buick & Dunlop 1990) provided a number of detailed contributions on areas in the eastern part of the GGT. Concurrently, Lowe (1982, 1983) and his students (DiMarco 1986; Lowe & Byerly 1986; DiMarco & Lowe 1989*a*, *b*; Yuan & Lowe 1990) carried out work on the Pilbara Supergroup of similarly high quality in contiguous locations. Such work has resulted in detailed environmental and depositional reconstructions for restricted parts of the succession (e.g. Groves

et al. 1981; Krapez & Barley 1987), but these have largely been limited to the eastern part of the northern GGT, and major stratigraphic units (e.g. the Mosquito Creek Formation) still await specialist sedimentological attention. None of the stratigraphic units of the Pilbara Supergroup has received the intensity of sedimentological study applied to the Witwatersrand Supergroup of the Kaapvaal Craton.

Palaeobiology. The Pilbara Craton has been well served by palaeobiological studies, largely due to the early, and independent, interest shown by both Elso Barghoorn and Preston Cloud. Although neither of these published papers on the palaeobiology of the Pilbara Craton the extent of later work owes much to their influence and inspiration. The Warrawoona Group is one of two roughly coeval units which contain the oldest physical evidence of life on Earth, the other being the Onverwacht Group of the Kaapvaal craton. The discovery of microfossils in the North Pole area (Fig. 3) was first announced by Dunlop et al. (1978) and extensive work has since been carried out both on these and on *c.* 3.5 Ga stromatolites from the same and neighbouring localities (Walter *et al.* 1980; Buick *et al.* 1981; Awramik *et al.* 1983; Buick 1985; Schopf & Packer 1987; Schopf 1993). Walter (1983) and Schopf & Walter (1983) discuss in detail the general palaeobiological significance of work prior to that year.

Physical and chemical volcanology. Apart from work on sedimentological aspects of some sedimentary rocks with a volcanic association (e.g. Dunlop & Buick 1981; DiMarco & Lowe 1989*a*, *b*) almost nothing has been published on the physical volcanology of the Pilbara Supergroup, and nothing by a specialist volcanologist; the massive thicknesses of ultramafic, mafic and felsic volcanic rocks in the Pilbara Supergroup remain a major challenge for specialist palaeovolcanological study. Detailed geochemical studies of the volcanic rocks began comparatively late. A systematic programme of analysis was initiated jointly by the GSWA and Bureau of Mineral Resources (BMR — now renamed the Australian Geological Survey Organisation) in 1975, largely through the personal enthusiasm of Andrew Glikson. Systematic sampling of mafic and felsic volcanic units of the Pilbara Supergroup, across and along strike, took place in the following years, and all of the 500+ samples were analysed for major oxides and a wide range of trace elements. The results appeared in 1981 (Glikson & Hickman 1981*a*, *b*), and additional data and interpretations followed (Jahn *et al.* 1981; Glikson *et al.* 1986*a*, *b*, 1987). While the results form a substantial geochemical

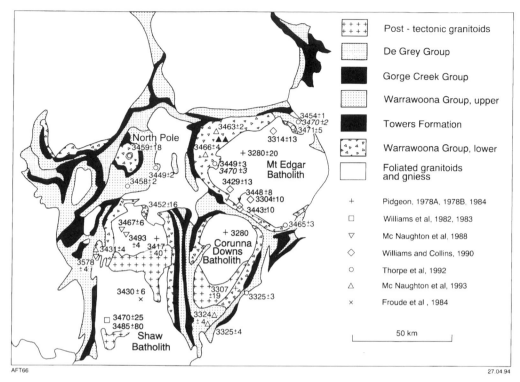

Fig. 3. Simplified geological map of the eastern part of the north Pilbara granite–greenstone terrane, modified from Hickman (1980). Locations and ages of all published zircon U–Pb ages from the area are marked, with sources cited. Italic figures below upright numbers adjacent to some symbols indicate ages from the same sample believed to be from xenocrystic zircons; reference should be made to the cited papers for more detailed interpretation of the ages shown. Solid symbols without adjacent ages indicate locations of samples which yielded equivocal ages. Most ages obtained from zircon U–Pb analyses using the Research School of Earth Sciences SHRIMP in Canberra. Exceptions are those of Pidgeon, which are conventional multi-grain TIMS results, also carried out at the RSES, and those of Thorpe *et al.* (1992) which are single- or multiple-grain analyses carried out at either the Jack Satterly laboratories of the Royal Ontario Museum or the geochronological laboratories of the Geological Survey of Canada.

database it seems fair comment that they produced few surprises, or major and unequivocal petrogenetic insights. Notwithstanding this comment, interesting results included the demonstration of significant chemical differences between the upper and lower parts of the supergroup, and systematic vertical trends within some formations, which confirmed their correlation on the basis of field mapping and limited the possibilities of structural repetition. The UWA carried out more locally directed work at the same time (Barley 1980; Barley *et al.* 1984).

Granitoids. Two main points need emphasis concerning knowledge of the granitoid complexes: (i) those which have been most closely examined have been found to consist of multiple plutons, between or within which there is a wide variety of gneisses and migmatites with structurally complex

relationships, and (ii) of the 24 granitoid bodies separately named by Hickman (1983) within the north Pilbara GGT, very few have received closer attention than that associated with reconnaissance 1 : 250 000 scale mapping. Hickman (1983) gives a good review of knowledge to that date. One of the first studies specifically concerned with granitoid chemistry was the relatively recent one of Blockley (1980), who summarized and discussed the results of analyses of 123 samples from the Mount Edgar, Shaw, and Yule batholiths. These, together with the Corunna Downs batholith (Fig. 3), are the only bodies which have received detailed attention. Davy & Lewis (1981, 1986) carried out a reconnaissance geochemical study of the Mount Edgar batholith, and supplemented it with petrological data; Davy (1988) also reported similar work on Corunna Downs. A UWA study of the Shaw batholith resulted in a substantial amount of geochemical, geochronological and structural work on that body (Bettenay *et al.* 1981; Bickle *et al.* 1983, 1985, 1989, 1993), while the detailed work of Collins and others (Collins 1983, 1989, 1990, 1993; Collins & Gray 1990; Williams & Collins 1990; Teyssier & Collins 1990) on the Mount Edgar batholith has brought knowledge of that body to about the same level as that of the Shaw batholith; the papers by Bickle *et al.* (1993) and Collins (1993) represent the most detailed discussions so far of granitoid petrogenesis.

Structure and tectonic development. Like work of any detail on granitoid domes, structural studies have tended to be restricted to the eastern part of the north Pilbara GGT, and most have inevitably been directed towards the associated problem of the tectonic relationship between granitoid complexes and greenstone belts, a relationship which is central to the main issue of the tectonic development of the entire GGT. Hickman (1983, 1984, 1992) was the first to suggest a staged sequence of crustal events from work within the Pilbara craton, and to introduce a modern structural nomenclature for recognizably differing fold and foliation ages; the 1984 contribution of Hickman remains the most comprehensive model for the development of the north Pilbara GGT. The papers on the Shaw and Mount Edgar batholiths cited under the preceding heading also address the question of the relationship between tectonic development and batholith emplacement. It is noteworthy that UWA authors emphasize lateral stress in tectonic schemes akin to modern plate tectonics for the Shaw batholith, whereas Collins and co-workers prefer an interpretation based on diapirism and vertical movements for the adjacent Mount Edgar batholith; if all the major granitoid complexes of the north Pilbara GGT have a common emplacement mechanism then much more work is clearly needed to establish its nature; conversely, if adjacent complexes have radically disparate emplacement modes, then close attention to the abundant unstudied bodies throughout the GGT is urgently needed. Krapez (1993) has given an overall tectonostratigraphic interpretation of GGT evolution, but this does not include such necessary components of a complete conceptual model as mechanisms of magma emplacement and petrogenesis.

Metamorphism. There have been almost no specialist studies of metamorphism in the north Pilbara GGT, although it is well known to be generally low grade; the work of Morant (1984), published more accessibly in Bickle *et al.* (1985), is virtually the only contribution to this topic, although Wijbrans & MacDougall's (1987) geochronological work also has strong relevance

for metamorphic timing. Such work is badly needed, in view of the constraints it might place on the thermal structure of the craton, and of the earth, during GGT development.

Mount Bruce Supergroup

Trendall (1983) reviewed work on the Mount Bruce Supergroup to that date. The following paragraphs therefore concentrate on later additions, but provide an outline of earlier work to put them in context. The BIFs of the Hamersley basin are well established in the literature as some of the best known on Earth, a circumstance due directly or indirectly to the presence within them of major deposits of crocidolite and high-grade iron ore.

The GSWA 1 : 250 000 mapping of the Pilbara craton, already noted, also covered the entire outcrop area of the Mount Bruce Supergroup. Mapping of the Mount Bruce Supergroup was lithostratigraphic from the start, and the formal lithostratigraphic subdivisions established by MacLeod *et al.* (1963), and modified by Trendall (1979), have served well as a framework for continuing research. That stratigraphic scheme includes the most detailed type section, for the Dales Gorge Member of the Brockman Iron Formation, ever published for any formally defined lithostratigraphic unit, of any age, on any continent (Trendall & Blockley 1968). Revision of these 1 : 250 000 maps (e.g. Seymour *et al.* 1988; Tyler *et al.* 1990), and remapping of selected areas within them at a scale of 1 : 100 000 (e.g. Tyler 1990; Thorne & Tyler 1994) have continued since 1983, and are ongoing (Clarke 1992). Blake (1990) has made an important contribution to 1 : 100 000 map coverage of the lower part of the Fortescue Group.

Other highlights of post-1983 work include the continuing Commonwealth Scientific and Industrial Research Organisation (CSIRO) work on the genesis of iron ore (Morris 1985), which has proved to have important feedback for the depositional environment of the host Hamersley Group BIFs (Morris 1993), the appearance of sufficient new geochronological data to place the Mount Bruce Supergroup in a more secure time frame (Pidgeon 1984; Arndt *et al.* 1991; Hassler 1991; Pidgeon & Horwitz 1991), and the discovery of the oldest known palaeosol, developed over a basalt flow of the Fortescue Group (Macfarlane & Holland 1990; Macfarlane *et al.* 1994*a, b*). A welcome addition to sedimentological knowledge of the Hamersley Group has been the continuing work of Bruce Simonson and his co-workers on the Wittenoom Formation, renamed from Wittenoom Dolomite (e.g. Simonson *et al.* 1993*a, b*); and detailed isotope chemistry on the Marra Mamba Iron Formation has provided a constructive challenge to some earlier concepts both on the chronology of BIF diagenesis, and on the genesis of the S macrobands of the Dales Gorge Member (Alibert & McCulloch 1990, 1993).

Blake (1984) has made a detailed and important contribution to knowledge of Fortescue Group sedimentology, and has related it to episodes of continental rifting. It seems unfortunate, though, that he has now elected to synthesize his results (Blake 1993) within the framework of a sequence stratigraphic scheme, first set out by Blake & Barley (1992), which is based, like that of Krapez (1993) for the Pilbara Supergroup, on the concepts of Haq et al. (1988). The scheme has also been adopted by Barley et al. (1992); it appears to represent, in essence, a preferred regrouping of the existing formally defined stratigraphy, rearranged without benefit of the discipline of any code, and unsupported, in respect to the Hamersley Group, by significant new field data.

Geophysics

Geochronology. The age of the new mineral 'pilbarite' from Wodgina, in the eastern part of the Pilbara craton, was determined by the chemical helium method and published (Simpson 1910) only four years after Rutherford had suggested the possibility of this approach (see de Laeter & Trendall 1979). Simpson's published analyses of this and other Wodgina minerals were later used by Cotton (1926) to calculate chemical lead ages; Holmes & Lawson (1927) recalculated these, and one result, of 1260 Ma, was accepted by Holmes (1927) as the oldest known mineral age, even though 'not very well-established'. This early start to geochronological work on the Pilbara Craton led to continuing attention as each new method was developed, as a result of which it is, in terms of an accurately determined time framework for development, one of the best known cratons on any continent. The ages of the main events in the 1000+ billion years of crustal history recorded in the craton have already been noted; the substantial body of published data to 1983 has been reviewed by Trendall (1983) and Blake & McNaughton (1984). Soon after the appearance of those reviews the zircon U–Pb method became accepted as the 'industry standard' for regional geochronological definition, and Pidgeon's (1978a, b) pioneering application of this method was quickly followed by others, both by 'conventional' multi-grain and later by single-grain (SHRIMP) work. Because two maps (Bickle et al. 1993, fig. 2; McNaughton et al. 1993, fig. 1) in the recent Pilbara special issue of *Precambrian Research* (Blake & Meakins 1993) either show published results in incorrect localities, or omit available data, Figure 3 is included here to show the locations and sources of all published zircon U–Pb ages from the eastern part of the northern GGT. Inspection clearly shows the coeval development of granitoid complexes and the contiguous greenstone belts. The only published zircon U–Pb age from the north Pilbara GGT, not included on Fig. 3, is that of Horwitz & Pidgeon (1993). Their result, of 3112 ± 6 Ma, came from five single zircons recovered from a felsic tuff in the Sholl Belt, some 400 km west of the area of Fig. 3; the unit was correlated by Hickman (1983) with the Duffer Formation of the Warrawoona Group, the age of which is now well established at c. 3.46 Ga (Pidgeon 1984; Thorpe et al. 1992; McNaughton et al. 1993). This unresolved problem emphasizes the immense scope for further work, not least by methods able to establish the timing of metamorphic and tectonic events later than primary rock ages (e.g. Wijbrans & Macdougall 1987).

Recent zircon U–Pb work from the Hamersley basin, and underpinning the time frame mentioned in the earlier summary, includes the papers of Arndt et al. (1991), Pidgeon & Horwitz (1991) and Hassler (1991).

Gravity. The BMR carried out a relatively widely spaced (11 km) gravity survey of the craton and adjacent areas in 1969–1972 (Fraser 1973, 1976, 1980). The resultant Bouguer anomaly map (Fraser 1980, plate B) shows that the northern GGT of the Pilbara Craton and the outcrop area of the Mount Bruce Supergroup share a closely similar anomaly pattern which Fraser calls the Fortescue Regional Gravity Complex. It seems a reasonable supposition that the boundaries of this approximate to the real edges of the craton. Additional gravity data were added during the deep crustal structure project (see below) and a revised, but essentially similar Bouguer gravity map was produced by Drummond & Shelley (1981, fig. 1).

Deep crustal structure. A seismic study of the crustal structure of the Pilbara Craton and the northern margin of the neighbouring Yilgarn Craton, was carried out jointly by the Australian National University and the BMR in 1977, using large blast explosions from iron ore mines on the craton. The main result (Drummond 1979, 1981, 1983) of this work was to show that the two cratons have different crustal structures; the crust of the Pilbara Craton has two layers, and a total thickness of 28–33 km, while that of the Yilgarn Craton has three layers, and is over 50 km thick. It also supported the interpretation from gravity patterns that the southern edge of the Pilbara Craton lies not far south of the southern edge of present exposure. *Palaeomagnetic studies.* Early observations (Irving & Green 1958) were too few to do more than expose problems for future work, while some following work (Porath 1967; Porath & Chamalaun 1968) was essentially an experiment in the use of palaeomagnetic methods to determine the age of iron ore formation, an impossible task at that time through inadequate definition of polar wander paths. Later work of high quality directed at this problem (Embleton 1978; Embleton *et al.* 1979; Schmidt & Embleton 1985) was confined to the volcanic rocks of the Fortescue Group, some possible feeder dykes, and a single major mafic intrusion in the western part of the northern GGT. Although valuable, these studies also proved to be premature, in that they preceded an adequate framework of field understanding and geochronological control; more such palaeomagnetic work, tightly controlled within a framework of high-precision geochronology, is now urgently needed both for a proper understanding of the evolution of the craton, and even more importantly to establish its position in relation to other similar foci of crustal growth on the early Earth. McElhinny & Senanayake (1980) distinguished three separate magnetic components in samples of Duffer Formation (lower Warrawoona Group; Figs 2 & 3) of the Marble Bar area. They used their results both to suggest an Archaean (3.5–2.6 Ga) apparent polar wander path for Australia, and to argue that the Earth's magnetic field existed at 3.5 Ga.

Heat flow. The Pilbara Craton shares the low heat flow characteristic of other ancient cratonic areas of the Earth. Three reliable measurements reported by Cull & Denham (1979) all fall in the range 40–50 mW m^{-2}, in contrast to the general range in the younger tectonic units of eastern Australia of *c.* 70–100 mW m^{-2}.

Aeromagnetic mapping. Aeromagnetic mapping, with varying specifications, has been carried out by the BMR, and all fifteen 1 : 250 000 sheets are available as total magnetic intensity maps from that organization, in some cases incorporating other available data. All data, contributing to those maps, have been integrated into the magnetic anomaly map of Australia compiled by Tarlowski *et al.* (1993), but no integrated anomaly map of the Pilbara Craton at a larger scale than the 1 : 5 000 000 of that map is available.

Tectonic evolution

No reasoned and integrated account of the tectonic evolution of the Pilbara craton, from the oldest volcanic units of the Pilbara Supergroup at *c.* 3.5 Ga to the completion of deposition of the Turee Creek Group at *c.* 2.4 Ga, has yet been published. Distinction is made here between integrated

concepts of GGT development (e.g. Hickman 1983, 1984) and separate conceptual schemes for all or part of the Mount Bruce Supergroup (e.g. Blake 1993; Blake & Barley 1992), which accept as a departure point the existence of an extensive region of continental crust, of which the GGT component of the Pilbara Craton represents a post-fragmentation remnant. Trendall (1984) has pointed out the continuity of magmatic and depositional activity throughout the history of the crustal entity now defined as the Pilbara Craton, and it is hardly conceivable that the deposition of the Mount Bruce Supergroup is unrelated to the evolution of the underlying GGT, but the key to that relationship has yet to be found.

Application of paradigms to the Pilbara craton

Concepts for the development of all tectonic units are most usefully expressed as dynamic three-dimensional models. The key difference between such models and paradigms, as already defined, is that models start from the totality of available evidence, and attempt to incorporate it into a self-consistent hypothesis based on general physical principles, whereas paradigms simply seek to identify some set of features present in the supposedly better understood unit elsewhere; if these features are present in the tectonic unit under study, understanding is implied by attaching to it the label of the paradigm. At worst, the logic of paradigms follows the argument: all cats have four paws; your pet has four paws; therefore yout pet is a cat. The 'X' in the earlier definition of a paradigm here becomes 'cat' and we have a 'cat paradigm' which, in terms of paradigmatic thought, satisfactorily labels and explains all four-footed pets.

In those terms, the following are some of the paradigms in terms of which the Pilbara craton, or some component of it, has been interpreted:

- the Great Unconformity paradigm;
- the gregarious batholith paradigm;
- the horizontal tectonics paradigm;
- the impact paradigm;
- the terrane paradigm;
- the rift-and-drift paradigm;
- the CFBP paradigm;
- the sequence stratigraphy paradigm;
- the island arc paradigm.

The use of paradigms characteristically diverts debate from the totality of evidence, and concentrates it instead onto the paradigm; and such debate is often conducted using paradigm-based nomenclature, destroying the objectivity of the investigative approach. Yet another problem is that paradigms usually involve implicit assumptions, which are in consequence excluded from discussion. Some examples, selected from the list above, are used below to illustrate these and other points.

The Great Unconformity, or Archaean–Proterozoic, paradigm is a particular good example of their inhibiting effect on geoscientific thought (Trendall 1984). This paradigm grew from an early dogma that the basal unconformity of the relatively undisturbed Huronian strata over crystalline rocks of the Superior Province in Canada was coeval with other great unconformities on other continents; and that all these were related to a unique event in earth history, separating 'Archaean' time before it from 'Proterozoic' time after. These paradigm-based names became widely accepted internationally, and have been used in the Pilbara Craton, often with the name Nullagine instead of Proterozoic. Although the Subcommission on Precambrian Stratigraphy (Plumb & James 1986) has formally destroyed the paradigm by defining Archaean and Proterozoic as purely chronometric terms (that is, as names for arbitrary divisions of measured time), the retention of the names still causes confusion through their original associations. From currently available geochronological results the best estimate of the position of the Archaean–Proterozoic boundary, defined by the Subcommission at 2500 Ma, is within the undisturbed BIF of the Dales Gorge Member of the Brockman Iron Formation of the Hamersley Group; the Pilbara Craton was clearly enjoying a period of relatively stable and continuous development at that time, so that the names serve no useful purpose, are unrelated to events in the craton, and should be abandoned in view of their derivation from a discredited paradigm.

The implicit assumption of the horizontal tectonics paradigm is that if horizontally directed forces can be shown to have operated within an ancient tectonic unit, then that unit was a component of a coherent plate, and the plate tectonic model consensually applied to the modern earth operated in the same way during its development. The paradigm has been applied to a part of the north Pilbara GGT (Bickle *et al.* 1980; Boulter *et al.* 1987), but the area mapped was miniscule in relation to the size of the craton, and others have demonstrated major vertical movements in similar areas nearby (Collins 1983; Teyssier & Collins 1990). The elevation of horizontal structures into a paradigm of plate tectonics applicable to the whole craton is clearly inappropriate at this stage of tectonic understanding.

The dogma of the sequence stratigraphy paradigm is that the entire sedimentary record of the earth is divisible, through the influence of regularly spaced sea level changes, into globally correlated packages of strata, separated by hiatal surfaces. The paradigm has recently been applied to the sedimentary record of both the north Pilbara GGT (Krapez 1993) and the Mount Bruce Supergroup (Blake & Barley 1992; Blake 1993). The resultant use of a stratigraphic nomenclature based on this paradigm, already commented on, is a prime example of the danger of using a paradigm-based terminology to discuss the paradigm itself.

An island arc paradigm has recently been applied to parts of the Pilbara Craton (Isozaki *et al.* 1991; Kimura *et al.* 1991), and although details of the concept have yet to appear the approach seems close to the paradigmatic logic already discussed: some features of island arcs are believed to be identifiable in the Pilbara craton, so the craton must be such an arc.

Conclusions

The Pilbara Craton is a superb natural laboratory for the study of early crustal development. From a logistic viewpoint there are no political or security problems for field work, and the area is well provided with regular commercial air services. There is enough relief to make exposure generally good, without impeding ground access by four-wheel drive vehicle; the population is sparse, so that access is not restricted by settlement or close agriculture. Most of the area is arid bushland, so that there is neither forest nor glacial cover, and the general depth of weathering is quite small. The dry climate in the (southern) winter months could hardly be improved upon for geological field work, and good air photos and satellite imagery are available, as well as both topographic and geological maps.

Enough work has been done on the geological development of the Pilbara Craton to know that the rocks record a complex and near-continuous history of sedimentation, volcanism, magmatism and tectonism over more than a billion years; but sufficient major problems remain to challenge the capabilities of both specialists and generalists within geoscience. Some examples of unanswered questions that need to be addressed by further work within the craton include:

- why is low-grade metamorphism prevalent in the GGT, at the time of whose development radiogenic heat production was several times higher than at present?

- were volcanism and sedimentation precisely synchronous in different greenstone belts, and continuous between them, or were the rocks of individual belts, especially in the eastern and western parts of the GGT, formed by local and diachronous events?

- what exactly took place to cause the basal unconformity of the Fortescue Group, and where did the material stripped off the underlying GGT go?

- what was the physical environment of eruption of the lowermost lavas (North Star Basalt) of the north Pilbara GGT, and what material were they extruded through and over?
- what is the precise petrogenetic connection between the volcanic rocks of the greenstone belts and the adjacent and penecontemporaneous granitoid complexes?

These, and many other general questions, can best be addressed by careful and intensive continuing study of the Pilbara Craton itself, rather than by the facile application of fashionable paradigms. Such work is easy to identify, and examples have already been noted in the course of this paper. An obvious priority is the production of more detailed systematic geological maps than the presently available 1 : 250 000 scale sheets; the resolution of many current stratigraphic and structural controversies will flow automatically from such work. Specialized structural studies should be integrated closely with the mapping, rather than be carried out independently. Although there is great scope for high-resolution isotope geochronological work, this would best be carried out, especially in the GGT, after more detailed maps are available. Future geochronology also needs close integration with more extensive and systematic palaeo-magnetic work.

These are only a few selected priorities from the vast range of further work needed to achieve the degree of understanding of the Pilbara Craton which is possible by the use of currently established geoscientific methods; whether the financial resources will be available for the necessary work to be accomplished within a reasonable time frame is another matter, and outside the scope of this paper. It is important that such future work should approach the Pilbara Craton in an objective way, rather than with a determination to fit its geology within the confines of some fashionable paradigm derived from the present state of the Earth.

John Sutton accompanied me during his first visit to the Pilbara Craton in 1970, and I am therefore grateful to Mike Coward for his invitation to participate in the John Sutton Memorial Meeting at Imperial College in September 1993, and for the consequent opportunity to contribute this paper in his memory. Bob Pidgeon kindly supplied information on the positions of some geochronological samples in Fig. 3. A first draft was improved as a result of constructive comments from John Blockley, Arthur Hickman, and Alan Thorne; the improved version was later substantially condensed following comment from an anonymous referee.

References

ACSN 1964, Australian Code of Stratigraphic Nomenclature. *Geological Society of Australia Journal,* **11**, 165–171.

—— 1973. Australian Code of Stratigraphic Nomenclature (4th Edition, reprinted with corrigenda and additional notes). *Geological Society of Australia Journal,* **20**, 105–112.

ALIBERT, C. & McCULLOCH, M. T. 1990. The Hamersley Basin revisited: the Sm–Nd isotopic systematics of banded iron-formations and associated "shales" and implications for the composition of the early Proterozoic seawater. *Third International Archaean Symposium, Extended Abstracts Volume,* 327–328.

—— & —— 1993. Rare earth element and neodymium isotopic compositions of the banded iron-formations and associated shales from amersley, Western Australia. *Geochimica et Cosmochimica Acta,* **57**, 187–204.

ARNDT, N. T., NELSON, D. R., COMPSTON, W., TRENDALL, A. F. & THORNE, A. M. 1991. The age of the Fortescue Group, Hamersley Basin, Western Australia, from ion microprobe zircon U–Pb results. *Australian Journal of Earth Sciences,* **38**, 261–281.

AWRAMIK, S. M., SCHOPF, J. W. & WALTER, M. R. 1983. Filamentous fossil bacteria 3.5×10^9 years old from the Archean of Western Australia. *Precambrian Research,* **20**, 357–374.

BARLEY, M. E. 1980. *The evolution of Archaean calc-alkaline volcanics: a study of the Kelly Greenstone Belt and McPhee Dome, eastern Pilbara Block*

Western Australia. PhD Thesis, University of Western Australia.

——, BLAKE, T. S. & GROVES, D. I. 1992. The Mount Bruce Megasequence Set and eastern Yilgarn Craton: examples of Late Archaean to Early Proterozoic divergent and convergent craton margins and controls on mineralization. *Precambrian Research,* **58**, 55–70.

——, DUNLOP, J. S. R., GLOVER, J. E. & GROVES, D. I. 1979. Sedimentary evidence for an Archean shallow-water volcanic-sedimentary facies, eastern Pilbara Block, Western Australia. *Earth and Planetary Science Letters,* **43**, 74–84.

——, SYLVESTER, G. C., GROVES, D. I., BARLEY, G. D. & ROGERS, N. 1984. Archaean calc-alkaline volcanism in the Pilbara Block, Western Australia. *Precambrian Research,* **24**, 285–320.

BATES, R. L. & JACKSON, J. A. (eds) 1987. *Glossary of Geology,* Third Edition. american Geophysical Institute, Falls Church, Virginia.

BETTENAY, L. F., BICKLE, M. J., BOULTER, C. A., GROVES, D. I., MORANT, P., BLAKE, T. S. & JAMES, B. A. 1981. Evolution of the Shaw Batholith — an Archaean granitoid–gneiss dome in the eastern Pilbara, Western Australia. *In:* GLOVER, J. E. & GROVES, D. I. (eds) *Archaean Geology.* Geological society of Australia, Special Publications, **7**, 361–372.

BICKLE, M. J., BETTENAY, L. F., BARLEY, M. E., CHAPMAN, H. J., GROVES, D. I., CAMPBELL, I. H. & DE LAETER, J. R. 1983. A 3500 Ma plutonic and calc-alkaline province in the Archaean East Pilbara Block. *Contributions to Mineralogy and Petrology*, **84**, 25–35.

——, ——, CHAPMAN, H. J., GROVES, D. I., MCNAUGHTON, N. J. CAMPBELL, I. H. & DE LAETER, J. R. 1989. The age and origin of younger granitic plutons of the Shaw Batholith in the Archean Pilbara Block, Western Australia. *Contributions to Mineralogy and Petrology*, **101**, 361–376.

——, ——, ——, ——, ——, MCNAUGHTON, N. J., CAMPBELL, I. H. & DE LAETER, J. R. 1993. Origin of the 3500–3300 Ma calc-alkaline rocks in the Pilbara Archaean: isotopic and chemical constraints from the Shaw Batholith. *In*: BLAKE, T. S. & MEAKINS, A. (eds) *Archaean and Early Proterozoic geology of the Pilbara region, Western Australia. Precambrian Research*, **60** (Special issue), 117–149.

——, ——, BOULTER, C. A., GROVES, D. I. & MORANT, P. 1980. Horizontal tectonic interaction of an Archean gneiss belt and greenstones, Pilbara Block, Western Australia. *Geology*, **8**, 525–529.

——, MORANT, P., BETTENAY, L. F., BOULTER, C. A., BLAKE, T. S. & GROVES, D. I. 1985. Archaean tectonics of the Shaw Batholith, Pilbara Block, Western Australia: structural and metamorphic tests of the batholith concept. *In*: AYRES, L. D., THURSTON, P. C., CARD, K. D. & WEBER, W. (eds) *Evolution of Archean Supracrustal Sequences*. Geological Association of Canada, Special Publications, **28**, 325–341.

BLAKE, T. S. 1984. The lower Fortescue Group of the northern Pilbara Craton: stratigraphy and palaeogeography. *In*: MUHLING, J. R., GROVES, D. I. & BLAKE, T. S. (eds) *Archaean and Proterozoic Basins of the Pilbara, Western Australia — Evolution and Mineralization Potential*. University of Western Australia, Geology Department and University Extension Publications, **9**, 123–143.

—— 1990. *Bedrock geology of the Fortescue Group of the northern Pilbara Craton*. Key Centre for Strategic Mineral Deposits, Geology Department, and University Extension, University of Western Australia, Map Series 1, 10 sheets at 1 : 100 000 scale.

—— 1993. Late Archaean crustal extension, sedimentary basin formation, flood basalt volcanism and continental rifting: the Nullagine and Mount Jope Supersequences, Western Australia. *Precambrian Research*, **60**, 185–241.

—— & BARLEY, M. E. 1992. Tectonic evolution of the Late Archaean to Early Proterozoic Mount Bruce Megasequence Set, Western Australia. *Tectonics*, **11**, 1415–1425.

—— & MEAKINS, A. (eds) 1993. *Archaean and Early Proterozoic geology of the Pilbara region, Western Australia. Precambrian Research*, **60** (Special Issue).

—— & MCNAUGHTON, N. J. 1984. A geochronological framework for the Pilbara region. *In*: MUHLING, J. R., GROVES, D. I. & BLAKE, T. S. (eds) *Archaean and Proterozoic Basins of the Pilbara, Western Australia — Evolution and Mineralization Potential*. University of Western Australia, Geology

Department and University Extension Publications, **9**, 1–22.

BLOCKLEY, J. G. 1980. *Tin deposits of Western Australia with special reference to the associated granites*. Western Australia Geological Survey, Mineral Resources Bulletin 12.

BOULTER, C. A., BICKLE, M. J., GIBSON, B. & WRIGHT, R. K. 1987. Horizontal tectonics pre-dating upper Gorge Creek Group sedimentation, Pilbara Block, Western Australia. *Precambrian Research*, **36**, 241–258.

BUICK, R. 1985. *Life and conditions in the early Archaean: evidence from 3500 M.Y. old shallow-water sediments in the Warrawoona Group, North Pole, Western Australia*. PhD Thesis, University of Western Australia, Nedlands.

—— & DUNLOP, J. S. R. 1990. Evaporitic sediments of early Archaean age from the Warrawoona Group, North Pole, Western Australia. *Sedimentology*, **37**, 247–277.

——, —— & GROVES, D. I. 1981. Stromatolite recognition in ancient rocks: an appraisal of irregularly laminated structures in an Early Archaean chert-barite unit from North Pole, Western Australia. *Alcheringa*, **5**, 161–181.

CLARKE, J. 1992. *Summary of progress of the Geological Survey of Western Australia from July 1991 to June 1992 and plans for 1992/93 to 1996/97*. Western Australia Geological Survey, Record 1992/1.

COLLINS, W. J. 1983. *Geological evolution of an Archaean batholith*. PhD Thesis, La Trobe University.

—— 1989. Polydiapirism of the Archaean Mount Edgar Batholith, Pilbara Block, Western Australia. *Precambrian Research*, **43**, 47–62.

—— 1990. Genesis of Archaean granitoids from the Mount Edgar Batholith: implications for crustal evolution, Pilbara Block, Western Australia. *third International Archaean Symposium, Extended Abstracts Volume*, 189–190.

—— 1993. Melting of Archaean sialic crust under high P_{H_2O} conditions: genesis of 3300 Ma Na-rich granitoids, Pilbara Block, Western Australia. *In*: BLAKE, T. S. & MEAKINS, A. (eds) *Archaean and Early Proterozoic Geology of the Pilbara Region, Precambrian Research*, **60** (Special issue), 151–174.

—— & GRAY, C. M. 1990. Rb-Sr systematics of an Archaean granite-gneiss terrain: the Mount Edgar Batholith, Pilbara Block, Western Australia. *Australian Journal of Earth Sciences*, **37**, 9–22.

COTTON, L. A. 1926. Age of certain radium-bearing rocks in Australia. *American Journal of Science*, **12**, 42–45.

CULL, J. P. & DENHAM, D. 1979. Regional variations in Australian heat flow. *BMR Journal of Australian Geology and Geophysics*, **4**, 1–13.

DAVY, R. 1988. Geochemical patterns in granitoids of the Corunna Downs Batholith, Western Australia. *Western Australia Geological Survey, Reports*, **23**, 51–84.

—— & LEWIS, J. D. 1981. The geochemistry of the Mount Edgar Batholith, Pilbara area, Western Australia. *In*: GLOVER, J. E. & GROVES, D. I. (eds)

Archaean Geology. Geological society of Australia, Special Publications, **7**, 373–384.

—— & —— 1986. *The Mount Edgar Batholith, Pilbara area, Western Australia: geochemistry and petrology.* Western Australia Geological Survey, Report 17.

DE LAETER, J. R. & TRENDALL, A. F. 1979. The contribution of geochronology to Precambrian studies in Western Australia. *Royal Society of Western Australia, Journal,* **62**, 21–31.

DIMARCO, M. J. 1986. *Stratigraphy, sedimentology and sedimentary petrology of the Early Archean Coongan Formation, Warrawoona Group, eastern Pilbara Block, Western Australia.* PhD Thesis, Louisiana State University, Baton Rouge.

—— & LOWE, D. R. 1989*a.* Stratigraphy and sedimentology of an early Archean felsic volcanic sequence, eastern Pilbara Block, Western Australia, with special reference to the Duffer Formation and implications for crustal evolution. *Precambrian Research,* **44**, 147–169.

—— & —— 1989*b.* Petrography and provenance of silicified early Archaean volcaniclastic sandstones, eastern Pilbara Block, Western Australia. *Sedimentology,* **36**, 821–836.

DRUMMOND, B. J. 1979. A crustal profile across the Archaean Pilbara and northern Yilgarn Cratons, northwest Australia. *BMR Journal of Australian Geology and Geophysics,* **4**, 171–180.

—— 1981. Crustal structure of the Precambrian terrains of northwest Australia from seismic refraction data. *BMR Journal of Australian Geology and Geophysics,* **6**, 123–135.

—— 1983. Detailed seismic velocity/depth models of the upper lithosphere of the Pilbara Craton, northwest Australia. *BMR Journal of Australian Geology and Geophysics,* **8**, 35–51.

—— & SHELLEY, H. M. 1981. Isostasy and structure of the lower crust and upper mantle in the Precambrian terrains of northwest Australia, from regional gravity studies. *BMR Journal of Geology and Geophysics,* **6**, 137–143.

DUNLOP, J. S. R. 1978. Shallow-water sedimentation at North Pole, Pilbara, Western Australia. *In*: GLOVER, J. E. & GROVES, D. I. (eds) *Archaean cherty metasediments: their sedimentology, micropalaeontology, biogeochemistry and relation to mineralization.* University of Western Australia Extension service, 30–38.

—— & BUICK, R. 1981. Archaean epiclastic sediments derived from mafic volcanics, North Pole, Pilbara Block, Western Australia. *In*: GLOVER, J. E. & GROVES, D. I. (eds) *Archaean Geology.* Geological Society of Australia Special Publications, **7**, 225–234.

——, MUIR, M. D., MILNE, V. A. & GROVES, D. I. 1978. A new microfossil assemblage from the Archaean of Western Australia. *Nature,* **274**, 676–678.

EMBLETON, B. J. J. 1978. The palaeomagnetism of 2400 m.y. old rocks from the Australian Pilbara Craton and its relation to Archæan-Proterozoic tectonics. *Precambrian Research,* **6**, 275–291.

——, ROBERTSON, W. A. & SCHMIDT, P. W. 1979. *A survey of magnetic properties of some rocks from*

northwestern Australia. Commonwealth Scientific and Industrial Research Organisation (CSIRO), Division of Mineral Physics, Investigations Report **129**.

ERIKSSON, K. A. 1981. Archaean platform-to-trough sedimentation, east Pilbara Block, Australia. *In*: GLOVER, J. E. & GROVES, D. I. (eds) *Archaean Geology.* Geological Society of Australia Special Publications, **7**, 235–244.

—— 1982. Geometry and internal characteristics of Archaean submarine channel deposits, Pilbara Block, Western Australia. *Journal of Sedimentary Petrology,* **52**, 383–393.

—— 1983. Siliciclastic-hosted iron-formations in the early Archaean Barberton and Pilbara sequences. *Journal of the Geological Society of Australia,* **30**, 473–482.

FITTON, M. J., HORWITZ, R. C. & SYLVESTER, G. 1975. *Stratigraphy of the Early Precambrian in the west Pilbara.* Commonwealth Scientific and Industrial research Organisation (CSIRO), Mineral Research Laboratory Report **FP 11**.

FLEMING, B. S. 1982. *The sedimentology, structure and stratigraphy of the Gorge Creek Group, northern Lalla Rookh Syncline, eastern Pilbara.* BSc (Hons) Thesis, University of Western Australia.

FRASER, A. R. 1973. *A discussion on the gravity anomalies of the Precambrian Shield of Western Australia.* Bureau of Mineral Resources, Geology and Geophysics (BMR), Record 1973/105 (unpublished).

—— 1976. Gravity provinces and their nomenclature. *BMR Journal of Australian Geology and Geophysics,* **1**, 350–352.

—— 1980. Reconnaissance gravity surveys in Western Australia and South Australia, 1969–1972. *In*: FRASER, A. R. & PETTIFER, G. R. (eds) *Reconnaissance gravity surveys in Western Australia and South Australia, 1969–1972.* Australia Bureau of Mineral Resources, Geology and Geophysics, Bulletin **196**, Part B, 13–25.

FROUDE, D. E., WIJBRANS, J. R. & WILLIAMS, I. S. 1984. 3400 and 3430 Ma ages from U–Pb analyses in the Western Shaw Belt, Pilbara Block. *Australian National University, Research School of Earth Sciences Annual Report for 1983,* 126–128.

GEE, R. D. 1979. Structure and style of the Western Australian shield. *Tectonophysics,* **58**, 327–369.

GLIKSON, A. Y. & HICKMAN, A. H. 1981*a. Geochemistry of Archaean volcanic successions, eastern Pilbara Block, Western Australia.* Bureau of Mineral Resources, Geology and Geophysics (BMR), Record 1981/36 (unpublished).

—— & —— 1981*b.* Geochemical stratigraphy and petrogenesis of Archaean basic–ultrabasic volcanic units, eastern Pilbara Block, Western Australia. *In*: GLOVER, J. E. & GROVES, D. I. (eds) *Archaean Geology.* Geological Society of Australia Special Publications, **7**, 287–300.

——, DAVY, R. & HICKMAN, A. H. 1986*a. Geochemical data files of Archaean volcanic rocks, Pilbara Block, Western Australia.* Bureau of Mineral Resources, Geology and Geophysics, Record 1986/14.

——, ——, ——, PRIDE, C. & JAHN, B.-M. 1987. *Trace element geochemistry and petrogenesis of Archaean felsic igneous units, Pilbara Block, Western Australia.* Bureau of Mineral Resources, Record 1987/30.

——, PRIDE, C., JAHN, B., DAVY, R. & HICKMAN, A. H. 1986B. *RE and HFS (Ti, Zr, Nb, P, Y) element evolution of Archaean mafic–ultramafic volcanic suites, Pilbara Block, Western Australia.* Bureau of Mineral Resources, Geology and Geophysics, Record 1986/6.

GROVES, D. I., DUNLOP, J. S. R. & BUICK, R. 1981. An early habitat of life. *Scientific American*, **245**, 64–73.

GSWA 1975. *The Geology of Western Australia.* Western Australia Geological Survey, Memoir **2**.

—— 1990. *Geology and Mineral Resources of Western Australia.* Western Australia Geological Survey, Memoir **3**.

HAQ, B. U., HARDENBOL, J. & VAIL, P. R. 1988. Mesozoic and Cenozoic chronostratigraphy and cycles of sea-level change. *In*: WILGUS, C. E. *et al.* (eds) *Sea-level change: an integrated approach.* Society of Economic Paleontologists and Mineralogists, Special Publications, **42**, 71–108.

HALLBERG, J. A. & GLIKSON, A. Y. 1981. Archaean granite-greenstone terranes of Western Australia. *In*: HUNTER, D. R. (ed.) *Precambrian of the Southern Hemisphere.* Elsevier, Amsterdam, 111–136.

HASSLER, S. W. 1991. *Depositional processes, paleogeography, and tectonic setting of the Main Tuff Interval of the Wittenoom Dolomite, late Archean Hamersley Group, Western Australia.* PhD Thesis, University of California, Santa Barbara, Geological Sciences.

HICKMAN, A. H. 1977a. New and revised definitions of rock units in the Warrawoona Group, Pilbara Block. *Western Australia Geological Survey, Annual Report for 1976*, 53.

—— 1977b. Stratigraphic relations of rocks within the Whim Creek Belt. *Western Australia Geological Survey, Annual Report for 1976*, 53–56.

—— 1980. *Lithological map and stratigraphic interpretation of the Pilbara Block, 1 : 1 000 000 scale.* Western Australia Geological Survey.

—— 1993. *Geology of the Pilbara Block and its environs.* Western Australia Geological Survey, Bulletin 127.

—— 1984. Archaean diapirism in the Pilbara Block, Western Australia. *In*: KRÖNER, A. & GREILING, R. (eds) *Precambrian Tectonics illustrated.* E. Schweizerbart'sche, Stuttgart, Germany, 113–127.

—— 1990. *Pilbara Excursion Guide.* Third International Archaean Symposium, Perth.

—— 1992. A 3000–2900 Ma change in the evolution of granite-greenstone terrains: was this the beginning of plate tectonics? *29th International Geological Congress, Kyoto, Japan, Abstracts*, **1**, 49.

HOLMES, A. 1927. *The Age of the Earth.* Benn's Sixpenny Library No. **102**, Ernest Benn Ltd, London.

—— & LAWSON, R. W. 1927. Factors involved in the calculation of the ages of radioactive minerals. *American Journal of Science*, **13**, 327–344.

HORWITZ, R. C. 1979. The Whim Creek Group, a discussion. *Journal of the Royal Society of Western Australia*, **61**, 67–72.

—— 1986. Unconformities and volcanic provinces of the Pilbara Block, Western Australia. *Twelfth International Sedimentological Conference, Canberra, Abstracts*, 143.

—— 1987. The unconformity in the Kelly Belt, east Pilbara Craton. *Journal of the Royal Society of Western Australia*, **70**, 49–53.

—— 1990a. Palaeogeographic and tectonic evolution of the Pilbara Craton, Northwestern Australia. *Precambrian Research*, **48**, 327–340.

—— 1990b. *The Archaean unconformity at Shay Gap, northeastern Pilbara Craton.* Commonwealth Scientific and Industrial Research Organisation (CSIRO), Division of Exploration Geoscience, Restricted Report 123R.

—— 1990c. Evolution of the Pilbara Craton. *Third International Archaean Symposium, Extended Abstracts Volume*, 79–80.

—— & PIDGEON, R. T. 1993. 3.1 Ga tuff from the Sholl Belt in the West Pilbara: further evidence for diachronous volcanism in the Pilbara Craton of Western Australia. *Precambrian Research*, **60**, 175–183.

HUGHES, F. E. 1990. *Geology of the Mineral Deposits of Australia and Papua and New Guinea.* Australasian Institute of Mining and Metallurgy, Melbourne.

INGRAM, P. A. J. 1977. A summary of the geology of a portion of the Pilbara Goldfield, Western Australia. *In*: MCCALL, G. J. H. (ed.) *The Archaean, Search for the Beginning.* Dowden, Hutchison & Ross, Stoudsburg, Pennsylvania, 208–216.

IRVING, E. & GREEN, R. 1958. Polar movement relative to Australia. *Geophysical Journal of the Royal Astronomical Society*, **1**, 64–72.

ISOZAKI, Y., MARUYAMA, S. & KIMURA, G. 1991. Middle Archean (3.3 Ga) Cleaverville accretionary complex in northwestern Pilbara Block, Western Australia. *Eos* (Program and Abstracts, Fall Meeting, American Geophysical Union), **72**, No. 44 Supplement, 542.

JAHN, B. M., GLIKSON, A. Y., PEUCAT, J. J. & HICKMAN, A. H. 1981. REE geochemistry and isotopic data of Archean silicic volcanics and granitoids from the Pilbara Block, Western Australia: implications for the early crustal evolution. *Geochimica et Cosmochimica Acta*, **45**, 1633–1652.

KIMURA, G., MARUYAMA, S. & ISOZAKI, Y. 1991. Early Archean accretionary complex in the eastern Pilbara craton, Western Australia. *Eos* (Program and Abstracts, Fall Meeting, American Geophysical Union) **72**, No. 44 Supplement, 542.

KNIGHT, C. L. (ed.) 1975. *Economic Geology of Australia and Papua New Guinea. 1. Metals.* Australasian Institute of Mining and Metallurgy, Monographs 5.

KRAPEZ, B. 1984. Sedimentation in a small, fault-bounded basin — the Ialla Rookh Sandstone, East Pilbara Block. *In*: MUHLING, J. R., GROVES, D. I. & BLAKE, T. S. (eds) *Archaean and Proterozoic Basins of the Pilbara, Western Australia — Evolution and Mineralization Potential.* University of Western Australia, Geology Department and University Extension Publications, **9**, 89–110.

—— 1989. *Depositional styles and geotectonic settings of Archean metasedimentary sequences: evidence from the Ialla Rookh Basin, Pilbara Block, Western*

Australia. PhD Thesis, University of Western Australia.

—— 1993. Sequence stratigraphy of the Archaean supracrustal belts of the Pilbara Block, Western Australia. *In*: BLAKE, T. S. & MEAKINS, A. (eds) *Archaean and Early Proterozoic Geology of the Pilbara Region, Precambrian Research,* **60** (Special issue), 1–45.

—— & BARLEY, M. E. 1987. Archaean strike-slip faulting and related ensialic basins: evidence from the Pilbara Block, Western Australia. *Geological Magazine,* **124**, 555–567.

LIPPLE, S. L. 1975. Definitions of new and revised units of the eastern Pilbara region. *Western Australia Geological Survey, Annual Report for 1974,* 58–63.

LOWE, D. R. 1982. Comparative sedimentology of the principal volcanic sequences of Archaean greenstone belts in South Africa, Western Australia and Canada — implications for crustal evolution. *Precambrian Research,* **17**, 1–29.

—— 1983. Restricted shallow-water sedimentation of early Archaean stromatolitic and evaporitic strata of the Strelley Pool Chert, Pilbara Block, Western Australia. *Precambrian Research,* **19**, 239–283.

—— & BYERLY, G. R. 1986. Early Archean silicate spherules of probable impact origin, South Africa and Western Australia. *Geology,* **14**, 83–86.

MACFARLANE, A. W. & HOLLAND, H. D. 1990. Paleoweathering of the Mt. Roe Basalt, Fortescue Group, and the oxygen content of the late Archean atmosphere. *Third International Archaean Symposium, Extended Abstracts Volume,* 289–291.

——, DANIELSON, A. & HOLLAND, H. D. 1994a. Geology and major and trace element chemistry of late Archean weathering profiles in the Fortescue Group, Western Australia: implications for atmospheric P_{O_2}. *Precambrian Research,* **65**, 297–317.

——, ——, —— & JACOBSEN, S. B. 1994b. REE chemistry and Sm-Nd systematics of late Archean weathering profiles in the Fortescue Group, Western Australia. *Geochimica et Cosmochimica Acta,* **58**, 1777–1794.

MACLEOD, W. N., DE LA HUNTY, L. E., JONES, W. R. & HALLIGAN, R. 1963. A preliminary report on the Hamersley Iron Province, North West Division. *Western Australia Geological Survey, Annual Report for 1962,* 44–54.

MAITLAND, A. G. 1909. *Geological investigations in the country lying between 21° 30′ and 25° 30′ S lat and 113° 30′ and 118° 30′ E long, embracing parts of the Gascoyne, Ashburton and West Pilbara Goldfields.* Western Australia Geological Survey, Bulletin 33.

MCELHINNY, M. W. & SENANAYAKE, W. E. 1980. Paleomagnetic evidence for the existence of the geomagnetic field 3.5 Ga ago. *Journal of Geophysical Research,* **85**, 3523–3528.

MCNAUGHTON, N. J., COMPSTON, W. & BARLEY, M. E. 1993. Constraints on the age of the Warrawoona Group, eastern Pilbara Block, Western Australia. *Precambrian Research,* **60**, 69–98.

——, GREEN, M. D., COMPSTON, W. & WILLIAMS, I. S. 1988. Are anorthositic rocks basement to the Pilbara

Craton? *Australian Conference on Geochronology, Abstracts,* 272–273.

MORANT, P. 1984. *Metamorphism of an Archaean granitoid–greenstone terrain, eastern Pilbara Block.* PhD Thesis, University of Western Australia.

MORRIS, R. C. 1985. Genesis of iron ore in banded iron-formation by supergene and supergene-metamorphic processes — a conceptual model. *In*: WOLF, K. H. (ed.) *Handbook of stratabound and stratiform ore deposits,* **13**, Elsevier, Amsterdam, 73–235.

—— 1993. Genetic modelling for banded iron-formation of the Hamersley Group, Pilbara Craton, Western Australia. *In*: BLAKE, T. S. & MEAKINS, A. (eds) *Archaean and Early Proterozoic Geology of the Pilbara Region, Western Australia. Precambrian Research,* **60**, 243–286.

MYERS, J. S. 1993. Precambrian history of the West Australian Craton and adjacent orogens. *Annual Reviews of Earth and Planetary Sciences,* **21**, 453–485.

PIDGEON, R. T. 1978a. Geochronological investigations of granite batholiths of the Archaean granite-greenstone terrain of the Pilbara Block, Western Australia. *In*: SMITH, I. E. M. & WILLIAMS, J. G. (eds) *Proceedings of the 1978 Archaean Geochemistry Conference.* University of Toronto, Ontario, 360–362.

—— 1978b. 3450 m.y.-old volcanics in the Archaean layered greenstone succession of the Pilbara Block, Western Australia. *Earth and Planetary Science Letters,* **37**, 421–428.

—— 1984. Geochronological constraints on early volcanic evolution of the Pilbara Block, Western Australia. *Journal of the Geological Society of Australia,* **31**, 237–242.

—— & HORWITZ, R. C. 1991. The origin of olistoliths in Proterozoic rocks of the Ashburton Trough, Western Australia, using zircon U–Pb isotopic characteristics. *Australian Journal of Earth Sciences,* **38**, 55–63.

PLUMB, K. A. & JAMES, H. L. 1986. Subdivisions of Precambrian time: recommendations and suggestions by the Subcommission on Precambrian Stratigraphy. *Precambrian Research,* **32**, 65–92.

PORATH, H. 1967. Palaeomagnetism and the age of Australian Haematite ore bodies. *Earth and Planetary Science Letters,* **2**, 409–414.

—— & CHAMALAUN, F. H. 1968. Palaeomagnetism of Australian haematite ore bodies, II — Western Australia. *Geophysical Journal of the Royal Astronomical Society,* **15**, 253–264.

RYAN, G. R. 1964. A reappraisal of the Archaean of the Pilbara Block. *Western Australia Geological Survey, Annual Report for 1963,* 25–28.

SCHMIDT, P. W. & EMBLETON, B. J. J. 1985. Prefolding and overprint magnetic signatures in Precambrian (~2.9–2.7 Ga) igneous rocks from the Pilbara Craton and Hamersley Basin, NW Australia. *Journal of Geophysical Research,* **90**, 2967–2984.

SCHOPF, J. W. 1993. Microfossils of the Early Archean Apex Chert: new evidence of the antiquity of life. *Science,* **260**, 640–646.

—— & PACKER, B. M. 1987. Early Archaean (3.3 billion to 3.6 billion-year-old) microfossils from

Warrawoona Group, Australia. *Science*, **237**, 70–73.

—— & WALTER, M. R. 1983. Archean microfossils: new evidence of ancient microbes. *In*: SCHOPF, J. W. (ed.) *Earth's earliest biosphere, its origin and evolution.* Princeton University Press, Princeton, New Jersey, 214–239.

SEYMOUR, D. B., THORNE, A. M. & BLIGHT, D. F. 1988. *Wyloo, Western Australia* (Second Edition). Western Australia Geological Survey, 1 : 250 000 Geological Series Explanatory Notes.

SIMONSON, B. M., HASSLER, S. W. & SCHUBEL, K. A. 1993*a*. *Lithology and proposed revisions in stratigraphic nomenclature of the Wittenoom Formation (Dolomite) and overlying formations, Hamersley Group, Western Australia.* Western Australia Geological Survey, Professional Papers 34.

——, SCHUBEL, K. A. & HASSLER, S. W. 1993*b*. Carbonate sedimentology of the early Precambrian Hamersley Group of Western Australia. *Precambrian Research*, **60**, 287–335.

SIMPSON, E. S. 1910. Pilbarite, a new mineral from the Pilbara goldfields. Verbatim report of paper read to Natural History and Science Society of Western Australia, published in *West Australian* newspaper of 17 August 1910 under heading 'A new radioactive mineral'.

TARLOWSKI, C., SIMONIS, F. & MILLIGAN, P. 1993. *Magnetic anomaly map of Australia*, Scale 1 : 5 000 000. Australian Geological Survey Organization.

TEYSSIER, C. & COLLINS, W. J. 1990. Strain and kinematics during the diapiric emplacement of the Mount Edgar Batholith and Warrawoona Syncline, Pilbara. *Third International Archaean Symposium, Extended Abstracts Volume*, 481–483.

THORNE, A. M. & TYLER, I. M. 1994. *Geology of the Paraburdoo 1 : 100 000 Sheet, Western Australia.* Western Australia Geological Survey, Record 1992/11.

THORPE, R. I., HICKMAN, A. H., DAVIS, D. W., MORTENSEN, J. K. & TRENDALL, A. F. 1992. Conventional U-Pb zircon geochronology of Archaean felsic units in the Marble Bar region, Pilbara Craton, Western Australia. *Precambrian Research*, **56**, 169–189.

TRENDALL, A. F. 1979. A revision of the Mount Bruce Supergroup. *Western Australia Geological Survey, Annual Report for 1978*, 63–71.

—— 1983. The Hamersley Basin. *In*: TRENDALL, A. F. & MORRIS, R. C. (eds) *Iron-formation: Facts and Problems.* Elsevier, Amsterdam, 69–129.

—— 1984. The Archean/Proterozoic transition as a geological event — a view from Australian evidence. *In*: HOLLAND, H. D. & TRENDALL, A. F. (eds) *Patterns of change in earth evolution.* Springer-Verlag, Berlin, 243–259.

—— 1992. The Woongarra Volcanics, a giant lavalike felsic sheet (GLFs) of the *c.* 2.8–2.4 Ga Hamersley Basin, Pilbara Craton, Western Australia. *29th International Geological Congress, Kyoto, Japan, Abstracts*, **2**, 495.

—— (in press). *Bibliography of the Precambrian geology of the Pilbara Craton, with a historical commentary.* Western Australia Geological Survey, Record 1995/?

—— & BLOCKLEY, J. G. 1968. Stratigraphy of the Dales Gorge Member of the Brockman Iron Formation, in the Precambrian Hamersley Group of Western Australia. *Western Australia Geological Survey, Annual Report for 1967*, 48–53.

TYLER, I. M. 1990. *Geology of the Newman 1 : 100 000 Sheet, Western Australia.* Western Australia Geological Survey, Record 1990/3.

——, HUNTER, W. M. & WILLIAMS, I. R. 1990. *Newman, Western Australia* (Second Edition). Western Australia Geological Survey, Record 1989/6.

WALTER, M. R. 1983. Archean stromatolites: evidence of the earth's earliest benthos. *In*: SCHOPF, J. W. (ed.) *Earth's earliest biosphere, its origin and evolution.* Princeton University Press, Princeton, New Jersey, 187–213.

——, BUICK, R. & DUNLOP, J. S. R. 1980. Stromatolites 3400–3500 Myr old from the North Pole area, Western Australia. *Nature*, **284**, 443–445.

WIJBRANS, J. R. & McDOUGALL, I. 1987. On the metamorphic history of an Archaean granitoid greenstone terrain, east Pilbara, Western Australia , using the $^{40}Ar/^{39}Ar$ age spectrum technique. *Earth and Planetary Science Letters*, **84**, 226–242.

WILHELMIJ, H. R. 1986. *Depositional history of the middle Archaean sedimentary sequences in the Pilbara Block, Western Australia: a genetic stratigraphic analysis of the terrigenous rocks of the Pilgangoora Syncline.* PhD Thesis, University of Western Australia.

—— & DUNLOP, J. S. R. 1984. A genetic stratigraphic investigation of the Gorge Creek Group in the Pilgangoora Syncline. *In*: MUHLING, J. R., GROVES, D. I. & BLAKE, T. S. (eds) *Archaean and Proterozoic Basins of the Pilbara, Western Australia — Evolution and Mineralization Potential.* University of Western Australia, Geology Department and University Extension Publications, **9**, 68–88.

WILLIAMS, I. S. & COLLINS, W. J. 1990. Granite-greenstone terranes in the Pilbara Block, Australia, as coeval volcano-plutonic complexes: evidence from U-Pb zircon dating of the Mount Edgar Batholith. *Earth and Planetary Science Letters*, **97**, 41–53.

——, PAGE, R. W., FOSTER, J. J., COMPSTON, W., COLLERSON, K. D. & McCULLOCH, M. T. 1982. Zircon U-Pb ages from the Shaw Batholith, Pilbara Block, determined by ion microprobe. Australian National University, Research School of Earth Sciences, Annual Report 1982, 199–203.

——, ——, FROUDE, D., FOSTER, J. J. & COMPSTON, W. 1983. Early crustal components in the Western Australian Archaean: zircon U-Pb ages by ion microprobe analysis from the Shaw Batholith and Narryer metamorphic belt. In Sixth Australian Geological Convention — Lithosphere Dynamics and Evolution of Continental Crust, Abstracts Volume, Geological Society of Australia, Abstracts 9, 169.

YUAN, P. B. & LOWE, D. R. 1990. Facies changes in the Strelley Pool Chert, Western Australia, and its implications. *Third International Archaean Symposium, Perth, Extended Abstracts Volume*, 321–322.

The generation and assembly of an Archaean supercontinent: evidence from the Yilgarn craton, Western Australia

JOHN S. MYERS

Geological Survey of Western Australia, 100 Plain Street, East Perth, Western Australia 6004

Abstract: The Yilgarn craton consists of granites, greenstones and granitic gneisses ranging from at least 3730 to 2550 Ma. These rocks can be divided into a number of tectonostratigraphic terranes that comprise distinct rock units with different geological histories. However all the terranes, regardless of their age and composition, show evidence of intense tectonic, volcanic, plutonic, and metamorphic activity between 2780 and 2630 Ma. This is interpreted as a major episode of plate tectonic activity which swept together and amalgamated a number of diverse crustal fragments (including volcanic arcs, back-arc basins and microcontinents) to form a super-continent, of which the Yilgarn craton is a fragment. The main features of the Narryer, Murchison and Kalgoorlie terranes are described as examples of the diversity of crustal components within the craton and the nature of the processes by which these components were combined to form the Yilgarn craton.

One of John Sutton's many assets was his infectious enthusiasm about geology at all scales from outcrops to continents, and his ability to simplify and synthesize diverse aspects of geology into broad hypotheses of crustal evolution.

As large numbers of radiometric age determinations first became available during the 1950s, it became apparent that most K–Ar and Rb–Sr ages clustered into a small number of peaks (Gastil 1960*a*, *b*). This led John to propose the concept of long-term cyclicity in the formation of continents (Sutton 1963). He called them chelogenic (shield-forming) cycles. This name has not been widely used in the geological literature, but the basis for the concept, that of periodicity in the peaks of radiometrically determined geological events, has been supported and enhanced by subsequent geochronology, involving the whole range of techniques old and new (Moorbath 1977, 1978; Condie 1989).

John Sutton related this periodicity in major crust-forming events to patterns of mantle convection. There is still debate about the ultimate cause of this periodicity and the kind of mantle processes involved (Le Pichon & Huchon 1984; Gurnis & Davis 1986; Gurnis 1988; Condie 1989; Hoffman 1989*a*). However at the Earth's surface, these events appear to reflect the formation of supercontinents (Hoffman 1989*a*; Kröner 1991). Well-known episodes of supercontinent formation occurred in the Early Proterozoic (Laurentia), Mid Proterozoic (Rodinia), Late Proterozoic (Gondwanaland) and Late Palaeozoic (Pangaea) (Dalziel 1991; Gaal 1992; Hoffman 1988, 1989*b*;

Stoeser & Camp 1985; Van Kranendonk *et al.* 1993), but these largely involved older crustal fragments that formed during an intense episode of continent formation in the Late Archaean. Evidence of this Late Archaean episode of continent formation is described below from the Yilgarn craton of Western Australia.

Yilgarn craton

The Yilgarn craton is bounded by faults and orogenic belts which truncate the structures within the craton (Fig. 1). The craton thus appears to be a remnant of a formerly larger Late Archaean continent which was subsequently fragmented by rifting, and then successively amalgamated with other fragments of continental crust during the Proterozoic (Myers 1993). The faults and orogenic belts which bound the Yilgarn craton have remained major lines of crustal weakness, and have repeatedly been reworked as zones of continental disruption and amalgamation, and intracontinental deformation.

The Archaean crustal remnant called the Yilgarn craton has remained intact and free of significant internal distortion since *c.* 2500 Ma. It has of course moved around the surface of the Earth, and has moved up and down, but vertical movements did not exceed *c.* 5 km, because rocks that formed at, or near, the Earth's surface are still at low metamorphic grade.

This relative stability near sea level throughout Proterozoic and Phanerozoic time resulted in the smooth surface that we see today. The flatness and

From COWARD, M. P. & RIES, A. C. (eds), 1995, *Early Precambrian Processes*, Geological Society Special Publication No. 95, pp. 143–154.

relative stability of this large raft of crust for all this time is, in some ways, even more impressive than more prominent, short-lived mountain ranges. The surface was covered by a thin blanket of Permian glacial, fluvioglacial and clastic sedimentary deposits which were largely removed before the Tertiary. It was subsequently covered by weathering products, several metres thick, during a Miocene–Pliocene episode of increasing aridity. Recent uplift to form an elevated plateau (mainly 250–500 m altitude), and associated erosion, have extensively reduced or removed this mantle of detritus and weathered rock, revealing fresh rock and rubble.

The Yilgarn craton (Fig. 1) mainly consists of granite and greenstones that formed between 3000 and 2600 Ma. Most of these rocks were metamorphosed at low grade and have always formed an upper part of the crust. Deeper Late Archaean crustal levels are exposed in the southwest and northwest. In the southwest (Dumbleyung and Hyden terranes), granites and greenstones were metamorphosed in granulite facies at c. 2640 Ma (Nemchin et al. 1994). The rocks were rapidly elevated and overlying crust eroded before c. 2600 Ma when they were intruded by high-level granites and partly recrystallized in greenschist facies (Myers 1990a). In the northwest (Narryer terrane), the c. 3700–3300 Ma Narryer gneiss complex was extensively intruded by granite, repeatedly deformed, and recrystallized in granulite facies at c. 2700 Ma. It remained at a deep crustal level until c. 1800–1600 Ma when it was rapidly elevated during collision and amalgamation of the Pilbara and Yilgarn cratons (Muhling 1990; Myers 1990b, part 1).

Rocks and structures are superficially similar over large parts of the craton, but in detail they can be subdivided into units with distinct geological histories which range in age from 3730 to 2550 Ma. These units form a number of tectonostratigraphic terranes (Fig. 1) that, regardless of their age, all indicate intense tectonic, volcanic, plutonic and metamorphic activity between 2780 and 2630 Ma. This activity is interpreted as a major episode of plate tectonic movements which swept together and amalgamated a number of diverse crustal fragments (including volcanic arcs, back-arc basins and microcontinents) to form the Yilgarn craton at this time (Myers 1990c).

This conclusion contrasts with most previous interpretations of the Yilgarn craton. During the 1960s and 1970s the greenstones were interpreted as remnants of a primordial simatic crust that was deformed during the diapiric emplacement of steep-sided granite batholiths (Glikson 1979). In the 1980s most models interpreted the greenstones and granites as products of rifting within older continental crust (Gee et al. 1981; Groves & Batt 1984; Groves et al. 1985, 1987; Hallberg 1986; Blake & Groves 1987). Groves et al. (1978) and Blake & Groves (1987) concluded that there was no evidence of subduction-related processes in either the granites or greenstones. However Barley & McNaughton (1988) interpreted the Norseman–Wiluna belt (approximately the Kalgoorlie, Gindalbie and Jubilee terranes of Fig. 1) in terms of a westward-dipping subduction zone, and this was further elaborated by Barley et al. (1989) and Morris (1993). In contrast, Campbell & Hill (1988) and Hill et al. (1991) proposed that the rocks and structures in the southern part of this belt reflect plume tectonic rather than plate tectonic processes, and Glikson (1993) suggested that peaks of granite–greenstone generation were related to meteorite impacts.

The Yilgarn craton was formerly divided into the Eastern Goldfields, Southern Cross and Murchison Provinces and the Western Gneiss Terrain (Gee et al. 1981). These were broad regions thought to be characterized by different proportions of major rock types. Subsequent mapping indicates that the Western Gneiss Terrain contains gneisses with diverse crustal histories, and some of the gneisses are integral parts of the granite–greenstone terrains (Myers 1990a). The granite–greenstone provinces comprise a number of distinct tectonostratigraphic terranes, and previously loosely defined subdivision is here replaced by terrane nomenclature based on recently mapped fault-bounded tectonostratigraphic units.

Three terranes are described below as examples of the diversity of the crustal components of the craton and the nature of the processes by which these components were combined to form the Yilgarn craton. All the ages referred to are approximate and, unless otherwise indicated, are based on U–Pb ages of zircons (Table 1).

Narryer terrane

The Narryer terrane forms the northwestern part of the Yilgarn craton (Fig. 1), and contains the oldest known components of the craton (De Laeter et al. 1981; Fletcher et al. 1988; Kinny et al. 1988; Myers 1988, 1990b, parts 1 & 2; Nutman et al. 1991). It consists of heterogeneous high-grade gneiss derived from granite and minor amounts of granodiorite, tonalite, layered basic intrusions and metasedimentary rocks. They include Meeberrie gneiss, that formed from 3730 Ma tonalite and 3680–3600 Ma granite, and the Manfred complex of 3730 Ma layered anorthosite, gabbro and ultramafic rocks. After, deformation and metamorphism these rocks were intruded by granite, between 3400 and 3300 Ma ago, that forms the Dugel gneiss. This

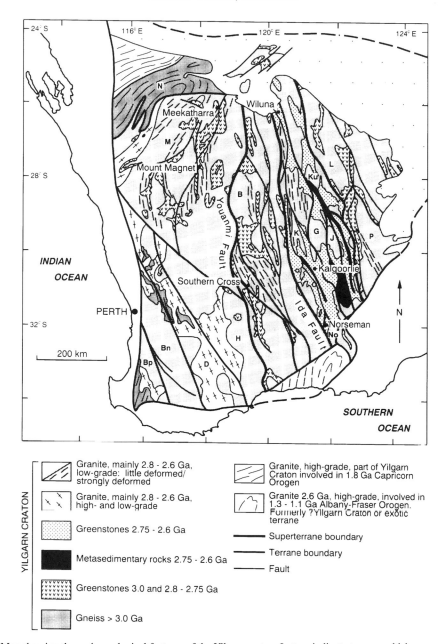

Fig. 1. Map showing the main geological features of the Yilgarn craton. Letters indicate terranes which are grouped into three superterranes: 1. west Yilgarn superterrane comprising N – Narryer, M – Murchison, Bp – Balingup, Bn – Boddington, D – Dumbleyung and H – Hyden terranes; 2. central Yilgarn superterrane comprising B – Barlee and Y – Yellowdine terranes; 3. east Yilgarn superterrane comprising K – Kalgoorlie, Ku – Kurnalpi, L – Laverton, G – Gindalbie, J – Jubilee, P – Pinjin and No – Norseman terranes. The open stipple indicates Yilgarn craton obscured by a cover of younger rocks.

second major plutonic episode culminated in high-grade metamorphism at 3300 Ma marked by melt patches and veins, and overgrowths on zircons.

Sedimentary rocks, now mainly quartzite, pelite and BIF (banded iron formation), were deposited on the older rocks between 3100 and 2700 Ma, and contain detrital zircons ranging back to 4270 Ma that are the oldest known fragments of terrestrial

Table 1. *The main sequence of events in four different terranes of the Yilgarn craton showing the times of their amalgamation*

NARRYER	MURCHISON	BARLEE	KALGOORLIE
Granite 2.64 - 2.62	Granite 2.64 - 2.60		Granite 2.64 - 2.60
COMPLETE ASSEMBLY OF YILGARN TERRANES 2.65			
DEFORMATION AND PEAK METAMORPHISM 2.68 - 2.65 During assembly of Narryer and Murchison terranes		**SYNTECTONIC GRANITE 2.66** During assembly of Barlee and Kalgoorlie terranes	
Granite 2.78 - 2.68 Greenstones c. 2.75	Granite 2.71 - 2.68 Greenstones 2.8 - 2.75 Granite 2.9 Greenstones 3.0 - 2.94	Granite 2.69 - 2.85 Greenstones 2.73	Granite 2.685 Greenstones 2.69
Granite 3.4 - 3.3 Granite 3.68 - 3.60 Tonalite , Gabbro Anorthosite 3.73	Composite Basement 3.5 - 3.1		

Ages in Ga are based on U–Pb studies of zircons published by Campbell & Hill (1988), Claoué-Long *et al.* (1988), Hill *et al.* (1992), Kinny *et al.* (1988), Myers (1990*b*, part 2), Nemchin *et al.* (1994), Nutman *et al.* (1991), Pidgeon & Wilde (1990), and Wiedenbeck & Watkins (1993).

rocks (Froude *et al.* 1983; Compston & Pidgeon 1986).

A third plutonic episode is marked by the widespread emplacement of granite sheets and plutons between 2780 and 2630 Ma. This episode included intense deformation and high-grade metamorphism related to tectonic interleaving by thrust stacking of adjacent parts of the Narryer and Murchison terranes. These terranes collided and amalgamated at 2680–2650 Ma (Figs 2b & 3b) and were intruded by sheets of granite.

There appears to be a hiatus in the zircon ages at *c.* 2680 Ma which may mark the onset of collision between the Narryer and Murchison terranes. Granite emplaced as sheets at 2800–2680 Ma may have been generated from oceanic crust that was subducted during ocean closure as the two slabs of continental crust converged (Figs 2a & 3a). Strongly deformed 2650 Ma granites contain 2900 Ma zircon xenocrysts (Kinny *et al.* 1990) that may reflect melting of subducted Murchison terrane during early stages of collision (Fig. 3b). Granites, dated at 2630 Ma, post-date high-grade metamorphism and may have been derived by melting of thickened continental crust during extensional collapse and erosion of the amalgamated continents.

The main structures within the Narryer terrane are upright fold interference structures. D1 recumbent folds and thrusts were refolded successively by upright folds with E–W (D2) (Fig. 2b) and then N–S (D3) axes (Fig. 2c).

The boundary between the Narryer and Murchison terranes is marked by the Yalgar fault. It is a zone of intense late Archaean and early

Proterozoic deformation that truncates the structures in both terranes. Sheets of late Archaean granite were emplaced along this zone and partly converted to mylonite. The fault zone is still active.

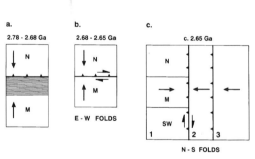

Fig. 2. Cartoon maps illustrating the main sequence of events leading to the amalgamation of terranes and superterranes. (**a**) Oceanic crust (horizontal lines) is consumed below the Narryer terrane (N) during the approach of the Murchison terrane (M) between 2.78 and 2.68 Ga. (**b**) Collision and amalgamation of the Narryer and Murchison terranes between 2.68 and 2.65 Ga led to sub-horizontal tectonic interleaving followed by upright E–W folds in both terranes, and dextral transcurrent movements between the terranes. (**c**) After removal of an eastern portion of the combined Narryer terrane (N), Murchison terrane (M) and southwestern Balingup, Boddington, Dumbleyung and Hyden terranes (collectively marked as SW) by rifting or transcurrent faulting, the west Yilgarn superterrane (1) was amalgamated with the central (2) and east (3) Yilgarn superterranes at *c.* 2.65 Ga. This resulted in upright north–south folds in all the terranes, and dextral transcurrent movement along the suture between the west and central Yilgarn superterranes.

ASSEMBLY OF NARRYER AND MURCHISON TERRANES

a. 2.80 - 2.68 Ga

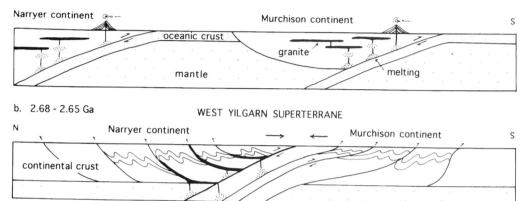

b. 2.68 - 2.65 Ga   WEST YILGARN SUPERTERRANE

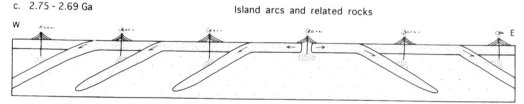

GENERATION AND ASSEMBLY OF EAST AND CENTRAL YILGARN SUPERTERRANES

c. 2.75 - 2.69 Ga   Island arcs and related rocks

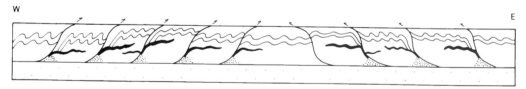

d. 2.69 - 2.66 Ga

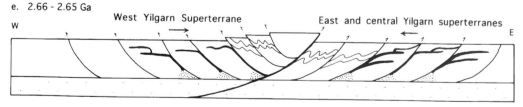

ASSEMBLY OF YILGARN SUPERTERRANES

e. 2.66 - 2.65 Ga

Fig. 3. Cartoon sections illustrating some features of the main sequence of events leading to the amalgamation of terranes and superterranes in the Yilgarn craton. (**a**) and (**b**) Assembly of the Narryer and Murchison terranes to form part of the west Yilgarn superterrane. (**c**) and (**d**) Generation and assembly of the east and central Yilgarn superterranes. (**e**) Assembly of the west Yilgarn superterrane and combined central and east Yilgarn superterranes.

Recent fault scarps may have formed in association with an earthquake recorded in 1885 (Williams 1979), and another earthquake in 1941 (Richter magnitude 6.8) was the largest and deepest (focal depth 33 km) ever recorded on the Australian ontinent (Denham 1976).

,vlurchison terrane

In contrast to the Narryer terrane, most of the Murchison terrane comprises granite and greenstones metamorphosed in prograde greenschist facies (Fig. 1). Deeper levels of the terrane are exposed in the west and southwest where the rocks are metamorphosed in amphibolite facies and include extensive belts of gneiss (Murgoo gneiss) derived from heterogeneous granites with emplacement ages of 3490, 3440, 2900 and 2780 Ma (Myers 1990a, b, part 2; Pidgeon & Wilde 1990; Nutman et al. 1991; Wiedenbeck & Watkins 1993). Much of the Murchison terrane has recently been mapped and described in detail by Watkins & Hickman (1990) and this work is summarized below.

Two major stratigraphic groups are recognized in the greenstones of the Murchison terrane. The lower part of the older, 3000 Ma (Pidgeon & Wilde 1990), Luke Creek Group consists of two sequences of submarine tholeiitic and high-Mg basalt lava flows overlain by BIF, that formed during two episodes of lava-plain volcanism. Most of the lavas are massive and contain very few pillow lava structures. The upper part of the Luke Creek Group consists of interlayered basalt and silicic volcanic and volcaniclastic rocks. There are fundamental differences in the geochemistry of the volcanic rocks in the northern and southern parts of the Luke Creek Group which suggest that the volcanic rocks were derived from different upper mantle sources.

The Luke Creek Group was intruded by sheets of granite at 2900 Ma and, together with the granite, was deformed in a sub-horizontal tectonic regime, D1. The granite was converted into gneiss that forms part of the Murgoo gneiss complex. The younger 2800 Ma Mount Farmer Group unconformably overlies the Luke Creek Group and consists of remnants of nine distinct volcanic centres and one epiclastic sedimentary basin, all of local extent. Parts of large layered basic intrusions occur at or near the eastern margin of the Murchison terrane (Fig. 1).

Thick sheets of monzogranite were intruded into the volcanic rocks and gneissose monzogranite at 2690–2680 Ma (Fig. 3b). They are compositionally similar to the older gneissose monzogranite and may have ben derived by similar processes from similar lower crustal source rocks.

All these rocks were deformed and repeatedly folded between 2680 and 2640 Ma. Fold interference structures resulted from older subhorizontal D1 thrusts and fold limbs being refolded successively by upright folds with E–W (D2) (Fig. 2b) and then N–S (D3) axes (Fig. 2c) (Myers & Watkins 1985). The rocks were recrystallized in prehnite–pumpellyite, greenschist or low amphibolite facies except in the southwest and west where the regional metamorphic grade is in amphibolite facies.

The last stages of ductile deformation (D4) in the Murchison terrane formed major N–S shear zones. These shear zones formed in a continuation of the regional D3 stress regime and spanned the emplacement of 2630 Ma plutons of granite, granodiorite and tonalite, and the peak of regional metamorphism. The plutons form two distinct groups that have different compositions, suggesting that they were derived from different crustal sources: older siliceous crust in the south and younger mafic crust in the north. This also appears to reflect two distinct kinds of lithosphere that were already juxtaposed to form the basement of the Murchison terrane before the eruption of the Luke Creek Group of volcanic rocks (Watkins et al. 1991). The 3490 and 3440 Ma components of the Murgoo gneiss complex may have been part of this basement.

Kalgoorlie terrane

The Kalgoorlie terrane (Fig. 1) is a fault-bounded unit of granite and greenstones with a greenstone stratigraphy and deformation history that are distinct from adjacent mapable granite–greenstone units (Swager et al. 1992). The components of the terrane formed immediately prior to the assembly of the craton. This contrasts with both the Murchison and Narryer terranes which largely consist of continental crust that is much older than the Late Archaean assembly of the Yilgarn craton.

The greenstones mostly formed at 2700–2690 Ma (Claoué-Long et al. 1988; Swager et al. 1992) and comprise a lower basalt unit of high-Mg basalt that passes upwards into tholeiite. This is followed by a unit of komatiite in which discontinuous bodies of dunite (olivine orthocumulate and adcumulate) are overlain by komatiite flows characterized by olivine spinifex textures. Thin layers of high-Mg basalt with variolitic textures occur towards the top of the komatiite unit which is overlain by an upper unit of tholeiitic and high-Mg basalt. This was followed by the deposition of silicic volcanic and sedimentary rocks. The silicic volcanic rocks are predominantly dacitic but range from rhyolite to andesite, and include lava and pyroclastic flows and waterlain tuff. They are interbedded with siltstone and sandstone. The upper part of this unit is dominated by clastic sedimentary rocks largely derived from silicic volcanic centres. The sequence is unconformably overlain by coarse clastic rocks (marked on Fig. 1 as metasedimentary rocks 2.75–2.6 Ga) that were deposited along major strike-slip fault zones during early stages of (D2) deformation.

The greenstone stratigraphy was stacked by thrusts and recumbent folds (D1), and then deformed by upright folds with NNW–SSE axes (D2). These structures were modified by sinistral transcurrent faults, generally sub-parallel to NW-striking terrane boundaries, and associated en echelon folds and dextral NNE-striking reverse faults (D3) (Swager et al. 1992).

Hammond & Nisbet (1992) and Williams & Whitaker (1993) proposed that both greenstone deposition and early granite emplacement were controlled by N–S crustal extension. These proposals are based on the interpretation of deformed granite domes as metamorphic core complexes. However, no sialic rocks have yet been identified as a basement, older than the greenstones, and most granites appear to have been emplaced as subhorizontal sheets into the greenstones, before or during deformation.

All three episodes of deformation occurred between 2680 and 2650 Ma. They led to the pronounced NNW–SSE tectonic grain of the region, and deformed and reactivated the terrane boundary faults. The terrane is internally disrupted by faults and shear zones into tectonic domains which contain various parts of the regional stratigraphy. Swager et al. (1992) consider that these faults may be reactivated early, syn-volcanic or D1 tear faults.

An E–W seismic reflection profile across the Kalgoorlie terrane, just north of Kalgoorlie (Goleby et al. 1993), indicates that the Ida fault, which forms the western boundary of the Kalgoorlie terrane (Fig. 1), dips 30° E to a depth of 25–30 km. It shows normal displacement of sub-horizontal structures, including the flat base of the greenstones. The greenstones appear to be truncated at a depth of 4–7 km by sub-horizontal sheets of granitic rocks and heterogeneous flat-lying deformation structures.

Monzogranite was emplaced in three pulses: before and during D2, during D3, and late during D3 and D4 (Witt & Swager 1989) at 2680, 2660 and 2620–2600 Ma (Campbell & Hill 1988; Hill et al. 1991, 1992). The granites are generally little deformed and weakly foliated, but were converted into gneiss in narrow belts of intense deformation. Although most of the granites have long been considered steep-sided diapiric structures (Glikson 1979; Witt & Swager 1989), the seismic reflection profiling of Goleby et al. (1993) indicates that the

granites within the greenstones crossed by this seismic line are flat sheet-like bodies.

Most rocks indicate regional prograde greenschist-facies metamorphism, except along the western margin of the terrane where metamorphism reached amphibolite facies.

The Kalgoorlie terrane may represent a strip of island-arc or back-arc crust (Fig. 3c) that was deformed, intruded by granite and metamorphosed, during collision and amalgamation with other greenstone terranes and microcontinents (Fig. 3d).

Tectonic assembly of the craton

The terranes of the Yilgarn craton can be divided into three major groups called superterranes (Fig. 1). The western part of the Yilgarn craton (Narryer, Murchison, Balingup, Boddington, Dumbleyung and Hyden terranes, here collectively called the west Yilgarn superterrane) (Figs 1 & 2c 1), contains a substantial amount of sialic crust older than 3000 Ma which, in the Murchison terrane, was blanketed by flood basalts at 3000 Ma. The central part of the craton (Barlee and Yellowdine terranes) here called the central Yilgarn (or Southern Cross) superterrane (Figs 1 & 2c 2), contains greenstone sequences and granite that formed at 2800–2700 Ma. The eastern part of the craton, east of the Ida fault (Fig. 1) here called the east Yilgarn (or Eastern Goldfields) superterrane (Fig. 2c 3), consists of greenstones and granite, and a minor amount of granitic gneiss, that all appears to have formed between 2730 and 2600 Ma.

The west Yilgarn superterrane was assembled between 2680 and 2650 Ma (Fig. 2a,b & Table 1). The structures within the superterrane are truncated to the east by the suture which joins it to the central Yilgarn superterrane (Fig. 2c 2). Therefore an eastern portion of the west Yilgarn superterrane appears to have been removed, perhaps by rifting or transcurrent faulting, before amalgamation with the central Yilgarn superterrane at 2650 Ma (Figs 2c, 3e). The central and east Yilgarn superterranes (Fig. 2c, 2 & 3) also appear to have been amalgamated at about 2650 Ma (Fig. 3e), but it is not known whether they were joined before collision with the west Yilgarn superterrane, or whether they were sequentially amalgamated with the west Yilgarn superterrane.

After, and perhaps during, assembly of these superterranes at 2650–2600 Ma, there were major dextral transcurrent movements along NNW–SSE faults and terrane boundaries within the central and east Yilgarn superterranes, and along the suture between the west and central Yilgarn superterranes (Figs 1 & 2). Large volumes of fluids passed through these fault zones and widely deposited gold during a late stage of these tectonic movements.

Similar evidence from other cratons

Other Archaean cratons such as the Superior
Province of Canada, the North Atlantic craton of
Canada and Greenland, and the Kaapvaal and
Zimbabwe cratons of southern Africa show similar
kinds of broadly contemporaneous events which
may also reflect the formation of a supercontinent,
or supercontinents, at this time.

In the Superior province (Fig. 4) a combination
of detailed mapping and a substantial amount of
zircon geochronology have established the timing
of both the sequence of events within the terranes
and the amalgamation of the terranes (Card 1990;
Corfu & Davis 1992; Hoffman 1989b; Williams
et al. 1992).

As in the Yilgarn craton, the terranes have
diverse ages and early geological histories and the
tectonic grain, marked by elongate rock bodies and
fold and fault structures, is sub-parallel to the
terrane and superterrane boundaries (Hoffman
1989b; Card 1990). The Minnesota gneiss in
the south (Fig. 4) is part of an older continent
(analogous to the Narryer terrane), to which the
granite–greenstone terranes were accreted from
the north. The Wawa–Abitibi terrane is largely

composed of juvenile, 2750–2710 Ma, island-arc
material (comparable with the Kalgoorlie terrane
and east Yilgarn superterrane). The Wabigoon
terrane to the north is also dominated by
2780–2710 Ma volcanic and plutonic rocks, but in
addition contains older 3000–2800 Ma rocks (as in
the Murchison terrane). Three distinct episodes of
volcanism and plutonism are recognized in the
northern Uchi–Sachigo terrane at 3020–2930 Ma,
2880–2830 Ma and 2740–2730 Ma, and the rocks
were deformed at 2730–2700 Ma (Hoffman 1989b;
Card 1990).

The terranes appear to have been amalgamated
sequentially from north to south, with the
termination of deformation and metamorphism at
2725 Ma in the Uchi–Sachigo terrane, 2705 Ma
in the Wabigoon terrane and 2695 Ma in the
Wawa–Abitibi terrane (Langford & Morin 1976;
Hoffman 1989b; Card 1990) (Fig. 4). The suture
zones and associated belts of metasedimentary
rocks were deformed by dextral transpression,
similar to widespread dextral transpression along
terrane and superterrane boundaries within the
Yilgarn craton.

Late Archaean terrane assembly has also been
described from the gneiss complex in West

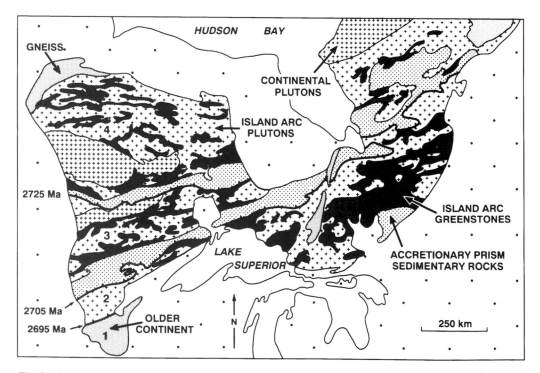

Fig. 4. Map showing the main geological features of the Superior Province (simplified from Hoffman 1989b
and Card 1990). The numbers refer to: 1, Minnesota gneiss; 2, Wawa–Abitibi terrane; 3, Wabigoon terrane;
4, Uchi–Sachigo terrane. Thick lines with barbs indicate major suture zones with age of suturing in Ma.

Greenland that forms part of the North Atlantic craton by Friend *et al.* (1988) and Nutman *et al.* (1989). These authors recognize four terranes: Færingehavn terrane of 3820–3600 Ma tonalitic gneiss; Akia terrane of 3070–2980 Ma dioritic and tonalitic gneiss, intruded by 2980–2940 Ma granite; Tasiusarsuaq terrane of 3000–2900 Ma metavolcanic amphibolite and megacrystic anorthosite which was intruded by tonalite and subsequently recrystallized in granulite facies during and after the emplacement of granite at *c.* 2800 Ma; Tre Brødre terrane of 2800–2750 Ma amphibolite and anorthosite intruded by granodiorite, recrystallized in prograde amphibolite facies. Friend *et al.* (1988) and Nutman *et al.* (1989) determined that the terranes were assembled, deformed and metamorphosed to amphibolite facies between 2750 and 2650 Ma.

De Wit *et al.* (1992) describe a much longer sequence of crustal growth by the successive amalgamation of crustal fragments between 3700 and 2700 Ma in the Kaapvaal craton of southern Africa. Extensive Mid-Archaean continental growth was followed by a major episode of Late Archaean continental collision in which the Kaapvaal craton was combined with the Zimbabwe craton along the Limpopo orogen at 2680 Ma. This led to substantial crustal thickening, uplift and erosion along the orogen.

Conclusions

The Late Archaean episode of supercontinent formation evident from the Yilgarn craton and the Superior Province differs from those that followed in the Proterozoic and Phanerozoic. It appears to have involved a much larger number of small crustal fragments. Many were newly-formed crust, and many consisted of island-arc and back-arc material. Large volumes of tonalite and granite were emplaced as sheets during tectonic stacking of their host rocks.

Over wide areas the amalgamation of crustal units did not lead to substantial crustal thickening and the formation of mountain belts. In the Yilgarn craton, relief remained relatively subdued across extensive regions (up to 1×10^6 km^2). However, some broad regions (200 km^2) were substantially elevated and eroded (such as the southwest part of the Yilgarn craton) during supercontinent assembly. Uplift and erosion also occurred in narrow belts (10×500 km) along normal extensional faults immediately after, and/or during late stages of, major crustal amalgamation.

In other cratons such as the North Atlantic craton, there is evidence of substantial and extensive crustal thickening during Late Archaean terrane assembly. The Late Archaean collision of the Kaapvaal and Zimbabwe cratons also led to substantial crustal thickening in the Limpopo orogen.

Once formed, the supercontinent (or supercontinents), of which the Yilgarn craton and Superior Province may be remnants, was flushed by the upward passage of enormous volumes of fluids. This led to the widespread deposition of gold in shear zones during the last tectonic adjustments within the supercontinent.

Thereafter the supercontinent, or supercontinents, became coherent rafts with generally low relief close to sea level.

John Sutton and Janet Watson in their last paper, presented by John at the 1985 Lewisian conference in Leicester a few days before Janet's death, concluded that 'the most critical problem' in the Lewisian complex was 'an understanding of how the accretion of the Scourian took place'. They considered that 'this was not a local event, but can be paralleled in many parts of the world'. 'It was an epoch-making phenomenon as important in understanding continental evolution as sea-floor spreading is now recognized to be in the formation of oceanic crust' (Sutton & Watson 1987). The Scourian appears to have been a small portion of this Late Archaean supercontinent(s).

John Sutton's 1963, pre-plate tectonic, conclusion that 'the continents have throughout geological time been periodically dispersed ... and then reassembled once more' is supported by subsequent investigations of the Earth's crust, at least as far back as the Late Archaean. Although this process may not have been inherently cyclic, nor always produced two supercontinents, as John suggested, supercontinents do appear to have been formed periodically and then fragmented, dispersed and reassembled in different combinations. The plate tectonic assembly of supercontinents appears to have been largely random, but was probably driven by less random evolution of the mantle.

The intensity and nature of Late Archaean supercontinent formation have never been matched since, and this may reflect the ending of a regime of higher heat flow and thicker oceanic crust, indicated by komatiites and megacrystic anorthosites, and a more ductile lithosphere.

References

BARLEY, M. E. & McNAUGHTON, N. J. 1988. The tectonic evolution of greenstone belts and setting of Archaean gold mineralization in Western Australia: geochronological constraints on conceptual models. *In::* Ho, S. E. & GROVES, D. I. (eds) *Advances in understanding Precambrian Gold Deposits*, **2**. Geology Department (Key Centre) and University Extension, The University of Western Australia, Publication, **12**, 23–40.

——, EISENLOHR, B. N., GROVES, D. I., PERRING, C. S. & VEARNCOMBE, J. R. 1989. Late Archean convergent margin tectonics and gold mineralization — a new look at the Norseman–Wiluna belt, Western Australia. *Geology*, **17**, 826–829.

BICKLE, M. J., CHAPMAN, H. J., BETTENAY, L. F., GROVES, D. I. & DE LAETER, J. R. 1983. Lead ages, reset Rb-Sr ages and implications for Archaean crustal evolution of the Diemals area, Western Australia. *Geochimica et Cosmochimica Acta*, **47**, 907–914.

BLAKE, T. S. & GROVES, D. I. 1987. Continental rifting and the Archean-Proterozoic transition. *Geology*, **15**, 229–232.

CAMPBELL, I. H. & HILL, R. I. 1988. A two-stage model for the formation of the granite-greenstone terrains of the Kalgoorlie-Norseman area, Western Australia. *Earth and Planetary Science Letters*, **90**, 11–25.

CARD, K. D. 1990. A review of the Superior Province of the Canadian Shield, a product of Archean accretion. *Precambrian Research*, **48**, 99–156.

CLAOUÉ-LONG, J. C., COMPSTON, W. & COWDEN, A. 1988. The age of the Kambalda greenstones resolved by ion-microprobe — implications for Archaean dating methods. *Earth and Planetary Science Letters*, **89**, 239–259.

COMPSTON, W. & PIDGEON, R. T. 1986. Jack Hills, evidence of more very old detrital zircons in Western Australia. *Nature*, **321**, 766–769.

CONDIE, K. C. 1989. *Plate tectonics and crustal evolution* (3rd edn). Pergamon Press, New York.

CORFU, F. & DAVIES, D. W. 1992. A U-Pb geochronological framework for the western Superior Province, Ontario. *In::* THURSTON, P. C., WILLIAMS, H. R., SUTCLIFFE, R. H. & STOTT, G. M. (EDS) *Geology of Ontario*. Ontario Geological Survey Special Volumes, **4**, 1135–1346.

DALZIEL, I. W. D. 1991. Pacific margins of Laurentia and East Antarctica-Australia as a conjugate rift pair: evidence and implications for an Eocambrian supercontinent. *Geology*, **19**, 598–601.

DE LAETER, J. R., FLETCHER, I. R., ROSMAN, K. J. R., WILLIAMS, I. R., GEE, R. D. & LIBBY, W. G. 1981. Early Archaean gneisses from the Yilgarn Block, Western Australia. *Nature*, **292**, 322–324.

DENHAM, P. 1976. *Earthquake hazard in Australia*. Australian Bureau of Mineral Resources Record 1976/31.

DE WIT, M. J., ROERING, C., HART, R. J., ARMSTRONG, R. A., DE RONDE, C. E. J., GREEN, R. W. E., TREDOUX, M., PEBERDY, E. & HART, R. A. 1992. Formation of an Archaean Continent. *Nature*, **357**, 553–562.

FLETCHER, I. R., ROSMAN, K. J. R. & LIBBY, W. G. 1988. Sm-Nd, Pb-Pb and Rb-Sr geochronology of the Manfred Complex, Mount Narryer, Western Australia. *Precambrian Research*, **38**, 343–354.

FRIEND, C. R. L., NUTMAN, A. P. & McGREGOR, V. R. 1988. Late Archaean terrane accretion in the Godthab region, southern West Greenland. *Nature*, **335**, 535–538.

FROUDE, D. O., COMPSTON, W. & WILLIAMS, I. S. 1983. Early Archaean zircon analyses from the central Yilgarn Block. *Australian National University, Canberra, Research School of Earth Sciences, Annual Report 1983*, 124–126.

GAAL, G. 1992. Global Proterozoic tectonic cycles and Early Proterozoic metallogeny. *South African Journal of Geology*, **95**, 80–87.

GASTIL, G. 1960a. Continents and mobile belts in the light of mineral dating. *21st International Geological Convention, Norden, 1960. Proceedings part 9*, 162–169.

—— 1960b. The distribution of mineral dates in time and space. *American Journal of Science*, **258**, 1–35.

GEE, R. D., BAXTER, J. L., WILDE, S. A. & WILLIAMS, I. R. 1981. Crustal development in the Yilgarn Block, Western Australia. *In::* GLOVER, J. E. & GROVES, D. I. (eds) *Archaean Geology*. Second International Archaean Symposium, Perth, Western Australia, 1980, Proceedings; Geological Society of Australia, Special Publications, **7**, 43–56.

GLIKSON, A. Y. 1979. Early Precambrian tonalite-trondhjemite sialic nucleii. *Earth Science Reviews*, **15**, 1–73.

—— 1993. Asteroids and early Precambrian crustal evolution. *Earth Science Reviews*, **35**, 285–319.

GOLEBY, B. R., RATTENBURY, M. S., SWAGER, C. P., DRUMMOND, B. J., WILLIAMS, P. R., SHERATON, J. W. & HEINRICH, C. A. 1993. *Archaean crustal structure from seismic reflection profiling, Eastern Goldfields, Western Australia*. Australian Geological Survey Organisation, Record 1993/15.

GROVES, D. I. & BATT, W. D. 1984. Spatial and temporal variations of Archaean metallogenic associations in terms of evolution of granitoid–greenstone terrains with particular emphasis on Western Australia. *In::* KRÖNER, A., HANSON, G. N. & GOODWIN, A. M. (eds) *Archaean Geochemistry*. Springer-Verlag, Berlin, 73–98.

——, ARCHIBALD, N. J., BETTENAY, L. F. & BINNS, R. A. 1978. Greenstone belts as ancient marginal basins or ensialic rift zones. *Nature*, **273**, 460–461.

——, PHILLIPS, G. N., Ho, S. E. & HOUSTOUN, S. M. 1985. The nature, genesis and regional controls of gold mineralization in Archaean greenstone belts of the Western Australian Shield: a brief review. *Geological Society of South Africa Transactions*, **88**, 135–148.

——, ——, ——, —— & STANDING, C. A. 1987. Craton-scale distribution of Archaean greenstone gold deposits: predictive capacity of the metamorphic model. *Economic Geology*, **82**, 2045–2058.

GURNIS, M. 1988. Large-scale mantle convection and the aggregation and dispersal of supercontinents. *Nature*, **332**, 695–699.

—— & DAVIES, G. F. 1986. Apparent episodic crustal growth arising from a smoothly evolving mantle. *Geology*, **14**, 396–399.

HALLBERG, J. A. 1986. Archaean basin development and crustal extension in the northeastern Yilgarn Block, Western Australia. *Precambrian Research*, **31**, 133–156.

HAMMOND, R. L. & NISBET, B. W. 1992. Towards a structural and tectonic framework for the central Norseman-Wiluna greenstone belt, Western Australia. *In:*: GLOVER, J. E. & HO, S. E. (eds) *The Archaean: Terrains, processes and metallogeny.* Geology Department (Key Centre) and University Extension, University of Western Australia Publication, **22**, 39–49.

HILL, R. I., CHAPPELL, B. W. & CAMPBELL, I. H. 1992. Late Archaean granites of the southeastern Yilgarn Block, Western Australia: age geochemistry, and origin. *Transactions Royal Society of Edinburgh: Earth Sciences*, **83**, 211–226.

——, CAMPBELL, I. H. & GRIFFITHS, R. W. 1991. Plume tectonics and the development of stable continental crust. *Exploration Geophysics*, **22**, 185–188.

HOFFMAN, P. F. 1988. United Plates of America, the birth of a craton: Early Proterozoic assembly of Laurentia. *Annual Reviews of Earth and Planetary Sciences*, **16**, 543–603.

—— 1989a. Speculations on Laurentia's first gigayear (2.0 to 1.0 Ga). *Geology*, **127**, 135–138.

—— 1989b. Precambrian geology and tectonic history of North America. *In:*: BALLY, A. W. & PALMER, A. R. (eds) *The geology of North America — an overview.* Geological Society of America, The geology of North America, **A**, 447–512.

KINNY, P. D., WIJBRANS, J. R., FROUDE, D. O., WILLIAMS, I. S. & COMPSTON, W. 1990. Age constraints on the evolution of the Narryer Gneiss Complex, Western Australia. *Australian Journal of Earth Sciences*, **37**, 51–69.

——, WILLIAMS, I. S., FROUDE, D. O., IRELAND, T. R. & COMPSTON, W. 1988. Early Archaean zircon ages from orthogneisses and anorthosites at Mount Narryer, Western Australia. *Precambrian Research*, **38**, 325–341.

KRÖNER, A. 1991. Tectonic evolution in the Archaean and Proterozoic. *Tectonophysics*, **187**, 393–410.

LANGFORD, F. F. & MORIN, J. A. 1976. The development of the Superior Province of northwestern Ontario by merging island arcs. *American Journal of Science*, **276**, 1023–1034.

LE PICHON, X. & HUCHON, P. 1984. Geoid, Pangea, and convection. *Earth and Planetary Science Letters*, **67**, 123–135.

MOORBATH, S. 1977. Ages, isotopes and evolution of the Precambrian continental crust. *Chemical Geology*, **20**, 151–187.

—— 1978. Age and isotopic evidence for the evolution of continental crust. *Philosophical Transactions of the Royal Society of London*, **A288**, 401–413.

MORRIS, P. A. 1993. *Archaean mafic and ultramafic volcanic rocks, Menzies to Norseman, Western Australia.* Geological Survey of Western Australia Report **36**.

MUHLING, J. R. 1990. The Narryer gneiss complex of the Yilgarn Block, Western Australia: a segment of Archaean lower crust uplifted during Proterozoic orogeny. *Journal of Metamorphic Geology*, **8**, 47–64.

MYERS, J. S. 1988. Early Archaean Narryer Gneiss Complex, Yilgarn Craton, Western Australia. *Precambrian Research*, **38**, 297–308.

—— 1990a. Western Gneiss Terrane. *In: Geology and Mineral Resources of Western Australia.* Geological Survey of Western Australia Memoirs, **3**, p. 13–31.

—— 1990b (compiler). Excursion 1: Narryer Gneiss Complex. *In*: HO, S. E., GLOVER, J. E., MYERS, J. S. & MUHLING, J. R. (eds) *Third International Archaean Symposium, Perth, Western Australia, 1990, excursion guidebook*, 61–95.

—— 1990c. Precambrian tectonic evolution of part of Gondwana, southwestern Australia. *Geology*, **18**, 537–540.

—— 1992. Tectonic evolution of the Yilgarn Craton, Western Australia. *In*: GLOVER, J. E. & HO, S. E. (eds) *The Archaean: Terrains, processes and metallogeny.* Geology Department (Key Centre) and University Extension, University of Western Australia Publication, **22**, 265–273.

—— 1993. Precambrian history of the West Australia Craton and adjacent orogens. *Annual Review of Earth and Planetary Sciences*, **21**, 453–485.

—— & WATKINS, K. P. 1985. Origin of granite-greenstone patterns, Yilgarn Block, Western Australia. *Geology*, **13**, 778–780.

NEMCHIN, A. A., PIDGEON, R. T. & WILDE, S. A. 1994. Timing of Late Archaean granulite facies metamorphism and the evolution of the southwestern Yilgarn Craton of Western Australia: evidence from U-Pb ages of zircons from mafic granulites. *Precambrian Research*, **68**, 307–321..

NUTMAN, A. P., FRIEND, C. R. L., BAADSGAARD, H. & MCGREGOR, V. R. 1989. Evolution and assembly of Archean gneiss terranes in the Godhåbsfjord region, southern West Greenland: structural, metamorphic, and isotopic evidence. *Tectonics*, **8**, 573–589.

——, KINNY, P. D., COMPSTON, W. & WILLIAMS, I. S. 1991. SHRIMP U-Pb zircon geochronology of the Narryer Gneiss Complex, Western Australia. *Precambrian Research*, **52**, 275–300.

PIDGEON, R. T. & WILDE, S. A. 1990. The distribution of 3.0 Ga and 2.7 Ga volcanic episodes in the Yilgarn Craton of Western Australia. *Precambrian Research*, **48**, 309–325.

STOESER, D. B. & CAMP, V. E. 1985. Pan-African microplate accretion of the Arabian Shield. *Geological Society of America Bulletin*, **96**, 817–826.

SUTTON, J. 1963. Long-term cycles in the evolution of the continents. *Nature*, **198**, 731–735.

—— & WATSON, J. 1987. The Lewisian complex: questions for the future. *In*: PARK, R. G. & TARNEY, J. (eds) *Evolution of the Lewisian and comparable Precambrian high grade terrains.* Geological Society, London, Special Publications, **27**, 7–11.

SWAGER, C. P., WITT, W. K., GRIFFIN, T. J., AHMAT, A. L., HUNTER, W. M., MCGOLDRICK, P. J. & WYCHE, S. 1992. Late Archaean granite–greenstones of the Kalgoorlie Terrane, Western Australia. *In*: GLOVER, J. E. & HO, S. E. (eds) *The Archaean: Terrains, processes and metallogeny*. Geology Department (Key centre) and University Extension, University of Western Australia, Publication **22**, 107–122.

VAN KRANENDONK, M. J., ST-ONGE, M. R. & HENDERSON, J. R. 1993. Paleoproterozoic tectonic assembly of northeast Laurentia through multiple indentations. *Precambrian Research*, **63**, 325–347.

WATKINS, K. P. & HICKMAN, A. H. 1990. *Geological evolution and mineralization of the Murchison Province, Western Australia*. Geological survey of Western Australia Bulletin **137**.

——, FLETCHER, I. R. & DE LAETER, J. R. 1991. Crustal evolution of Archaean granitoids in the Murchison Province, Western Australia. *Precambrian Research*, **50**, 311–336.

WIEDENBECK, M. & WATKINS, K. P. 1993. A time scale for granitoid emplacement in the Archaean Murchison Province, Western Australia, by single zircon geochronology. *Precambrian Research*, **61**, 1–26.

WILLIAMS, H. R., STOTT, G. M., THURSTON, P. C., SUTCLIFFE, R. H., BENNETT, G., EASTON, R. M. & ARMSTRONG, D. K. 1992. Tectonic evolution of Ontario: summary and synthesis. *In*: THURSTON, P. C., WILLIAMS, H. R., SUTCLIFFE, R. H. & STOTT, G. M. (eds) *Geology of Ontario*. Ontario Geological Survey Special Volumes, **4**, 1255–1314.

WILLIAMS, I. R., 1979. Recent fault scarps in the Mount Narryer area, Byro 1 : 250 000 sheet. *Geological Survey of Western Australia Annual Report for 1978*, 51–55.

WILLIAMS, P. R., & WHITAKER, A. J. 1993. Gneiss domes and extensional deformation in the highly mineralized Archaean Eastern Goldfields Province, Western Australia. *Ore Geology Reviews*, **8**, 141–162.

WITT, W. K. & SWAGER, C. P. 1989. Structural setting and geochemistry of Archaean I-type granites in the Bardoc–Coolgardie area of the Norseman–Wiluna belt, Western Australia. *Precambrian Research*, **44**, 323–351.

Lode-gold deposits of the Yilgarn block: products of Late Archaean crustal-scale overpressured hydrothermal systems

D. I. GROVES[1], J. R. RIDLEY[1], E. M. J. BLOEM[1], M. GEBRE-MARIAM[1],
S. G. HAGEMANN[1], J. M. A. HRONSKY[1], J. T. KNIGHT[1], N. J. MCNAUGHTON[1],
J. OJALA[1], R. M. VIELREICHER[1], T. C. MCCUAIG[2] & P. W. HOLYLAND[3]

[1]*Key Centre for Strategic Mineral Deposits, Department of Geology and Geophysics,
The University of Western Australia, Nedlands, Western Australia, Australia 6009*
[2]*Department of Geological Sciences, University of Saskatchewan, Saskatoon,
Saskatchewan, Canada S7N OWO*
[3]*Terra Sancta, 48 Peoples Avenue, Gooseberry Hill, Western Australia 6076*

Abstract: Although the lode-gold deposits of the Yilgarn block are hosted by a variety of rocks, and their structural style, associated alteration and ore mineralogy are also variable, common parameters suggest that they represent a coherent group of epigenetic deposits, most of which formed during a widespread (500 000 km^2) and broadly synchronous (2635 ± 10 Ma) hydrothermal event in the closing stages of the Late Archaean tectonothermal evolution of the host granitoid–greenstone terrains. Progressive variations in deposit parameters can be correlated with the metamorphic grade of the enclosing greenstone successions. These systematic variations, combined with evidence for their timing and the *P–T* conditions of their formation, indicate that the deposits form a continuum in which gold deposition took place from < 5 km depth (1 kbar, 180°C) to > 15 km depth (> 5 kbar, 700°C), marking hydrothermal fluid flow and fluid evolution through the middle and upper crust.

The primary ore fluid appears to have been an overpressured, low salinity $H_2O–CO_2–CH_4$ fluid originating from a deep source. Upward fluid advection was strongly channelized along vertically extensive conduits. Although there is a gross regional association between clusters of gold deposits and craton- or greenstone-scale deformation zones, these do not appear to have been the primary fluid conduits, at least during their major phase of structural and magmatic activity. Further, all kinematic types of lower-order structure are mineralized, and the fossil fluid conduits exposed at the mine scale are extremely variable, some with no obvious fault or shear control. A potentially unifying hypothesis that can explain this extreme variability in the nature of the conduits is that fluid flow was focussed into zones of low mean stress in the granitoid–greenstone terrains. This can explain the selective occurrence of gold deposits adjacent to irregular or fault-bounded granitoid contacts in some goldfields, the selective mineralization of competent units (e.g. dolerites) in elongate greenstone belts that contain such units and are oriented sub-perpendicular to the far-field compressive stress, and the lack of mineralization in major, largely planar, shear zones undergoing simple shear. The orientation of the greenstone belts with respect to the far-field compressive stress appears to be a crucial factor in defining the potential for strike-extensive zones of low mean stress. This potentially can explain why granitoid–greenstone belts with a high density of sub-parallel craton- and greenstone-scale deformation zones are most highly mineralized and commonly contain the giant gold deposits which, themselves, are located in geometrically anomalous zones within these greenstone belts.

Lode-gold deposits are a widespread feature of most Archaean granitoid–greenstone terrains worldwide (e.g. Canada, Colvine 1989; Australia, Ho *et al.* 1990; Zimbabwe, Foster 1989), and, after the giant Witwatersrand goldfields, contain some of the world's largest gold deposits (e.g. Timmins deposits, Ontario; Golden Mile, Western Australia). The deposits record a major hydrothermal fluid event, with a minimum flux calculated for the whole Yilgarn block on the basis of gold and quartz solubility of 10^{10} t fluid (after Hronsky 1993).

As a group the deposits are diverse, and are sited in a wide range of structures in almost the entire spectrum of lithologies in the hosting terrains. As such, their genesis has been the subject of considerable debate, with early controversy over syngenetic (± remobilization) versus epigenetic hypotheses for their origin, particularly for deposits hosted by sedimentary rocks (e.g. BIF). These have been largely resolved in favour of epigenetic models involving fluid and gold sources distal to the depositional sites, although more local

From COWARD, M. P. & RIES, A. C. (eds), 1995, *Early Precambrian Processes*,
Geological Society Special Publication No. 95, pp. 155–172.

remobilization models are still proposed (e.g. Hutchinson, 1993). Much of the recent discussion has centred on defining their geodynamic setting (e.g. Wyman & Kerrich 1988; Barley *et al.* 1989), the range of crustal environments in which they occur (e.g. Colvine 1989; Groves 1993), their structural controls (e.g. Sibson *et al.* 1988; Hodgson 1989), and the source of the ore fluids and ore solutes that formed the deposits: for example, see reviews of models by Kerrich and Fryer (1988) and in Ho *et al.* (1990). Emphasis has been placed in recent years on developing a holistic model by integrating evidence from a variety of scales (craton- to mine-scale) and sources (structural, petrological, geochemical and isotopic data). Large-scale fluid focusing is clearly required to produce large lode-gold deposits, and it is the scale (first-order versus lower-order structures) and nature of the fluid focusing process that has been one of the more difficult genetic problems to resolve (see Ridley 1993).

This paper examines the problem of the scale and nature of fluid focusing in the generation of Archaean lode-gold deposits of the Yilgarn block, Western Australia, by: (1) showing that the deposits represent a coherent genetic group formed late in the tectonic evolution of the craton, more or less contemporaneously on a craton scale over the entire spectrum of crustal depths represented by exposed greenstone belts, (2) summarizing evidence for deep fluid and solute sources, (3) examining the nature of exposed hydrothermal conduits at a variety of fossil crustal depths, and (4) reviewing the nature of fluid focusing processes that can satisfy the spatial and temporal constraints imposed by the integrated sum of the geological evidence. It is essentially the physical, not chemical, aspects of Archaean lode-gold deposits that are reviewed and discussed here.

Crustal continuum of lode-gold deposits

Although most of the well-documented Archaean lode-gold deposits are hosted in greenschist-facies terrains, and have been considered as 'meso-thermal' deposits (e.g. Groves *et al.* 1989; Kerrich & Wyman 1990), there has been a growing realization that the lode-gold deposits are sited in a variety of metamorphic settings ranging from sub-greenschist to upper amphibolite and even lower granulite facies (e.g. Colvine 1989; Foster 1989). Barnicoat *et al.* (1991), Groves *et al.* (1992) and Groves (1993) have taken this further by suggesting that the deposits actually formed penecontempor-aneously over a range of crustal depths approxi-mated by the prograde metamorphic conditions of the domains in which the deposits are now sited.

The evidence for such a crustal continuum model is briefly discussed below. Groves *et al.* (1992) and Groves (1993) provide a more comprehensive discussion. Analysis of the model is vital for estab-lishing the dimensions of the hydrothermal systems responsible for Yilgarn gold mineralization.

Coherent genetic group

Before data from any group of deposits can be integrated into a unifying genetic model, it must be demonstrated that they form a coherent genetic group. This is particularly important for the Archaean lode-gold deposits as superficially they appear as a heterogeneous group in terms of structure, host rock, mineralogy of wallrock alteration, and ore paragenesis.

Features that do, however, show that the majority of these deposits are a coherent group include: (1) they are epigenetic and controlled by structures that developed or were reactivated late in the structural evolution of the belts, (2) they are consistently enriched in Au, Ag, As and W, with variable enrichments in B, Bi, Pb, Sb and Te, and generally lack Cu and Zn, (3) their alteration zones are commonly enriched in CO_2, S, K and other LILE ($\pm$ Na $\pm$ Ca) despite contrasting alteration assemblages that correlate with the metamorphic grade of the host rocks (see below), (4) lateral alteration zoning is on the scale of centimetres to tens of metres whereas vertical zoning is normally on the scale of hundreds of metres (e.g. Mikucki *et al.* 1990), and (5) one component of the ore fluid in all deposits was a deeply sourced low-salinity H_2O–CO_2–CH_4 fluid, although phase-separation and/or fluid mixing contributed to the variety of fluids trapped as fluid inclusions in auriferous quartz veins.

The few available precise ages of mineralization based on robust geochronology of alteration minerals show that gold mineralization occurred at 2635 $\pm$ 10 Ma (Wang *et al.* 1993; Barnicoat *et al.* 1991; Clark *et al.* 1989) over a wide geographic range (Reedys, Griffins Find, Victory, from west to east on Fig. 1) and over a range of metamorphic environments. This is supported by Pb isotope model ages of least-radiogenic ore-related sulphides which are essentially within error ($\pm$ 30 Ma) of each other over the Yilgarn block (Browning *et al.* 1987; McNaughton *et al.* 1990), despite evidence for variable timing of mineraliz-ation with respect to peak metamorphism, with pre-peak (rare or absent), syn-peak (common) and post-peak (abundant) all being documented.

Thus, available evidence strongly supports the hypothesis that the Yilgarn lode-gold deposits are a broadly synchronous coherent genetic group,

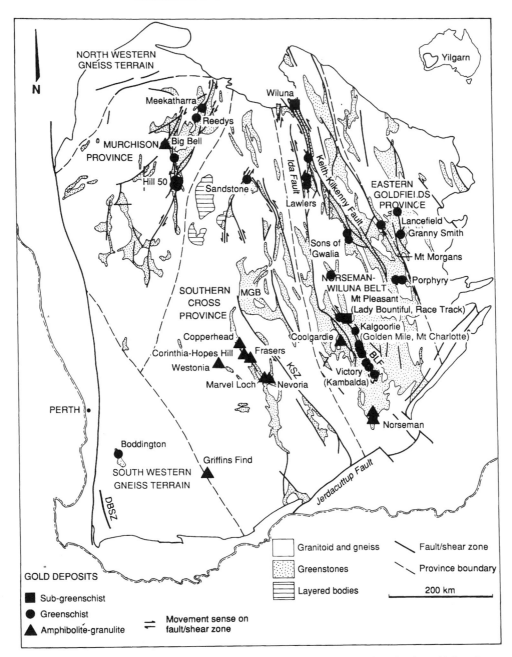

Fig. 1. Geological map of the Yilgarn block, Western Australia, showing the distribution of major gold deposits (>10t Au) in relation to greenstone belts and transcraton deformation zones. Deposits discussed in the text are shown in terms of their metamorphic grade. Adapted from Groves (1993).

although differences in age of up to 20–30 Ma cannot be discounted from available dating, and deposits styles that overprint each other may be present in the same gold field (e.g. Mt Charlotte style overprinting Golden Mile style at Kalgoorlie).

Crustal depth variation

If the lode-gold deposits are a coherent group that are sited in a variety of metamorphic settings, they potentially formed over a spectrum of crustal depths equivalent to that represented by the host metamorphic rocks. Evidence for the existence of such a spectrum is provided by gross trends with increasing metamorphic grade (cf. Colvine 1989; Foster 1989; Witt 1993): (1) from more brittle structures to more ductile structures (Fig. 2), (2) from breccias and discordant quartz veins to foliation-parallel veins and shear-zone replacement deposits, and (3) from plumose, comb, cockade and vughy quartz textures (Gebre-Mariam *et al.* 1993) through massive or laminated veins (Knight *et al.* 1993; McCuaig *et al.* 1993) to veins with coarse grained granoblastic quartz (Barnicoat *et al.* 1991).

Proximal wallrock alteration assemblages in mafic/ultramafic host rocks show similar progressive variations (Fig. 3) from ankerite/dolomite–white mica–chlorite at the lowest metamorphic grades, through ankerite/dolomite–white mica–biotite (phlogopite)–chlorite ± albite, to amphibole–biotite–plagioclase and diopside–biotite–garnet ± K-feldspar assemblages at the highest metamorphic grades (e.g. Mueller & Groves 1991; Witt 1991). The opaque mineral assemblages of the deposits show a sympathetic trend from S-rich

assemblages of pyrite ± arsenopyrite ± pyrrhotite through typically pyrrhotite ± arsenopyrite to pyrrhotite ± arsenopyrite ± loellingite with increasing metamorphic grade (Fig. 3). Tellurides, stibnite, sulphosalts, tetrahedrite and electrum are more common at low metamorphic grades (e.g. Gebre-Mariam *et al.* 1993), and, in sympathy, Ag, Sb, Pb and S are more enriched at such grades, whereas Bi and Cu appear enriched in deposits in higher metamorphic grade settings (e.g. Perring *et al.* 1991).

It should be noted that these broad trends are based on observations on widely spaced gold deposits in the Yilgarn block. However, there is some evidence for transitions reflecting these gross trends on the district scale at Norseman (McCuaig *et al.* 1993), on the mine scale at Mt Charlotte (Mikucki & Heinrich 1993) and elsewhere (Mikucki *et al.* 1990).

Collectively, available fluid-inclusion data and mineral equilibria calculations, summarized by Groves *et al.* (1992), suggest that these variations in structure, vein styles, vein textures, alteration mineralogy and ore mineralogy reflect *P–T* conditions ranging from 180°C at < 1 kbar to 700°C at 5 kbar.

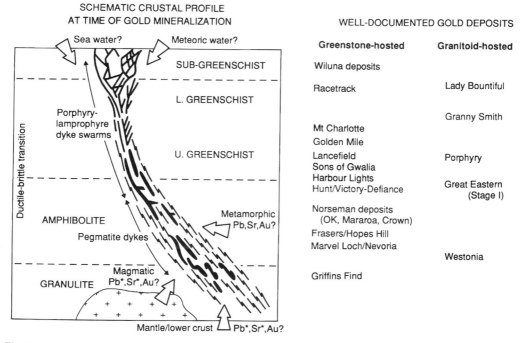

Fig. 2. Schematic reconstruction of a hypothetical hydrothermal system extending over a crustal range of 20–25 km showing potential fluid and ore solute sources (arrows). The probable vertical extent of gold mineralization in specific segments of the Yilgarn block is shown, together with examples of well studied deposits. Adapted from Groves (1993).

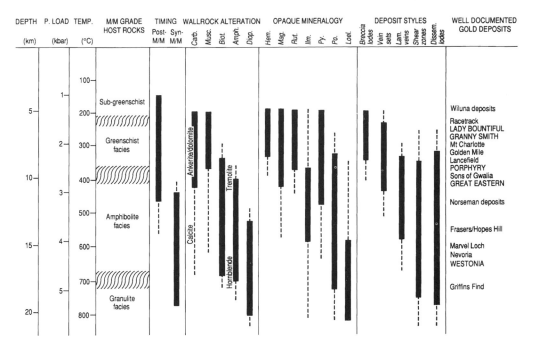

Fig. 3. Summary of various features of Late Archaean lode-gold deposits over the crustal continuum of their deposition. Note that the *P–T*-depth conditions shown do not correspond to those of peak metamorphism but to the conditions *at the time of gold mineralization*. The metamorphic grades shown are for peak metamorphic conditions of the host rocks to mineralization. The characteristics shown are based on a small number of well-documented examples and are, of necessity, generalized. Individual alteration minerals are shown for simplicity although it is recognized that wallrock alteration assemblages are the *critical* features characterizing the different $P–T–X_{CO_2}$ conditions of alteration. Typical examples of well documented gold deposits are listed, with granitoid hosted deposits in capitals.

Timing of gold mineralization in amphibolite-hosted deposits

There is a strong body of evidence from Archaean lode-gold deposits worldwide that mineralization was slightly post-peak metamorphism where the so-called 'mesothermal' deposits are sited in greenschist- to lowermost amphibolite-facies domains (e.g. Clark *et al.* 1989). In such cases, there is no argument that the *P–T* conditions determined from auriferous veins and related wallrock alteration are those of gold deposition. However, in higher-grade metamorphic settings it is necessary to demonstrate that the determined *P–T* conditions are not simply those of metamorphism superimposed on previously formed mineralization as suggested, for example, by Phillips (1985) for the Big Bell deposit (Fig. 1).

There are now a number of carefully constrained studies of gold mineralization in amphibolite-facies settings in Western Australia (e.g. Barnicoat *et al.* 1991; Mueller & Groves 1991; Mueller *et al.* 1991b; Knight *et al.* 1993; McCuaig *et al.* 1993;

Mikucki & Ridley 1993; Neumayr *et al.* 1993*a, b*). These show that gold mineralization was broadly contemporaneous with peak metamorphism because: (1) the deposits are hosted in syn-metamorphic structures, (2) the alteration assemblages comprise high temperature minerals and have a high thermodynamic variance, indicating a high *T*, open metasomatic system, (3) vein minerals preserve fine oscillatory zoning indicative of growth under conditions of high fluid flux, (4) there are deformed and annealed to undeformed vein minerals, with the latter oriented perpendicular to vein walls in places, and (5) there is sub-microscopic gold in the earliest recognizable sulpharsenide phases in the ore paragenesis (e.g. Neumayr *et al.* 1993*a*).

Mid to upper crustal deposit continuum

Evidence from gold the majority of deposits Yilgarn wide thus supports the concept of a continuum of lode-gold deposits formed broadly synchronously throughout the Yilgarn block (at 2635 ± 10 Ma from available published data) over a range of crustal depths as shown schematically in Fig. 2. It must be pointed out that nowhere has the entire range of metamorphic/crustal environments been documented from a single greenstone belt. However, Witt (1991) has mapped metasomatic isograds, based on proximal alteration assemblages, that broadly parallel metamorphic grade variations in the Menzies–Kambalda area of the Norseman–Wiluna Belt, and Knight *et al.* (1993) and McCuaig *et al.* (1993) present further evidence for broadly coincident metasomatic and metamorphic isograds from Coolgardie and Norseman, respectively, providing further support for a deposit continuum.

Implications for fluid and solute sources

A major implication of the model is that not only were hydrothermal systems developed broadly synchronously over the entire craton (Fig. 1), but that the dimensions of individual hydrothermal systems were potentially very large with scope for vertically extensive flow through a variety of potential fluid and solute reservoirs and extensive opportunity for fluid modification via fluid–wallrock interaction (Mikucki & Ridley 1993).

Importantly, the deposits in the higher grade metamorphic settings cannot be derived solely from devolatilization of greenstone successions at the greenschist–amphibolite transition as proposed by several authors (e.g. Fryer *et al.* 1979; Phillips & Groves 1983; Groves & Phillips 1987) and recently reviewed by Phillips & Powell (1993). Many of these deposits are sited in higher-grade metamorphic settings and may be only a few hundred metres from the contact between host greenstones and granitoid intrusions that are earlier than, or synchronous with, peak metamorphism (e.g. Barnicoat *et al.* 1991; Knight *et al.* 1993).

Evidence for deeply sourced fluids

The fluid inclusion and mineral equilibria evidence for precursor low salinity H_2O–CO_2–CH_4 fluids with XCO_2 between 0.1 and 0.2 (e.g. Ho *et al.* 1985; Clark *et al.* 1989) is itself suggestive of a deeply sourced metamorphic or magmatic (e.g. Holloway 1976) fluid. Available stable isotope data from deposits in upper greenschist-facies and higher metamorphic grade environments are non-definitive (e.g. Golding *et al.* 1989), but are consistent with derivation of various fluid components from a variety of crustal reservoirs intersected by fluids advecting along vertically

extensive conduits as predicted by the crustal continuum model.

Radiogenic isotope data provide more definitive evidence that deep crustal fluids were derived from below the greenstone volcanic piles. For example, the least-radiogenic Pb isotope ratios of ore-related pyrite and galena from a representative coverage of Yilgarn gold deposits, are sufficiently radiogenic to implicate a Pb component sourced from older granitic crust or granitoids derived from it (Browning *et al.* 1987; McNaughton *et al.* 1990). Importantly, there are regional scale variations in Pb isotope ratios of these sulphides that correspond broadly to the initial Pb isotope ratios of neighbouring crustally derived granitoids (Fig. 4A), implicating derivation of Pb from the terrains below the greenstone belts (McNaughton *et al.* 1990, 1993). In addition, initial $^{87}Sr/^{86}Sr$ ratios of ore-related scheelites from some deposits (Fig. 4B) are higher than the estimated initial ratios of greenstone belts at the time of gold mineralization (Mueller *et al.* 1991a, Knight, 1994), implicating a Sr contribution from old crust or anatectic granitoids (cf. King & Kerrich 1989).

Ridley (1990) and Mikucki & Ridley (1993) argue that the deeply sourced ore-fluid component of the auriferous hydrothermal systems was saturated with respect to quartz and a K-bearing phase, such as biotite or K-feldspar, providing a further indication that the fluid equilibrated with granitic rocks. Although the exact source of this fluid is still open to debate, it is suffice for the objectives of this paper to illustrate that a deeply sourced fluid is involved and that upward fluid movement along vertically extensive fluid conduits is implicated.

Surface water at high crustal levels

The recognition that some deposits (e.g. Wiluna, Mt Pleasant, Fig. 1) are hosted by brittle structures, are sited in low-grade metamorphic terrains, and have some of the textural and mineralogical characteristics of epithermal gold deposits (e.g. Clout *et al.* 1990; Gebre-Mariam *et al.* 1993) raises the possibility that surface water may have been involved in the hydrothermal systems at high crustal levels. Gebre-Mariam *et al.* (1993) and Hagemann *et al.* (1994) present fluid inclusion and stable isotope data that demonstrate significant differences between components of the interpreted ore fluid at these crustal levels from those at inferred deeper levels. These data, together with Pb isotope ratios of gold-related sulphides, are best interpreted to indicate dual hydrological fluid systems, with incursion of surface water that mixed with a deeply sourced fluid similar to that of the deposits formed at deeper crustal levels.

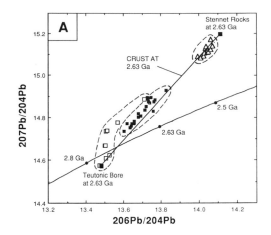

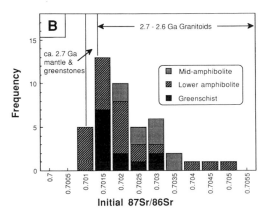

Fig. 4. Summary diagram of radiogenic isotope compositions of hydrothermal minerals from Yilgarn lode-gold deposits. (**A**) ^{207}Pb/^{204}Pb v. ^{206}Pb/^{204}Pb diagram showing the least radiogenic Pb isotope ratios of ore-related galenas and pyrites from gold deposits in the Norseman–Wiluna belt, referenced to the growth curve of Stacey & Kramers (1975). The Pb from the Norseman area (triangles) in the southern part of the belt, together with the Pb from the Kambalda–Kalgoorlie–Menzies area (small filled squares) in the central part of the belt, and from the Leonora–Wiluna area (large open squares) in the northern part of the belt, is compared with the estimated composition of coeval crustal Pb estimated from the end members at Stennet Rocks and Teutonic Bore (from McNaughton *et al.* 1990). (**B**) Frequency histogram showing the initial ^{87}Sr/^{86}Sr ratios of ore-related scheelites from Yilgarn gold deposits in greenschist, lower- and mid-amphibolite facies settings, compared to the ranges in initial ^{87}Sr/^{86}Sr ratios for coeval mantle-derived greenstones and granitoids derived mostly from older crust (data from Mueller *et al.* 1991*a*, *b*; Knight 1994).

Fluid conduits

The discussion above demonstrates that the Yilgarn gold deposits were formed over a spectrum of crustal depths which, assuming that calculated *P* conditions equal lithostatic pressure, ranged from < 5 km to about 20 km (Fig. 3), and were deposited largely from deeply sourced ore fluids. Through combining data obtained from gold deposits exposed at different structural levels, important constraints can be placed on the nature of the fluid conduits down to the base of the greenstone belts, and the major controls on upward fluid flow can be determined. The approach taken below is to briefly discuss evidence that the fluids were overpressured and hence generally upward moving, to examine the types of structures with which the deposits are associated, and to establish any controls common to the deposits. This provides significant constraints on the mechanism(s) of fluid focusing through the vertically extensive hydrothermal system that controlled the location of gold deposits in the Yilgarn block.

Fluid overpressuring

A number of authors have argued that the vein structures and textures characteristic of lode-gold deposits worldwide imply vein formation under conditions of high fluid pressures, and, at least transiently, under supralithostatic fluid pressures (e.g. Robert & Brown 1986; Sibson *et al.* 1988; Cox *et al.* 1991). Supralithostatic pressures are indicated in particular by 'flat' veins in which crack seal textures indicate purely extensional vein growth (cf. Sibson *et al.* 1988), and by the geometry of vein arrays (Fyfe *et al.* 1978). High fluid pressures are indicated by evidence from mineral zonation or habit of growth into open, fluid-filled space, and by hydraulic breccias (cf. Phillips 1972). In the Yilgarn block, evidence for high or supralithostatic pressures is present in deposits formed at all crustal depths. Hydraulic breccias showing multiple fracturing and veining events have been recorded in a number of low-temperature deposits (e.g. Gebre-Mariam *et al.* 1993), as shown in Fig. 5A and D. Subhorizontal extensional vein arrays within and immediately adjacent to wallrock alteration zones, commonly in shear zones (e.g. Victory Defiance; Clark *et al.* 1989) as illustrated in Fig. 5E, are recorded at all crustal levels. Crack-seal textures and open-space filling growth textures are most commonly recorded in relatively low-temperature deposits (e.g. Hagemann *et al.* 1992), but euhedral growth zoning in hydrothermal diopside has been recognized through SEM imagery at the amphibolite-facies Frasers deposit (Barnicoat *et al.* 1991).

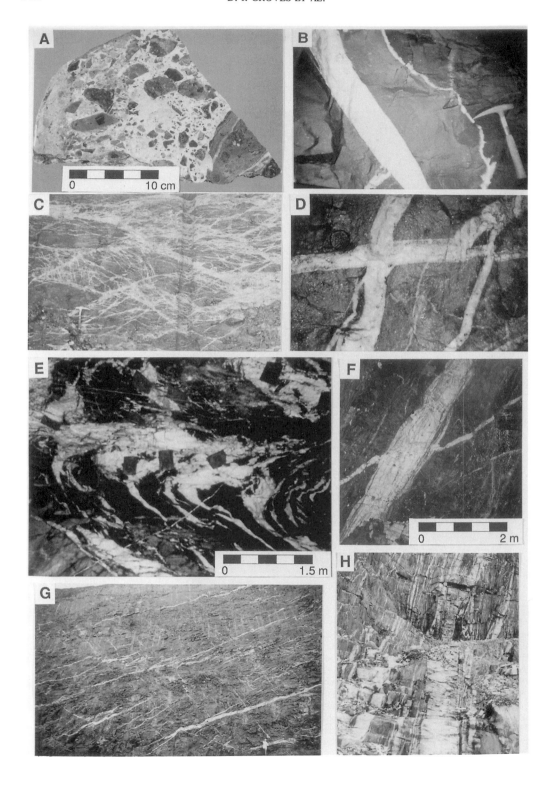

That fluids were locally at supralithostatic pressures, and more generally at high pressures, means that they were overpressured with respect to hypothetical hydrostatic pressure at all depths. An important implication of this is that fluid flow will generally be outwards, away from a fluid source. In particular there will be an upward directed hydraulic gradient forcing bulk upward flow of the fluid (Fyfe *et al*. 1978).

Nature of fluid conduits

As noted above, the nature of the depositional sites varies broadly with crustal level of mineralization, with a progressive change from brittle through brittle–ductile to dominantly ductile hosting structures with depth. However, this is an over-simplification as the nature of the hosting structures, the competency contrasts between adjacent lithologies, and the degree of regional strain in those lithologies outside of fluid conduits, are all important factors in controlling the nature of structural sites for gold mineralization, as illustrated below.

The best described deposits, hosted in amphibolite-facies terrains, are from the Southern Cross (Bloem *et al*. 1994) and Coolgardie (Knight *et al*. 1993) goldfields.

The Southern Cross greenstone belt is a narrow attenuated belt flanked by granitoids (Fig. 1) in which all components of the greenstone succession have penetrative planar and/or linear metamorphic fabrics (i.e. high strain state). Dolerite sills, the host to many of the large deposits in the Norseman–Wiluna Belt (e.g. Kalgoorlie and Kambalda gold-fields, Fig. 1), are rare. In this belt, the hosting structures for gold deposits (e.g. Corinthia-Hopes Hill, Frasers; Bloem *et al*. 1994; Barnicoat *et al*. 1991) are ductile shear zones sited along litho-logical contacts. Gold deposits occur along specific segments of the shear zones, and the majority of gold-bearing quartz veins and alteration zones are parallel to the margins of the zones (Fig. 5H). In contrast, the Coolgardie gold deposits, although in terrains that are at similar metamorphic grade to those at Southern Cross and with similar *P–T* conditions of mineralization (Knight *et al*. 1993), are hosted by essentially brittle–ductile structures in a terrain that shows overall low bulk strain and has a stratigraphic succession similar to that at Kalgoorlie and Kambalda. Here, some deposits are located in laminated quartz veins along brittle–ductile faults (e.g. Kings Cross: Fig. 5F) and strongly resemble those at Norseman. In other cases, particularly for deposits in dolerite sills, mineralization is associated with steeply inclined zones containing brittle vein arrays in which individual veins have both subhorizontal and subvertical orientations (e.g. Three Mile Hill), and there are no obvious connecting shear zones or faults that could have acted as fluid conduits (Fig. 5G).

In lower metamorphic-grade domains, under greenschist and greenschist–amphibolite transition conditions, the classic brittle–ductile shear-zone hosted deposits, such as those described by Clark *et al*. (1986) from Victory-Defiance (Fig. 5E), may be developed. In such deposits, shear zone parallel quartz veins and alteration zones may be accompanied by arrays of extensional or shear veins that are oblique to the shear zones. The most competent rocks in the succession, commonly dolerites (e.g. deposits at Kalgoorlie, Kambalda) or BIF units (deposits at Hill 50, Mt Morgans: Vielreicher *et al*. 1994), tend to be selectively mineralized (e.g. Phillips & Groves 1983). In some cases (e.g. Golden Mile), there is evidence that more ductile structures (e.g. Boulter *et al*. 1987) and strongly

---

Fig. 5. Plate showing the nature of auriferous mineralization sites at different crustal levels. (**A**) Mineralized breccia from the low-greenschist facies, porphyritic basalt-hosted Racetrack deposit, Mt Pleasant. Note the variable size and angular to subrounded shape of the breccia fragments, and that some fragments contain earlier mineralization, indicating multiple stages of hydrothermal activity. (**B**) Simple brittle quartz vein with a well developed proximal alteration halo comprising muscovite–ankerite–pyrite, and distal alteration of chlorite and calcite, North–South Shoot, Golden Kilometre, Mt Pleasant. Note the lack of fabric development in the adjacent wallrock domains, indicating the overall brittle nature of deformation. This deposit is hosted by a quartz gabbro zone within the differentiated Mt Pleasant sill. (**C**) Mineralized, conjugate, brittle vein-sets and related carbonate-sericite alteration envelopes at the greenschist facies, granitoid-hosted Granny Smith mine, Laverton. (**D**) Brittle quartz vein-sets at the greenschist facies, dolerite-hosted Mt Charlotte mine, Kalgoorlie. (**E**) Sigmoidal quartz vein-array in a brittle-ductile shear zone at the upper-greenschist facies, dolerite-hosted Junction mine, Kambalda. (**F**) Steeply dipping, boudinaged and laminated quartz reef sited in a brittle-ductile shear zone, and hosted by amphibolite-facies basalts at Kings Cross mine, Coolgardie. Note also the shallowly-dipping quartz vein which truncates the main quartz reef, but is deformed by the host shear zone fabric. (**G**) Subhorizontal, brittle quartz vein-sets hosted by a differentiated unit of a gabbro sill at the amphibolite-facies Three Mile Hill mine, Coolgardie. (**H**) Laminated to massive quartz veins sited in a broad ductile-brittle shear zone at the basalt-hosted, mid-amphibolite-facies Hopes Hill mine, Southern Cross.

mineralized brittle structures (e.g. Clout *et al.* 1990) may represent temporally distinct events. Mineralized structures are commonly reactivated, and hence classification of host structures is difficult. At this crustal level, the classic vein-array (commonly termed stockwork) deposits, such as those at Mt Charlotte (Fig. 5D), are developed, commonly in highly competent granophyric units in dolerite sills. Although bounded by faults, recent studies (Mengler 1993) show that mineralization was not centred around these faults and the faults are largely unmineralized. They appear to have acted largely as caps or valves (not conduits) to fluid flow. The fracture sets within unfaulted granophyre, now filled with quartz veins (Fig. 5D), were the zones of primary fluid flow.

At interpreted higher crustal levels (e.g. Wiluna, Mt Pleasant), the deposits are generally located along brittle-ductile to brittle, commonly strike-slip, fault zones (e.g. Hagemann *et al.* 1992) and the deposits are located in extensional veins or breccias (Fig. 5A, B), commonly in the more competent units, in extensional sites along the

fault zones (Hagemann *et al.* 1992; Gebre-Mariam *et al.* 1993).

Thus, there is an array of structures that host gold mineralization. These structures are controlled by a combination of crustal depth, regional strain state and lithology of hosting successions, and contrasts in rock competency. At all depths, however, a spectrum is recognized between situations where fluid flow is dominantly within a fault or shear zone, to where it is dominantly within a subvertical zone of fractured rock within a single lithological unit. There is no obvious trend of increasing importance of major fault or shear zone structures with depth. Figure 6 shows, schematically, the nature of fluid conduits in faults and shear zones, and of lithologically controlled conduits, for each crustal depth based on well-studied examples of the lode-gold deposits.

Structures controlling deposit location

The extreme variation in the nature of hosting structures, and inferred fluid conduits, at the mine

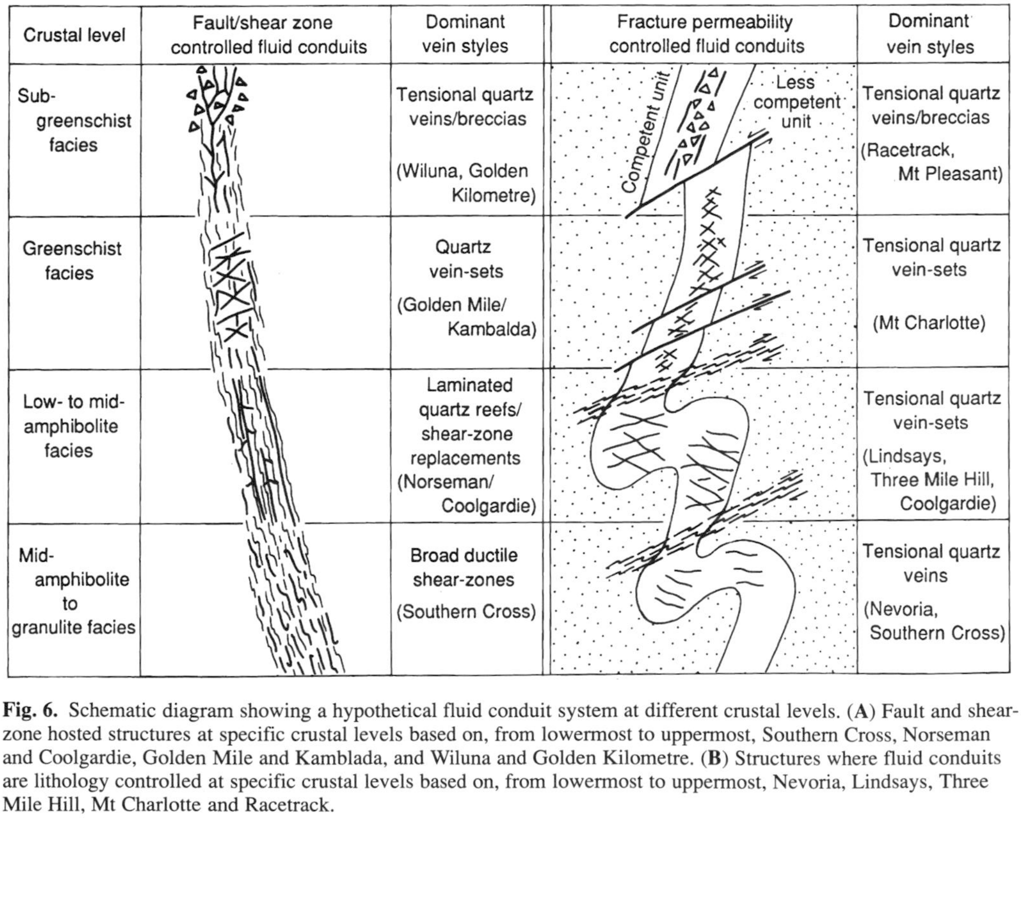

Crustal level	Fault/shear zone controlled fluid conduits	Dominant vein styles	Fracture permeability controlled fluid conduits	Dominant vein styles
Sub-greenschist facies		Tensional quartz veins/breccias (Wiluna, Golden Kilometre)		Tensional quartz veins/breccias (Racetrack, Mt Pleasant)
Greenschist facies		Quartz vein-sets (Golden Mile/ Kambalda)		Tensional quartz vein-sets (Mt Charlotte)
Low- to mid-amphibolite facies		Laminated quartz reefs/ shear-zone replacements (Norseman/ Coolgardie)		Tensional quartz vein-sets (Lindsays, Three Mile Hill, Coolgardie)
Mid-amphibolite to granulite facies		Broad ductile shear-zones (Southern Cross)		Tensional quartz veins (Nevoria, Southern Cross)

Fig. 6. Schematic diagram showing a hypothetical fluid conduit system at different crustal levels. (**A**) Fault and shear-zone hosted structures at specific crustal levels based on, from lowermost to uppermost, Southern Cross, Norseman and Coolgardie, Golden Mile and Kambalda, and Wiluna and Golden Kilometre. (**B**) Structures where fluid conduits are lithology controlled at specific crustal levels based on, from lowermost to uppermost, Nevoria, Lindsays, Three Mile Hill, Mt Charlotte and Racetrack.

scale, combined with the craton-scale distribution and vertically extensive nature of the gold deposits, raises the question of the overall nature and scale of structural control on the deposits.

On the craton scale (Fig. 1), there is a gross relationship between the location of gold deposits and crustal- or greenstone belt-scale deformation zones, and a similar relationship is noted with the so-called 'breaks' in the Superior Province, Canada. Large gold deposits appear to be particularly well developed where there are swarms of lamprophyre and porphyry dykes associated with these deformation zones (e.g. Wyman & Kerrich 1988; Perring et al. 1989; Vielreicher et al. 1994). In granitoid–greenstone belts, which lack such large deformation zones, there is a lower incidence of gold deposits (e.g. Groves & Phillips 1987). Despite this large-scale association, the first-order structures rarely host gold deposits at any structural level in the Yilgarn block. The deposits are, instead, hosted in second-, third- and lower-order geometric structures which may or may not show a geometric relationship to the larger scale deformation zones (e.g. Eisenlohr et al. 1987; Witt 1993). That the major structures are rarely mineralized at any level suggests that they are not feeders to the deposit-scale conduits. Rather, it is suggested in the discussion below that there may be an indirect rather than direct structural and temporal relationship between focused fluid flow and the major episode of structural and magmatic activity on the first-order structures.

On a lower, greenstone belt scale, there is also a broad spatial correlation between regional-scale (tens to hundreds of kilometres strike-length) faults and shear zones and gold-deposit location but, again, most of the deposits are sited at some distance from these structures. There is also no specific association between the deposits and structures of a particular kinematic group (e.g. Hronsky et al. 1990; Witt 1993). At inferred upper crustal levels, the gold deposits are mostly associated with strike-slip fault zones (e.g. Hagemann et al. 1992; Gebre-Mariam et al. 1993), but at inferred lower crustal levels strike-slip, reverse and normal faults may all be important locally (Hronsky et al. 1990), even in a single goldfield.

Although much gold mineralization appears to have formed synchronously with primary fault movement, an important factor in the control of many deposits is the reactivation of earlier formed fault or shear zones, with lode-gold structures superimposed on pre-existing structures (e.g. Grigson et al. 1990). At Wiluna, for instance, ore zones are sited along faults with displacements of up to 1 km, yet ore is essentially undeformed and the siting of the deposits can be understood in terms of the present (post-faulting) geometry of units and

structures (Hagemann et al. 1992). Gold mineralization is thus commonly associated with the latest significant movements on these structures.

Heterogeneous stress orientation

The lode-gold deposits clearly formed late in the structural history of the greenstone terrains when the far-field compression is widely accepted to have been broadly E–W to ENE–WSW (e.g. Swager et al. 1990; Libby et al. 1991). Two detailed structural studies have been carried out at Granny Smith (Ojala et al. 1993) and Coolgardie (Knight et al. 1993) to determine the theoretical maximum compressive principal palaeostress orientation for gold-hosting structures: the results are shown schematically in Fig. 7A and B and compared to a theoretical model in Fig. 7A.

At Granny Smith, a deposit in a greenschist-facies greenstone domain, conjugate vein arrays are consistently developed in the mineralized granitoid pluton (Fig. 5C). On the assumption that the bisector of the acute angle between the conjugate sets is the direction of the maximum compressive principal stress and the line of intersection of the sets is the axis of the intermediate stress (e.g. Hobbs et al. 1976), most vein sets indicate the predicted ENE–WSW compression. However, there are variations of up to 90° in the stress orientation about both vertical and horizontal axes (Fig. 7C), particularly where the granitoid contact is most irregular and concave with respect to the far-field stress. The zones of highest ore grade correlate strongly with these irregularities (Ojala et al. 1993).

At Coolgardie, similar heterogeneous stress orientation (Fig. 7B) is noted in a higher-grade (amphibolite-facies) metamorphic terrain around a granitoid body on a much larger scale (Knight et al. 1993). Here, a variety of mineralized structures whose kinematics are well constrained have been used to determine the stress orientation at the time of mineralization. Although the ore-confining structures have variable orientations, they are considered to be contemporaneous based on the metamorphic assemblages that define their fabrics and their domainal development (Knight et al. 1993). The heterogeneous stress orientation at Coolgardie is, as at Granny Smith, related to the shape of the adjacent granitoid, with greatest heterogeneity occurring where the granitoid margin and/or the geometry of the greenstone belt is most complex.

In both cases, the heterogeneities in the stress field have influenced the form and nature of the deposits. Other authors have pointed out a similar influence of stress heterogeneities on deposit formation: for example, Dubé et al. (1989) at the

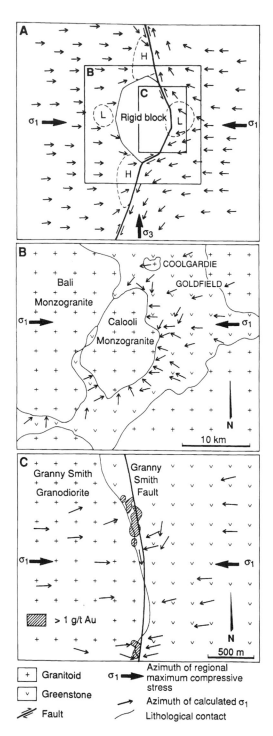

Norbeau gold mine, Quebec, Canada, and Oliver *et al.* (1990) at the polymetallic uranium mine at Mary Kathleen, Australia.

Fluid focusing

The discussion above provides a number of important constraints on fluid focusing: (1) craton-scale deformation zones are unlikely to have been the major fluid conduits, at least during their major phase of structural and magmatic activity, (2) structures that form the fluid conduits are not uniquely reverse, normal or strike-slip structures, (3) there is no unique type of high-permeability zone that acted as fluid conduits, and (4) at least in some instances, local- to regional-scale hetero-geneous stress orientation appears to be coincident with zones of high ore-fluid flux. The potential relationship between fluid flow and mean rock stress is discussed below. It is stressed from the outset that this discussion is generalized due to space restrictions on this review.

Mean rock stress and fluid flow

Ridley (1993), following the approaches suggested by Oliver *et al.* (1990) and Holyland (1990), has examined the requirements for fluid focusing, and the relationships between fluid focussing, fluid pressure and stress fields: the theoretical con-siderations are not repeated here. Variations in mean rock stress will inevitably develop in an inhomogeneous rock pile undergoing deformation (e.g. Dewers & Ortoleva 1989). If fluid pressures are buffered at close to rock pressures, fluids will flow upwards and towards sites of low mean rock stress. This focusing will have the effect of raising fluid pressures, and hence decreasing effective

---

Fig. 7. Summary diagram showing the azimuth of the calculated maximum principal compressive palaeostress direction based on the orientations and kinematics of mineralized structures. (**A**) Theoretical displacements about a rigid inclusion (e.g. granitoid) in an inhomo-geneous matrix with visco-plastic rheology (e.g. layered greenstones) which includes a zone of early slip (e.g. shear zone) wrapping one side of the inclusion. Note the position of low mean stress zones. From Ridley (1993). Compare this theoretical model with the examples of the Coolgardie goldfield and Granny Smith mine shown in Figs 7B and 7C, respectively. (**B**) Coolgardie goldfield showing the gross granitoid-greenstone geometry and the azimuth of σ_1, determined from mineralized and unmineralized structures across the goldfield. Simplified from Knight *et al.* (1993). (**C**) Granny Smith mine, Laverton, showing the azimuth of calculated σ_1, based on the measurement of mineralized conjugate vein sets. Simplified from Ojala *et al.* (1993).

pressures, in the sites of low mean rock stress. Sites of fluid focusing are thus predicted to be sites of relatively low effective pressures, as is implied from vein structures described above. Importantly, potential sites of low mean stress strongly resemble the sites of gold deposition in the greenstone belts. For example, the position of dilatant, low mean-stress sites associated with irregularities along faults (e.g. Sibson 1987; Hodgson 1989) include dilatational jogs, releasing bends and Riedel shears, the precise sites of gold mineralization in the upper-crustal brittle regimes at Wiluna (Hagemann et al. 1992) and Mt Pleasant (Gebre-Mariam et al. 1993). Hronsky et al. (1991) have shown that only very minor variations in orientation of a shear zone are necessary to incur significant variations in mean stress and, hence, fluid flux.

Rigid bodies may also be sites of variable mean stress, with the precise siting of low mean stress zones dependant on the shape of the rigid body with respect to the regional stress field and the nature of the contact (faulted or sheared versus undeformed) with less competent units. Theoretical models of low mean stress distribution (Ridley 1993) compare well with the distribution of heterogeneous stress and higher-grade gold mineralization in greenstone belts adjacent to rigid granitoid bodies, such as described by Ojala et al. (1993) and Knight et al. (1993). For example, the theoretical model shown in Fig. 7A is almost identical to the natural example at Granny Smith shown in Fig. 7C.

In layered sequences comprising interlayered rocks with contrasting competency, there are systematic differences in stress fields dependant on the orientation of the layering with respect to the maximum principal palaeostress direction (Casey 1980). Ridley (1993) has shown that where the stress field is such that there is compression perpendicular to the layering, and hence extension parallel to that layering, mean stress will be lower in the competent layers. In contrast, where there is compression parallel to layering, the most in-competent layers will have lower mean stress, and where there is compression at 45° to layering there will be minimal variation in mean stress. In most of the greenstone belts in the Yilgarn, layering is oriented between NNW–SSE and NNE–SSW, whereas the far field stress at the time of gold mineralization was oriented broadly E–W. In support of this model, it is normally the most competent units in the succession (e.g. dolerite sills) that are best mineralized where they are part of the lithostratigraphy (e.g. Phillips & Groves 1983; Groves & Phillips 1987; Witt 1993).

The above discussion indicates that focusing through zones of low mean stress in the granitoid–greenstone terrains potentially provides an embracing model that can explain the variability of structural controls of the gold deposits in the terrain. In particular, the model can explain the fact that some fluid conduits in competent units, for example, at Mt Charlotte (Clout et al. 1990), appear to be fracture zones restricted to those units and without any visible hydraulic connection to synchronously active discrete faults or shear zones along which fluid might have been channelled.

Discussion and conclusions

The majority of lode-gold deposits of the Yilgarn block, although varying in their host rocks, structural style, alteration and ore mineralogy, represent a coherent genetic group of epigenetic deposits that formed during a major hydro-thermal event. This event occurred very late in the tectonometamorphic evolution of the host granitoid–greenstone terrains. Available data suggest that most gold mineralization occurred over a wide area (500 000 km^2) of the Yilgarn block, more or less synchronously in the Late Archaean at c. 2635 ± 10 Ma.

The deposits show progressive variations in structural style, proximal wallrock alteration assemblages, vein textures, and ore mineralogy in parallel with the metamorphic grade of the enclosing greenstone successions. Deposits in sub-amphibolite-facies terrains appear to post-date peak metamorphism, whereas those in higher grade metamorphic terrains are broadly synchronous with peak metamorphism. Overall, the features of the deposits and the P–T conditions of their formation, as calculated from fluid inclusions and/or alteration assemblages, are consistent with the concept that the deposits represent a crustal continuum with gold mineralization formed more or less syn-chronously under conditions that ranged from < 1 kbar at 180° to > 5 kbar at 700°C, i.e. at crustal depths ranging from < 5 km to about 20 km.

There is evidence from all inferred crustal levels that the low salinity H_2O–CO_2–CH_4 ore fluids were deeply sourced and overpressured. Although craton- to regional- scale deformation zones are broadly coincident with major clusters of lode-gold deposits, mantle-derived lamprophyres and crustal-derived felsic porphyries, available evidence suggests that they were not the major ore-fluid conduits. At all crustal levels, most deposits are related to second-, third- or even lower-order faults and shear zones, although no particular kinematic type of structure is preferentially mineralized; strike-slip, oblique-slip, reverse and normal structures are all mineralized in various parts of the terrain. In some instances, there is no obvious control on mineralization by fault or shear zones, and fluid conduits appear to be pipe-like fracture zones, commonly in the more competent units in the terrain. It is thus difficult to define a structural model for fluid focusing that is applicable to all lode-gold deposits in the terrain.

The concept that zones of low mean rock stress could have focused fluid flow, and hence controlled the siting of lode-gold deposits, appears the most likely to have universal application. Not only can it explain the location of important gold deposits, such as Granny Smith and the deposits in the Coolgardie Goldfield, but it can potentially explain the wide variety of fluid conduits in differently oriented greenstone belts and at different crustal levels. For example, Ridley (1993) has argued that the occurrence of deposits such as Nevoria, south of Southern Cross, and Mt Pleasant (Fig. 1) within pressure shadow zones of granitoids, is consistent with low stress distribution around a rigid body undergoing E–W compression, whereas low mean stress distribution at Granny Smith (Fig. 7A, C) is strongly controlled by the shear zone along the eastern margin of the granitoid body. On a more regional scale, most of the gold deposits are sited in competent units, or at the contacts between units of contrasting competency, in greenstone belts that are sub-perpendicular to the inferred E–W compression, again consistent with predicted low mean stress distribution. Conversely, incompetent units are rarely mineralized, except in fold hinges where layering is locally sub-parallel to the far field maximum compressive stress, and greenstone belts that are not sub-perpendicular to the compressive stress (e.g. in the western part of the Murchison province, Fig. 1) are essentially unmineralized.

Ridley (1993) has also pointed out that the lack of mean stress variations between different rock layers in shear zones undergoing simple shear parallel to the layering provides a possible explanation for the observation that the craton- to regional-scale shear zones in the Yilgarn block, and elsewhere, are essentially devoid of gold mineralization. Variations in mean stress will only occur where there are irregularities in the shear zones such as jogs or the lensing out of specific stratigraphic units.

Various authors (e.g. Kerrich 1986; Groves *et al.* 1989) have shown that most gold production has come from greenschist-facies terrains, and it is widely considered that this is due to enhancement of gold deposition during fluid pressure fluctuations resulting from episodic fault movement around the frictional–quasiplastic (brittle–ductile) transition at the base of the seismogenic zone. However, Ridley (1993) has pointed out that variations in mean stress vary linearly with deviatoric stress, and hence that the enhanced fluid focusing, due to larger mean stress variations at this level, is an alternative explanation of the gold deposit distribution.

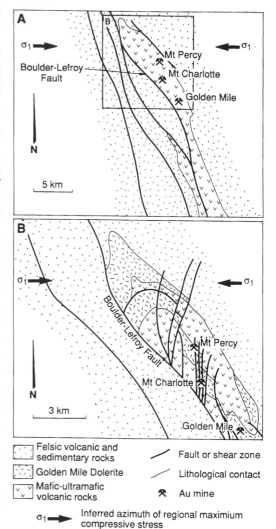

Fig. 8. Geological sketch map of the Kalgoorlie district showing: (**A**) the distribution of relatively competent mafic-ultramafic sequences virtually surrounded by less competent metasedimentary rocks on the gross scale, and (**B**) fault-bounded blocks of competent Golden Mile dolerite surrounded by less competent rocks on a smaller scale. All mineralized zones are in parts of the sequence oriented at a high angle to the inferred regional stress field. Adapted from Swager & Griffin (1990).

Finally, the selective concentrations of gold deposits in terrains with a high density of sub-parallel craton- to regional-scale deformation zones can potentially be explained in terms of the model involving fluid focusing into low mean stress zones. Such terrains have a very large proportion of their greenstone belts oriented at a high angle to the far field stress and, in addition to the normal

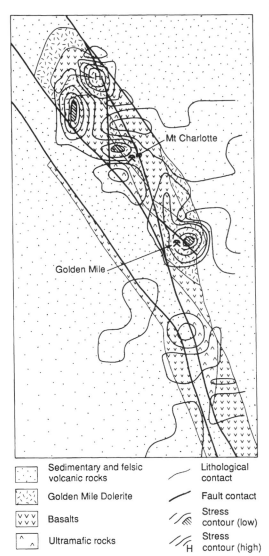

Sedimentary and felsic volcanic rocks

Golden Mile Dolerite

Basalts

Ultramafic rocks

Lithological contact

Fault contact

Stress contour (low)

Stress contour (high)

Fig. 9. Simplified geological map of the Kalgoorlie area prepared for application of stress mapping technology as detailed by Holyland (1990). Simplified contours showing variations in magnitude of mean stress coincide with the two major areas of gold mineralization at the Golden Mile and Mt Charlotte.

heterogeneities within the greenstone sequences, there are commonly mismatches, albeit subtle in some cases, in orientation and stratigraphy across the deformation zones which provide further inhomogeneities in the greenstone successions and bordering granitoids. Such greenstone belts have an ideal geometry for the generation of gold deposits in competent rocks along considerable strike lengths (e.g. mineralized dolerite sills from south of Kambalda to Kalgoorlie, Fig. 1). Within such greenstone belts, it is commonly the zones of greatest inhomogeneity that are mineralized. For example, the giant Kalgoorlie deposits are sited in a refolded thrusted sequence (Swager 1989) whose geometry has produced a distribution of competent dolerite units almost completely enclosed by less competent units at a variety of scales, and oriented at a high angle to the Late Archaean far field stress field (Fig. 8A, B). The geometry of the giant Timmins goldfield in the Superior Province, Canada, is remarkably similar (e.g. Hodgson 1989).

If this model is valid, calculation of stress fields is a tool for targeting potential sites of gold mineralization through prediction of sites of enhanced fluid flow. Sites of low mean stress are controlled by a combination of inhomogeneities in the granitoid-greenstone belts and their gross orientation and geometry with respect to the far field stress. They are therefore highly variable in accord with the observed variation in structural style of the lode-gold deposits. Numerical computation methods have been applied to the calculation of two-dimensional stress fields in inhomogeneous material (e.g. Holyland 1990), and the good correlation between calculated low mean stress zones and mineralized areas (e.g. Kalgoorlie area; Fig. 9) lends considerable support to the model of fluid focussing as a result of variations in mean stress at a variety of scales.

We are grateful to our colleagues in the Key Centre for Strategic Mineral Deposits for stimulating discussion. The research described in this paper was supported by many mining and exploration companies, MERIWA, ARC and UWA. We are most grateful for this support. E.B., M.G.-M., S.H., J.K., J.O. and R.V. all acknowledge receipt of OPRS and URS scholarships. T.C.M. acknowledges NSERC and U of S postgraduate scholarships. Terra Sancta acknowledges the use of Itasca Inc. UDEC code for the numerical modelling.

References

BARLEY, M. E., EISENLOHR, B., GROVES, D. I., PERRING, C. S., VEARNCOMBE, J. R. 1989. Late Archean convergent margin tectonics and gold mineralization: a new look at the Norseman-Wiluna Belt, Western Australia. *Geology*, **17**, 826–829.

BARNICOAT, A. C., FARE, R. J., GROVES, D. I. & MCNAUGHTON, N. J. 1991. Syn-metamorphic lode-gold deposits in high-grade Archean settings. *Geology*, **19**, 921–924.

BLOEM, E. J. M., DALSTRA, H. J., GROVES, D. I. & RIDLEY, J. R. 1994. Metamorphic and structural setting of Archaean amphibolite-hosted gold deposits near Southern Cross, Southern Cross Province, Yilgarn

Block, Western Australia. *Ore Geology Reviews*, **9**, 183–208.

BOULTER, C. A., FOTIOS, M. G. & PHILLIPS, G. N. 1987. The Golden Mile, Kalgoorlie: a giant gold deposit localised in ductile shear zones by structurally induced infiltration of an auriferous metamorphic fluid. *Economic Geology*, **82**, 1661–1678.

BROWNING, P., GROVES, D. I., BLOCKELY, J. R. & ROSMAN, K. J. R. 1987. Lead isotopic constraints on the age and source of gold mineralization in the Archean Yilgarn Block, Western Australia. *Economic Geology*, **82**, 971–986.

CASEY, M. 1980. Mechanics of shear zones in isotropic dilatant materials. *Journal of Structural Geology*, **2**, 143–147.

CLARK, M. E., ARCHIBALD, N. J. & HODGSON, C. J. 1986. The structure and metamorphic setting of the Victory gold mine, Kambalda, Western Australia. *In*: MACDONALD, A. J. (ed.) *Gold '86*. An International Symposium on the Geology of Gold Deposits, Proceedings Volume, Toronto, 243–254.

——, CARMICHAEL, D. M., HODGSON, C. J. & FU, M. 1989. Wallrock alteration, Victory gold mine, Kambalda, Western Australia: processes and P-T-XCO$_2$ conditions of metasomatism. *In*: KEAYS, R. R., RAMSAY, W. R. H. & GROVES, D. I. (eds) *The Geology of Gold Deposits: The Perspectives in 1988*. Economic Geology, Monograph, **6**, 445–459.

CLOUT, J. M. F., CLEGHORN, J. H. & EATON, P. C. 1990. Geology of the Kalgoorlie Gold Field. *In*: HUGHES, F. E. (ed.) *Geology of the Mineral Deposits of Australia and Papua New Guinea*. Australasian Institute of Mining and Metallurgy, Melbourne, Monograph Series, **14**, 411–431.

COLVINE, A. C. 1989. An empirical model for the formation of Archean gold deposits: products of final cratonization of the Superior Province, Canada. *In*: KEAYS, R. R., RAMSAY, W. R. H. & GROVES, D. I. (eds) *The Geology of Gold Deposits: The Perspectives in 1988*. Economic Geology, Monograph, **6**, 37–53.

COX, S. F., WALL, V. J., ETHERIDGE, M. A. & POTTER, T. F. 1991. Deformational and metamorphic processes in the formation of mesothermal vein-hosted gold deposits—examples from the Lachlan fold belt in central Victoria, Australia. *Ore Geology Reviews*, **6**, 391–423.

DEWERS, T. & ORTOLEVA, P. J. 1989. Mechano-chemical coupling in stressed rocks. *Geochimica et Cosmochimica Acta*, **53**, 1243–1258.

DUBÉ, B., POULSEN, H. & GUHA, J. 1989. The effects of layer anisotropy on auriferous shear zones: The Norbeau mine, Quebec. *Economic Geology*, **84**, 871–878.

EISENLOHR, B. N., GROVES, D. I. & PARTINGTON, G. A. 1989. Crustal-scale shear zones and their significance to Archaean gold mineralization in Western Australia. *Mineralium Deposita*, **24**, 1–8.

FOSTER, R. P. 1989. Archean gold mineralization in Zimbabwe: Implications for metallogenesis and exploration. *In*: KEAYS, R. R., RAMSAY, W. R. H. & GROVES, D. I. (eds) *The Geology of Gold Deposits: The Perspectives in 1988*. Economic Geology, Monograph, **6**, 54–70.

FRYER, B. J., KERRICH, R., HUTCHINSON, R. W., PIERCE, M. G. & ROGERS, D. S. 1979. Archaean precious metal hydrothermal systems, Dome Mine, Abitibi greenstone belt. Part I. *Canadian Journal of Earth Sciences*, **16**, 421–439.

FYFE, W. S., PRICE, N. J. & THOMPSON, A. B., 1978. *Fluids in the Earth's Crust*. Elsevier, Amsterdam.

GEBRE-MARIAM, M., GROVES, D. I., MCNAUGHTON, N. J., MIKUCKI, E. J. & VEARNCOMBE, J. R. 1993. Archaean Au-Ag mineralization at Racetrack, near Kalgoorlie, Western Australia: a high crustal-level expression of the Archaean composite lode-gold system. *Mineralium Deposita*, **28**, 375–387.

GOLDING, S. D., MCNAUGHTON, N. J., BARLEY, M. E., GROVES, D. I., HO, S. E., ROCK, N. M. S. & TURNER, J. V. 1989. Archean carbon and oxygen reservoirs: their significance for fluid sources and circulation paths for Archean mesothermal gold deposits of the Norseman-Wiluna Belt, Western Australia. *In*: KEAYS, R. R., RAMSAY, W. R. H. & GROVES, D. I. (eds) *The Geology of Gold Deposits: The Perspectives in 1988*. Economic Geology, Monograph, **6**, 376–388.

GRIGSON, M. W., VEARNCOMBE, J. R. & GROVES, D. I. 1990. Structure of the Meekatharra-Mount Magnet Greenstone Belt, Murchison Province, Yilgarn Block, Western Australia: implications for regional controls on Archaean lode-gold mineralization. *In*: GLOVER, J. E. & HO, S. E. (comps) *Third International Archaean Symposium, Perth, 1990, Extended Abstracts Volume*, 411–413.

GROVES, D. I. 1993. The crustal continuum model for late-Archaean lode-gold deposits of the Yilgarn Block, Western Australia. *Mineralium Deposita*, **28**, 366–374.

—— & PHILLIPS, G. N. 1987. The genesis and tectonic control on Archaean gold deposits of the Western Australian Shield: a metamorphic-replacement model. *Ore Geology Reviews*, **2**, 287–322.

——, BARLEY, M. E. &.HO, S. E., 1989. Nature, genesis, and setting of mesothermal gold mineralization in the Yilgarn Block, Western Australia. *In*: KEAYS, R. R., RAMSAY, W. R. H. & GROVES, D. I. (eds) *The Geology of Gold Deposits: The Perspectives in 1988*. Economic Geology, Monograph, **6**, 71–85.

——, ——, BARNICOAT, A. C., CASSIDY, K. F., FARE, R. J., HAGEMANN, S. G., HO, S. E., HRONSKY, J. M. A., MIKUCKI, E. J., MUELLER, A. G., MCNAUGHTON, N. J., PERRING, C. S., RIDLEY, J. R. & VEARNCOMBE, J. R. 1992. Sub-greenschist to granulite-hosted Archaean lode-gold deposits of the Yilgarn Craton: a depositional continuum from deep-sourced hydrothermal fluids in crustal-scale plumbing systems. *In*: GLOVER, J. E. & HO, S. E. (eds) *The Archaean: Terranes, Processes and Metallogeny*. Geology Department (Key Centre) & University Extension, The University of Western Australia, Publication **22**, 325–337.

HAGEMANN, S. G., GEBRE-MARIAM, M. & GROVES, D. I. 1994. Surface-water influx in shallow-level Archean lode-gold deposits in Western Australia. *Geology*, **22**, 1067–1070.

——, GROVES, D. I., RIDLEY, J. R. & VEARNCOMBE, J. R. 1992. The Archean lode-gold deposit at Wiluna, high-level, brittle-style mineralization in a strike-slip regime. *Economic Geology*, **87**, 1022–1053.

Ho, S. E., GROVES, D. I. & BENNETT, J. M. (eds) 1990. *Gold deposits of the Archaean Yilgarn Block, Western Australia: Nature, Genesis and Exploration Guides*. Geology Department (Key Centre) & University Extension, The University of Western Australia, Publication, **20**.

——, —— & PHILLIPS, G. N. 1985. Fluid inclusions as indicators of the nature and source of ore fluids and ore depositional conditions for Archaean gold deposits of the Yilgarn Block, Western Australia. *Transactions of the Geological Society of South Africa*, **88**, 149–158.

HOBBS, B. E., MEANS, W. D. & WILLIAMS, P. F. 1976. *An Outline of Structural Geology*. John Wiley & Sons, Inc., New York.

HODGSON, C. J. 1989. The structure of shear-related, vein-type gold deposits: a review. *Ore Geology Reviews*, **4**, 231–273.

HOLLOWAY, J. R. 1976. Fluids in the evolution of granitic magmas: consequences of finite CO_2 solubility. *Bulletin of the Geological Society of America*, **87**, 1513–1518.

HOLYLAND, P. W. 1990. Targeting of epithermal ore deposits using stress mapping techniques. *Proceedings Pacific Rim Congress 1990*, **3**. Australasian Institute of Mining and Metallurgy, Melbourne, 337–341.

HRONSKY, J. M. A. 1993. *The role of physical and chemical processes in the formation of gold ore-shoots at the Lancefield gold deposit, Western Australia*. PhD thesis, University of Western Australia.

——, CASSIDY, K. F., GRIGSON, M. W., GROVES, D. I., HAGEMANN, S. G., MUELLER, A. G., RIDLEY, J. R., SKWARNECKI, M. S. & VEARNCOMBE, J. R. 1990. Deposit and mine-scale structure. *In*: Ho, S. E., GROVES, D. I . & BENNETT, J. M. (eds) *q.v.*, 38–54.

——, RIDLEY, J. R. & CATHCART, J. C. 1991. Structural controls on the distribution of ore shoots in a shear-zone hosted gold deposit: the Lancefield example. *In: Structural Geology in Mining and Exploration, Extended Abstracts*. Geology Department (Key Centre) & University Extension, The University of Western Australia, Publication, **25**, 147–149.

HUTCHINSON, R. W. 1993. A multi-stage, multi-process genetic hypothesis for greenstone-hosted gold lodes. *Ore Geology Reviews*, **8**, 349–382.

KERRICH, R. 1986. *Archaean lode gold deposits of Canada: Part II. Characteristics of the hydrothermal system and models of origin*. Economic Geology Research Unit, University Witwatersrand, Johannesburg, Information Circular, **183**.

—— & FRYER, B. J. 1988. Lithophile-element systematics of Archean greenstone belt Au–Ag vein deposits: implications for source processes. Canadian *Journal of Earth Sciences*, **25**, 945–953.

—— & WYMAN, D. 1990. The geodynamic setting of mesothermal gold deposits: an association with accretionary tectonic regimes. *Geology*, **18**, 882–885.

KING, R. W. & KERRICH, R. 1989. Strontium isotope compositions of tourmaline from lode gold deposits of the Archean Abitibi greenstone belt (Ontario-

Quebec, Canada): implications for source reservoirs. *Chemical Geology*, **79**, 225–240.

KNIGHT, J. T. 1994. *The geology and genesis of Archaean amphibolite-facies lode-gold deposits in the Coolgardie Goldfield, Western Australia, with special emphasis on the role of granitoids*. PhD thesis, University of Western Australia.

——, GROVES, D. I. & RIDLEY, J. R. 1993. The Coolgardie Goldfield, Western Australia: district-scale controls on an Archaean gold camp in an amphibolite facies terrane. *Mineralium Deposita*, **28**, 436–456.

LIBBY, J., GROVES, D. I. & VEARNCOMBE, J. R. 1991. The nature and tectonic significance of the crustal-scale Koolyanobbing shear zone, Yilgarn Craton, Western Australia. *Australian Journal of Earth Sciences*, **38**, 229–245.

McCUAIG, T. C., KERRICH, R., GROVES, D. I. & ARCHER, N. 1993. The nature and dimensions of regional and local gold-related hydrothermal alteration in tholeiitic metabasalts in the Norseman goldfields: the missing link in a crustal continuum of gold deposits? *Mineralium Deposita*, **28**, 420–435.

McNAUGHTON, N. J., CASSIDY, K. F., DAHL, N., GROVES, D. I., PERRING, C. S. & SANG, J. H. 1990. Sources of ore fluid and ore components: lead isotope studies. *In*: Ho, S. E., GROVES, D. I. & BENNETT, J. M. (eds) *q.v.* 226–236.

——, GROVES, D. I. & WITT, W. K. 1993. The source of lead in Archaean lode gold deposits of the Menzies–Kalgoorlie–Kambalda region, Yilgarn Block, Western Australia. *Mineralium Deposita*, **28**, 495–502.

MENGLER, F. C. 1993. *Structural and lithological context of stockwork-style gold mineralization, Mt Charlotte, Kalgoorlie, Western Australia*. BSc Honours thesis, University of Western Australia.

MIKUCKI, E. J. & HEINRICH, C. A. 1993. Vein- and mine-scale wall-rock alteration and gold mineralization in the Archaean Mount Charlotte deposit, Kalgoorlie, Western Australia. *In*: WILLIAMS, P. R. & HALDANE, J. A. (comps) *An International Conference on Crustal Evolution, Metallogeny and Exploration of the Eastern Goldfields, Extended Abstracts*. Australian Geological Survey Organisation, Record, 1993/54, 135–140.

—— & RIDLEY, J. R. 1993. The hydrothermal fluid of Archaean lode-gold deposits at different metamorphic grades: compositional constraints from ore and wallrock alteration assemblages. *Mineralium Deposita*, **28**, 469–481.

——, GROVES, D. I. & CASSIDY, K. F. 1990. Alteration patterns: wallrock alteration in sub-amphibolite facies gold deposits. *In*: Ho, S. E., GROVES, D. I. & BENNETT, J. M. (eds) *q.v.* 60–78.

MUELLER, A. G. & GROVES, D. I. 1991. The classification of Western Australian greenstone-hosted gold deposits according to wallrock alteration mineral assemblages. *Ore Geology Reviews*, **6**, 291–331.

——, DE LAETER, J. R. & GROVES, D. I. 1991a. Strontium isotope systematics of hydrothermal minerals from epigenetic Archean gold deposits in the Yilgarn Block, Western Australia. *Economic Geology*, **86**, 780–809.

——, Groves, D. I. & Delor, C. P. 1991*b*. The Savage Lode magnesian skarn in the Marvel Loch gold-silver mine, Southern Cross Greenstone Belt, Western Australia. Part II Pressure-temperature estimates and constraints on fluid sources. *Canadian Journal of Earth Sciences*, **28**, 686–705.

Neumayr, P., Cabri, L. J., Groves, D. I., Mikucki, E. J. & Jackman, J. 1993*a*. The mineralogical distribution of gold and relative timing of gold mineralization in two Archean settings of high metamorphic grades in Australia. *Canadian Mineralogist*, **31**, 711–725.

——, Groves, D. I., Ridley, J. R. & Koning, C. D. 1993*b*. Syn-amphibolite facies Archaean lode gold mineralization in the Mt. York District, Pilbara Block, Western Australia. *Mineralium Deposita*, **28**, 469–481.

Ojala, V. J., Ridley, J. R., Groves, D. I. & Hall, G. C. 1993. The Granny Smith Gold Deposit: the role of heterogeneous stress distribution at an irregular granitoid contact in a greenschist facies terrane. *Mineralium Deposita*, **28**, 409–419.

Oliver, N. H. S., Valenta, R. K. & Wall, V. J. 1990. The effect of heterogeneous stress and strain on metamorphic fluid flow, Mary Kathleen, Australia, and a model for large-scale fluid circulation. *Journal of Metamorphic Geology*, **8**, 311–331.

Perring, C. S., Barley, M. E., Cassidy, K. F., Groves, D. I., McNaughton, N. J ., Rock, N. M. S., Bettenay, L. F., Golding, S. D. & Hallberg, J. A. 1989. The association of linear orogenic belts, mantle-crustal magmatism, and Archean gold mineralization in the eastern Yilgarn Block of Western Australia. *In*: Keays, R. R., Ramsay, W. R. H. & Groves, D. I. (eds) *The Geology of Gold Deposits: The Perspectives in 1988.* Economic Geology, Monograph, **6**, 571–584.

——, Groves, D. I. & Shellabear, J. N. 1991. *The geochemistry of Archaean gold ores from the Yilgarn Block of Western Australia: Implications for gold metallogeny.* Mineral and Energy Research Institute of Western Australia, Reports, **82**.

Phillips, G. N. 1985. Interpretation of Big Bell/Hemlo-type gold deposits: precursors, metamorphism, melting and genetic constraints. Transactions of the *Geological Society of South Africa*, **88**, 159–173.

—— & Groves, D. I. 1983. The nature of Archaean gold-bearing fluids as deduced from gold deposits of Western Australia. *Journal of the Geological Society of Australia*, **30**, 25–39.

—— & Powell, R. 1993. Links between gold provinces. *Economic Geology*, **88**, 1084–1098.

Phillips, W. J., 1972. Hydraulic fracturing and mineralization. *Journal Geological Society London*, **128**, 337–359.

Ridley, J. R. 1990. Source of ore fluid and ore components: alteration assemblages. *In*: Ho, S. E., Groves, D. I. & Bennett, J. M. (eds) *q.v.* 268–272.

—— 1993. The relations between mean rock stress and fluid flow in the crust: With reference to vein- and lode-style gold deposits. *Ore Geology Reviews*, **8**, 23–37.

Robert, F. & Brown, A. C. 1986. Archaean gold-bearing quartz veins at the Sigma mine, Abitibi greenstone belt, Quebec: Part I. Geological relations and formation of the vein system. *Economic Geology*, **81**, 578–592.

Sibson, R. H. 1987. Earthquake rupturing as a hydro-thermal mineralizing agent in hydrothermal systems. *Geology*, **15**, 701–704.

——, Robert, F. & Poulsen, K. H. 1988. High-angle reverse faults, fluid pressure cycling and meso-thermal gold–quartz deposits. *Geology*, **16**, 551–555.

Stacy, J. S. & Kramers, J. D. 1975. Appproximation of terrestrial lead isotope evolution by a two-stage model. *Earth and Planetary Science Letters*, **26**, 207–221.

Swager, C. P. 1989. Structure of the Kalgoorlie green-stones – regional deformation history and implications for the structural setting of gold deposits within the Golden Mile. *Geological Survey of Western Australia, Report*, **25**, 59–84.

—— & Griffin, T. J. 1990. *Geology of the Archaean Kalgoorlie Terrane (northern and southern sheets).* Western Australian Geological Survey, 1:250,000 geological series map.

——, Witt, W. K., Griffin, T. J., Ahmat, A. L., Hunter, W. M., McGoldrick, P. J. & Wyche, S. 1990. A regional overview of the Late Archaean Kalgoorlie granite-greenstones of the Kalgoorlie Terrane. *In*: Ho, S. E., Glover, J. E., Myers, J. S. & Muhling, J. R. (eds) *Third Archaean International Symposium, Perth, 1990, Excursion Guidebook.* Geology Department (Key Centre) and University Extension, University of Western Australia, Publication, **21**, 205–220.

Vielreicher, R. M., Groves, E. I., Ridley, J. R. & McNaughton, N. J. 1994. A replacement origin for the BIF-hosted gold deposit at Mt Morgans, Yilgarn Block, Western Australia. *Ore Geology Reviews*, **9**, 325–347.

Wang, L. G., McNaughton, N. J. & Groves, D. I. 1993. An overview of the relationship between granitoid intrusions and gold mineralization in the Archaean Murchison Province, Western Australia. *Mineralium Deposita*, **28**, 482–494.

Witt, W. K. 1991. Regional metamorphic controls on alteration associated with gold mineralization in the Eastern Goldfields Province Western Australia: Implications for the timing and origin of Archean lode-gold deposits. *Geology*, **19**, 982–985.

—— 1993. Lithological and structural controls on gold mineralization in the Archaean Menzies-Kambalda area, Western Australia. *Australian Journal of Earth Sciences*, **40**, 65–86.

Wyman, D. & Kerrich, R. 1988. Alkaline magmatism, major structures and gold deposits: implications for greenstone belt metallogeny. *Economic Geology*, **83**, 454–461.

Late Archaean structure and gold mineralization in the Kadoma region of the Midlands greenstone belt, Zimbabwe

R. J. HERRINGTON

Department of Mineralogy, The Natural History Museum, London SW7 5BD, UK

Abstract: Major Archaean greenstone gold deposits are located in distinctive structural settings, and Zimbabwe is a good example with over 90% of its gold production being derived from structurally controlled veisns, shear zones and probably epigenetic replacement BIF hosted deposits.

The structural framework of the Zimbabwe craton is dominated by the evolution of the *c.* 2700 Ma Limpopo belt. Strongly developed, major shear zones were initiated at this time. These major shear zones focused strain, which in the Midlands greenstone belt, evolved from pure N–S compression, to lateral extrusion of the main Rhodesdale gneiss block westwards.

The structural features seen in the gold deposits of the Midlands greenstone belt are consistent with such a model of NNW–SSE compression linked to lateral extrusion of the Rhodesdale gneiss block westwards. Initiation of structures during NNW–SSE compression resulted in early folding such as the Kadoma anticline, together with initiation of the major mineralized shear systerms as thrust faults. Simple shear evolved to transpressive shear with development of oblique dextral and sinistral mineralized shear systems. Rotation of the principal stress direction clockwise to an ENE–WSW orientation led to evolution of dominantly dextral transpressive shear on the major Lily, Munyati and Rhodesdale boundary shear systems, probably in response to the lateral extrusion of the Rhodesdale gneiss block westwards. This even is linked to waning gold mineralization and accounts for the major dextral offsets along the Lily fault and the Munyati shear close to Battlefields.

The Archaean Midlands greenstone belt of Zimbabwe has been a significant gold producer on the world scale, with at least eight deposits having produced over 10 tonnes of gold and two (Cam & Motor, Globe & Phoenix) having produced over 100 tonnes of gold (Foster *et al.* 1986). Probably all of these deposits are structurally controlled (Foster *et al.* 1986; Foster 1989) and most are hosted within units of the Bulawayan Group (Fig. 1).

North of Kadoma, the Basaltic Unit of the Bullawayan is host to a number of important gold deposits, notably the major Dalny gold deposit (>50 t Au produced) and the deposits around Golden Valley. The Dalny deposit is developed in a shear zone array within tholeiitic lavas which contain minor intercalated black shales (Carter 1991). The structurally controlled Patchway (20 t Au produced) and Golden Valley (*c.* 35 t Au produced), together with the minor Erin and Hamburg deposits, comprise dominantly discrete veins and minor shear zone disseminations hosted within iron-rich tholeiites of the Upper Greenstones (Herrington 1991).

Close to Kadoma, the Cam & Motor gold deposit, historically Zimbabwe's largest gold producer (145 t Au), is located at the steeply dipping contact between basaltic flow units and a mixed metasedimentary sequence (Collender 1964). The style is one of an anastomosing and branching series of veins, clearly structurally controlled. South of Kadoma, the Brompton deposit rather unusually occurs within the Rhodesdale Gneiss complex, and forms several parallel, moderately dipping veins developed within thin felsic intrusive bodies intrudeed along narrow shear zones (Tomlinson 1982).

Farther south, the Indarama and Broomstock deposits occur within mafic volcanics and BIF of the Bulawayan and are also structurally controlled (Nutt 1985; Nutt *et al.* 1988). The area immediately around Kwekwe is one of the most productive regions (Foster *et al.* 1986) with the Globe & Phoenix deposit having produced some 120 t of gold to date (Foster 1989). Mineralization is developed where altered and carbonated ultramafic rocks are in contact with granitic gneisses of the Rhodesdale batholith. Veins are hosted in both carbonate-rich ultramafiic rocks and granitic gneiss, pinching out into talc-schist units (Porter & Foster 1991).

Recent aeromagnetic mapping (Anon 1984) and structural analyses of some of these deposits (e.g. Carter 1991; Herrington 1991; Porter & Foster 1991), together with new regional studies (e.g. Pitfield & Campbell 1989) and older published studies (e.g. Bliss 1970; Robertson 1976; Catchpole 1987; Nutt *et al.* 1988), provide good information which can be synthesized into a regional structural

From COWARD, M. P. & RIES, A. C. (eds), 1995, *Early Precambrian Processes*, Geological Society Special Publication No. 95, pp. 173–191.

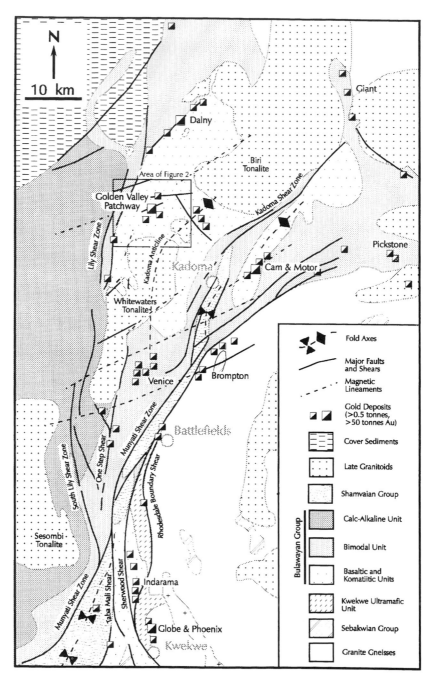

Fig. 1. Geological map of the Midlands greenstone belt, Zimbabwe with interpreted structural information from aeromagnetic, air photographic and remote sensing data (compiled from Bliss 1970; Robertson 1976; Anon 1984; Foster *et al.* 1986; Pitfield & Campbell 1990; Herrington 1991)

chronology. Such a chronology can be placed in the broader craton-scale deformation models which have been proposed.

This paper reports on detailed structural information collected by the author at the Patchway and Golden Valley deposits, a key area to the north of Kadoma close to the major Lily fault, which shows structural features consistent with the model for evolution of the greenstone belt as a whole when integrated with other deposit scale studies and the author's own observations from other deposits.

Structural framework of the Zimbabwe craton

It is clear that structural evolution of the Zimbabwe craton was linked to that of the Limpopo belt and Kaapvaal craton (Coward 1976). Strong compressional tectonics were active across the Limpopo belt and structures well into the Zimbabwe craton were influenced by this. Studies such as Coward *et al.* (1976) and Key & Hutton (1976) recognized the regional aspect to deformation patterns and the influence of the craton-bounding mobile belts. This work, compiled and expanded by Coward (1980), outlined three broad structural domains in Zimbabwe: (1) craton, (2) Northern Marginal Zone (of the Limpopo belt), and (3) Central Zone (of the Limpopo belt).

Domain 1, the main craton in which all the areas discussed in this paper lie, forms a domain dominated by strongly arcuate foliation, often focused into restricted structural zones. In the southwest portion of the area, which covers the Bulawayo, Gwanda, Shangani and Mberengwa greenstone belts, a NW–SE-trending upright to inclined tectonic fabric is developed in granitic gneisses and greenstones. Farther east this fabric is deflected to form steep NE–SW shear zones. Some of these shears show evidence of extensive strike-slip motion (Stowe 1974). In the KweKwe area of the Midlands greenstone belt, the fabric is mainly steep and arcuate around the older gneiss complex at its eastern margin. The northern greenstone belts of Mount Darwin and Harare are again dominated by NE–SW structures. The eastern part of the craton is dominated by structures developed in response to some form of NNW–SSE imposed shortening across the craton, whilst structure in the more western part indicate some lateral shortening in a WSW–ENE direction to the west of the main Rhodesdale gneiss block (Treloar *et al.* 1992).

Geological framework of the Midlands greenstone belt

The geological mapping of the greenstone belt (Bliss 1970; Robertson 1976) has been recently complemented by acquisition of new aeromagnetic data produced at 1 : 1 000 000 (Anon 1984). This new data, together with recently published air-photo interpretations of Pitfield & Campbell (1989) and the author's own air-photo study of the area NW of Kadoma (Herrington 1991), allows more interpretation of regional structure to be added to the geological maps. This is presented as an interpretive compilation which is shown on Fig. 1.

The strongest feature in the Patchway–Golden Valley area is the Lily fault, described by Bliss (1970) as a major strike-slip structure (Fig. 1). In the north the structure branches towards Dalny with continuation north to form the boundary between the Archaean and Proterozoic cover. Southwards, the South Lily shear follows the western margin of the Whitewaters tonalite body, joining the Munyati shear, west of Indarama. The fault is well defined in the northern part of the belt, specifically to the west of the Patchway–Golden Valley region where it forms the boundary between the Calc-alkaline and Bimodal units of the greenstones. Towards the north, a NE–SW-trending splay runs through the Chakari region with the development of parallel N–S structures, accommodating movement around a large infold of Proterozoic rocks. In the south the Lily fault steps southeast to join the Munyati shear system in the south. A large dextral displacement is recorded across the Lily fault (Bliss 1970), apparently late in the structural evolution of the belt.

The Munyati shear clearly defines the contact between the Bulawayan and Shamvaian Groups in the Battlefields area. Northwards the shear branches northwestwards to form the Kadoma shear zone and eastwards to join the Rhodesdale boundary shear. Southwards, the Munyati shear branches into the Munyati, Taba–Mali and Sherwood shear zones around Indarama. In the southern part of the Midlands the Munyati shear forms the boundary between the Shamvaian and the greenstones of the Bulawayan Group. In the north it merges with the Rhodesdale shear in the Battlefields area. Again evidence for both early sinistral and later dextral components to shearing on this structure are recorded (Pitfield & Campbell 1990).

The Rhodesdale boundary shear forms the eastern boundary of the Midlands greenstone belt merging with the Sherwood shear in the south and the Munyati system, south of Kadoma. The shear zone forms the eastern margin of the greenstone belt, trending roughly N–S in the KweKwe area, curving around to NE–SW towards Chegutu. The shear is marked by a zone of distinctly foliated granitic gneiss in the Rhodesdale gneiss unit which contains zones of locally silicified and carbonatized rocks. Evidence for both early sinistral and later dextral strike-slip components to shearing are indicated (Stowe 1980).

Stowe (1980) had previously identified the Lily, Rhodesdale and Munyati structures as major strike-slip shears with early sinistral and later dextral shear sense. In addition, the zone between the major Lily and Munyati–Rhodesdale structures in the Kadoma region shows some linear features in the magnetic data which are largely sub-parallel to, or splays off, the major first order structures. One such magnetic feature extrapolates through the Cam & Motor region. A broad area of elevated magnetic response is indicated over the White-waters and Biri tonalite bodies. The relevance of this is not clearly understood, but it may be due to the presence of magnetite in these intrusives.

These observations show the focus of intense deformation in the KweKwe to Battlefields area which to the north opens into more complex splay structures branching northwards to Dalny and northeastwards to Kadoma.

Geology of the Patchway–Golden Valley area

The gold deposits of the Golden Valley region are hosted within the Upper Bulawayan sequence of rocks of the Midlands greenstone belt (Fig. 1). These rocks can be compared to similar rocks in the Mberengwa area. The contrast of metamorphic grade between the Rhodesdale gneiss complex (moderate) and the greenstones (low) suggests that the gneiss forms the basement sequence to the greenstone succession (Robertson 1976). Whole-rock Pb–Pb age dates corroborate this, with the gneiss dated at 2976^{+121}_{-132} Ma and the most reliable Pb–Pb dates from the Upper Greenstone sequence indicating an age of 2659^{+38}_{-39} Ma (Taylor et al. 1991). More recently, SHRIMP U–Pb data derived from zircons return precise dates of 2702 ± 6 Ma for the Maliyami Formation and 2696 ± 7 Ma for a rhyodacite of the Upper Bulawayan (Wilson et al. this volume).

All the deposits studied in the Golden Valley area are hosted in the Basaltic Unit of the Upper Bulawayan which is the most important stratigraphic unit for gold production in the craton (Foster & Wilson 1984). The Basaltic Unit forms the core of the major Kadoma anticline and extends from Chegutu in the north to an area to the south-west of Kadoma in the south. The unit is dominated by flow rocks, mainly of basaltic and andesitic composition and often pillowed (Bliss 1970; Foster et al. 1979). These compositions have been confirmed by the author's own analyses in the Golden Valley region (Herrington 1991). Subordinate inter-flow sediments and volcano-sediments are also recorded, seen underground at Dalny (Carter 1991). The package of flow rocks is accompanied by numerous thin quartz porphyry intrusives (termed 'felsites' locally), probably high-level dykes or sills, which show variable age relationships but the majority apparently intruded later than the original mafic flow rocks. These intrusives are recorded at a number of mines in the Kadoma region (Bliss 1970), including the Golden Valley, Patchway and Erin mines. The intrusives are of tonalitic to grandodioritic composition (Bliss 1970; Herrington 1991), making them similar in chemistry to the larger Sesombi suite to intrusives and the small stocks of Marizani, Chadshunt and Polperro near Chakan.

The greenstone sequence is intruded by the Sesombi tonalite suite (Bliss 1970). Recent Pb–Pb dates of 2579^{+154}_{-173} Ma (Taylor et al. 1991) span the previously quoted dates and are compatible with the intrusives closely post-dating the Upper Greenstones. Mapping suggests that the White-waters complex intruded in two stages as an early sodic phase which was deformed before a later more potassic phase, possibly represented by the Grandeur potassic granite, located at the southern tip of the Whitewaters body tonalite (Bliss 1970).

Clearly post-dating all these rocks are a series of generally NE–SW-trending dolerite dykes. Recorded extensively in the belt (Bliss 1970), these are seen to cut the Lion Hill stock and Whitewaters tonalite bodies. Macgregor (1930) proposes an Upper Karoo age for these dykes, although the undoubted deformation of some of the intrusives seen at Patchway (Herrington 1991) would appear to indicate that at least some of the intrusives are much earlier than this, possibly part of the Proterozoic Plumtree swarm of 2.15 Ga (Wilson et al. 1987).

Accurate dating of the gold mineralization itself has recently been attempted by Sm–Nd determination of scheelite which is a ubiquitous accessory of the veins in the Kadoma region (Bliss 1970). Scheelite from a number of gold deposits' which included Patchway and Golden Valley (S. D. G. Campbell pers. comm.), gave an isochron age of 2668 ± 64 Ma.

Description of gold-bearing structures in the Golden Valley region

The gold deposits in the region northwest of Kadoma are almost entirely hosted within vein structures of two clearly defined orientations of strike closely approximating to 020° and 070° (see Fig. 2a) which fits well with the bulk of linear features observed from the interpretation of aerial photos of the region (Fig. 2b).

(a)

(b)

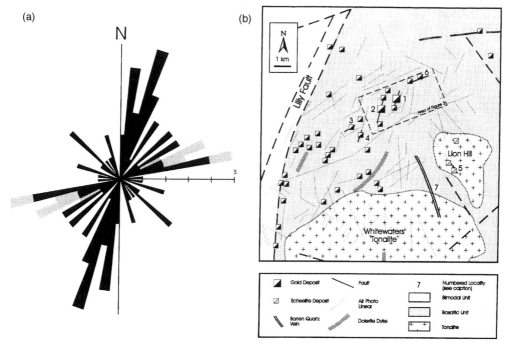

Fig. 2. (a) Rose diagram of strike orientation of gold-bearing mineralized structures (in black, *n* = 50) and major cross-cutting shear-zones (in grey, *n* = 5) in the Golden Valley–Patchway region (compiled from Bliss 1970). (b) Summary of aerial photograph interpretation of the Golden Valley–Patchway region (Herrington 1991). 1, Patchway mine; 2, Golden Valley mine; 3, Erin mine; 4, Hamburg mine; 5, Lion Hill scheelite prospect; 6, Rouge mine; 7, Lion Hill barren quartz vein. Geology modified from Bliss (1960).

Mineralization at Patchway is confined to a single complex vein within an envelope of sheared mafic volcanics. Bifurcations and duplications are common, particularly where the vein-hosting shear is also intruded by so-called 'felsites' (Bliss 1970). The strike of the vein varies from 005° in the south to 025° in the northern part and dip is within the range 25–30° westwards. Golden Valley shows a very similar style of mineralization. Mineralization at Patchway, Golden Valley and sub-parallel structures are very similar texturally, all forming classic veins of ribboned appearance (Herrington *et al.* 1993). At Patchway actual quartz vein widths rarely exceed 2.5 m, whilst at Golden Valley 5 m widths are recorded. Where the mineralization takes the form of multiple quartz veins at Patchway, common when felsite intrudes the hosting shear zone, an aggregate vein thickness of over 5 m is known. At Golden Valley, mineralization is generally restricted to a single, complex vein and almost without exception at Golden Valley a felsite intrusive forms the immediate hanging wall to the vein.

Mineralization at Patchway is divided into two orebodies, separated by a zone where no significant

quartz veining is present. This is termed the 'Central Shear Zone' (Herrington 1991) and it extends to Golden Valley, where it also represents a zone in the mine where only highly deformed slivers of gold–quartz veining are present. The traces of the intercepts of this zone with the orebodies are shown clearly as parallel, E–W-trending lines on the plan of mined areas on Fig. 3, and a generally E–W-trending alignment of individual mining blocks is a strong feature of other areas of the mine forming a principal structural control to economic mineralization.

In addition to the E–W trend, the Patchway northern orebody clearly shows a break between mineralization on the upper and lower mine levels. A 100 m zone with an 030° trend, where mineralization is again deformed and highly faulted, is present. In other areas of the mine, minor sub-parallel faults cause disruption to mineralization and the intercept of these faults and the orebody, constitutes the second structural control to economic mineralization. Also apparent from the plan, particularly on the upper parts of the northern orebody at Patcvhway, are a series of roughly N–S trending zones where mineralization is missing

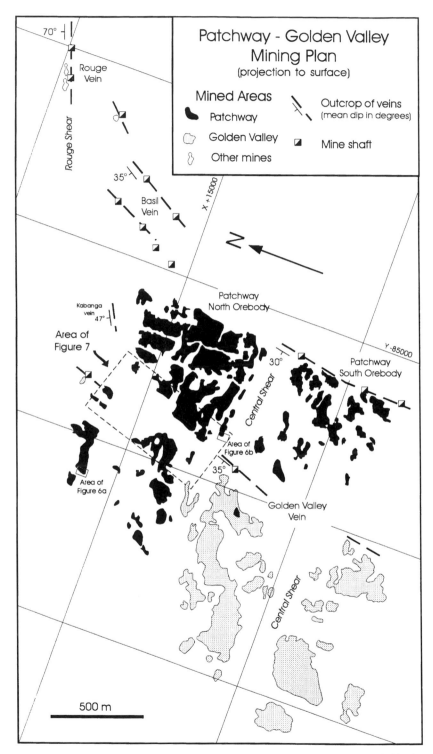

Fig. 3. Mine plans projected to surface for the Pathway, Golden Valley and Rouge mines (compiled by the author).

across normal faults, which simply displace the mineralization. Many of these normal faults are intruded by dolerite dykes. The Golden Valley orebodies show a very similar spatial pattern to those at Patchway. The Golden Valley mineralization has further extensions to north and south of the two main orebodies, where limited mining has been carried out. The deeper levels of the Golden Valley mine have similar normal fault disruptions to mineralization to that seen at Patchway, which are also associated with late dolerite dykes. The Golden Valley workings have reached one such structure named the 'Terminal Fault' at the level of deepest workings. The fault offset of the mineralization has not been intersected below this point although similar faults at Patchway show a variable normal throw suggesting thjat extension to the Golden Valley mineralization may occur at an unspecified depth.

At the northern end of both Golden Valley and Patchway, the Rouge mine is developed on a 070° structural trend, dipping 55–60° NNW. Mineralization style is clearly different to that of the Patchway and Golden Valley veins, being described by Bliss (1970) as: 'schist impregnated with pyrite, pyrrhotite and minor amounts of quartz in lenses and stringers. The zone is normally wide, varying from 44 inches to over 100 inches but the distribution of ore was irregular'. The description of mineralization is somewhat similar to the style and trend of mineralization displayed at the nearby Erin mine, which is also developed on an 070° structure.

The other minor deposits, such as Basil (see Fig. 3), resemble the Patchway mineralization, the vein having a typical ribboned appearance and a similar strike (Bliss 1970). Finally, low-grade mineralization, known as the Kabanga vein system, has been investigated at the northwestern limit of the mine (see Fig. 3) which has a roughly NE–SW trend.

Structural geology of the Patchway–Golden Valley deposits

Early folding

The earliest recorded deformation event in the greenstone belt as described earlier is the development of large scale regional folding (Bliss 1970). The folding has led to formation of the Kadoma anticline (Fig. 1) and the Patchway area would lie on the northwest limb of this feature. At Patchway primary bedding features are rare, but there is indication of a generally shallow northwest dip to the package of flow rocks, with units being the right way up in agreement with other authors (Bliss 1970; Ward 1964).

D2 deformation

The next deformation event, here termed D2, produced the structures which host the main gold–quartz mineralization. The structures themselves are represented by ductile to brittle–ductile zones which can be traced for over 1.5 m of strike. Down-dip the structures are known to a depth of 1 km which at 30° dip means a true extent of at least 2 km. Within these structures, ribboned quartz–gold veins themselves form semi-continuous elements of the structures, which at Patchway measure up to 300 m along strike with a down dip extent of up to 250 m. At Golden Valley the continuous sections of vein may be over 300 m in strike length with dip extensions of up to 500 m. The Basil vein (Fig. 3) is a similar sub-parallel structure partially investigated by underground workings at Patchway. Texturally the vein shows identical textures to the Patchway main vein (Herrington 1991), although low gold grades do not currently support mining. In addition, the Hamburg structure, which lies south of Patchway, follows a similar structure, and is considered to have formed contemporaneously. From the broad relationships the veins would all appear to represent a shear vein, parallel to the main failure of the zone. Figure 4a shows a composite plot for poles to gold-bearing veins in the Patchway, Basil, Golden Valley and Hamburg deposits. The great circle plotted is drawn about the central point of the >30% contour plot of all the poles to veins. The data show a relatively small spread, consistent with all the deposits being developed on parallel, regionally significant, structures. In all cases there is strong development of a shear fabric marginal to the quartz veins and this has a generally consistent relationship with the main vein. The angular relationship of this S-fabric versus the main failure plane, C is supportive of a reverse plus sinistral strike-slip movement on the C failure plane (Fig. 4b). Other macroscopic kinematic indicators are rare, but there is a suggestion of minor reverse thrusting in pillow lava flow tops on the mine's 10 level. Reverse plus strike-slip movement accompanied the periodic hydrothermal fluid flow through the main failure surface which led to formation of the ribbon quartz veins (Herrington et al. 1993). Evidence for the broadly compressive nature of the stress field on the main mineralized structure is seen in the intense flattening of pillows in the shear zone and within the quartz vein itself where a variable development of stylolitization of the veins and wallrock septa is seen across the vein. However it is clear that for the main failure plane, the quartz veining is evidence for volume inflation across the structure.

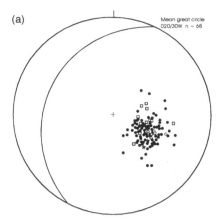

(a)

Mean great circle
020/30W n = 68

Pi poles to D2 gold-quartz veins
•   Patchway mine
□   Golden Valley Mine
*   Hamburg Mine

D3 effects on D2 mineralization

The D3 deformation is marked by a second fabric at Patchway (S3) which is developed at variable intensity throughout the deposit within the vein hosting shear zone (Fig. 5a), but more importantly in discrete cross-cutting zones which disrupt gold mineralization (Herrington 1991).

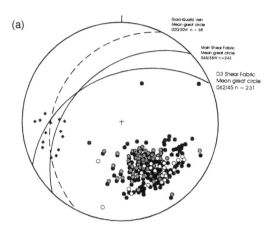

(a)

Gold-Quartz Vein
Mean great circle
020/30W n = 68

Main Shear Fabric
Mean great circle
048/35W n=243

D3 Shear Fabric
Mean great circle
062/45 n = 231

Pi poles to D3 shear fabric
•   Patchway North Orebody
⊚   Patchway "Central Shear"
○   Erin Mine

♦   Plunge of early boudinage and
    minor folding of D2 veining

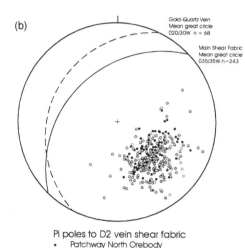

(b)

Gold-Quartz Vein
Mean great circle
020/30W n = 68

Main Shear Fabric
Mean great circle
035/35W n=243

Pi poles to D2 vein shear fabric
•   Patchway North Orebody
⊚   Patchway South Orebody
○   Golden Valley

Fig. 4. (**a**) Pi plot of D2 gold–quartz veins from Patchway, Golden Valley and Hamburg mines. Great circle plotted about mean of data, n = 68. (**b**) Pi plot of shear fabric related to D2 gold-quartz veining at Patchway and Golden Valley. Great circle plotted about mean of data, n = 243.

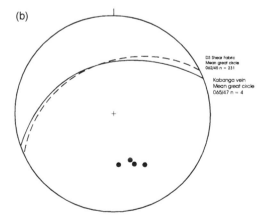

(b)

D3 Shear Fabric
Mean great circle
062/45 n = 231

Kabanga vein
Mean great circle
065/47 n = 4

Pi poles to Kabanga Veins

D3 deformation

The second major structural event during the development of the gold deposits in the region around Patchway is the development of a second major trend of structures striking approximately 070°. This trend is parallel to: (a) the 'Central shear zone' at Patchway and similar zones at Golden Valley, (b) the 'Rouge shear' and associated Rouge orebody (see Fig. 3) and (c) the Erin orebody.

Fig. 5. (**a**) Pi plot of D3 shear fabric at Patchway and Erin mines. Great circles plotted about data means. Also plotted are plunges to boudinage crests and minor folds in D2 gold–quartz veins. (**b**) Pi plot of D3 Kabanga veins measured at Patchway mine. Great circle plotted about mean of 4 data.

In general, D3 is characterized by boudinage and thinning of the vein. The boudinage is indicative of volume loss across the structures, again indicative of a compressive regime. Examination of both the ribbon quartz vein and the felsite intrusive body, also caught in the later deformation zone, indicates a repetition of these units at the northern part of the mine, consistent with a sinistral strike-slip component to shear (Fig. 6a). Parts of the vein shows minor folding, particularly in the more intensely deformed areas. Plunges of the axial traces of these minor folds is parallel to the plunge of boudin crests, consistent with formation during the same deformation event (Fig. 5a).

In the region of the Central shear zone thin remnants of earlier D2 mineralization are present as boudinaged sections scattered throughout the zone, sometimes only present as thin lenses. To the north of the mine the D3 deformation is largely confined to discrete but intense cross-cutting zones which clearly disrupt mineralization, complicating mining operations. Again there is clear evidence of sinistral strike-slip motion on these structures, which is accompanied by the apparent rotation of the faulted segments of the quartz–gold veins to closer parallelism with the strike of the D3 zones in this area (Fig. 6b).

Erin mine structure

The main host structure for mineralization at Erin is parallel to the main D3 structures mapped at Patchway (Fig. 5a). The deformation style of the Erin structure is quite different to that seen surrounding the main veins at Patchway and Golden Valley. Alteration takes the form of intense carbonate flooding of opened shear foliation, lending the mineralized zone a characteristic striped appearance. Minor folds in the mineralized zone are common, strikingly open in form in some places. Mapping at Erin in this structure indicates that a similar sense of shear can be deduced from all the kinematic indicator elements, showing dominantly sinistral strike-slip motion (Fig. 6c). Such a shear sense supports the view that the Erin structure is a D3 structure, with similar sense to that at Patchway.

Other minor veins

The so called Kabanga reef, intercepted at the northwest end of Patchway mine, also appears to be located on D3 structures (Fig. 3). Figure 5b shows a stereonet projection of the data collected from this area indicating its association with the D3 trend. The mineralized structures examined underground show some strong similarities with the alteration style at Erin (Herrington 1991).

Good evidence for quite late deformation on the D3 structures is provided from the literature. Bliss (1970) notes that the Rouge shear, host to the Rouge gold deposit (Fig. 3), which has an identical trend to the Patchway central shear and the Erin structure, in fact truncates the late dolerite dyke intrusions which are ascribed to the D4 event (see below). Although this cannot be seen at Patchway, a clearly more protracted period of deformation post-dating mineralization is recorded on the D3 structures.

D4 deformation

The latest major deformation event recorded at the Patchway mine is expressed as a major period of extension which was oriented in an almost E–W direction across the region. The deformation is largely manifest as a distinct event of normal dip-slip on the main vein, leading to another period of stretching and boudinage of mineralization. Accompanying this normal faulting of the mineralization is the intrusion of later dolerite dykes. This deformation event is responsible for the fault disruption and attenuation seen between the upper and lower sections of the mine at Patchway.

D4 is characterized by common asymmetric boudinage with deformation later evolving to more brittle normal faulting. This seems to have been closely accompanied by the intrusion of the dolerite dykes. Stereonet projection of data shows the late faults, veining and dyke intrusions (Fig. 7), which is suggesative of a strong Riedel geometric control to the elements, with faults developed along 'R' and 'T' planes, infilled with quartz and tourmaline, and in the case of some the 'T' orientation much later doleritic magma which in places cuts the quartz–tourmaline veining (Herrington 1991).

Figure 8 is a perspective view of part of the North Orebody at Patchway which demonstrates the combined effects of superimposed D3 and D4 features. These features can be referred back to the plan of mined areas (Fig. 3) where the westward plunging terminated to mineralization caused by D3 on the D2 vein and the roughly N–S- and NNW–SSE-trending termination to mineralization caused by D4 on the D2 vein can be observed. Complex ore blocks are thus defined by superimposition of these two deformation events.

Very late regional features

Veins hosted in the large tonalite bodies have produced scheelite and are largely barren of gold (Bliss 1970). These veins probably post-date the D2 gold veins, since the Sesombi tonalite is seen to have a metamorphic aureole overprinting gold mineralization in shear zones of ages presumed

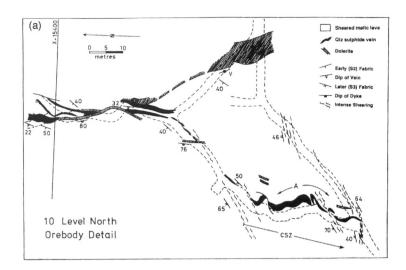

(a)

Sheared mafic lava
Qtz sulphide vein
Dolerite

Early (S2) Fabric
Dip of Vein
Later (S3) Fabric
Dip of Dyke
Intense Shearing

10 Level North
Orebody Detail

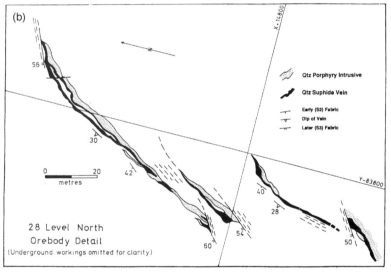

(b)

Qtz Porphyry Intrusive
Qtz Suphide Vein

Early (S2) Fabric
Dip of Vein
Later (S3) Fabric

28 Level North
Orebody Detail
(Underground workings omitted for clarity)

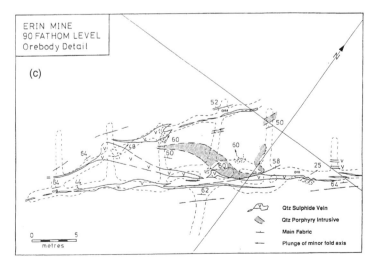

ERIN MINE
90 FATHOM LEVEL
Orebody Detail

(c)

Qtz Sulphide Vein
Qtz Porphyry Intrusive
Main Fabric
Plunge of minor fold axis

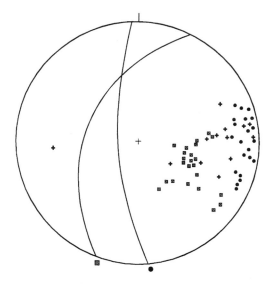

Pi poles to D4 features
- Late extension veins
- Late normal faults
- Dolerite dykes

Fig. 7. Pi plot of D4 features at Patchway mine. Great circles plotted about mean of data.

to be contemporaneous with D2 (Robertson 1976). Finally, late faults trending NW–SE cut all lithologies in the region (Bliss 1970). These faults host 'barren' quartz mineralization when striking 150–160° in the Patchway area. One such vein was examined at Lion Hill (Fig. 2) and it is believed to represent very late deformation since the faults would appear to cross-cut all regional geological features.

Structural summaries of the Midland gold deposits

Golden Valley–Patchway

Figure 9 summarizes the Golden Valley–Patchway events, where early oblique sinistral reverse-slip on NNE–SSW-trending structures evolved to sinistral

slip in linked ENE structures and was accompanied by gold mineralization. During this event, the NNE–SSW-trending structures are clearly more favourably oriented for dilation and hence are the best mineralized. The ENE structures continued to focus deformation after cessation of gold mineralization and are seen to deform the NNE gold veins. Intrusion of the tonalities belonging to the Sesombi suite probably occurred late in the deformation–mineralization event. This deformation is consistent with initial shortening directed in a roughly N–S direction, later modified to a NNE–SSW-directed principal stress direction. Much later E–W extension resulted in faulting and dolerite dyke intrusion (Herrington 1991).

Dalny mine

Gold mineralization is largely confined to an array of gold-bearing structures, related to a 10 km long NE–SW-trending shear zone hosted in tholeiites with minor shales and BIFs. A complex history of deformation is indicated where drag folding is indicative of sinistral strike-slip shear (Leigh 1964) with shear fabrics supportive of dominantly reverse dip-slip on the main Dalny shear during mineralization followed by later dextral strike-slip reactivation (Carter & Foster 1990; Carter 1991). Mineralization is then truncated by NNW–SSE-trending faults with fault offsets consistent with either sinistral shear or normal faulting. These events are postdated by the intrusion of NNE–SSW-oriented dolerite dykes.

Cam & Motor mine

The shear zone hosting the main mineralized vein lies along the contact between greenstones and metasediments which at lower levels takes the form of a drag fold, the shape of which suggests a early sinistral strike-slip shear during folding (Collender 1964). The Motor vein and related Cam spur vein truncate the E–W-trending Cam veins at depth suggesting a complex linked vein system which may have evolved during later ENE–WSW-directed dextral strike-slip shearing (Fig. 10a). This is compatible with the present orientation of the vein system. It is suggested by Bliss (1970) that

Fig. 6. (a) Level plan of 10 Level North, Patchway mine (see Fig. 3 for location). Plan shows intense D3 deformation to D2 quartz veining (in black) close to the northern margin of the ;Central Shear Zone'. (b) Level plan at 28 Level North, Patchway mine (see Fig. 3 for location). Plan shows fault repetition of the D2 quartz veining anf felsite intrusive close to the current northern margin of current mine workings. (c) Level plan of 90fm Level, Erin mine (see Fig. 2). Plan shows deformed felsite intrusive and quartz veining. Minor structures are consistent with dominantly sinistral strike-slip shearing.

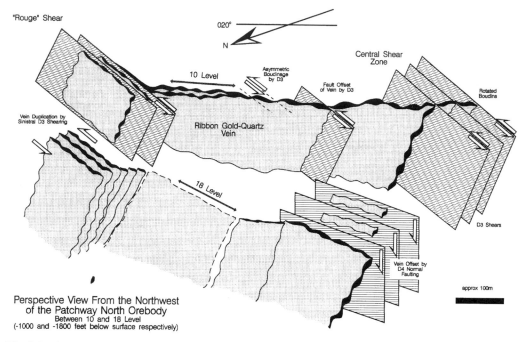

Fig. 8. Partly schematic perspective view of part of the Patchway North Orebody (area shown on Fig. 3). Black areas represent D2 gold–quartz veining which is cut by D3 (wavy lines) and D4 (cross hatched) structures. D3 leads to thinning, boudinage and shear repetition of the D2 veining. D4 leads to boudinage and normal faulting of the D2 veining.

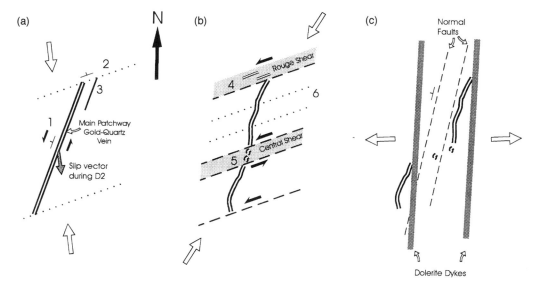

Fig. 9. Schematic plan of the deformation and mineralization events seen at Patchway. (**a**) D2 deformation focused on 020° trending structure in response to NNW–SSE applied stress. 1, Gold–quartz mineralization during oblique sinistral plus reverse slip. 2, Cross structures, later important during D3, are unfavourably oriented during D2. 3, Sub-parallel veins, such as Basil, also form. (**b**) D3 deformation focused on 060–070° structures during rotation of principal stress direction clockwise. Rouge and Central shears focus much deformation with formation of minor gold-bearing veining at Kabanga (4). D2 veining is fault repeated and boudinaged in this event (5). The main areas of the mine see development of minor D3 boudinage and shearing (6). (**c**) D4 deformation in response to E–W extension. Normal faulting and dyke intrusion disrupt the D2 mineralization.

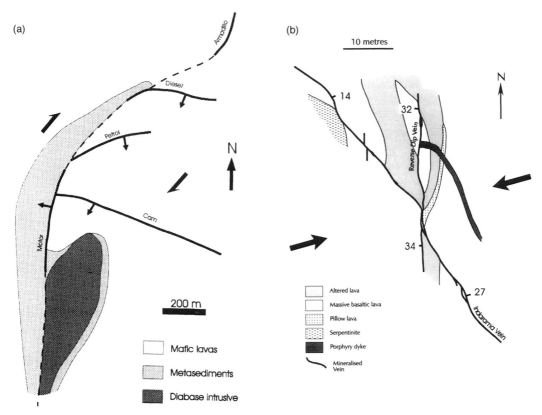

Fig. 10. (**a**) Surface plan of the Cam & Motor gold deposit (after Collender 1964). Later deformation on the mineralized structures, notably clockwise rotation of the main Motor shear, is consistent with dextral strike-slip shearing. (**b**) Plan of the veins on 4 Level, Indarama gold deposits (after Nutt 1985). The flat-lying veins form a conjugate pair with generally reverse sense of displacement on the veins, consistent with ENE–WSW directed compression.

the veins themselves are folded after mineralization implying such a protracted deformation–mineralization event. Andesite lavas make the best host rocks for mineralization; pillow lavas, greywackes and sediments make poor host rocks which may be a function of the rheological properties of the lithologies.

Undergroun, there is good evidence for dominantly reverse-sense shearing during mineralization with overthrusting of the main Brompton vein via minor duplex structures. Mineralization is disrupted by later faulting on cross structures which show both sinistral and dextral offsets in different areas (Tomlinson 1982).

Brompton mine

This district contains the currently producing Brompton mine and related Revie and Pink deposits. The main host rocks are somewhat unusually the basement Rhodesdale gneiss complex which is locally intruded by mafic dykes, now altered to amphibolites. Mineralization is related to broadly NNE–SSW-trending linements which probably represent parallel structures which host thin felsite intrusions and ribbon quartz veins.

Venice–What Cheer mines

Mineralization in this region occurs in two distinct orientations. Steeply easterly dipping NNE–SSW-trending veins contrast with shallow southerly dipping WNW–ESE-trtending veins, both interpreted as forming under reverse thrust faulting (Nutt et al. 1988). Catchpole (1987) suggests that thrusting on the shallowly dipping veins post-dates the steeper set. Formation of both vein sets is consistent with NNE–SSW-directed compression.

Battlefields area

The productive deposits in the Battlefields area (Invincible etc.) strike NNW to WNW and dip mainly southwestwards. Mineralization appears to be related to early thrusting on these structured which are cut later by ENE–WSW-trending sinistral faults. Farther south, the Washington mine has NNE–SSW-trending veining with evidence for both reverse sense thrusting and perhaps later oblique dextral shear sense (Pitfield & Campbell 1990). These shear senses are consistent with dominantly NNE–SSW compression followed by later rotation to a roughly E–W oriented principal stress.

Kwekwe area

At the Indarama mine, north of Kwekwe, conjugate low-angle veins are developed with a roughly N–S orientation consistent with simple shear developed in response to a roughly ENE–WSW compression event (Fig. 10b). The associated Broom stock deposit, lying to the north, occurs in the N–S-trending fold nose of oxide facies BIF which is highly fractured at the fold closure. This fold closure is also consistent with roughly E–W-directed compression.

Farther south, the major Globe & Phoenix deposits are hosted within the so called Kwekwe ultramafic unit and the adjoining Rhodesdale gneiss. The deposits comprise three main vein sets (Foster et al. 1991). Early carbonization and silicification of the ultramafics was crucial ground preparation, probably during development of the major Rhodesdale boundary shear complex. Mineralization occurred during reverse sense motion on NNW vein-infilled shears and supports a model of E–W- or NE–SW-directed compression, possibly coincident with dextral shear along NNW striking shears (Porter & Foster 1991).

Deformation model for the Kadoma region with reference to regional events

Summarizing the observations above with the regional work of previous authors (Stowe 1971, 1974; James 1975; Coward 1976; Coward et al. 1976; Coward & Fairhead 1980; Treloar et al. 1992), the proposed sequence of late Archaean regional deformation, which led to the gold-mineralization events across the Midlands area, can be linked to deformation over the Zimbabwe craton and can be broadly divided into two stages: early and later deformation.

Early deformation

Early deformation in the Shurugwi greenstone belt in the south of the cratonic area probably predates the bulk of the deformation elsewhere (Stowe 1974). After this event NNW–SSE-directed shortening across the belt probably led to the regional folding events in the 2700 Ma greenstone belts, which largely predated intrusion of the Sesombi suite of granitoids (Coward 1976; Coward et al. 1976). This appears to precede gold mineralization.

Later deformation

Development of the observed WSW–ENE shortening in the Midlands and Bulawayo greenstone belts cannot be explained by a simple compression event. McCourt & Vearncombe (1987) propose a component of an E–W collision event lying to the east to account for the westerly directed thrusting of the Limpopo Central Zone, since it is clear that the cratonic area moved some 200 km west with respect to the Limpopo belt (Coward 1976). This event led to major shortening across the south-west greenstone belts which are elongated in a N–S orientation (Coward 1980).

The whole of the cratonic area was probably not displaced, however, as dextral strike-slip movement is recorded on some of the NE–SW-trending shears on the northern margin of the Rhodesdale gneiss block (Pitfield & Campbell 1989). Therefore the Rhodesdale gneiss block alone may have moved westwards with strike-slip motion occurring on the bounding shears both north and south. This type of event could lead to a complex series major strike-slip movement on the previously developed NE–SW thrust faults.

The recent models for emplacement of the Great Dyke point to crustal extension during a period of pure shear (Wilson & Prendegast 1989) with development of the sub-parallel Popoteke Fault set. Light (1982) relates this event with the collision of the Zimbabwe and Kaapvaal cratons, which infers a continued period of NNW–SSE-oriented compression right up to the end of the Archaean. In this case the only way for the Rhodesdale gneiss block to have moved westwards, is in some kind of lateral extrusion model, where the central portions of the largely undeformed Rhodesdale gneiss are squeezed westwards (Treloar et al. 1992). Support for this lateral movement of the block in the central part of the craton comes from the observation of dominantly reverse thrust movement on N–S-trending mineralized structures in the KweKwe area (Nutt et al. 1988; Porter & Foster 1991).

The structural features seen in the gold deposits of the Midlands greenstone belt are consistent with such a model of NNW–SSE compression linked to

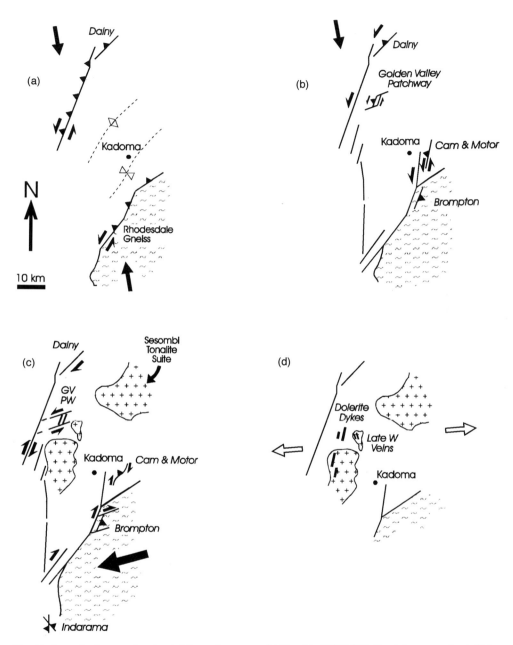

Fig. 11. Synoptic diagram of regional deformation events. (**a**) Roughly NNW–SSE-directed compression initiates major structures at greenstone belt boundaries. Initiation of mine structures, possibly at culmination of major regional folding (D1). (**b**) Continued NNW–SSE compression focused on structures. Gold mineralization during transpressional motion on lower order structures related to the 'craton-scale' shear zones (D2). (**c**) Switch of deformation to cross structures in Pathway–Golden Valley area may relate to lateral extrusion of Rhodesdale gneiss block westwards. Strike-slip component on most structures in the Kadoma region reverses. Stress regime becomes more favourable for emplacement of Sesombi tonalite suite along apex of Kadoma anticline (D3). (**d**) Late extension results in normal faulting and dolerite dyke swarm intrusion (possibly post-Archaean?) (D4).

lateral extrusion of the Rhodesdale gneiss block westwards (Fig. 11 and Table 1). Initiation of structures during NNW–SSE compression resulted in early folding such as the Kadoma anticline (Bliss 1970), together with initiation of the major mineralized shear systerms as thrust faults (Stage A). Simple shear evolved to transpressive shear with development of oblique dextral and sinistral mineralized shear systems (Stage B). At Patchway, reverse plus sinistral shearing is evident on the main gold-bearing structures at this time. In the Cam & Motor region, the Motor shear, which was probably initiated as a roughly N–S wrench fault, was likely to have been favourably oriented for formation of the Cam and associated vein systems as the branches of a compressive flower structure (Woodcock & Fischer 1986) under sinistral strike-slip shear (Fig. 12). Rotation of the principal stress direction clockwise to an ENE–WSW orientation led to evolution of dominantly dextral transpressive shear on the major Lily, Munyati and Rhodesdale

boundary shear systems, probably in response to the lateral extrusion of the Rhodesdale gneiss block westwards (Stage C). This phase of deformation accounts for the major dextral offsets along the Lily fault (Bliss 1970) and the Munyati shear close to Battlefields (Pitfield & Campbell 1989). The main part of this deformation at Patchway is shown to be dominantly reverse plus a component of sinistral strike-slip. Such a movement sense suggests continued compression, probably from the northwest. Carter & Foster (1990) concludes a similar slip vector for the Dalny structure during the bulk of the deformation during mineralization. This event is likely to have led to the current morphology of the Cam & Motor system where later dextral strike-slip shearing is evident (Fig. 10a). Pitfield & Campbell (1990) note stretching lineations on the Rhodesdale boundary shear-zone pitching to the NE, again consistent with the movement sense on the Lily fault associated structures. These authors further suggest mineralization in the

Table 1. *Tabulated summary of the deformation events in Midland greenstone belt during the Late Archaean, with particular reference to the Kadoma region*

Regional-scale structural evolution		Deposit-scale structural evolution		
Stage (see Fig. 11)	Deformation events	Stage (see Fig. 11)	Deformation events	Mineralization
A	Early NNW–SSE compression, major folds, initiation of major faults?	D1	Metamorphism in greenstone sequence, large-scale folding	Pillow margin quartz infillings
B	Continued NNW–SSE compression, reverse plus sinistral strike-slip motion on first-order structures	D2 (A on Fig. 9)	Reverse plus sinistral strike-slip faulting on second-order structures. Gold–quartz mineralization in 020° structures at Patchway and Brompton; mineralization in 040–045° structures at Cam & Motor and Dalny	Quartz–sulphide gold veins
C	Intrusion of Sesombi tonalite suite linked to expulsion of Rhodesdale gneiss block westwards. Switch to dextral strike-slip shearing on the first order structures	D3 (B on Fig. 9)	Sinistral strike-slip motion on 070° structures in the Golden Valley–Patchway area. Attenuation and faulting of earlier D2 gold-bearing veins, waning phase of gold mineralization on 070° structures (e.g. Patchway and Erin mine), folding of Cam & Motor veins?	Quartz–tourmaline (± sulphides, gold)
D	Regional E–W extension culminating in emplace-ment of dolerite dyke swarm	D4 (C on Fig. 9)	N–S-oriented normal faults, late gold-poor extension veins, scheelite veins in tonalite bodies, lastly intrusion of dolerite dykes	Quartz–tourmaline extension veins, quartz–scheelite veins
Late	Late brittle fractures, dominantly striking 150–160°, barren quartz veins			'Barren' quartz veins

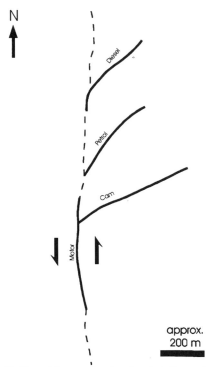

N

approx.
200 m

Fig. 12. Plan of the postulated early orientation and sense of displacement on the Cam & Motor structures during 'B' phase of deformation shown on Fig. 11. Current orientation of the Cam & Motor structures (Fig. 10a) is due to modification during Phase 'C' of the deformation.

Battlefields area may be related to this early thrust plus strike-slip faulting. Nutt *et al.* (1988) report on deposits in the KweKwe region farther south where high angle reverse faulting accompanies mineralization. Since the mineralization is likely to be of similar age to that in the Kadoma region, it supports the proposed lateral extrusion of the Rhodesdale block westwards at this time. If this event occurred at the same time as the generally applied NE–SW compression, as seems likely, then rather complex displacement relationships might be expected.

The bulk of gold mineralization links with the switch from Stage A deformation, the early simple shear during NNW–SSE-directed shortening, to Stage B, the later transpressive shearing.

Stage C was probably accompanied by waning

gold mineralization, largely along 070° structures with the intrusion of the larger tonalite bodies such as the Whitewaters stock (and regionally the Sesombi tonalite suite). Orientation of the stress field in the Kadoma region into an ENE–WSW direction would be conducive to intrusion of these bodies parallel to the local σ1 at this time. The strong geochemical affinity of these tonalite diapirs and the mine felsite intrusives is quite clear, and the intrusion of the diapirs would be a logical result of the large body of magma rising higher into the upper crust. These tonalities show that they have also suffered some deformation, particularly in the marginal zones, which is strong evidence for a syntectonic intrusion event. At the end of this event the brittle tonalite hosted scheelite-bearing veins probably formed.

Any regional model must take account of the similar, but easterly dipping structures are seen to the east of Lion Hill which are host to small gold deposits (Bliss 1970). Anhaeusser (1976) shows how these deposits are distributed in arcuate fashion around the tonalite intrusives and thus alludes to a genetic relationship between the veins and the tonalites. However, initiation of the host structures to these deposits can be explained if these initiated as the conjugate thrusts to the Patchway–Golden Valley veins and later structural modification by the foreceful emplacement of the Whitewaters tonalite could explain the observed arcuate distribution.

Lastly, Stage D is coincident with the intrusion of the regionally extensive vein swarms of the Proterozoic (Wilson *et al.* 1987), coincident with continued N–S compression after the end of the Archaean.

The author wishes to acknowledge the support of an RTZ bursary for PhD studies at Imperial College on which the paper is largely based. I should also like to acknowledge Royal Society support for a study visit to the University of Zimbabwe in 1993, kindly permitted by The Natural History Museum. I would like to thank my hosts and various friends there, particularly T. Blenkinsop, for much help and discussion but would not like to hold them responsible for my views. Many thanks also to RTZ and Falcon mines in particular, for access to the various mines. I must thank my colleagues at the NHM for their review of early drafts of the manuscript and to two anonymous reviewers whose comments much improved the paper.

References

DANHAEUSSER, C. R. 1976. The nature and distribution of Archaean gold mineralisation in southern Africa. *Minerals Science and Engineering*, **3**, 46–84.

ANON 1984. *Magnetic anomaly map of part of Zimbabwe. 1:10000 000 Scale.* Geological Survey of Zimbabwe, Ministry of Mines.

BLISS, N. W. 1970. *The geology of the country around Gatooma, Rhodesia. Rhodesian Geological Survey Bulletin,* **64**.

CARTER, A. H. C. 1991. Fluid-rock interaction and gold deposition within a late mArchaean shear zone, Dalny mine, Zimbabwe. *In:* LADEIRA, E. A. (ed.)

Proceedings of Brazil Gold '91, An International Symposium on the Geology of Gold Belo Horizonte, 1991. A. A. Balkema, Rotterdam, 263–268.

—— & FOSTER, R. P. 1990. Tectonic, thermal and chemical evolution of a late Archaean auriferous shear zone, Dalny mine, Zimbabwe. *GAC-MAC Annual Meeting, Program with Abstracts*, A22.

CATCHPOLE, S. J. 1987. Gold mineralisation related to shear zones in the Venice Group of mines near Kadoma, Zimbabwe. *In*: *African Mining*. Institute of Mining and Metallurgy, 71–88.

COLLENDER, F. D. 1964. The geology of the Cam & Motor mine, southern Rhodesia. *In*: HAUGHTON, S. H. (ed.) *The geology of some ore deposits in southern Africa*. Geological Society of South Africa, Special Publications, **2**, 15–27.

COWARD, M. P. 1976. Archaean deformation patterns in South Africa. *Philosophical Transactions of the Royal Society, London*, **A283**, 313–331.

—— 1980. Shear zones in the Precambrian crust of southern Africa. *Journal of Structural Geology*, **2**, 19–27.

—— & FAIRHEAD, D. 1980. Gravity and structural evidence for the deep structure of the Limpopo belt, southern Africa. *Tectonophysics*, **68**, 31–43.

——, JAMES, P. R. & WRIGHT, L. I. 1976. The movement pattern across the northern margin of the Limpopo mobile belt, southern Africa. *Geological Society of America Bulletin*, **87**, 601–611.

—— & WILSON, J. F. 1984. Geological setting of Archaean gold deposits in Zimbabwe. *In*: FOSTER, R. P. (ed.) *Gold '82: Geology, geochemistry and genesis of gold deposits*. A. A. Balkema, Rotterdam, 521–551.

——, FISHER, N. J., PORTER, C. W., FABIANI, W. M. B. & CARTER, A. H. C. 1991. The tectonic and magmatic framework of Archaean lode-gold mineralisation in the Midlands greenstone belt, Zimbabwe. *In*: LADEIRA, E. A. (ed.) *Proceedings of Brazil Gold '91, An International Symposium on the Geology of Gold, Belo Horizonte, 1991.* A. A. Balkema, Rotterdam, 359–366.

——, MANN, A. G., MILLER, R. G. & SMITH, P. J. R. 1979. Genesis of Archaean gold mineralisation with reference to three deposits in the Gatooma area, Rhodesia. Geological Society of South Africa, Special Publications, **5**, 25–38.

——, ——, STOWE, C. W. & WILSON, J. F. 1986. Archaean gold mineralisation in Zimbabwe. *In*: Anhaeusser, C. R. & Maske, S. (eds). *Mineral Deposits of Southern Africa*, Geological Society of South Africa, **1**, 43–112.

HERRINGTON, R. J. 1991. *The relationship between fluids and structure at the Patchway gold mine, Zimbabwe*. PhD Thesis, Univ. London.

——, RANKIN, A. H. & SHEPHERD, T. J. 1993. Fluid chemical and structural evolution of gold-quartz veins, Patchway mine, Zimbabwe. *In*: MAURICE, Y. (ed.) *Proceedings of the 8th Quadrennial IAGOD Symposium*. E. Schweizerbart'sche Verlagsbuchhandlung (Nägele u. Obermiller), D-7000 Stuttgart 1, 681–694.

JAMES, P. R. 1975. A deformation study of the northern margin of the Limpopo belt, Rhodesia. PhD Thesis, Univ. Leeds.

KEY, R. M. & HUTTON, S. M. 1976. The tectonic generation of the Limpopo Mobile Belt, and a definition of its western extremity. *Precambrian Research*, **8**, 375–413.

LEIGH, R. W. 1964. The geology of the Dalny mine. *In*: *The geology of some ore deposits of Southern Africa*, Geological Society of South Africa, **2**, 29–40.

LIGHT, M. P. R. 1982. The Limpopo mobile belt: a result of continental colllision. *Tectonics*, **1**, 325–342.

MACGREGOR, A. M. 1930. T*he geology of the country between Gatooma and Battlefields*. Geological Survey of Southern Rhodesian Bulletin, **38**.

McCOURT, S. & VEARNCOMBE, J. R. 1987. Crustal-scale shear zones in the Limpopo belt, southern Africa. *Geological Society of Canada, Programs with Abstracts*, 12.

NUTT, T. H. C. 1984. Gold mineralisation in the Broomstock Extension and Indarama mines and their bearing on the genesis of the Kwekwe–Sherwood Block goldfield. *In*: BRODERICK, T. J. (ed.) *Annals of the Geological Survey of Zimbabwe*, **10**, Harare, 107–120.

——, McCOURT, S. & VEARNCOMBE, J. R. 1988. Structure of some antimony-gold deposits from the Kaapvaal and Zimbabwe cratons. *In*: HO, S. E. & GROVES, D. I. (eds) *Advances in understanding Precambrian gold deposits*. **2**. Geol. Dept. & Univ. Extension, Univ. Western Australia, Publications, **12**, 63–80.

PITFIELD, P. E. J. & CAMPBELL, S. D. G. 1990. Integrated exploration, Midlands goldfield project: Preliminary results. *In*: BAGLOW, N. (ed.) *Annals of the Zimbabwe Geological Survey*, **15**, Harare, 21–33.

PORTER, C. W. & FOSTER, R. P. 1991. Multi-phase ductile-brittle deformation and the role of Archaean thrust tectonics in the evolution of the Globe and Phoenix gold deposit, Zimbabwe. *In*: LADEIRA, E. A. (ed.) *Proceedings of Brazil Gold '91, An International Symposium on the Geology of Gold, Belo Horizonte, 1991.* A. A. Balkema, Rotterdam, 665–671.

ROBERTSON, I. D. M. 1976. *The geology of the country around Battlefields, Gatooma district*. Geological Survey of Rhodesia Bulletin, **76**.

STOWE, C. W. 1971. Summary of the tectonics of the Rhodesian Archaean craton. *In*: GLOVER, J. E. (ed.) *Symposium on Archaean Rocks*. Special Publication of the Geological Society of Australia, **3**, 373–383.

—— 1974. Alpine-type structures in the Rhodesian basement at Selukwe. *Journal of the Geological Society , London*, **130**, 411–425.

—— 1980. Wrench tectonics in the Archaean Rhodesian Craton. *Transactions of the Geological Society of South Africa*, **83**, 193–205.

TAYLOR, P. N., KRAMERS, J. D., MOORBATH, S., WILSON, J. F., ORPEN, J. L. & MARTIN, A. 1991. Pb/Pb, Sm-Nd and Rb-Sr geochronology in the Archaean Craton of Zimbabwe. *Chemical Geology*, **87**, 175–196.

TOMLINSON, L. A. 1982. Brompton mine. *In*: *GOLD '82 Excursion Guidebook*. Geological Society of Zimbabwe, Harare, 43–45.

TRELOAR, P. J., COWARD, M. P. & HARRIS, N. W. 1992. Himalayan-Tibetan analogies for the evolution of the Zimbabwe Craton and Limpopo Belt. *Precambrian Research*, **55**, 571–587.

WARD, J. W. H. 1968. Geological aspects of the Patchway gold mine, Rhodesia. Transaction of the Geological Society of South Africa, **71**, Annex, 147–158.

WILSON, A. H. & PRENDEGAST, M. D. 1989. The Great Dyke of Zimbabwe–I: tectonic setting, stratigraphy, petrology structure, emplacement and crystallisation. *In*: PRENDEGAST, M. D. & JONES, M. J. (eds) *Magmatic Sulphides–The Zimbabwe Volume*, Institution of Mining and Metallurgy, London, 1–20.

WILSON, J. F., JONES, D. L. & KRAMERS, J. D. 1987. Mafic dyke swarms in Zimbabwe. In: HALLS, H. C. & FAHRIG, W. F. (eds) *Mafic dyke swarms*. Geological Association of Canada, Special Papers, **34**, 433–44.

——, NESBITT, R. W. & FANNING, C. M. 1995. Ziron geochronology of Arachaean felsic sequencies in the Zimbabwe craton: a revision of greenstone stratigraphy and a model for constant growth. *This volume*.

WOODCOCK, N. H. & FISCHER, M. 1986. Strike-slip duplexes. *Journal of Structural Geology*, **8**, 725–735.

Early Proterozoic mafic dykes in the North Atlantic and Baltic cratons: field setting and chemistry of distinctive dyke swarms

DAVID BRIDGWATER[1], FLEMMING MENGEL[2], BRIAN FRYER[3], PAUL WAGNER[4] & SØREN CLAUDIUS HANSEN[1]

[1]*Geological Museum, Øster Voldgade 5-7 Copenhagen K, DK 1350 Denmark*
[2]*Danish Lithosphere Centre, Øster Voldgade 10 Copenhagen K, DK 1350 Denmark*
[3]*Department of Earth Sciences, University of Windsor, Ontario, Canada*
[4]*Department of Earth Sciences, University of Alberta, Edmonton, Canada*

Abstract: Widespread swarms of basic dykes intruded the Archaean North Atlantic craton (NAC) and the NW Baltic shield between 2.5 and 1.9 Ga. The oldest Mg-rich dykes and sills are 2.4–2.5 Ga and are associated with rift controlled acid and basic magmatism. In SW Greenland the main high-Mg swarms are *c.* 2.2 Ga. The high-Mg dykes have high Si, LILE and LREE and are low in Ca, Al, Ti and Nb. Isotopic compositions (Sr, Nd and Pb) demonstrate a high degree of crustal contamination. Individual dykes are remarkably constant in composition but there are large variations between adjacent dykes. Tholeiitic swarms were emplaced at *c.* 2.2 Ga in the southern part of the craton in SW and SE Greenland and Labrador. The majority are normal moderately evolved tholeiites with no marked LILE or LREE enrichment, a few *c.* 2.2 Ga comparatively mafic tholeiitic dykes show a LILE and LREE enrichment. Dense swarms of distinctive Fe-enriched hornblende-bearing tholeiitic dykes (the *c.* 2.0 Ga Kangâmiut dykes) intruded the northern part of the NAC in Greenland. Many are composite with intermediate centres. The Kangâmiut dykes were emplaced at depth during shearing in a regional N–S compressional regime and developed metamorphic mineral assemblages at the time of dyke injection. The 1.83 Ga Avayalik dykes from the eastern foreland of the Torngat orogen, northern Labrador show comparable syn-tectonic emplacement and autometamorphic assemblages. Emplacement was controlled by regional sinistral shearing after the first calc-alkaline igneous activity had occurred within the orogen.

John Sutton and Janet Watson's landmark paper in 1951 on the use of the Scourie dykes to divide the Lewisian of NW Scotland in time and space into areas where the dykes were unmetamorphosed (Scourian) and areas where equivalent dykes were strongly deformed and metamorphosed (Laxfordian) formed the basis on which many of those present at this memorial symposium started their careers of unraveling Precambrian shield areas. At the same time as Sutton and Watson presented their ideas at the 1948 International Geological Congress in London, Noe-Nygaard (1948, 1952) and Ramberg (1949) pioneered the use of comparable dykes to divide the Precambrian of SW Greenland into the Nagssugtoqidian, a metamorphic and metamorphic complex affected by post-dyke orogenic events, from an earlier complex in which there is little or no post-dyke tectonism and metamorphism (pre-Nagssugtoqidian). These relatively simple field criteria, which use the intrusion and deformation of basic dyke swarms both as relative time markers and also to mark the spatial limits of orogenic belts, have served

Precambrian geology well, and have been one of the most successful methods used for first order divisions of shield areas as far back in time as the Early Archaean (McGregor 1973; Bridgwater *et al.* 1975). Intrusion and transport of magma from deeper levels through the crust is commonly assumed to represent crustal extension (Tarney & Weaver 1987). In extreme cases, such as during the Tertiary rifting of the continental margin of SE Greenland, the volume of dykes may approach 100% of the total outcrop. However, the intrusion of large swarms of basic dykes is not an isolated geological event restricted to one particular tectonic regime. When examined in detail many dyke swarms are syn-kinematic with respect to major orogenic movements and are, for example, emplaced in conjugate sets of shear zones related to contractional deformation. Without an understanding of the tectonic regime into which dykes were emplaced, their use to separate plutonic episodes, either in time or space, or to define a particular stage in the evolution of an ideal orogeny can give problems to the unwary. Properly studied

From COWARD, M. P. & RIES, A. C. (eds), 1995, *Early Precambrian Processes,*
Geological Society Special Publication No. 95, pp. 193–210.

dyke swarms serve both as important indicators of the timing, orientation and magnitude of paleostress and, where they are involved in later tectonism, the degree of subsequent crustal deformation. It is one of the main purposes of the present review to look in a little more detail than was possible nearly 50 years ago at the character of some of the dykes broadly equivalent to the Scourie dykes which intrude the North Atlantic and Baltic cratons.

Sutton & Watson's division of the Lewisian and Ramberg and Noe Nygaard's comparable division of the Greenland shield were fully vindicated when isotopic measurements showed that in both cases the structural and metamorphic state of dykes could be used to separate areas of pre-2.5 Ga Archaean gneiss, essentially unaffected by later events, from areas where there was a strong tectonic and metamorphic overprint in the period 1.65–1.95 Ga. Treated as a group the Scourie dykes are part of the numerous dyke swarms which intrude the Archaean North Atlantic craton (NAC) in Greenland and eastern Canada and the dykes which intrude the Archaean rocks of the northern Baltic shield in the period 2.5–1.9 Ga. In detail it is now known that both the Scourie dykes and the dykes in the adjacent Archaean cratons were intruded during several episodes (see for example, Heaman & Tarney 1989; Tarney 1992) and each episode may have marked a distinct thermal and tectonic regime in the crust. Apart from the dykes, much of the direct evidence for events between 2.5 Ga and 1.9 Ga has been lost in the North Atlantic craton because of deep erosion, and there has been a general tendency to assume that the shield was essentially stable during this period. In areas such as the northern Baltic shield, which is generally not as deeply eroded as the NAC, it can be shown the period between 2.5 Ga and 2.0 Ga was one of considerable mantle activity and instability in the crust. This was marked by the deposition of at least two sequences of sediments and volcanics separated by periods of folding, metamorphism and the intrusion of basic plutonic suites and dykes. In the more deeply eroded North Atlantic shield dykes may be the only direct evidence that this area was also affected by regional tectonic events in the period from *c.* 2.0–2.5 Ga. The present field and geochemical data suggest that the conditions in which each of the swarms was emplaced varied through time and from place to place. In this paper two distinctive types of dyke are highlighted to see what evidence they can give about this period of supposed relative quiescence in the continental crust. These are the high-Mg dykes and swarms of more Fe-enriched tholeiitic syn-tectonic dykes concentrated near the margins of the Early Proterozoic mobile belts.

High-Mg dykes

Early Proterozoic dykes and layered bodies with high Mg/Fe ratios at a given Si content (Mg numbers generally between 65–80) form distinctive swarms in the Karelian and North Atlantic cratons and are an important component in the Scourie dyke swarm (Tarney & Weaver 1987). They are often associated with, and indistinguishable in age from, tholeiites intruded in parallel swarms. Some of the most primitive tholeiites have Mg/Fe ratios which overlap those seen in the high-Mg dyke swarm (Fig. 1) but can be distinguished from these by their comparatively high Ca and Al contents (Table 1).

The high-Mg intrusions vary in age from 2.4–2.5 Ga in Kola and Karelia (Mitrofanov & Balashov 1990) and NW Scotland (Heaman & Tarney 1989) to 2.2 Ga (SW Greenland). In some areas, for example SE and locally SW Greenland, there is more than one generation of high-Mg dykes, a locally developed early E–W swarm (which may correlate with the 2.4 Ga Scourie dykes) and a later 2.2 Ga approximately N–S swarm which is dominant in SW Greenland. In Greenland there is some confusion about the nomenclature used. The main swarm of *c.* N–S

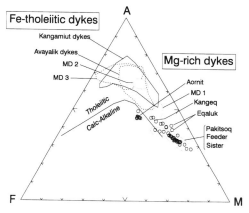

Fig. 1. AFM diagram showing the main fractionation trends and the difference in Fe/Mg ratios between the Mg-rich suite and the other dyke swarms cutting the Archaean craton in SW Greenland. The E–W Eqaluk and Kangeq dykes and the Si enriched N–S Aornit dyke from Fiskefjord are markedly less Mg-rich than other members of the swarm. The N–S MD1 dykes from Frederikshåb are intermediate between the Mg-rich swarms and the younger MD 2 and 3 swarms and the more iron enriched Kangâmiut dykes and the Avayalik swarm in Labrador. The Aornit dyke falls in the field of calc-alkaline rocks rather than tholeiites possibly reflecting the addition of alkalis by contamination rather than by fractionation processes.

Table 1. *Representative analyses from the dyke swarms described in the text*

Sample	No.	SiO_2	TiO_2	Al_2O_3	FeO	MnO	MgO	CaO	Na_2O	K_2O	P_2O_5	LOI	Mg No.	Cr
Pakitsoq	125501	53.22	0.50	11.82	9.44	0.20	14.15	7.45	2.22	0.84	0.15	0.83	73.00	1290
Isua	h292289	53.50	0.60	9.19	9.47	0.20	15.93	7.05	1.85	1.01	0.06	n.d.	72.90	793
Aornit	VM/88/8	56.20	0.56	15.10	8.17	0.13	6.08	8.28	2.82	1.03	0.11	n.d.	63.00	275.00
Sister	14917	50.24	0.40	6.38	9.92	0.20	23.31	5.59	1.38	0.42	0.03	1.56	82.90	3881
Equaluk	125506	51.01	0.75	7.00	10.97	0.19	14.75	11.28	1.99	0.58	0.03	1.16	73.50	1633
Kangâmiut	158074	50.61	1.79	12.74	14.08	0.25	5.29	10.43	2.53	0.38	0.17	1.03	43.60	128
Kangâmiut	158078	58.98	1.64	12.10	11.41	0.18	2.22	6.32	3.36	1.37	0.41	0.95	28.00	53.00
MD1	85431	50.42	0.58	16.98	8.30	0.15	7.99	11.34	2.05	0.28	0.08	0.89	66.50	305
Avayalik	DB92-97A	49.73	1.15	13.93	13.17	0.22	6.35	11.14	1.82	0.30	0.08	1.64	47.00	70.80
Avayalik	DB92-100D	56.19	2.02	13.44	13.66	0.19	4.09	7.26	2.56	1.92	0.40	1.51	29.80	36.50
MD2	a243095	49.70	2.24	12.60	16.18	0.25	4.32	9.35	2.35	0.99	0.20	1.69	32.00	48
MD2	m243095	56.54	1.20	16.45	10.70	0.18	1.15	7.52	3.51	0.75	0.34	1.13	18.10	50

Sample	No.	Ni	V	Cu	Pb	Zn	Bi	Rb	Cs	Ba	Sr	Tl	Ga	Li
Pakitsoq	125501	373	187	43	5.60	59	0.15	22.85	0.87	298	260	0.19	12.08	12.28
Isua	h292289	167	181	78	12	109	n.d.	23	0.82	296	188	0.17	14	23
Aornit	VM/88/8	59.00	n.d.	n.d.	5.40	n.d.	n.d.	26.20	0.63	344.02	303.28	0.13	n.d.	12.21
Sister	14917	1419	n.d.	n.d.	n.d.	n.d.	n.d.	13.00	n.d.	164.00	130.00	n.d.	n.d.	n.d.
Equaluk	125506	523	n.d.	n.d.	n.d.	n.d.	n.d.	12.00	n.d.	222.00	216.00	n.d.	n.d.	n.d.
Kangâmiut	158074	43	438	75	5.32	132.00	0.04	6.43	0.12	118.34	172.73	0.04	18.00	6.37
Kangâmiut	158078	10.00	175	32	8.41	145.00	0.06	32.73	0.64	433.97	274.41	0.13	28.00	11.57
MD1	85431	228	214	n.d.	4.12	n.d.	0.03	6.49	0.23	116.00	187.00	0.05	n.d.	8.97
Avayalik	DB92-97A	n.d.	377.12	131.06	6.15	120.02	n.d.	9.32	n.d.	93.17	162.94	n.d.	17.80	n.d.
Avayalik	DB92-100D	49.65	n.d.	n.d.	n.d.	166.00	n.d.	57.20	n.d.	n.d.	252.00	n.d.	n.d.	n.d.
MD2	a243095	65	340	251	n.d.	n.d.	n.d.	19.50	n.d.	406	91.80	n.d.	n.d.	n.d.
MD2	m243095	39	66	70	n.d.	n.d.	n.d.	21.10	n.d.	408	229	n.d.	n.d.	n.d.

Sample	No.	Sc	Nb	Zr	Hf	Y	Th	U	La	Ce	Pr	Nd	Sm	Eu
Pakitsoq	125501	n.d.	3.30	64	n.d.	10.97	1.98	1.70	13.69	31.61	0.00	14.20	3.22	1.01
Isua	h292289	30	3.60	76	1.87	12.00	2.19	0.41	13.28	28.12	3.33	13.00	2.72	0.74
Aornit	VM/88/8	22.78	2.61	46.72	1.56	11.30	2.74	0.51	15.22	29.99	3.67	13.92	2.75	0.76
Sister	14917	n.d.	n.d.	44.00	n.d.	n.d.	n.d.	n.d.	n.d.	n.d.	n.d.	n.d.	n.d.	n.d.
Equaluk	125506	n.d.	n.d.	54.00	n.d.	n.d.	n.d.	n.d.	n.d.	n.d.	n.d.	n.d.	n.d.	n.d.
Kangâmiut	158074	43.87	8.46	69.46	2.33	24.66	1.09	0.30	9.05	22.50	3.91	13.99	3.82	1.33
Kangâmiut	158078	19.83	19.58	73.82	2.09	30.06	4.81	1.07	32.69	73.51	9.34	39.30	8.35	2.40
MD1	85431	34.00	2.05	33.00	0.90	12.22	0.59	0.11	6.15	13.74	1.79	7.72	1.82	0.65
Avayalik	DB92-97A	49.70	6.28	67.55	2.08	23.93	1.78	n.d.	9.42	19.68	n.d.	11.14	2.69	1.03
Avayalik	DB92-100D	n.d.	n.d.	20.14	n.d.	n.d.	2.31	n.d.	51.50	112.00	n.d.	55.60	9.36	2.67
MD2	a243095	n.d.	14	144	n.d.	40	n.d.	n.d.	10.90	26.50	n.d.	n.d.	5.09	1.53
MD2	m243095	n.d.	13	226	n.d.	59	n.d.	n.d.	19.60	50.2	n.d.	n.d.	8.75	2.51

Sample	No.	Gd	Tb	Dy	Ho	Er	Tm	Yb	Lu	Ta	$\varepsilon_{Nd}(2.2)$	$T_{(DM)}G$
Pakitsoq	125501	2.72	n.d.	2.32	n.d.	1.01	n.d.	n.d.	n.d.	n.d.	−3.90	2.83
Isua	h292289	2.89	0.38	2.30	0.44	1.27	0.18	1.24	0.18	1.87	−7.03	3.14
Aornit	VM/88/8	2.42	0.38	2.30	0.49	1.36	0.19	1.23	0.19	0.29	−6.14	3.04
Sister	14917	n.d.	n.d.	n.d.	n.d.	n.d.	n.d.	n.d.	n.d.	n.d.		
Equaluk	125506	n.d.	n.d.	n.d.	n.d.	n.d.	n.d.	n.d.	n.d.	n.d.		
Kangâmiut	158074	4.35	0.74	4.73	1.64	2.85	0.40	2.66	0.40	0.96		
Kangâmiut	158078	7.82	1.09	6.21	3.17	3.20	0.43	2.53	0.36	1.42		
MD1	85431	2.18	0.34	2.25	0.47	1.45	0.29	1.36	0.28	0.13		
Avayalik	DB92-97A	n.d.	0.43	n.d.	n.d.	n.d.	n.d.	n.d.	n.d.	n.d.		
Avayalik	DB92-100D	n.d.	n.d.	n.d.	n.d.	n.d.	n.d.	n.d.	n.d.	n.d.		
MD2	a243095	n.d.	n.d.	n.d.	n.d.	n.d.	n.d.	3.41	n.d.	n.d.		
MD2	m243095	n.d.	n.d.	n.d.	n.d.	n.d.	n.d.	5.38	n.d.	n.d.		

Samples with five and six figure numbers are the property of the Geological Survey of Greenland. Analytical details available from D. Bridgwater.

dykes includes the majority of the high Mg-dykes. These dykes were originally called the MD1 swarm, based on a field chronology from the southern part of the craton which showed they were the earliest regionally developed dykes (Rivalenti 1975). This chronology was later applied throughout the NAC in Greenland (Bridgwater *et al.* 1976, 1985), a correlation which has been confirmed by isotopic age determinations. The high concentration of Mg-rich minerals and high-Mg bulk chemistry for many of the MD1 dykes was recognized by many geologists (Berthelsen & Bridgwater 1960; Rivalenti 1975, Bridgwater *et al.* 1976, 1985; Hall *et al.* 1985). These dykes were interpreted as being derived from distinctive Mg-rich magmas by Bridgwater *et al.* (1985) and Hall *et al.* (1987). The term 'boninitic-norite (BN)' was introduced by Hall & Hughes (1987) and Hall *et al.* (1987) to emphasize the chemical similarity between the high-Mg members of the N–S dyke swarm with modern boninites. The present authors are, however, unable to make so clean a distinction between the tholeiitic suite and 'BN' suite as Hall & Hughes (1987), either on field evidence or on chemical grounds. The term MD1 is used in this paper in the original field chronological sense for the earliest N–S regional swarms in SW Greenland and the term high-Mg dykes is used for the distinctive Mg-rich, high SiO_2, LILE and LREE and low Ca, Al, Ti and Nb dykes which form part of (but which do not belong exclusively to) the MD1 swarm. Although the high-Mg dykes have some chemical characters in common both with modern boninites and with the Late Archaean – Early Proterozoic layered complexes such as Stillwater and Bushveld (see also Schiøtte 1983, 1988; Bridgwater *et al.* 1985; Tarney 1992), the geological setting of the high-Mg dykes and the mechanisms by which they acquired their chemistry is so different to that of boninites that it is here considered that the term 'boninitic norite' should be dropped. The term siliceous high-magnesium basalt (Sun *et al.* 1989) does not have the genetic implications of boninitic norite but still does not cover the most silica-rich suite of the dykes described here (see the Aornit dyke, Table 1).

The main high-Mg dyke swarm in Greenland occurs in the northern Sukkertoppen region, where it trends NNW–SSE to NNE–SSW. Single dykes with the same trend and chemical character occur as far south as Ivigtut and on the southeast coast near Tingmiarmiut. The dykes vary in thickness from a few metres up to 200 m and are widely spaced. Individual bodies can be traced for up to 80 km (en echelon dyke trains up to 100 km). Locally the dykes have diffuse margins with the local high grade gneiss country rocks (the Feeder dyke, Berthelsen & Bridgwater 1960, and a 200 m

dyke cutting the Isua supracrustal rocks, Wagner 1982; Nutman *et al.* 1995). Except for these very rare examples the present authors agree with Tarney (1992) that the high-Mg dykes show remarkably little reaction with their immediate country rocks. They have chilled, sometimes glassy margins, set with up to 15% phenocrysts and xenocrysts of olivine and pyroxene. Individual dykes are chemically relatively homogeneous both across and along their length but adjacent dykes may show a marked range in chemistry, mineralogy and petrology which makes it difficult to classify the suite under one heading in strict chemical terms. An almost full range of the rock types found in the dykes was described by Berthelsen & Bridgwater (1960), from the Fiskefjord area where two early E–W dykes, the Kangeq and Eqaluk dykes with chemical characters intermediate between tholeiites and the high Mg-swarm, are cut by five NNW–SSE dykes (from west to east the Pakitsoq, Feeder, West Sister, East Sister and Aornit). Representative chemical analyses from the dykes described are given in Table 1.

The Pakitsoq dyke is representative of the most common rock type found in the high-Mg dykes in SW Greenland. It is a norite with phenocrystic orthopyroxene and clinopyroxene (commonly Ca poor) enclosed by intermediate plagioclase oikocrysts. These are often cloudy. Olivine (Fo87–80) is found as partly resorbed xenocrysts in the chilled margins and rarely in the coarser grained dyke centres. More calcic clinopyroxene is typically late, and occurs intergrown with both orthopyroxene, sub-calcic clinopyroxene and plagioclase (see Hall *et al.* 1985 for a detailed description of pyroxenes from comparable dykes). Biotite, K-feldspar and quartz are found as interstitial material.

The Sister dykes show more extreme Mg-rich bulk compositions. Olivine and clinopyroxene form up to 90% of the dyke mass, interpreted to have been intruded as a crystal mush. The interstitial plagioclase is comparatively calcium poor (<An55) for a rock with a Mg number > 75 emphasizing the low Ca and Al content of these rocks. Chilled contacts have not been reported.

The Aornit dyke is only classified as a high-Mg dyke because of its field association with more typical members of the suite and because its Mg number (65) is high when taking the SiO_2 content of up to 58% into account. The Aornit dyke also differs from typical members of the swarm in that olivine is absent, orthopyroxene is only found as partly resorbed phenocrysts and a relatively Ca-rich plagioclase (>An65) crystallized early and forms about 50% of the rock as laths. This illustrates a paradoxical feature of the high-Mg dykes as a suite in the Fiskefjord area in that the Ca content of the

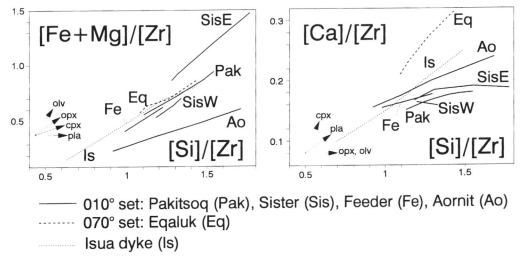

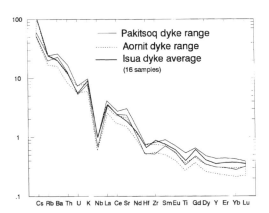

Fig. 2. Element ratio diagrams (Pearce 1968) illustrating that the high-Mg dykes in the Fiskefjord – Isua area show ranges in compositions determined by a mineral assemblage which is unique for each individual dyke.

feldspars is highest in those dykes which have the most Fe-rich mafic minerals (Berthelsen & Bridgwater 1960). Total CaO and Al_2O_3 contents are comparable to normal tholeiites (Table 1). The clinopyroxenes are complex mixtures of pigeonite and augite (Berthelsen & Bridgwater 1960). There is a large amount of interstitial quartz–K-feldspar intergrowths.

Chemistry

The total chemical variation within individual dykes is small and while major element fractionation trends are well defined they only show a limited range. The bulk compositions and the minerals controlling these trends vary from dyke to dyke (Fig. 2). There is a very marked LILE and LREE enrichment and negative Nb and Ti anomalies in all the high-Mg dykes. The trace element patterns are remarkably constant both within individual dykes and between dykes in spite of large bulk chemical differences (Fig. 3). There is little evidence that incompatible elements, such as the LREE, are fractionated by the main rock-forming minerals crystallizing in the dykes. This suggests that they are concentrated in late minor phases.

Isotope chemistry

Rb–Sr studies on the Pakitsoq dyke yielded a 2110 ± 85 Ma isochron with an ISr of 0.70365 + 32 (Fig. 4, analytical details and full data for all isotopic determinations available from

D. Bridgwater and will be published in a subsequent paper). This is 100 Ma younger than a U/Pb age of 2214 ±10 Ma obtained on zircons from the 200 m high-Mg dyke cutting the Isua supracrustal rocks (Nutman *et al.* 1995). Wagner (1982) obtained a scatter from the same dyke at Isua,

Fig. 3. MORB-normalized element contents showing the range in compositions in the Pakitsoq and Aornit dykes together with the average of 16 analyses from the Isua dyke. These are closely similar although mineralogically there is marked variation between the three dykes (see Fig. 2). This similarity between dykes is marked in spite of the massive contamination seen in the Isua dyke in outcrop whereas the Pakitsoq and Aornit dykes show no contamination at the present erosion level. This supports the contention that contamination of the high-Mg swarm occurred at depth.

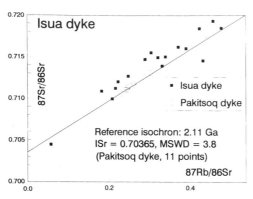

Fig. 4. Rb–Sr results from the 200 m N–S dyke at Isua
with the Rb–Sr isochron from the Pakitsoq dyke as
reference. Note that the majority of data points lie
above the reference isochron. This is consistent with
contamination from a source with a higher amount
of radiogenic Sr at the time of intrusion such as the
c. 3.7 Ga country rocks. The scatter is interpreted as
caused by lack of homogenization between the dyke
magma and contaminants from the country rock at the
time of intrusion. The immediate country rocks to the
dyke vary from *c.* 3.6–3.7 Ga Rb-rich gneisses to 3.8 Ga
Rb-depleted basic rocks and would have developed a
large range of $^{87}Sr/^{86}Sr$ ratios at 2.2 Ga. The dyke can
be traced for at least 80 km south of the Isua supracrustal belt, where it intrudes younger (*c.* 3.0 Ga)
gneisses. Lateral transport of contaminated material
along the fissure would mix different crustal sources.
Analytical details available from D. Bridgwater.

which falls largely above the 2.1 Ga isochron from
the Pakitsoq dyke (Fig. 4) suggesting the Isua dyke
contained more radiogenic Sr than the Pakitsoq
dyke at the time of crystallization. The lack of
variation in whole-rock Sm/Nd ratios means that
Sm–Nd isochrons cannot be obtained from these
dykes. Nd (DM) model ages range between 2.65–
2.8 Ga (Pakitsoq) and 2.95–3.1 Ga (Aornit and
Isua). The difference in Nd model ages between the
Pakitsoq and Isua dykes can be correlated with the
difference in age between the *c.* 2.8 Ga age of
the country gneisses in outer Fiskefjord area and
the > 3.5 Ga gneisses from the Isua area. Pb isotopic compositions vary from dyke to dyke but are
consistently less radiogenic than the composition of
suites derived from a moderately depleted mantle at
2.2 Ga. The Pb isotopic compositions for the Isua
and Aornit dykes plot along mixing lines between
Pb derived from a source with a mantle-like
composition at 2.2 Ga ($\mu1 = 7.7$) and Pb with an
isotopic composition comparable to a mixture of
the units making up the local crust (Figs 5 & 6).
The Pb isotopic compositions from the Aornit
dyke like that of the Nd suggest contamination by
a crustal component which is older than 3.0 Ga.

High-Mg tholeiites enriched in LIL elements

In the Fiskenæsset area the MD1 swarm contains
both low Ca and Al, Mg-rich dykes and olivine-
and orthopyroxene-bearing moderately Mg-rich
dolerites with relatively high Ca and Al contents
concentrated in Ca plagioclase. Like the high-Mg
dyke swarm these rather primitive tholeiites are
also enriched in Si and LIL elements (Table 1,
Fig. 7). In many papers before the recognition
of a distinct Mg-rich magma type (Bridgwater
et al. 1985; Hall *et al.* 1987), these dykes with
apparently transitional characters between the
high-Mg types and the younger tholeiitic rocks in
the same area were used to suggest that the various
generations of Early Proterozoic 'MD' dykes were
co-genetic and derived from a single tholeiitic
source. This was based on comparatively simple
geochemical arguments such as a regular increase
in Fe/Mg ratio from the oldest to youngest dyke
generations (Fig. 1). More sophisticated trace
element and REE studies show that the early more
Mg-rich tholeiites are more enriched in LIL and
LRE elements than the younger Fe-enriched dykes
in the same area and a simple line of descent from
early Mg-rich to later Fe-rich dykes is unlikely. The
dykes with compositions intermediate between the
high-Mg suite and the normal tholeiites could
indicate either different mantle sources for the
different generations of tholeiite or possibly
tholeiites with the same mantle source, in which
the most primitive and hotter magmas have been
modified by strong crustal contamination so that
their LILE and LREE chemistry are dominated
by a crustal signature. The Aornit dyke, with its
unusual combination of high-Mg number (65), high
SiO_2 (56%) and high LILE/LREE contents but with
much higher CaO and Al_2O_3 than is seen in the
other high-Mg dykes in the same swarm, could
be interpreted as the result of mixing a normal
tholeiite magma with high-Mg magma which had
been highly contaminated with material from the
lower continental crust.

Interpretation of the mechanics by which the high-Mg dykes developed their composition and implications for their tectonic setting

There is a marked disagreement in the literature
about the nature of the primary magma from which
the high-Mg dykes were formed and in particular
the reason for the unusual combination of high Mg
and high LILE contents. One school, represented
by Hall & Hughes (1967) and Hall *et al.* (1987),
considers the LILE and LREE enrichment and the

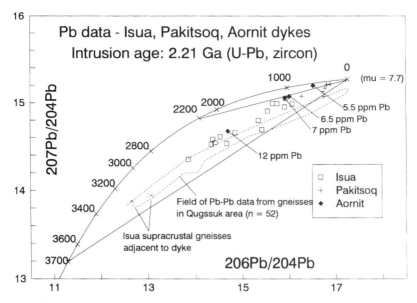

Fig. 5. Whole rock $^{207}Pb/^{204}Pb-^{206}Pb/^{204}Pb$ ratios from the Isua, Pakitsoq and Aornit dykes. Note (1) All except one data point (the least Pb-enriched sample from the Aornit dyke) lie below a reference isochron drawn from 2.1 Ga to the Present assuming a μl value of 7.7 for an upper mantle source. This implies a source which was depleted, rather than enriched, in U before 2.1 Ga. (2) The Aornit and Isua dykes show Pb isotopic ratios which scatter in narrow fields between a plausible composition for Pb derived from a depleted mantle source at 2.2 Ga and the average crust in the inner Fiskefjord – Isua area. There is a good correlation between the absolute amount of Pb in individual samples from the Aornit dyke and their isotopic composition. This is not seen in the Isua dyke, possibly reflecting a much more heterogeneous host at 2.2 Ga. Data from the regional country rocks in the Qugssuk area from Garde (1988).

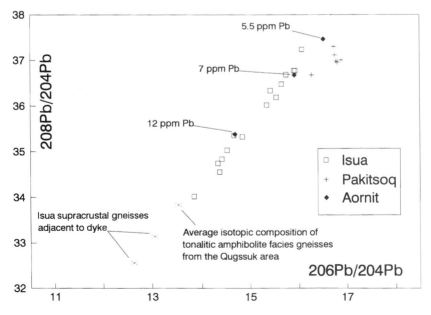

Fig. 6. Whole-rock $^{208}Pb/^{204}Pb-^{206}Pb/^{204}Pb$ isotope plots showing a similar mixing line between values approaching those of uncontaminated mantle derived material at 2.2 Ga, the Aornit sample with the least total Pb, and the local crust in SW Greenland.

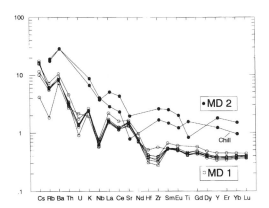

Fig. 7. MORB-normalized element patterns (open circles) from the olivine-bearing tholeiitic MD1 dykes, described by Rivalenti (1975) contrasted with element patterns from a zoned medium to highly fractionated MD2 dyke (solid circles) from SW Greenland (Bridgwater *et al.* 1985). The olivine dolerite shows similar element patterns to those from the Aornit Pakitsoq and Isua dykes but at lower levels of LREE and LIL enrichment. If it is accepted that the Isua dyke shows LREE and LIL enrichment due to crustal contamination, it can be argued that this must also be the case for the olivine dolerites from the MD1 swarm. Conversely if these patterns reflect differences in mantle sources, the high Mg dykes and the MD1 olivine-bearing tholeiites appear closer related than the MD1 tholeiites and the MD 2 and 3 swarms. The chilled basic margin of the MD2 dyke shows no negative Nb and Ti anomaly while the highly fractionated centre shows large negative anomalies. This is interpreted as due to the removal of a Fe–Ti oxide phase during fractionation.

high ISr and Archaean model Nd ages indicate derivation from 'a depleted harzburgitic mantle' which was 'metasomatically enriched in LIL elements by fluids derived from Archaean oceanic and lower continental crust material returned to the mantle during Late Archaean crustal thickening', that is to say approximately 500 Ma before dyke injection. A second school, represented by some of the present writers (Bridgwater *et al.* 1985), have suggested that the LILE enrichment and isotopic character of the dykes are more easily explained by contamination of a parent primitive high-Mg magma with components selectively extracted from the lower crust. Part of the divergence of views is perhaps due to differences in opinion on how and where contamination between basic magma and the continental crust can take place. Weaver & Tarney (1983) and Tarney & Weaver (1987) have pointed out that, in the case of similar enrichments seen in the Mg-rich Scourie dykes, the concentrations of elements such as Rb and Th in the immediate country rocks are too low to give rise to the higher

concentrations in the dykes cutting them by assimilation of country rocks in the dyke magma. Bridgwater *et al.* (1985), however, specifically invoked selective transfer between basic magmas and partially melted crust at depth for the dykes in Greenland rather that the simple assimilation of continental crust in a mantle-derived, basic magma. A discussion of selective transfer between basic magmas and acid melts from the adjacent gneisses is given by Blichert-Toft *et al.* (1992). Furthermore, while in this paper it is argued that the isotopic composition of the dykes is at least in part controlled by the crust, through which the magmas feeding dykes were intruded, evidence for local assimilation of the adjacent country rocks is restricted to two examples. Even in these cases the acid and basic melts do not appear to mix together physically to any extent. Contamination, if it occurred in other dykes, must have taken place at depth and the composition of the rocks immediately adjacent to individual bodies, need have little bearing on the crust with which the dykes reacted. When the overall layered nature of the NAC Archaean gneiss complex, in which depleted granulites frequently overlie undepleted amphibolite-facies gneisses, is taken into account there are no a priori grounds to assume that the source of contamination was as depleted as the rocks intruded by the dykes at the present surface.

The question whether secondary processes, such as crustal contamination and mixing between magmas at different stages in their evolution, control the chemistry of the dykes rather than long lasting differences in lithosphere compositions can be discussed using the following observations.

(1) The dykes cause local massive remelting of the country rocks which suggests that they were hot magmas (see the Feeder dyke, Berthelsen & Bridgwater 1960, fig. 5). This phenomenon is particularly well documented where the 200 m wide Isua dyke intrudes supracrustal rocks containing high percentages of hydrous minerals, such as hornblende and biotite. At this locality the margins of the dyke is diffuse over tens of metres and there is a complex mixture of noritic dyke material and pods and diffuse patches of leucocratic material, locally containing feldspar xenocrysts derived from Archaean pegmatites in the country rock. The bulk chemistry of the Isua and Feeder dykes, which show field evidence for exchange with the country rocks, does not differ markedly from dykes where there is no evidence of mixing at the present erosion level (except there is a greater scatter in such parameters as ISr isotope ratios of individual samples calculated at the time of intrusion).

(2) Whole-rock chemical compositions in all

the individual dykes studied, except the Isua dyke, show little variation suggesting efficient mixing. This is consistent with a hot, low viscosity magma which possibly developed turbulent flow. Such a magma is more liable to contamination when it comes into contact with crustal rocks than a tholeiitic magma. In contrast, adjacent dykes may have markedly different chemical and isotopic compositions. This either requires a range of mantle sources or the modification of a primary magma by a secondary process.

(3) Where chilled margins are developed, these, plus their population of high-Mg olivine (Fo > 86) microxenocrysts and pyroxene phenocrysts, have identical bulk chemical compositions to the coarse-grained centres of the dykes (Berthelsen & Bridgwater 1960; Hall *et al.* 1987). Quartz-rich interstitial material is present in all dykes except the very Mg-rich Sister dykes. The Mg-rich olivine is unstable in the magma from which the SiO_2 rich (>52% SiO_2) ground mass in the chilled margins crystallized and is rare in dyke centres where quartz and orthopyroxene crystallize. Orthopyroxene is stable in all the dykes except the Aornit dyke where it is found as corroded xenocrysts surrounded by Ca-rich minerals, such as augite and labradorite. The high-Mg olivines, the high Cr and Ni contents, the low-Ca pyroxenes and the overall low Ca and Al contents found in all the dykes, except the Aornit dyke, suggest that the parent magma was mafic and probably charged with olivine xenocrysts or fragments of mantle restite material. The preservation of high-Mg olivine, as partly resorbed crystals, almost exclusively in the chilled margins, strongly suggests that the oversaturation in Si seen in these rocks must have occurred late in the development of the magmas and have been comparatively short lived. There appears to be no convincing mineral crystallization path by which a single parent magma, which crystallized olivine with Fo > 86, could evolve to a series of rocks with Mg numbers > 75 but with silica contents of > 52%.

(4) There is local evidence of more than one magma in individual sills from the Rinkian fold belt (Schiøtte 1988) and high-Mg dykes from SE Greenland. The order of intrusion differs from dyke to dyke so that in one case the most Mg-enriched magma intruded first, in a second last. Unless each pulse is seen as derived from a separate mantle source, the presence of two chemically related magmas in one dyke requires an intermediate stage in the formation of the magmas which fed the dyke.

(5) Positive ε_{Sr} and negative ε_{Nd} values, with respect to a depleted mantle at 2.2 Ga can either be explained by contamination from an older crustal source or derivation from a mantle reservoir which had been enriched in Rb and Nd in the Late

Archaean, that is, several hundred million years before 2.2 Ga (Hall *et al.* 1987). The variation in isotopic composition from dyke to dyke, which can be partly correlated with variations in the isotopic composition of gneiss terranes through which the dykes intruded, suggests crustal contamination unless it is accepted that the mantle beneath the North Atlantic craton varies laterally over distances of a few tens of kilometres and mirrors the composition of the overlying continent. The unradiogenic Pb isotope compositions, characteristic of these dykes, preclude derivation of Pb from a mantle enriched in U in the Late Archaean but are consistent with contamination from the depleted Archaean lower continental crust through which the dykes were intruded. The extremely unradiogenic nature of the Pb and Nd in the Aornit dyke, which resembles that seen in the Isua dyke, suggests that the Aornit dyke may have passed through pre-3.0 Ga crust at depth. This process has been demonstrated from the tholeiitic dyke swarms farther south in the craton (Kalsbeek & Taylor 1985) which contain extremely unradiogenic Pb, not present in the gneisses intruded by the dykes at the present surface. It is argued that if contamination on this scale can occur when a normal tholeiite magma is intruded through cold crust, the effects of intruding very hot Mg-rich magmas to form larger dykes would have been even greater. Because of the distinctive unradiogenic character of the Early Archaean continental crust in parts of West Greenland and Labrador, Pb isotopes are a particularly powerful tool to show crustal contamination. The Pb isotope compositions of the Scourie dykes, determined by Waters *et al.* (1990), are more radiogenic than those described here, and the surrounding crust, and do not offer the same unique constraints about the source of the Pb. Using the same arguments as those above, this may indicate that the Lewisian granulite gneiss terranes overlie a crust which is less depleted in U.

It is therefore suggested that the petrology, chemistry and isotope chemistry of the dykes is best explained by mixing processes. The most plausible source for the enrichment in Si, LIL and LRE elements is by massive contamination of hot mafic magmas when they came into contact with the lower crust shortly before injection of the magma into individual fissures. The evidence for several phases of injection and differences between individual dykes, in particular the different order of crystallization seen between the Pakitsoq and Aornit dykes, are not easily explained in terms of a straight forward contamination of a single high-Mg magma with variable amounts of continental crust followed by crystal fractionation. One possibility is that the original mantle derived magmas feeding the dykes ranged in composition between mafic high-Mg basalt and normal tholeiites. Part of the variation between dykes could represent different proportions of two sources mixed before intrusion. This idea has some support in the observation that high-Mg dykes are contemporaneous with tholeiitic dykes and that there are some dykes, such as the Aornit dyke, with chemical characters between the two end members. Such a mixing could take place in the mantle or in a magma chamber at depth in the crust. In view of the isotopic evidence that at least some crustal contamination occurred before dyke injection, the second idea is preferred by the present authors. Both Berthelsen & Bridgwater (1960) and Schiøtte (1983, 1988) suggest that the magmas, which formed the high-Mg dykes, are derived from one or more stratified magma reservoirs in the lower crust comparable, for example, to the Stillwater or Bushveld complexes or layered bodies in the Kola–Karelian shield. Similar ideas were developed by Kalsbeek & Taylor (1985) to explain the zoned nature of tholeiitic dykes in the MD swarms farther south. If different levels of such a chamber were tapped when it was partly differentiated, the range of magma types seen in the dykes could be obtained without calling on a different mantle source for each dyke. Such a magma chamber would furthermore provide a better opportunity for melting the adjacent crust and for chemical interchange between continental crust and dykes than that provided by single, rapidly injected, small intrusions.

The tectonic setting in which the high-Mg dykes were emplaced is not known. The high-Mg nature of the magmas requires an unusual primary melt from the mantle. The breadth and length of single dykes and the lack of internal structures within the dykes, the presence of which could indicate differential movements along the dyke fissures, suggests that they were emplaced in an extensional regime. The presence of Early Proterozoic layered sills derived from similar magmas in the Rinkian belt farther north in West Greenland, the 2.5–2.4 Ga layered complexes in Kola–Karelia and the 2.4 Ga high-Mg Scourie dykes, which are all interpreted as intruded in rift environments, suggest the period between 2.5 to 2.0 Ga was one of considerable crustal extension in the North Atlantic - Baltic shield area. The ultimate control of this activity must lie in the mantle. Unfortunately until the role of crustal contamination can be measured, many of the geochemical parametres used to define the type of mantle, from which the high-Mg magmas were derived, are of uncertain value. The consistent low Nb and Ti contents of these dykes, their high Mg/Fe ratios and their low Ca and Al contents must however reflect the mantle source of the magmas.

Early Proterozoic dyke swarms intruded into active shear zones

Many of the Early Proterozoic dyke swarms emplaced into the North Atlantic and Baltic cratons show clear evidence of being syn-tectonic in the sense that they were emplaced in the same time interval during which regional shearing occurred in the surrounding gneisses. In some cases, episodes of shearing separate different members of the same regional dyke swarm or different pulses of magma in the same fissure, in others the textures and mineralogy within single dykes show that shearing occurred during their crystallization and they can be regarded as true syn-tectonic intrusions. The timing and the tectonic settings which control the shearing differ from area to area. Three examples are briefly described.

2.4 –2.45 Ga dykes swarms with high - pressure metamorphic assemblages cutting the Kolvitsa anorthosite, Belomorian belt (White Sea area)

The Archaean gneisses of northern Karelia and the White Sea coast of the Kola Peninsula are cut by numerous dykes many of which show the so called drusite textures (garnet overgrowths on clinopyroxene/hornblende cores). This was originally interpreted by Federov (1896) to indicate the slow cooling of igneous bodies at depth, an idea which many (including the present authors) have used to interpret the assemblages in the Early Proterozoic dykes in the North Atlantic craton. Sudovikov (1939) used the 'drusites' as time markers to separate early periods of folding and migmatization in the Belomorian (Archaean) country rocks from post-dyke metamorphism and folding (Proterozoic) in the same way that Sutton & Watson used the metamorphic and structural state of the Scourie dykes. More recent work has shown that at least three suites of dykes intruded the Baltic shield in the period 1.9 to 2.5 Ga. These are partly affected by tectonic and metamorphic overprinting between 1.7 and 1.9 Ga, (V. Glebovitsky, pers. comm. 1992). In the western arm of the White sea (Kandalaksha Bay) both high-Mg dykes and high-Al tholeiites were emplaced between 2.4 and 2.45 Ga. The latter are spectacularly exposed cutting the Kolvitsa anorthosite, dated at 2.45 Ga (Mitrofanov *et al.* 1993), but are cut by younger members of the same northosite–leucogabbro–diorite suite, also emplaced between 2.4 and 2.45 Ga. The dykes, which were described by Balagansky & Kozlova (1987), form an anastomosing complex, in which many individuals are composite. Coarse gabbroic anorthosite centres are common. The composite bodies frequently show a strong internal foliation which in some dykes is restricted to one phase only. Foliated dykes locally cut less foliated dykes and visa versa. A large number of the dykes appear to have been torn apart during extension rather than deformed in a compressional regime. Many of the dykes develop high-pressure clinopyroxene–garnet mineral assemblages, some by static replacement of a pre-existing hornblende and plagioclase-bearing igneous or earlier metamorphic assemblage, others during shear movements parallel to the dykes. As individual minerals forming the high pressure assemblages, which form the foliation in one dyke, are cut by later dykes in the same swarm with the same high pressure assemblages, the metamorphic minerals are interpreted to have formed during the same time interval as dyke injection, between 2.4 Ga and 2.45 Ga. The shearing is concentrated in 2.4–2.45 Ga igneous rocks, which include both acid and basic members and show the chemical characteristics of rift-controlled magmatism. This rather unusual combination of shearing, high-pressure metamorphism and magmatism, with the chemical characteristics of rift regimes, has lead Bridgwater et al. (1994) to suggest that the dykes are part of an anorthosite leucogabbro–granite suite emplaced in the deep crust during extensional shearing. The shearing may be related to the tectonic escape of lower crust, thickened and heated by the emplacement of major basic igneous complexes. Similar shear structures, essentially contemporaneous with intrusion, have been noted from anorthosites in the Grenville province (Van Breemen & Higgins 1993).

Syn-kinematic dykes with autometamorphic assemblages in the margins of Proterozoic mobile belts

The Kangâmiut dyke swarm. The Kangâmiut dykes were originally described by Ramberg (1949) and Noe Nygaard (1952) and a regional map of them was published by Escher et al. (1975). Closely comparable dykes occur throughout the Nagssugtoqidian of SE Greenland (except in younger Early Proterozoic sedimentary belts and intrusions) and are regarded as the continuation of the same swarm (Bridgwater & Gormsen 1968; Escher et al. 1976; Bridgwater et al. 1990). Field relationships in SW Greenland show that dykes in the Fiskefjord area with which the Kangâmiut swarm are correlated, are younger than the 2.21 Ga high-Mg dykes. They are separated in time from the high-Mg dykes by movements along prominent NE trending faults (Berthelsen & Bridgwater 1960). Two dykes have yielded identical Rb–Sr ages of 1950 ± 60 Ma (Kalsbeek et al. 1978). Recent U/Pb ion-probe determinations on zircons from one of the quartz diorite samples, used for the Rb–Sr studies, has yielded an age of 2046 ± 8 Ma.(A. Nutman, pers. comm. June 1994) confirming the dykes are 100–200 Ma younger than either the 2.21 Ga high-Mg dykes or tholeiitic dykes from the southern part of the craton which yield Rb–Sr ages of 2.13 ± 65 Ga (Kalsbeek & Taylor 1985). The latter may correlate with the 2.235 ± 2 Ga Kikkartavak dykes (U–Pb on baddeleyite) which intrude the southernmost section of the Archaean craton in Labrador (Cadman et al. 1993).

Dykes from the Kangâmiut swarm occur in the Archaean gneiss complex for approximately 150 km south of the thrust zones traditionally taken as marking the the southern Nagssugtoqidian Front in SW Greenland and continue as deformed bodies in the reworked Archaean gneisses for approximately 50–100 km to the north of the front. In this foreland area, south and southwest of Søndre Strømfjord, the Kangâmiut dykes form a dense NNE-trending swarm of Fe-tholeiitic dykes (Escher et al. 1975). Individual dykes range in width from less than 10 to 50 m, with a few reaching 100 m wide. The swarm does not continue into the granulite-facies rocks forming an approximate 100 km wide belt towards the centre of the orogen. Recent field work by the Danish Lithosphere Centre (1994) shows that the main swarm does not cut supracrustal rocks regarded as Proterozoic cover sequences.

In detail the Kangâmiut dykes show regional changes in both intrusion form and in primary character over a distance of c. 100 km south of the Nagssugtoqidian Front. In the south near Sukkertoppen, where the dykes intrude dry granulite-facies Archaean gneisses, many of the dykes are composite. They have chilled margins with micro-porphyritic hornblende followed by zones of hornblende quartz dolerite, homogeneous amphibolite and foliated garnet amphibolite, with foliation increasing towards the centres of the dykes. Where dykes of the same swarm intersect, foliated younger dykes cut across the foliation in older dykes demonstrating that both the mineral assemblages and the foliation are developed in the same time interval as the emplacement of the swarm and are not superimposed during a later regional tectonic event (Windley 1970). About 50 km farther north, between Kangâmiut and Søndre Strømfjord, c. 10% of the dykes are zoned with hornblende–augite dolerite margins followed by zones of rocks with irregular wavy textures and patches of hornblendite and more leucocratic material. The latter are interpreted as formed by separating the last material to crystallize from a hornblende-rich crystal mush by filter pressing in the dyke fissure during deformation. The central zones of the most fractionated dykes consist of garnet–biotite–albite–epidote quartz-diorite which in places occupies over half the dyke fissures. Occasionally garnet–albite–quartz diorite pods develop in the unsheared margins of the dykes which implies that this assemblage is part of the late magmatic evolution of the dykes and not a late phenomenon connected with post-dyke shearing alone (Fahrig & Bridgwater 1976). The relation between the mineralogical/petrological zoning and shearing in the dykes is complex. Commonly the quartz–garnet-diorite centres coincide with the greatest amount of shearing in the dykes but unsheared quartz–garnet-diorite veins cut the outer zones of some dykes showing that differentiation of this material had started before the onset of shearing. Shearing concentrated in the centres of the dykes continued after the margins of the dykes solidified. Locally the shearing is found nearer one margin and post-dates the solidification of the dyke. Other dykes are intruded into sheared country rocks which were partly retrogressed by fluids before the injection of dyke magma. Scattered examples of comparable composite dykes occur at Søndre Strømfjord airport, 150 km to the NE. As the tectonic boundary of the Nagssugtoqidian is approached, between Søndre Strømfjord and Itivdleq, the dykes change primary orientation from NNE–SSW to NE–SW (Escher *et al.* 1976). The main dyke swarm swings from NNE to NE and a second subordinate set was intruded along shear zones, trending approximately ESE. The two directions of shearing and the two directions of dyke emplacement are essentially contemporaneous. Individual dykes turn sharply from ESE to NE. The dykes contain hornblende–garnet–orthopyroxene assemblages, interpreted as the result of autometamorphism as the dykes crystallized at depth (Jack 1978). Composite dykes of the type seen farther south become rare. Escher *et al.* (1976) interpret the complex intrusion patterns seen in the dykes to show that the dykes were emplaced along conjugate active shear fractures. North of Itivdleq the Kangâmiut dykes were emplaced in amphibolite facies with localized E–W-trending high strain zones. Pressure and temperature studies on the dykes suggest they experienced initial hydration followed by widespread dehydration (Mengel *et al.* 1993). Outside the high strain zones north of Itivdleq, the Kangâmiut dykes contain hornblende–orthopyroxene–garnet assemblages comparable to those recorded from the granulite-facies areas studied by Jack (1978).

In summary, the emplacement of the Kangâmiut dykes along the coast, south of Søndre Strømfjord, was intimately connected with shearing and the movement of aqueous fluids. Primary hornblende assemblages crystallized in the marginal zones of many dykes and suggest that the dykes crystallized from wet magmas. The dry nature of the granulite-facies country rocks precludes a local origin for the fluids. Quartz-diorite centres are found in many dykes 50–100 km south of the tectonic margin of the Nagssugtoqidian but are less common as the mobile belt is approached. Reconnaissance oxygen isotope studies have yielded $\delta^{18}O$ values between + 4 and + 6.5 for both the basaltic margins and the quartz-diorite centres (P. De Groot, pers. comm. 1994). This is slightly lower than many basalts but markedly higher than the O isotopic composition of the Scourie dykes (Cartwright & Valley 1991). In contrast with the Scourie dykes studied there is thus no evidence for a major input of water from a sedimentary source. The high concentration of dykes near the margin of the mobile belt, their emplacement into a compressional rather than extensional regime, their high water contents and the formation of dioritic magmas all suggest that dyke emplacement and the formation of the orogen are linked. One possibility is that the dykes were emplaced during a *c.* 2.0 Ga compressional regime in which amphibolite-facies crust was thrust from the north beneath the granulite-facies gneisses forming the Archaean craton to the south. The changes in the character of the dykes from north to south may be controlled by the depth at which water could be released from the lower thrust slice.

Equivalents of the Kangâmiut dykes occur throughout the 300 km wide Nagssugtoqidian mobile belt on the southeast coast of Greenland. They form a particularly dense swarm in the southern marginal zone of the belt extending 100 km south of Ammassalik (Bridgwater *et al.* 1990, fig. 3) where they cross-cut earlier high-Mg dykes. The dykes were emplaced into active shear zones which trend approximately E–W and are interpreted to be contemporaneous with the similar zones of deformation in the Nagssugtoqidian mobile belt on the west coast. Outside the shear zones individual dykes branch abruptly. Many of the dykes are composite with hornblende-rich basaltic margins and foliated garnet–quartz-diorite centres. These persist even where the dykes cut the undeformed and unmetamorphosed centres of high-Mg dykes and are interpreted as primary features comparable to those seen on the west coast. Both the dykes and the E–W shear zones are affected by later overthrusting from the N and NW correlated with the main cross-orogen shortening in the Nagssugtoqidian on the west coast. Neither the high-Mg dykes nor the characteristic composite hornblende-bearing dykes are known south of a major thrust at Pikiutdleq (64° N) which may mark a major tectonic break between different Proterozoic terranes. Dykes with similar characters to the Kangâmiut dykes form a major swarm in the northern border zone of the mobile belt for 100 km north of Ammassalik. Individual dykes are up to 100 m wide and can be traced for tens of kilometres. Many are composite bodies intruded into active shear zones in which the Archaean granulite-facies country rocks are retrogressed. In this zone of the mobile belt the dykes are essentially undeformed by Proterozoic movements and are discordant to earlier structures in the country rocks which show no clear evidence of regional post-dyke metamorphism. Two major shear directions controlling the dykes are recognized at c. 130° and 60° and are interpreted to be caused by a N–S compressional regime at the time of dyke injection. Some post-dyke movements are concentrated in the margin of individual dykes along the same shear zones which controlled their injection. Preliminary U/Pb ion-probe determinations from zircons extracted from the quartz-dioritic centre of one of the dykes have yielded ages within error of the 2.046 Ga U/Pb age from the Kangâmiut swarm from SW Greenland (A. Nutman, pers. comm. July 1994). The centres of the dykes contain garnet-orthopyroxene-clinopyroxene bearing assemblages yielding temperatures of c. 750°C at up to 10–12 kbar.

Garnet forms rims around magmatic plagioclase phenocrysts. The high pressure minerals are here interpreted as autometamorphic assemblages developed in dykes emplaced at depth comparable to the autometamorphic assemblages described by Jack (1978) and Mengel et al. (1993, 1994) from the west coast.

Geochemistry. When treated as a swarm the Kangâmiut dykes (*s.s.*) and their supposed equivalents in SE Greenland display distinct Fe-enrichment trends (Fig. 1) and have highly evolved centres (Fig 1, Table 1). They contain moderately high LILE concentrations and show marked LREE enrichment patterns (Fig. 8) which distinguish them from many of the c. 2.2 Ga tholeiitic swarms in the southern part of the craton. No significant differences were found between the chemistry of dykes, in which hornblende crystallized as a primary igneous mineral (south of Søndre Strømfjord), and those interpreted as having developed autometamorphic granulite-facies mineral assemblages, north of Søndre Strømfjord. The Fe-enrichment seen in Fig. 1 is not easily explained in terms of hornblende fractionation within the individual dyke fissures since crystallization of this mineral generally generates a calc-alkaline trend. However fractionation trends from individual dykes do not

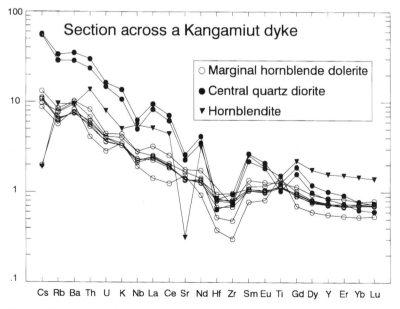

Fig. 8. MORB-normalized element plot from samples across a Kangâmiut dyke. Open circles, marginal hornblende dolerite; filled circles, garnet-bearing quartz-diorite centre; filled triangles, hornblendite lens from transition zone between margin and centre interpreted as formed by removal of late crystallizing magma by the filter pressing of a hornblende-rich crystal mush. Note that there are no marked Nb and Ti anomalies in the Kangâmiut dyke.

show marked Fe enrichment on a AFM plot. It is therefore argued that the main Fe enrichment occurred before dyke emplacement and that individual dykes were derived from lower crustal/subcrustal chambers which had already undergone marked differentiation, for example, by the extraction of olivine.

The Avayalik dykes, northernmost Labrador. The Avayalik dykes (Wardle *et al.* 1993) are a regional swarm of dolerites which occur throughout the northernmost extent of the North Atlantic craton in Labrador (Nain province) on the eastern margin of a major Early Proterozoic mobile belt (Torngat orogen). The Avayalik swarm occurs from Cape Chidley (60°30′N) at least as far south as Trout Trap Fiord (59°N). Their southern extent and field relationship to the *c.* 2.2 Ga tholeiitic dyke swarms, which occur farther south in the Nain province, is not known but rare N–S dykes, with a similar LILE and LREE enriched chemistry to Avayalik dykes, cut the dominant E–W swarm of tholeiites in the Saglek–Hebron area (58°15′N). The latter does not show such marked enrichment in LIL elements. In the southern part of the region recently investigated (59°N), *c.* E–W dykes are dominant but are not separable in time regionally from slightly less abundant NNE-trending dykes. Individual dykes change direction and branch abruptly. Single E–W dykes reach over 100 m in width and can be traced for tens of kilometres across the regional (Archaean) N–S strike of the country rocks. The E–W dykes are approximately at right angles to the tectonic and metamorphic front of the 1.95–1.7 Ga Torngat orogen which forms the centre of the Labrador peninsula to the west. A progressive metamorphic overprint can be traced in the dykes from east to west for approximately 50 km across strike. The main effect visible is a static high-pressure metamorphism in which first garnet and then garnet plus clinopyroxene progressively replaces primary igneous minerals as the Torngat orogen is approached. In some parts of the foreland the Archaean country rocks were already re-crystallized under Proterozoic high-pressure granulite-facies conditions before the injection of the dykes. In other parts of the area the dykes contain a high-pressure granulite-facies assemblage while the country rocks are apparently unaffected. The latter is comparable to the situation described above from the northern margin of the Nagssugtoqidian mobile belt in East Greenland. A few of the larger dykes farthest away from the orogen, contain small pegmatitic patches in which garnet overgrows plagioclase. These predate the development of garnet and hornblende-filled fractures and garnet–clinopyroxene replacement of the main mass of the dykes, both of which become

more noticeable as the Torngat orogen is approached. The early growth of garnet is interpreted as an autometamorphic effect suggesting the dykes were emplaced at depth (compare the Kangâmiut dyke swarm of SE and SW Greenland and the drusites in Karelia). On the outer coast farther north at *c.* 60°N, NE–SW trends become dominant. E–W dykes cut NE–SW-trending dykes and vice versa. Individual dykes branch, sometimes at right angles, and may swing from NE–SW to E–W and back again. As the Torngat orogen is approached the dyke swarm swings N–S and trends parallel to the Komatorvik shear zone, a major Early Proterozoic structure which was the locus of intense sinistral movement, retrogression and migmatization in the period *c.* 1.791–1.710 Ga (Scott & Machado 1994*a*). Part of the swing in dyke direction can be ascribed to post-dyke rotation into the sinistral shear zone during a regional transpressional regime. However, like the Kangâmiut dykes in SW Greenland, when the intrusion mechanics of the dykes are examined in detail the shearing and dyke injection can be shown to be part of a single, possibly protracted event. Individual dykes swing into N–S shear zones where they cut the early fabrics, and in places earlier dykes in the same swarm, but are themselves cut by later movements. Some dykes contain inclusions of mylonite which are only known in the northern Nain province from the Komatorvik shear zone. Many dykes have small apophyses which point in opposite directions on either side of the dykes. These are consistent with intrusion into fissures undergoing sinistral movement (compare the features described from the Kangâmiut dykes, Escher *et al.* 1976). The major movements dated from the Komatorvik shear zone occurred 50–100 Ma after the collisional event in the Torngat orogen which culminated in high grade metamorphism at *c.* 1.86 Ga (Bertrand *et al.* 1993) and after copious calc-alkaline igneous activity within the orogen and its eastern foreland zone between 1.93 and 1.86 Ga. The calc-alkaline plutons extend in a N–S zone along the eastern margin of the orogen at least as far south as 57°N and are interpreted as resulting from the subduction of Proterozoic ocean floor beneath the North Atlantic craton. No consistent field or petrological features have been found which can separate the dykes into different groups on a regional basis. Also no clear cross-cutting relations between the Avayalik dykes and the calc-alkaline intrusions, which could show that as a suite they were either older or younger than the major dioritic and tonalitic plutons have been found. The dykes vary from fine grained dolerites to medium grained dolerites with abundant black plagioclase megacrysts. Hornblende-bearing dykes occur close to the *c.* 1.9 Ga calc-alkaline plutons and we have been unable to separate these in time from more typical dolerites. Geochemically they range from tholeiites to quartz diorites. They are moderately highly fractionated with Mg numbers as low as 30 and TiO_2 and P_2O_5 as high as 3% and 0.5% respectively. They show moderate to high LILE and LREE enrichment and negative Nb and Th anomalies when compared to the adjacent elements on a spider diagram but have absolute Nb and Th contents above MORB (Fig. 9). Some samples show a small negative Sr anomaly, other samples with plagioclase phenocrysts show a small positive anomaly. In many respects the Avayalik dykes show similar, but slightly more LILE enriched, geochemical trends to those seen in the 2.2 Ga Kikkertavik dykes from the southern part of the Archaean craton (Cadman *et al.* 1993). The hornblende-bearing dykes show similar trace element distribution patterns. It is suggested that the Avayalik dykes may be part of the subduction-related calc-alkaline suite rather than representing early extension related magmatism before the onset of the collision which formed the Torngat orogen.

A U/Pb age of 1.835 Ga was obtained by Scott & Machado (1994*b*) on zircons extracted from a small basic pegmatite pod from the centre of a *c* 100 m wide E–W dyke from Cape Kikkiviak, (60°01′N). This dyke shows the formation of hornblende, clinopyroxene and garnet along fractures formed during the regional metamorphic overprint in the area but otherwise preserves igneous minerals in both the normal dyke rock and the pegmatite from which zircon was extracted. Monazites from the regional Archaean gneisses from western Avayalik island, 10 km farther to the north where the dykes show a similar metamorphic overprint, preserve a Late Archaean age of 2.691 ± 2 Ga (Scott & Machado 1994*b*). If the regional Proterozoic metamorphic overprint in this area had little effect on the U–Pb system in monazite, it is argued that it is unlikely to have completely reset the U/Pb system in zircons from an undeformed dyke and that the 1.835 Ga age should be regarded as the age of intrusion of the dyke. This is compatible with the regional syn-tectonic nature of the Avayalik dykes described above and implies that they were emplaced into the foreland of the Torngat orogen up to 70 Ma after the intrusion of voluminous calc-alkaline suites, interpreted as formed from subduction-related processes (Scott & Machado 1994*b*). The dykes thus do not represent a pre-orogenic extensional regime.

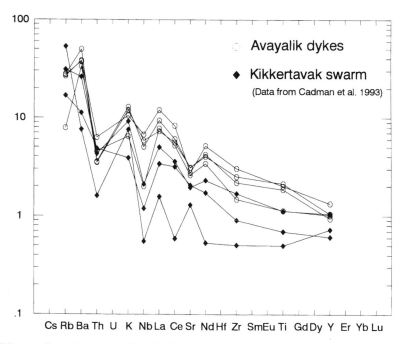

Fig. 9. MORB-normalized element plots from the Avayalik dykes, northernmost Labrador, contrasted with similar patterns from the Kikkertavik dykes *c.* 700 km farther south (Cadman *et al.* 1993). The marked negative Sr anomaly in the Avayalik dykes is thought to be controlled by removal of plagioclase phenocrysts.

Conclusions

The lateral equivalents to the Scourie dykes in the North Atlantic craton and Baltic shield have been used with a great deal of success as field criteria to separate Archaean from Proterozoic events on a regional basis. However as structural, geochronological and geochemical knowledge increases it has become clear that basic dykes are intruded under a variety of conditions and considerable care is needed before the assumption is made that regional dyke swarms always represent an extensional regime predating orogeny. In the North Atlantic and Baltic cratons the period between 2.5 Ga and 1.8 Ga was one of considerable tectonic and magmatic activity which varied both through time and from place to place. The basic dykes emplaced during this period can give us some of the best evidence how and when these changes took place. In particular the importance of the high-Mg dykes which show a range in ages between 2.5 and 2.2 Ga is emphasized. These appear to have been intruded under tensional conditions and crystallized from distinctive magmas, presumably related to equally distinctive conditions in the mantle and overlying crust. In contrast, a second group, the Kangâmiut and Avayalik dykes, was emplaced as regional swarms in active shear zones in the forelands to orogenic belts at the same time as calc-alkaline igneous activity within the belts. Provided that accurate age determinations, allied to modern structural and geochemical methods, are used to recognize the fact that dykes swarms, such as the Scourie dykes and their equivalents, can no longer be regarded as a single suite increases, rather than diminishes, their potential both as geological time markers and guides to the regional tectonic environment into which they were emplaced.

The authors wish to thank G. Rivalenti and V. R. McGregor for providing material from the MD1 and high-Mg dykes in West Greenland. We acknowledge funding from SNF and the Carlsberg foundation (Denmark), NSERC Canada (grants to T. Rivers, H. Baadsgaard and B. Fryer). In particular we would like to thank H. Baadsgaard who supervised the original MSc thesis by one of us (P.W.) and who collected the material used from the Isua dyke, J. Blichert-Toft and F. Albarede provided Sm–Nd isotopic data and suggested the importance of relict olivine in understanding the petrogenesis of the high Mg-dykes. We thank V. Balagansky for discussions and a translation of his description of the dykes in the Kolvitsa area and V. Glebovitsky for arranging field work in the area. The Pb isotopic data used in the figures from the Aornit and Pakitsoq dykes was obtained by M. Thirlwall, RHBNC on contract and will be published in full later. R. Wardle, Geological Survey of Newfoundland provided field support and encouragement. We thank J. Tarney and L. Heaman and M Thirlwall for constructive reviews. We acknowledge permission to publish this paper granted by the Geological Survey of Greenland.

References

BALAGANSKY, V. V. & KOZLOVA, N. YE. 1987. [A basic dyke complex in the Kochinny Cape area and its place in the development of the Kolvitsa zone.] *In*: IVANOVA, T. N., YEFIMOV, M. M., SMOL'KIN, V. F. & DUKUCHAEVA, V. S. (eds) [*Basic-hyperbasic magmatism in the major structural–formational zones of the Kola Peninsula.*] Apatity, Kola Science centre, 55–62. (in Russian).

BERTHELSEN, A. & BRIDGWATER, D. 1960. *On the field occurrence and petrography of some basic dykes of supposed Pre-Cambrian age from the southern Sukkertoppen District, western Greenland.* Meddelelser om Grønland, **123**.

BERTRAND, J.-M., RODDICK, J. C., VAN KRANENDONK, M. J. & ERMANOVICS, I. 1993. U-Pb geochronology of deformation and metamorphism across a central transect of the Early Proterozoic Torngat Orogen, North River map area, Labrador. *Canadian Journal of Earth Sciences,* **30**, 1470–1489.

BLICHERT-TOFT, J., LESHER, C. E. & ROSING, M. T. 1992. Selectively contaminated magmas of the Tertiary East Greenland macrodike complex. *Contributions to Mineralogy and Petrology,* **110**, 154–172.

BRIDGWATER, D., FRYER, B. J. & GORMAN, B. E. 1985. Proterozoic basic dykes in southern Greenland and the coast of Labrador; tectonic setting, intrusion forms and chemistry: *Extended Abstracts, International conference on mafic dyke swarms, University of Toronto, Canada*, 15–21.

—— & GORMSEN, K. 1968. Precambrian rocks of the Angmagssalik area, East Greenland. *Rapport Grønlands Geologiske Undersøgelse*, **15**, 61–71.

——, AUSTRHEIM, H., HANSEN, B. T., MENGEL, F., PEDERSEN, S. & WINTER, J. 1990. The Proterozoic Nagssugtoqidian mobile belt of southeast Greenland: A link between the eastern Canadian and Baltic shields. *Geoscience Canada*, **17**, 305–310.

——, COLLERSON, K. D., HURST, R. W. & JESSEAU, C. W. 1975. Field characters of the Early Precambrian rocks from Saglek, coast of Labrador. *Geological Survey of Canada, Paper 75-1 part A*, 287–296.

——, GLEBOVITSKY, V. A., SEDOVA, I., MILLER, J., ALEXEJEV, N., BOGDANOVA, M., YEPHIMOV, M. M., CHEKULAEV, V. P., ARESTOVA, N. A. & LOBACH ZHUCHENKO, S. 1994. Sub-horizontal stretching fabric and high grade metamorphic assemblages in c. 2.5–2.35 Ga syntectonic igneous suites from the Belomorian fold belt. Evidence of acid-basic igneous activity during extension in the deep crust. IGCP-Symposium 275/371. Nottingham. *Terra Nova*, **6**, abstract supplement 2, 4.

——, KETO, L., McGREGOR, V. R. & MYERS, J. S. 1976. Archaean gneiss complex of Greenland. *In*: ESCHER, A. & WATT, W. S. (eds) *Geology of Greenland*, Geological Survey of Greenland, Copenhagen, 18–75.

CADMAN, A. 1991. The petrogenesis and emplacement of Proterozoic dyke swarms, Part 3, Geochemistry and magmatic evolution of the mafic dyke swarms of the Hopedale Block, Labrador. *In*: *Current research (1991)*. Newfoundland Department of Mines and Energy, Report **91–1**, 191–204.

——, HEAMAN, L., TARNEY, J., WARDLE, R. J. & KROGH, T. E. 1993. U-Pb geochronology and geochemical variation within two Proterozoic mafic dyke swarms, Labrador. *Canadian Journal of Earth Sciences*, **30**, 1490–1504.

CARTWRIGHT, I. & VALLEY, J.W. 1991. Low-^{18}O Scourie dyke magmas from the Lewisian complex, northwest Scotland. *Geology*, **19**,578–581.

ESCHER, A., ESCHER, J. C. & WATTERSON, J. 1975. The reorientation of the Kangâmiut dyke swarm, West Greenland. *Canadian Journal of Earth Sciences*, **12**, 158–173.

——, JACK, S. & WATTERSON, J. 1976. Tectonics of the North Atlantic Proterozoic dyke swarm. *Philosophical Transactions of the Royal Society of London*, **A280**, 529–539.

FAHRIG, W. F. & BRIDGWATER, D. 1976. Late Archean – Early Proterozoic paleomagnetic pole positions from West Greenland. *In* WINDLEY, B. F. (ed.) *The early history of the Earth*. John Wiley and Sons, London. 427–439.

FEDEROV, E. S. 1896. [On a new group of igneous rocks.] *Izvestiya Moskovkogo selsko-khozyaistvennogo instituta (Transactions of the Agricultural institute of Moscow)*, **1**, 168-189 (in Russian).

GARDE, A., 1989 Retrogression and fluid movement across a granulite-amphibolite boundary in middle Archaean Nuk gneisses, Fiskefjord, southern West Greenland. *In*: BRIDGWATER, D. (ed.) *Fluid movements, element transport and the composition of the deep crust*. NATO ASI series C, **281**, 125–137.

HALL, R. P. & HUGHES, D. J. 1987. Noritic dykes of southern West Greenland: Early Proterozoic boninitic magmatism. *Contributions to Mineralogy and Petrology*, **97**, 169–182.

——, —— & FRIEND, C. R. L. 1985. Geochemical evolution and unusual pyroxene chemistry of the MD tholeiite dyke swarm from the Archaean craton of southern West Greenland. *Journal of Petrology*, **26**, 253–282.

——, ——, ——, & SNYDER, G. L. 1987. Proterozoic mantle heterogeneity: geochemical evidence from contrasting basic dykes. *In*: PHARAOH, T. C. BECKINSALE, R. D. & RICKARD, D. (eds) *Geochemistry and Mineralization of Proterozoic Volcanic Suites*. Geological Society, London, Special Publications, **33**. 9–21.

HEAMAN, L. M. & TARNEY, J. 1989. U-Pb baddeleyite ages for the Scourie dyke swarm, Scotland: evidence for two distinct intrusion events. *Nature*, **340**, 705–708.

JACK, S. M. B. 1978. *The North Atlantic Proterozoic dyke swarm*. Ph.D. thesis, Liverpool University.

KALSBEEK, F. & TAYLOR, P. N. 1985. Age and origin of early Proterozoic dolerite dykes in South-West Greenland. *Contributions to Mineralogy and Petrology*, **89**, 307–316.

——, BRIDGWATER, D. & ZECK, H. P. 1978. A 1950 ± 60 Ma Rb-Sr whole rock isochron age from two Kangâmiut dykes and the timing of the Nagssugtoqidian (Hudsonian) orogeny in West Greenland. *Canadian Journal of Earth Sciences*, **15**, 1122–1128.

McGREGOR, V. R. 1973. The early Precambrian gneisses of the Godthåb district, West Greenland. *Philosophical Transactions of the Royal Society of London*, **A273**, 343–358.

MENGEL, F., KORSTGÅRD, J. A. & BRIDGWATER, D. 1993. Post emplacement metamorphic reactions in mafic dykes: autometamorphism, hydration and dehydration-reactions. *Terra Abstracts*, **5**, 404–405.

——, —— & —— 1994. Metamorphism of the Kangâmiut dyke swarm, SW Greenland: Implications for development of the southern margin of the Nagssugtoqidian Orogen. IGCP-Symposium 275/371. Nottingham. *Terra nova*, **6**, abstract supplement 2, 12.

MITROFANOV, F. P. & BALASHOV, Yu. A. (eds) 1990. *Geochronology and genesis of layered basic intrusions, volcanites and granite gneisses of the Kola Peninsula*. Apatity Kola Science Centre of the USSR Academy of Sciences.

——, BALAGANSKY, V. V., BALASHOV, YU. A., GANNIBAL, L. F., DOKUCHAEVA, V. S., NEROVICH, L. I., RADCHENKO, M. K. & RYUNGENEN, G. I. 1993. U-Pb age of gabbro anorthosite massifs in the Lapland granulite belt. *Abstracts, First international Barent symposium, Kirkenes, Norway*.

NOE-NYGAARD, Λ. 1948. A new orogenic epoch in the Precambrian of Greenland. *18th International Geological Congress, Great Britain. 1948, Titles & Abstracts*, 100.

—— 1952. A new orogenic epoch in the Precambrian of Greenland. *Report of the 18th International Geological Congress, Great Britain, 1948*, Part 13, 199–204.

NUTMAN, A. P., HAGIYA, H. & MARUYAMA, S. 1995. SHRIMP U-Pb single zircon geochronology of a Proterozoic mafic dyke, Isukasia, southern West Greenland. *Bulletin Geological Society Denmark*, **42**, in press.

PEARCE, T. H. 1968. A contribution to the theory of variation diagrams. *Contributions to Mineralogy and Petrology*, **19**, 142–157.

RAMBERG, H. 1949. On the petrogenesis of the gneiss complexes between Sukkertoppen and Christianshaab, West Greenland. *Meddelelser dansk geologisk Forening*, **11**, 312-327.

RIVALENTI, G. 1975. Chemistry and differentiation of mafic dykes in an area near Fiskenaesset, West Greenland. *Canadian Journal of Earth Sciences*, **12**, 721-730.

SCHIØTTE, L. 1983. *Geology of the area between Satut and the Inland Ice Umanak Fjord, West Greenland*. MSc thesis, Copenhagen University.

—— 1988. *Field occurrence and petrology of deformed metabasic bodies in the Rinkian mobile belt,*

Umanak district, West Greenland. Rapport Grønlands geologiske Undersøgelse, **141**.

SCOTT, D. J. & MACHADO, N. 1994a. U-Pb Geochronology of the northern Torngat orogen: Results from work in 1993. *In*: WARDLE, R. J. & HALL, J. (eds) *The ECSOOT Report for 1993.* University of British Columbia, Lithoprobe Secretariat Reports, **36**, 141–185.

—— & —— 1994b. U-Pb geochronology of the northern Torngat orogen, Labrador, Canada. A record of Paleoproterozoic magmatism and deformation. *Precambrian Research*, **70**, 169–190.

SUTTON, J. & WATSON, J. 1951. The pre-Torridonian metamorphic history of Loch Torridon and Scourie areas in the north-west Highlands, and its bearing on the chronological classification of the Lewisian. *Quarterly Journal Geological Society of London*, **106**, 241–307.

SUDOVIKOV, N. G. 1939. *[On the petrology of the western Belomorian region (granitization of Belomorian rocks)]*. Trudy Leningradsogo geologischeskogo upraleniya [(Transactions of the Leningrad Geological administration.] **19** (in Russian).

SUN, S. S., NESBITT, R. W. & McCULLOCH, M. T. 1989. Geochemistry and petrogenesis of Archaean and early Proterozoic siliceous high-magnesium basalts. *In*: Crawford, A. J. (ed.) *Boninites*. Unwin Hyman, London, 149–173.

TARNEY, J. 1992. Geochemistry and significance of mafic dyke swarms in the Proterozoic. *In*: CONDIE, K. C. (ed.) *Proterozoic crustal evolution*. Elsevier, Amsterdam and New York, 151–179.

—— & WEAVER, B. L. 1987. Geochemistry and petrogenesis of Early Proterozoic dyke swarms. *In*: HALLS, H. C. & FAHRIG, W. H. (eds.) *Mafic dyke swarms*. Geological Association of Canada, Special Paper, **34**, 81–94.

VAN BREEMEN, O. & HIGGINS, M. D. 1993. U-Pb zircon age of the southwest lobe of the Havre – Saint-Pierre Anorthosite complex, Grenville Province, Canada. *Canadian Journal of Earth Sciences*, **30**, 1453–1457.

VOCKE, R. D. Jr. 1982. *Petrogenetic modeling in an Archean gneiss terrain, Saglek, Northern Labrador.* Ph.D. thesis, State University of New York, Stony Brook,.

WAGNER, P. A. 1982. *Geochronology of the Ameralik dykes at Isua, West Greenland.* MSc thesis, University of Alberta, Edmonton,

WARDLE, R. J., VAN KRANENDONK, M. J., MENGEL, F., SCOTT, D., SCHWARZ, S., RYAN, B. & BRIDGWATER, D. 1993. Geological mapping in the Torngat Orogen, northernmost Labrador. *In*: *Current research (1993).* Newfoundland Department of Mines and Energy, Geological Survey Branch, Report 93-1, 77–89.

WATERS, F. G., COHEN, A. S., O'NIONS, R. K. & O'HARA, M. J. 1990. Development of Archaean lithosphere deduced from chronology and isotope chemistry of Scourie dykes. *Earth and Planetary Science Letters*, **97**, 241–255.

WEAVER, B. L. & TARNEY, J. 1983. Chemistry of the subcontinental mantle: inferences from Archaean and Proterozoic dykes and continental flood basalts. *In*: HAWKESWORTH, C. J. & NORRY, M. J. (eds) *Continental flood basalts and mantle xenoliths.* Shiva, Nantwich, 575–577.

WINDLEY, B. F. 1970. Primary quartz ferro-dolerite/garnet amphibolite dykes in the Sukkertoppen region of West Greenland. *In*: NEWALL, G. & RAST, N. (eds) *Mechanisms of igneous intrusion. Geological Journal, Special Issue*, **2**, 79–92.

Palaeoproterozoic Laurentia–Baltica relationships: a view from the Lewisian

R. G. PARK

Department of Geology, Keele University, Staffordshire, ST5 5BG, UK

Abstract: Using a modified Patchett *et al.* (1978) pre-Grenville fit of Laurentia and Baltica, tectonic information from the Lewisian complex of Britain and from neighbouring Palaeoproterozoic belts of Laurentia and Baltica is reviewed. It is shown that a remarkable consistency in convergence directions existed across the whole region in the period 1.9–1.8 Ga suggesting that the reconstruction used may be correct, and that a common movement pattern can be inferred for the Palaeoproterozoic period within the North Atlantic region.

A speculative plate tectonic history for the region between *c.* 2.6 and *c.* 1.5 Ga is discussed in four stages. (1) 2.6–2.4 Ga: development of conjugate shear-zone systems in the North Atlantic craton. (2) 2.4–2.0 Ga: rifting and dyke emplacement in older cratons; creation of oceanic and intracontinental basins. (3) 2.0–1.8 Ga: subduction at active margins of older cratons with creation of magmatic arcs; collision of cratons accompanied by closure of intra-cratonic basins and accretion of arc terranes. (4) 1.8–1.5 Ga: development of new active margin discordant to the previous ones, with significant changes in convergence direction within the amalgamated continental assembly.

The tectonic history of the Lewisian complex of NW Scotland has been studied since the work of the Geological survey in the 1880s, and after John Sutton and Janet Watson rekindled interest in the complex (Sutton & Watson 1951), it has been the focus of intensive research for over forty years. Although many dates are poorly constrained, the chronology of tectonic events is now relatively well known (e.g. see Park & Tarney 1987; Park 1991 and Park *et al.* 1994 for recent summaries). However NW Scotland is only a small fragment of a much larger Palaeoproterozoic landmass, and in order to understand better and explain Lewisian tectonism, it is necessary to view this complex in its regional context. Major advances in knowledge of surrounding Palaeoproterozoic terrains of Laurentia and Baltica over the last decade have revealed remarkable similarities in nature and timing of Palaeoproterozoic events over the region (e.g. see Kalsbeek *et al.* 1993; Park 1994).

Interest in Proterozoic plate tectonics and continental assemblies has been stimulated by recent attempts by Moores (1991), Dalziell (1992) and Hoffman (1991) to establish the pattern of continental assembly, break-up and re-assembly during the later part of the Proterozoic, and Hoffman (1990) has proposed a model for the Palaeoproterozoic assembly of the various cratons of Laurentia. However there is no general agreement as to how Laurentia and Baltica should be fitted together during this period (e.g. compare Gower 1990 with Gorbatschev & Bogdanova 1993), and the evidence from the British Isles,

which occupies a key position along the join between the two, is often ignored.

The purpose of this paper is to establish a plausible geometric relationship between Laurentia and Baltica, and, using the well-known tectonic sequence in the Lewisian as a starting point, to review the tectonic sequences in the neighbouring belts of Laurentia and Baltica, surrounding the British Isles, in order to assess whether a common sequence and movement pattern can be established for the Palaeoproterozoic period.

Tectonic summary of the Lewisian terrain

The Lewisian complex, composed predominantly of Archaean granitoid gneisses, extensively reworked during the Palaeoproterozoic, forms the islands of the Outer Hebrides and the northwestern coastal strip of NW Scotland, west of the Caledonian front. It is also assumed to form the basement to northern Scotland, NW of the Great Glen fault and of the adjacent continental shelf. Recent summaries of Lewisian tectonic history are given in Park & Tarney (1987), Park (1991) and Park *et al.* (1994).

During the earliest Proterozoic, deformation assigned to the Inverian (late Scourian) event formed steep NW–SE shear zones in amphibolite facies reworking earlier granulite-facies Archaean gneisses. On the Scottish mainland, major Inverian shear zones occupy most of the Southern Region, and probably also the Northern Region, with a

From COWARD, M. P. & RIES, A. C. (eds), 1995, *Early Precambrian Processes*, Geological Society Special Publication No. 95, pp. 211–224.

211

combined outcrop width of between 50 and 70 km. This event is constrained rather imprecisely by an Sm–Nd age of 2.49 Ga (Humphries & Cliff 1982) for the end of the Archaean metamorphism and 2.4 Ga (Heaman & Tarney 1989) for the intrusion of the earliest dated Scourie dykes, which cut Inverian structures. Between 2.4 and 2.0 Ga (Cohen et al. 1988; Heaman & Tarney 1989), at least two phases of NW–SE mafic dyke injection took place (collectively known as the Scourie dyke swarm) partly influenced by the pre-existing Inverian fabric. Shearing continued locally during this period. Mafic volcanics and sediments of the Loch Maree Group were deposited probably around 2.1–2.0 Ga (O'Nions et al. 1983) in a basin considered to be ensialic, rather than oceanic, in nature (see Johnson et al. 1987).

Deformation post-dating the dyke swarm and the supracrustal rocks is assigned to the Laxfordian period, and may be divided into two main stages, the earlier of which, comprising D1 and D2 of the Laxfordian deformation sequence, took place under amphibolite- to granulite-facies conditions and is dated at 1870 ± 40 Ma in South Harris (Cliff et al. 1983). Widespread younger dates at c. 1700 Ma (e.g. Van Breemen et al. 1971) attributed to this event reflect post-tectonic granite and pegmatite emplacement. The D1/D2 deformation produced a strong, gently inclined or sub-horizontal, planar/linear fabric with the extension direction oriented NW–SE. Coward & Park (1987) relate this deformation to a major mid-crustal shear-zone network, separating and enclosing less deformed crustal blocks whose relative movements gave rise to the observed D1–D2 structures. Higher-level blocks are thought to have moved northwestwards relative to lower on a combination of low-angle shear zones and steep lateral ramps. Later, extensional, movements reverse the previous sense of movement (Wynn this volume).

A later Laxfordian deformation (D3) took place under greenschist- to amphibolite-facies conditions and resulted in refolding of the previous gently-inclined fabrics into upright NW–SE folds and steep NW–SE shear zones. This deformation was transpressional with a strong compressional component across the strike of the belt, combined with dextral shear. D3 has not been directly dated but can be bracketed between pre-D3 granite and pegmatite sheets at c. 1.7 Ga (Van Breemen et al. 1971; Lyon et al. 1973; Taylor et al. 1984) and widespread c. 1.5 Ga K–Ar reset ages (Moorbath & Park 1972). Further deformation episodes recorded in the Lewisian rocks are localized and took place in the brittle field at the present level of exposure.

The earlier part of the Lewisian palaeomagnetic record matches the contemporaneous Laurentian shield polar wander path for the period 1820–1750 Ma and is attributed to the closing stages of the main Laxfordian metamorphic event (Piper 1992). The later part of the Lewisian record is considered to relate to the D3 event and is correlated by Piper (1992) with the Laurentian APW path between 1700 and 1630 Ma. This later magnetic sequence is considered to post-date all major tectonism.

The Laxfordian tectonic record thus indicates an earlier period of NW–SE convergence at c. 1.9–1.8 Ga and a later period of approximately N–S convergence at c. 1.7–1.5 Ga. The critical change in convergence direction, which seems to correspond with a hiatus in the palaeomagnetic record, can be dated approximately at 1.75–1.70 Ga.

Division into tectonic stages

A summary of tectonic information on the Palaeoproterozoic belts of Laurentia/Baltica adjoining the Lewisian complex was assembled by Park (1994), who presents relevant geochronological, structural, metamorphic and igneous information for each belt, along with the inferred tectonic setting and movement vector (see Park 1994, table 1 for details). The dated events are divided into four stages, not all of which are represented in every belt. These stages may be thought of as tectonic periods within which several different kinds of tectonic event occurred, and between which marked regional changes may be discerned. These changes, if not synchronous over the region, appear to have taken place within perhaps 10 Ma or so. The stages may be summarized as follows (only the main or 'typical' events in each stage are specified):

(1) 2.6–2.4 Ga: initiation of conjugate shear zones in the North Atlantic craton (cf. Watterson 1978) (= the Inverian event of the Lewisian);

(2) 2.4–2.0 Ga: emplacement of mafic dyke swarms and volcanics (= the Scourie dyke swarm and the Loch Maree Group of the Lewisian);

(3) 2.0–1.8 Ga: production and emplacement of calc-alkaline granitoid plutons and related volcanics; formation of high-grade metamorphic belts accompanied by intense regional deformation (= Laxfordian D1–D2 events in the Lewisian);

(4) 1.8–1.5 Ga: further production and emplacement of calc-alkaline granitoid magmas; but accompanying structural and metamorphic effects generally more localized and less intense than in (3) (= Laxfordian granites and D3 deformation in the Lewisian).

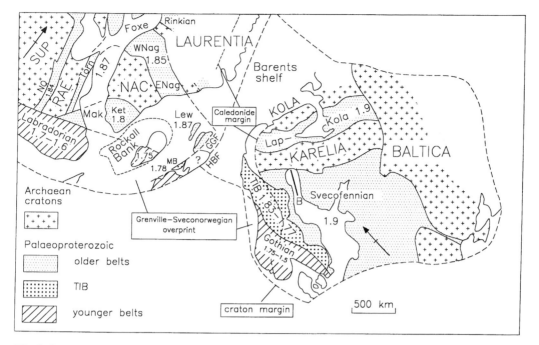

Fig. 1. Reconstruction of eastern Laurentia and Baltica for the period *c*. 1.9–1.5 Ga, based on the palaeomagnetically determined fit of Patchett *et al.* (1978), modified to accommodate the northern British Isles and Rockall Bank (see text). Dates give closest approximation to age of main tectonic activity. Note that the map is oriented parallel to present N in Britain; present N in Laurentia and Baltica is also shown. Abbreviations: NAC, N Atlantic (= Nain) craton; SUP, Superior craton; Mak, Makkovik belt; Ket, Ketilidian belt; W Nag, W Nagssugtoqidian belt; E Nag, E Nagssugtoqidian (or Ammassalik) belt; Lew, Lewisian; Lap-Kola, Lapland–Kola belt; TIB, Trans-Scandinavian igneous belt; GGF, Great Glen fault; HBF, Highland Boundary fault; B, BABEL deep seismic reflection line.

Laurentia–Baltica fit

The primary evidence for any Laurentia–Baltica reconstruction must be based on the palaeomagnetic data, and specifically on the existence of a common polar wander path for the two continents in the interval *c*. 1900–1250 Ma (Piper 1976). Major changes in plate movement pattern are considered to be reflected in prominent hairpin bends in the palaeomagnetic polar wander track. Three such changes are recorded by Elming *et al.* (1993) in Baltica at *c*. 2.6 Ga, *c*. 2.1 Ga and *c*. 1.75 Ga. The first corresponds to stage 1, above, the second to stage 2, and the third to stage 4. The major Svecokarelian collision event (stage 3) is marked by only a minor change in direction of the APW path at *c*. 1.9 Ga.

The common pole position for Laurentia and Baltica in the period 1260–1190 Ma is situated at long. 5°E, Lat. 10°S (e.g. see Patchett *et al.* 1978). Piper (1982) calculates a rotation of –41° about a Euler pole at 8°E, 10°N to bring Fennoscandia (Baltica) into its postulated pre-1190 Ma position adjacent to Laurentia. This reconstruction (see Stearn & Piper 1984) places Baltica adjacent to eastern Laurentia, contiguous to East Greenland (Fig. 1) but oriented about 90° anticlockwise from its orientation in the Mid- to Late Proterozoic (Patchett & Bylund 1977).

The Piper (1976) reconstruction, however, leaves no space for the British Isles and surrounding continental shelf. To accommodate this piece of continental crust, known to be Palaeoproterozoic or older, a more easterly position of Baltica relative to Laurentia has been obtained for Fig. 1, by a small clockwise rotation of Baltica about the same palaeopole. A further small adjustment was made to bring the Barents shelf margin of Baltica into juxtaposition with East Greenland, assuming that the shelf has doubled in width due to extensional thinning. The imprecision of the palaeomagnetic data of course means that the fit could be hundreds of kilometres in error. It should be noted that this fit is about 80° anticlockwise of the post-Grenville fit of Patchett & Bylund (1977) and of the Palaeoproterozoic restorations of Kalsbeek *et al.* (1993) and Gorbatschev & Bogdanova (1993) but is closer to that used by Gower (1990).

Using this reconstruction, the tectonic evidence from the Palaeoproterozoic belts surrounding the British Isles is reviewed (see Fig. 1) in order to assess whether a common movement pattern can be established for the Palaeoproterozoic period.

The 'Internal' belts of Laurentia

The Eastern Nagssugtoqidian or Ammassalik belt (SE Greenland)

The Eastern Nagssugtoqidian or Ammassalik belt (Kalsbeek 1989) has been compared directly with the Lewisian by Wright et al. (1973), Myers (1987) and more recently by Kalsbeek et al. (1993). The two terrains are adjacent if the effects of Atlantic ocean opening are removed (Fig. 1). Much of the 300 km wide belt consists, like the Lewisian, of reworked Archaean granitoid gneisses cut by mafic dykes correlated with the Lewisian Scourie dykes and by other intrusions. However, unlike the Lewisian, the central part of the belt contains abundant metasediments, deposited around 2.1–2.0 Ga, which were thought by Bridgwater et al. (1991) to represent the fill of an ensialic basin. However Kalsbeek et al. (1993) suggest a possible suture between the metasediments and the reworked Archaean gneisses to the south.

The structure of the belt is dominated by NE-dipping, S-vergent regional nappes and shear zones with N–S to NW–SE linear fabrics and by NW–SE upright folds; NE-plunging linear structures have also been recognized (Chadwick & Vasudev 1989). The Proterozoic metasediments are cut by deformed and metamorphosed intrusions of the Ammassalik igneous complex yielding a date of 1886 Ma (Hansen & Kalsbeek 1989).

The tectonic history thus appears to show considerable similarity to the Lewisian, with an early deformation accompanied by high-grade metamorphism (corresponding to stage 3) at c. 1.8 Ga involving movements sub-parallel or slightly oblique to the strike of the belt, followed (stage 4?) by deformation caused by convergence across the belt. Post-tectonic intrusions are dated at c. 1.68 Ga (Kalsbeek et al. 1993).

The Western Nagssugtoqidian and Foxe–Rinkian belts (SW Greenland)

The Western Nagssugtoqidian belt of SW Greenland is separated from the Eastern Nagssugtoqidian belt by the Greenland ice cap but its southern part, at least, is thought to be continuous with it. A summary of the belt is given by Korstgård et al. (1987). The belt is of similar width to the Eastern Nagssugtoqidian belt and contains reworked Archaean gneisses and Palaeoproterozoic paragneisses, metavolcanics and calc-alkaline plutonic intrusions. The metavolcanics yield a model age of 2.1–2.0 Ga (Kalsbeek et al. 1984, 1987, 1993). Isotopic data from the paragneisses indicate derivation from juvenile Proterozoic sources. The central plutonic zone has been interpreted by Kalsbeek et al. (1987) as a juvenile magmatic arc, suggesting a collisional origin for the belt.

The earliest recorded Proterozoic deformation (stage 1) produced steep dextral shear zones accompanied by amphibolite-facies retrogression of the granulite-facies Archaean basement near the southern boundary of the belt. A 200 km wide swarm of mafic dykes in the southern sector of the belt, which have been compared with the Lewisian Scourie dykes, post-date the shear zones and yield an Rb–Sr whole-rock age of 1.95 Ga (Kalsbeek et al. 1987); they may thus be assigned to stage 2. The dykes are affected by regional overthrusting towards the south on N-dipping shear zones, also in amphibolite facies.

In the central part of the belt, a major sinistral strike-slip shear zone formed in granulite-facies gneisses (Sorensen 1983) deforms a 1.92 Ga quartz-diorite intrusion and contains syn-tectonic pegmatites dated at c. 1.85 Ga (Hickman & Glassley 1984; Kalsbeek et al. 1984). This event corresponds to stage 3. The southwards overthrusting in the south may represent stage 4.

Only a small Archaean block separates the Western Nagssugtoqidian belt from the Rinkian belt, which continues westwards across Baffin island in Arctic Canada, where it is known as the Foxe zone (see Hoffman 1989). The Foxe–Rinkian belt is up to 430 km wide and is composed of Palaeoproterozoic metasediments and reworked Archaean gneisses, strongly deformed under high-pressure, low temperature metamorphism.

According to Grocott & Pulvertaft (1990) the main deformation in the Rinkian is associated with WNW–ESE to NW–SE elongation lineations, formed under high-temperature, low-pressure conditions. This deformation was followed by a phase of ductile overthrusting to the NW. The Proven igneous complex, which yielded an Rb–Sr date of c. 1860 Ma (Kalsbeek 1981), is considered to be syn-tectonic to the main high-temperature event; which thus correlates with stage 3.

The belt was interpreted by Grocott & Pulvertaft (1990) as a deformed back-arc extensional basin situated on the upper plate of a N-dipping subduction zone which was located within the Western Nagssugtoqidian belt to the south. However more recent work (Grocott & Davies 1994) indicates that the tectonic transport in this region is towards the north or northwest, raising the possibility that the

Rinkian represents the deformed passive margin of a northern Archaean craton, and that the southern Nagssugtoqidian margin may be backthrust onto the Nain craton.

The Eastern Churchill Province

The belt of Palaeoproterozoic activity continues southwards from the Western Nagssugtoqidian and Foxe–Rinkian belts along the west side of the North Atlantic craton, where it is known as the Eastern Churchill Province. This province is about 400 km wide, and consists of three separate units, the New Quebec, Rae and Torngat belts (e.g. see Wardle *et al.* 1990*b*). The New Quebec 'orogen' contains a belt of 2.1–1.8 Ga old supracrustal metasediments and volcanics which have been overthrust westwards onto the Superior craton. East of this belt are reworked Archaean rocks of the Rae Province, intruded by calc-alkaline granitoid plutons. This central belt is interpreted as the upper plate of a collision orogen. Initial transport directions were towards both the NW and the SW, but later major shear zones along the margin of the belt are dextral. Initial thrusting and collision are dated at 1.88–1.82 Ga although pegmatite intrusion and retrogressive deformation persisted until 1.77 Ga (Machado 1990). The calc-alkaline De Pas batholith within the Rae province, dated at 1.84–1.81 Ga, is interpreted as syn- to post-collisional by Wardle *et al.* (1990*b*).

East of the Rae Province lies the Torngat belt or 'orogen' (see e.g. Korstgård *et al.* 1987; Ermanovics & Van Kranendonk 1990; Wardle *et al.* 1990*b*, 1992) which is a Palaeoproterozoic belt, between 75 and 200 km wide, separating the North Atlantic craton to the east from Archaean rocks of the Rae province to the west. The belt consists of reworked Archaean rocks, Palaeoproterozoic granitic and dioritic plutons, mafic dykes and a band of high-grade metasediment (the Tasiuyak gneiss) which reaches a maximum width of 13 km. Calc-alkaline plutonic rocks in the belt are interpreted as part of a subduction-related magmatic arc (Wardle *et al.* 1990*b*). The western edge of the Nain craton is also intruded by a suite of primary, mantle-derived calc-alkaline plutons ranging in age from 1.91 to 1.87 Ga (Scott & Machado 1994*b*). These are held to indicate easterly subduction beneath the Nain craton. The Tasiuyak paragneiss, which has yielded detrital zircons with a maximum depositional age of 1.94 Ga, is thought to have been derived from an accretionary prism (Scott & Machado 1994*b*).

The earlier structures are west-vergent ductile thrusts or shear zones, whereas the later movements involve sinistral strike-slip movements along steep shear zones and east-vergent thrusts onto the Nain craton. Several precise Ub–Pb zircon ages ranging from 1.91 to 1.86 Ga are considered to date the main high-grade metamorphic/deformational event (stage 3) (Scott & Machado 1994*a*). The later movements (stage 4) have been dated at between 1.79 and 1.78 Ga from pegmatites within mylonite and are considered to date the final uplift of the belt (Bertrand *et al.* 1990). The earlier movements are attributed to oblique collision between the Nain (North Atlantic) and Rae cratons, the later to subsequent shortening across the belt (Wardle *et al.* 1990*b*) and backthrusting onto the Nain craton (Van Kranendonk & Wardle 1994).

According to Wardle & Van Kranendonk (1994), the Palaeoproterozoic evolution of the East Churchill Province was controlled by the successive indentation of the Nain and Superior cratons into the Rae hinterland, a prong of which was forced southwards to form the central zone of the orogen. Initial collision along the Torngat orogen took place at 1.87 Ga and was followed by collision of the Superior craton at 1.84 Ga to form the New Quebec orogen. Later movements in both orogens were dominated by strike-slip accommodation, sinistral in the Torngat, dextral in the New Quebec, to the northward progress of the two cratons relative to the Rae prong.

The Lapland–Kola belt of Baltica

The Lapland–Kola belt (for a fuller description, see Bridgwater *et al.* 1991) is between 200 and 400 km wide and separates the Archaean cratons of Kola in the north and Karelia in the south. The southern margin of the craton is marked by a narrow, S-dipping zone variously termed the Polmak, Pechenga, Imandra and Varguza belt (PPIV). It contains 2.4–2.0 Ga-old volcanic and sedimentary assemblages, including an andesitic volcanic arc sequence, and is interpreted by Berthelsen & Marker (1986) as a collisional suture zone. South of this zone lies the Lapland granulite belt, which is up to 100 km wide in the west but wedges out eastwards. It consists of high grade deformed Palaeoproterozoic metasedments intruded by a syntectonic plutonic suite, and has been compared with the central high-grade gneiss zones of the Nagssugtoqidian and Torngat belts. Major bounding thrusts, and internal ductile shear zones dip N and NE (Marker 1990). Peak metamorphism is dated at between 1.95 and 1.87 Ga (Bernard-Griffiths *et al.* 1984; Daly & Bogdanova 1991). Since the sediments appear to have been derived from older sialic crust, the belt has been interpreted as an ensialic back-arc basin on the upper plate of the subduction zone in the PPIV (Berthelsen & Marker 1986). An alternative interpretation is proposed by Campbell *et al.* (1994) who link these metasediments with similar rocks in the Nagssugtoqidian and Torngat belts and favour a continental margin setting for their derivation.

The southern part of the belt consists of NE-dipping thrust units of highly deformed Archaean gneisses interpreted as the reworked northern margin of the Karelian craton (see Gaal *et al.* 1989). The Archaean rocks are intruded by a suite of mafic dykes and layered intrusions dated at 2.45 Ga (Gaal *et al.* 1989). Farther west, the Archaean basement is overlain by an Palaeoproterozoic supracrustal sequence including a greenstone belt with komatiitic pillow lavas dated at 2.1–2.0 Ga (Often 1985). These rocks are intruded by syn- and post-tectonic granitoid plutons. The southwest marginal zone is characterized by SW-directed thrust movements (Marker 1990).

Summarizing the tectonic pattern of this belt, the thrust/shear zone systems in the central part of the belt and at the southern margin dip SW, parallel to the postulated subduction zone, and tectonic transport is inferred to be northeastwards. In the southern zone, overthrusting (backthrusting?) is to the SW towards the Karelian craton. The main thrust/shear zone movements are dated at c. 1.9–1.8 Ga by Petrov (1988) from metamorphic mineral ages and correspond to stage 3.

The 'External' or marginal belts

The Makkovik-Ketilidian belt

The belts situated on the southern side of Laurentia are interpreted as accretionary magmatic arc terrains on the active margin of the contemporary continent (e.g. see Hoffman 1989). The Makkovik–Ketilidian belt is characterized by a border zone of reworked Archaean basement with infolded remnants of Palaeoproterozoic meta-sedimentary and metavolcanic cover. In Greenland this zone is succeeded southwards by a median zone consisting of a calc-alkaline batholith and a southern zone of migmatitic gneisses. According to Patchett & Bridgwater (1984) the latter two zones appear to consist of juvenile Proterozoic crust, and a major shear zone separates the reworked border zone from the juvenile crust in the south. Structures in the supracrustal rocks to the south of the batholith are considered to result from southeast-ward overthrusting, and the batholith is interpreted as an arc terrane accreted obliquely onto the Nain craton (Chadwick & Garde 1994).

In Labrador, older granites in the north of the Makkovik belt are dated at 1.89 Ga and the main deformational/metamorphic event in the supra-crustal belt is bracketted between 1.81 Ga volcanics and 1.80 Ga post-tectonic plutons (Kerr *et al.* 1994).

The LITHOPROBE reflection profile across the Makkovik–Ketilidian belt in the Labrador Sea (Kerr *et al.* 1994) indicates prominent low-angle NW-dipping reflectors corresponding to the thrust structures in the southern part of the orogen, and the northern margin of the orogen is marked by a strong S-dipping reflector interpreted as the suture separating the arc terrane from reworked Archaean craton to the north.

Younger granitoid plutons in the Makkovik belt, dated at 1720 and 1650 Ga (Kerr *et al.* 1994) and 1.76 Ga post-tectonic rapakivi-type granites in Greenland (Kalsbeek & Taylor 1985), may be attributed to the Labradorian (see below).

The Svecofennian domain

The Karelian craton is bounded on its southwest side by a Palaeoproterozoic belt which is divided into two main zones, a northeastern zone (known as the Karelian belt or Karelides) consisting of reworked Archaean basement and Palaeo-proterozoic cover, and a southwestern zone (the Svecofennian *sensu stricto*) consisting of juvenile Palaeoproterozoic igneous and sedimentary rocks (see Gaal & Gorbatschev 1987; Gorbatschev & Bogdanova 1993). The upper parts of the supracrustal sequence in the Karelian zone (i.e. lying on the Karelian craton) are considered to have formed in a passive margin setting, but the south-west boundary of the craton is now marked by a collisional suture dipping NE. Tectonic movements in the collisional suture zone involve both NE-directed and SW-directed thrust movements. This suture has been imaged on deep seismic reflection profiles as cutting the Moho (BABEL Working Group 1990).

South of the suture, the Svecofennian comprises a series of magmatic arc terrains that have become welded to the Karelian craton (Gaal & Gorbatschev 1987). Early granitoid plutons of this arc terrain in Finland are dated at 1.89–1.88 Ga; these pre-date the main deformational/metamorphic event, which is post-dated by 1.87 Ga dykes (Ehlers *et al.* 1993) and falls in stage 3. The presence of zircons yielding dates of 1.88–2.13 Ga (Claesson *et al.* 1990) and of model Nd ages of 2.4–2.0 (Valbracht *et al.* 1994) indicate that parts of the Svecofennian terrain may have had a lengthy history prior to accretion onto Baltica.

A second generation of granites emplaced at 1.83–1.77 Ga (Anderson 1991) are attributed to the younger Gothian belt, i.e. stage 4 (see below). A later deformation produced NW–SE dextral strike-slip shear zones dated at between 1.85 Ga and 1.79 Ga, which have led to suggestions that convergence across the belt was oblique (Park 1985).

The main geological events of the Svecofennian belt (i.e. stage 3) are coeval with those recorded in the Lapland–Kola belt to the NE, and the Svecofennian belt thus represents part of the accretionary margin of the Palaeoproterozoic continent which had been assembled by 1.9 Ga.

The Labrador belt

The Makkovik orogen in Labrador appears to be crossed obliquely by the younger Labrador belt (Fig. 1). The latter belt is also interpreted as an accretionary magmatic arc terrain consisting of 1.71–1.62 Ga plutons and metamorphic rocks which form the basement to the mid-Proterozoic Grenville province (see Hoffman 1989; Gower 1990). The lack of older inherited ages in the magmatic and metasedimentary rocks indicates that the belt is primarily accretionary (Schärer & Gower 1988) and formed the southern active margin of Laurentia between c. 1.7 and c. 1.6 Ga, i.e. during stage 4.

Both the Makkovik and Labrador belts appear to cross-cut the Palaeoproterozoic Torngat and New Quebec belts and the intervening Rae province at a high angle (although younger rocks obscure the junction) suggesting that the Makkovik–Labrador belt may occupy the site of a former rifted or strike-slip margin (see Wardle et al. 1990a).

The Gothian belt

The Gothian belt (also known as the southwest Scandinavian domain) is summarized by Gaal & Gorbatschev (1987) and Gorbatschev & Bogdanova (1993). At the eastern margin of this belt is a 150–220 km wide zone of volcanic and plutonic igneous rocks known as the Trans-Scandinavian Igneous Belt (TIB). This belt is discordant to the gross structure of the Sveco-fennian belt and also truncates the boundary of the Karelian craton (although its precise course in the north is obscured by the younger Caledonian belt), but individual plutons within the belt display a late kinematic relationship to the Svecofennian (Wikstrom 1991).

The granitoid rocks display a more alkaline trend than the typically calc-alkaline Svecofennian plutons and have been attributed to extensive reworking of Svecofennian crust above an E-dipping subduction zone. Dates from the TIB rocks fall mainly between 1.83 and 1.77 Ga (e.g. see Skjöld 1984; Wilson et al. 1985; Patchett et al. 1987; Andersson 1991). The main part of the Gothian belt, west of the TIB, is composed of variably deformed and metamorphosed calc-alkaline plutons, dated at c. 1.75–1.5 Ga (Welin et al. 1982; Jacobsen & Heier 1978; Verschure

1985) emplaced into Palaeoproterozoic meta-sediments varying from shallow marine or continental in the east to marine in the west. Those in SW Sweden have been shown to lack inherited pre-Svecofennian crustal material (Ähäll & Daly 1989). The main deformation of the Gothian rocks, dated at c. 1580 Ma by Åhäll et al. (1990), involved localized SW-vergent movements on E-dipping shear zones in SW Sweden (Park et al. 1987) and in south Norway. Compressive movements with a component of dextral shear are noted by Starmer (1991).

The Gothian belt thus represents the active margin of Baltica between c. 1.8 Ga and 1.5 Ga. Because there is no consistent sense of younging across the belt, it is thought likely that arc terrains of various ages may have been accreted during this period (Ähäll & Persson 1992; Gorbatschev & Bogdanova 1993).

The Malin block and the Rockall plateau

The Malin block (Muir et al. 1989) comprises four inliers of Precambrian basement defining a NE–SW elongate block in the Malin Sea area immediately south of the Lewisian outcrop of Northwest Scotland. The easternmost part of the belt is exposed on the Rhinns of Islay and the westernmost on the island of Inishtrahull, off the north coast of Ireland. The block is tectonically isolated both from the Lewisian and from the Dalradian rocks to the south and east by splays of the Great Glen fault.

The rocks of the Malin block comprise a deformed and metamorphosed sub-alkaline igneous complex emplaced at 1.78 Ga (Marcantonio et al. 1988; Daly et al. 1991), i.e. during stage 4. Sm–Nd isotopic data (Daly et al. 1991) indicate that this is close to the age of crustal extraction and that, unlike the Lewisian complex to the north, the Malin rocks represent juvenile Proterozoic material rather than reworked Archaean crust. The age of deformation on Inishtrahull may be estimated from the meta-morphic age of 1710 Ma obtained by $^{39}Ar/^{40}Ar$ dating of amphibole (Roddick & Max 1983). The Malin block has been described as having affinities both with the Ketilidian of southern Greenland and the Svecofennian of Scandinavia (Muir et al. 1994). However, in age and chemistry, it is more comparable with the TIB in the Gothian belt.

The Rockall plateau to the west contains deformed granulite-facies orthogneisses (Morton & Taylor 1991) which differ from the Malin rocks in their more calc-alkaline composition, more intense deformation and much higher metamorphic grade. U–Pb zircon data from submarine drillcores on the southern Rockall Bank (Daly et al. 1994) yield a precise date of 1.75 Ga and confirm the juvenile

Table 1. *Significant events in the Palaeoproterozoic history of Laurentia/Baltica between 2.0 and 1.5 Ga*

Stage 3

1.92	Pre-tectonic plutons in W Nagssogtoqidian	Subduction
1.91–1.87	Plutons in Rae and W Nain cratons	Subduction
1.90–1.87	Main Lapland–Kola deformation	Collision
1.90–1.86	Svecofennian plutons	Subduction
1.89	Ammassalik igneous complex	?Subduction
1.89	Ketilidian plutons	Subduction
1.89–1.87	Main Svecokarelian deformation	Collision
1.87	Main Torngat deformation (Nain/Rae collision)	Collision
1.86	Main Rinkian deformation	Collision
1.85	Main W Nagssugtoqidian deformation	Collision
1.84	Main New Quebec deformation (Superior/Rae)	Collision
1.84–1.81	Emplacement of De Pas batholith in Rae	Post-collision

Stage 4

1.83–1.77	TIB plutons	Subduction
1.81–1.80	Main deformation in Makkovik	Collision
1.80–1.76	Main deformation of Ketilidian	Collision
1.79–1.78	Late deformation in Torngat	Collision
1.76	Post-tectonic plutons in Ketilidian	Post-collision
1.75–c. 1.5	Gothian plutons	Subduction
1.71–1.62	Labradorian plutons	Subduction
1.58	Gothian deformation	Accretion

References for individual dates are given in the text. Dating of tectonic events is estimated either by using ages of peak metamorphism or by bracketing deformation phases between pre-tectonic and post-tectonic intrusions, or in a few favourable cases by using ages of syntectonic intrusions. However many gaps in the geochronological coverage exist and the chronologies for most of the belts discussed must be regarded as provisional.

character of the Maein–Rockall terrane. The Malin–Rockall terrane is considered to form a link in a formerly continuous Palaeoproterozoic magmatic arc, embracing both the Labradorian and Gothian belts, along the southern active margin of Laurentia-Baltica (Fig. 1) between *c.* 1.8 and 1.6 Ga.

According to the currently available age data (see Table 1), the Makkovik–Ketilidian belt appears to be composed of rocks with a similar age range to the Svecofennian, but the date of accretion onto the Nain craton is considerably younger (by about 40 Ma) than the age of collision of the internal belts around the west, north and northeast

sides of the craton. This belt is therefore provisionally interpreted as part of the Svecofennian arc terrane formed during stage 3, but transported northwards and accreted to the craton during stage 4, at the beginning of the Gothian period.

Discussion and conclusions: Palaeoproterozoic kinematic history

Stage 1: 2.6–2.4 Ga

Little is known about this period in most of the belts under consideration — certainly in comparison to the amount of data available for the

Fig. 2. Speculative kinematic models for the periods 1.9–1.85 (**A**) and 1.83–1.55 (**B**) Ga based on the reconstruction of Fig. 1 showing distribution of cratons, tectonically active intracontinental belts and accretionary magmatic arc terrains at the active continental margin. In collisional or contractional belts, the relative movement vector between the blocks on either side (the convergence direction) has generally been established from the orientations of elongation lineations in the highly strained parts of major ductile shear zones (see Shackleton & Ries 1984; Coward & Daly 1984). Arrows represent presumed main directions of upward tectonic transport within the active belts. Abbreviations as Fig. 1; also EGC, E Greenland craton; KAR, Karelia craton. (**A**) 1.9–1.85 Ga. Note parallelism of convergence directions. (**B**) 1.83–1.55 Ga. Note that the site of the active continental margin has now shifted to the Labradorian and Gothian magmatic arc terrains, and that the Trans-Scandinavian igneous belt (TIB), at the eastern margin of the Gothian belt, is highly discordant to the older Palaeoproterozoic belts of Baltica. Directions of tectonic transport in the active intracontinental belts surrounding the North Atlantic craton indicate a significant clockwise change in convergence direction from (A).

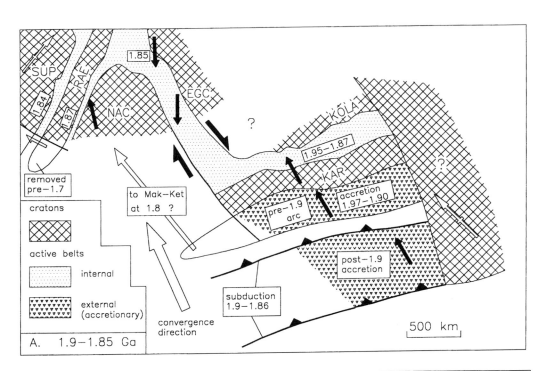

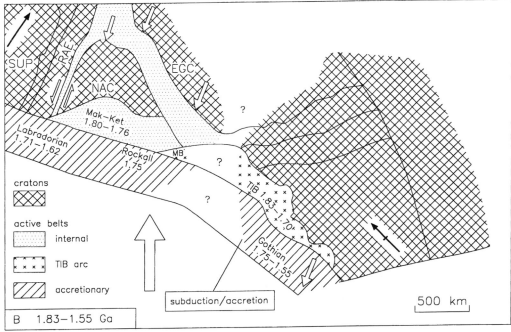

period 1.9–1.7 Ga. The relative positions of the various Archaean cratonic fragments prior to the 1.9–1.8 Ga continental assembly are highly uncertain, as the palaeomagnetic data do not allow us to decide whether the Rae, Kola and Karelia cratons were close to their 1.9 Ga relative positions. However it is thought that, at c. 2.6 Ga, the Archaean regions had become generally stable and were able to support uniform stress systems. In the North Atlantic craton, a set of conjugate major shear zones is attributed by Watterson (1978) to general N–S* compression throughout the craton at around this time.

In the Lewisian, the Inverian shear zone is considered to form part of this system and is attributed to dextral strike-slip movements between the North Atlantic and East Greenland cratons (see Park & Tarney 1987).

Stage 2: 2.4–2.0 Ga

During the period 2.45–1.95 Ga, basic dyke swarms are recorded in Scotland, Labrador, West and East Greenland and Baltica, indicating conditions of widespread extension or transtension. Between c. 2.1 and 2.0 Ga, ophiolites and mafic volcanics, interpreted as ocean-floor material, were formed in the areas of the new Quebec and Lapland–Kola belts, and at the southern margin of Karelia. There is evidence for the opening of an intracontinental extensional (or transtensional) basin at this time in the Lewisian (Loch Maree group) and possibly also in the East Nagssugtoqidian. Supracrustal rocks interpreted as passive-margin sequences and/or accretionary prisms occur in the New Quebec, Torngat, Rinkian and Lapland–Kola belts reinforcing their collisional status. In several cases, there appears to be a transition from oceanic to island-arc volcanic assemblages, indicating the commencement of subduction regimes that were to become widespread during stage 3.

Stage 3: 2.0–1.8 Ga (Fig. 2A)

Calc-alkaline plutonic suites were emplaced in all the Palaeoproterozoic belts considered above during the period 1.92–1.86 Ga, indicating widespread subduction around the west and north margins of the North Atlantic craton, and between the Kola and Karelia cratons. The accretionary regimes of the Svecofennian and Makkovik–Ketilidian belts are indistinguishable geochronologically from the other belts, and it would appear therefore that, at the southeastern margin of the

region, subduction created a series of magmatic island arcs which were subsequently welded to the Archaean cratons by arc collision.

During the same period, but possibly 20–30 Ma later than the onset of subduction, major convergent movements occurred across the Palaeoproterozoic belts surrounding the North Atlantic craton and in Baltica (assumed to be contiguous). Collision took place at 1.89–1.87 Ga in the Lapland–Kola belt, and along the southern margin of the Karelian craton (i.e. the Karelides, where the direction of overthrusting is consistent with those in the Lapland–Kola belt), after which subduction moved south to a position within the present Svecofennian belt to produce continued Svecofennian magmatism until about 1.86 Ga.

At 1.87 Ga, the North Atlantic craton collided with the Rae–North Greenland hinterland causing the major orogenies in the Torngat (1.87 Ga), Rinkian (1.86 Ga) and Western Nagssugtoqidian (1.85 Ga) belts, and about 1.84 Ga, the Superior craton also joined the assemblage. During the same period convergence took place across the Lewisian and Eastern Nagssugtoqidian terrains. It is clear from Fig. 2A that the movement directions in the ductile shear zones of the Torngat, Nagssugtoqidian, Rinkian, Lewisian, Lapland–Kola and Karelide belts are approximately parallel, which lends support to the Fig. 1 reconstruction and suggests that the region experienced a uniform plate movement system. It would seem, therefore, that the apparently continent-wide NW–SE convergent system at c. 1.9 Ga was not a single event, but embraced several separate collision events, commencing in Baltica and spreading west to the North Atlantic and finally to the Superior craton, and lasting between 30 and 50 Ma.

Stage 4: 1.8–1.5 Ga (Fig. 2B)

A new phase seems to have commenced at 1.83 Ga with the development of an active margin in Baltica which was discordant to the trend of the previous margin, lying about 40° clockwise of its previous position. An important implication of this geometrical arrangement is that the western part of pre-Gothian Baltica may have been removed prior to the initiation of Gothian subduction (see Fig. 2A). The apparent discordance of the Makkovik–Labradorian front is another indication of the same phenomenon (cf. Wardle et al. 1990a). Subduction along this new margin gave rise to the emplacement of the Trans-Scandinavian Igneous Belt at 1.83–1.77 Ga. At the same time (1.81–1.80 Ga) sections of the older arc system may have been displaced by dextral strike-slip movements to become accreted onto the Laurentian margin to form the Makkovik–Ketilidian belt. This

---

*Note that all orientations in this concluding section are referred to present north in the British Isles.

new belt cross-cuts the previously formed compressional structures of Baltica and Laurentia (Fig. 2B) and indicates an abrupt change in plate tectonic setting. This accretionary event seems to have been responsible for the later deformation in the Torngat orogen, and possibly also for the less well-dated later structures in the Nagssugtoqidian and Lewisian terrains.

Between 1.76 and 1.55 Ga, the locus of subduction shifted to the southern margin of Laurentia to form a series of magmatic arcs extending from the Labradorian to Rockall Bank, and probably linking with the Gothian belt, where magmatic arc-creation continued west of the TIB. The continuity of this belt forms a good test of the Laurentia–Baltica fit used in Fig. 1. The evidence for dextral transpression in the Gothian belt indicates that convergence across the southern margin of the supercontinent was oblique, and supports the suggestion that the missing continental fragments may have been removed by strike-slip terrane displacement (see Fig. 2A).

References

ÅHÄLL, K.-I. & DALY, J. S. 1989. Age, tectonic setting and provenance of Östfold-Marstrand belt supracrustals: westward crustal growth of the Baltic shield at 1760 Ma. *Precambrian Research*, **45**, 45–61.

—— & PERSSON. 1992. The Åmål-Horred belt: recognition of a c. 1.60 Ga calc-alkaline magmatic arc in SW Scandinavia. *Geologiska Föreningens, Stockholm Förhandlinger*, **114**, 448.

——, DALY, J. S. & SCHÖBERG, H. 1990. Geochronological constraints on mid-Proterozoic magmatism in the Östfold-Marstrand belt; implications for crustal evolution in Southwestern Sweden. *In*: GOWER, C. F., RIVERS, T. & RYAN, B. (eds) *Mid-Proterozoic Laurentia–Baltica*. Geological Association of Canada, Special Papers, **38**, 97–115.

ANDERSSON, U. B. 1991. Granitoid episodes and mafic-felsic magma interaction in the Svecofennian of the Fennoscandian shield, with main emphasis on the c. 1.8 Ga plutonics. *Precambrian Research*, **51**, 127–149.

BABEL WORKING GROUP. 1990. Evidence for Early Proterozoic plate tectonics from seismic reflection profiles in the Baltic shield. *Nature*, **348**, 34–38.

BERNARD-GRIFFITHS, J., PEUCAT, J. J., POSTAIRE, B., VIDAL, PH, CONVERT, PH & MOREAU, B. 1984. Isotopic data (U-Pb, Rb-Sr, Pb-Pb and Sm-Nd) on mafic granulites from Finnish Lapland. *Precambrian Research*, **23**, 325–348.

BERTHELSEN, A. & MARKER, M. 1986. Tectonics of the Kola collision suture and adjacent Archaean and Early Proterozoic terrains in the northeastern region of the Baltic shield. *Tectonophysics*, **126**, 31–55.

BERTRAND, J.-M., VAN KRANENDONK, M. J., HANMER, S., RODDICK, J.C. & ERMANOVICS, I.F. 1990. Structural and metamorphic geochronology of the Torngat orogen in the North River–Nutak Transect area, Labrador: preliminary results of U-Pb dating. *Geoscience Canada*, **17**, 297–301.

BRIDGWATER, D., MARKER, M. & MENGEL, F. 1991. The eastern extension of the Early Proterozoic Torngat orogenic zone across the Atlantic. *In*: WARDLE, R. J. & HALL, J. (eds) *Eastern Canadian Shield onshore-offshore transect (ECSOOT), Lithoprobe report 27*. Memorial Univ., Newfoundland, 76–91.

CAMPBELL, L. M., BRIDGWATER, D. & MARKER, M. 1994. A geochemical comparison of major Proterozoic units along the Torngat-Nagssugtoqidian-Lapland-Kola collisional belt. *Terra Abstracts*, **6**, 2, 5.

CHADWICK, B. & GARDE, A. A. 1994. Reappraisal of the Early Proterozoic Ketilidian belt of South Greenland. *Terra Abstracts*, **6**, 2, 5.

—— & VASUDEV, V. N. 1989. Some observations on the structure of the Early Proterozoic Ammassalik mobile belt in the Ammassalik region, South-East Greenland. *In*: KALSBEEK, F. (ed.) *Geology of the Ammassalik region, South-East Greenland*. Grønlands Geologiske Undersøgelse Rapports, **146**, 29–40.

CLAESSON, S., HUHMA, H., KINNY, P. & WILLIAMS, I. S. 1990. U-Pb dating of detrital zircons from Svecofennian metasediments. *Abstr 2nd Symp on Baltic Shield Lund IGCP 275*, 27.

CLIFF, R. A., GRAY, C. M. & HUHMA, H. 1983. A Sm-Nd isotopic study of the South Harris Igneous Complex, the Outer Hebrides. *Contributions to Mineralogy and Petrology*, **82**, 91 98.

COHEN, A. S., WATERS, F. G., O'NIONS, R. K. & O'HARA, M. J. 1988. A precise crystallisation age for the Scourie dykes and a new chronology for crustal development in North-west Scotland (abs). *Chemical Geology*, **70**, 19.

COWARD, M. P. & DALY, J. S. 1984. Crustal lineaments and shear zones in Africa: their relationship to plate movements. *Precambrian Research*, **24**, 27–445.

—— & PARK, R. G. 1987. The role of mid-crustal shear zones in the Early Proterozoic evolution of the Lewisian. *In*: PARK, R. G. & TARNEY, J. (eds) *Evolution of the Lewisian and comparable Precambrian high-grade terrains*. Geological Society, London, Special Publications, **27**, 127–138.

DALY, J. S. & BOGDANOVA, S. 1991. Timing of metamorphism in the Lapland granulite belt, Finland (abs). *In*: TUISKU, P. & LAAJOOKI, K. (eds) *Metamorphism, deformation and structure of the crust*. Report of joint meeting of IGCP projects 275 and 304. *Res Terrae*, University of Oulu, Finland.

——, FITZGERALD, R. C., BREWER, T. S., MENUGE, J. F., HEAMAN, L. M. & MORTON, A. C. 1994. Persistent geometry of Palaeoproterozoic juvenile crust on the southern margin of Laurentia–Baltica: evidence from Ireland and Rockall Bank. *Terra Abstracts*, **6**, 2, 5.

——, MUIR, R. J. & CLIFF, R. A. 1991. A precise U-Pb zircon age for the Inishrahull syenitic gneiss,

County Donegal, Ireland. *Journal of the Geological Society, London,* **148**, 639–642.

DALZIEL, I. W. D. 1992. Antarctica: a tale of two super-continents? *Annual Reviews of Earth and Planetary Science,* **20**, 501–526.

EHLERS, C., LINDROOS, A. & SELONEN, O. 1993. The late Svecofennian granite-migmatite zone of southern Finland – a belt of transpressive deformation and granite emplacement. *Precambrian Research,* **64**, 295–309.

ELMING, S.-A., Pesonen, L. J., LEINO, M. A. H., KHRAMOV, A. N., MIKHAILOVA, N. P., KRASNOVA, A. F., MERTANEN, S., BYLUND, G. & TERHO, M. 1993. The drift of the Fennoscandian and Ukrainian shields during the Precambrian: a palaeomagnetic analysis. *Teconophysics,* **223**, 177–198.

ERMANOVICS, I. F. & VAN KRANENDONK, M. J. 1990. The Torngat orogen in the North River – Nutak Transect area of Nain and Churchill provinces. *Geoscience Canada,* **17**, 297–301.

GAAL, G. & GORBATSCHEV, R. 1987. An outline of the Precambrian evolution of the Baltic shield. *Precambrian Research,* **35**, 15–52.

——, BERTHELSEN, A., GORBATSCHEV, R., KESOLA, R., LEHTONEN, M. I., MARKER, M. & RAASE, P. 1989. Structure and composition of the Precambrian crust along the POLAR profile in the northern Baltic Shield. *Teconophysics,* **162**, 1–25.

GORBATSCHEV, R. & BOGDANOVA, S. 1993. Frontiers in the Baltic shield. *Precambrian Research,* **64**, 3–21.

GOWER, C. F. 1990. Mid-Proterozoic evolution of the eastern Grenville Province, Canada. *Geologiska Föreningens, Stockholm Förhandlinger,* **112**, 127–139.

GROCOTT, J. & DAVIES, S. 1994. Deformation at the southern boundary of the late-Archaean Åta tonalite and the extent of the Burwell terrane in West Greenland. *Terra Abstracts,* **6**, 2, 7.

—— & PULVERTAFT, C. T. R. 1990. The Early Proterozoic Rinkian belt of central West Greenland. *In*: LEWRY, J. F. & STAUFFER, M. R. (eds) *The Early Proterozoic Trans-Hudson orogen.* Geological Association of Canada, Special Papers, **37**, 443–462.

HANSEN, B. T. & KALSBEEK, F. 1989. Precise age for the Ammassalik intrusive complex. *In*: KALSBEEK, F. (ed.) *Geology of the Ammassalik region, South-East Greenland.* Grønlands Geologiske Undersogelse Rapports, **146**, 46–47.

HEAMAN, L. M. & TARNEY, J. 1989. U-Pb baddelyite ages for the Scourie dyke swarm, Scotland: evidence for two distinct intrusion events. *Nature,* **340**, 705–708.

HICKMAN, M. H. & GLASSLEY, W. E. 1984. The role of metamorphic fluid transport in the Rb-Sr isotopic resetting of shear zones: evidence from Nordre Strømfjord, West Greenland. *Contributions to Mineralogy and Petrology,* **87**, 265–261.

HOFFMAN, P. F. 1989. Precambrian geology and tectonic history of North America. *In*: BALLY, A. W. & PALMER, A. R. (eds) *The Geology of North America – an overview.* Geological Society of America VA 447–512.

—— 1990. Dynamics of the tectonic assembly of north-east Laurentia in geon 18 (1.9–1.8 Ga). *Geoscience Canada,* **17**, 222–226.

—— 1991. Did the breakout of Laurentia turn Gondwanaland inside-out? *Science,* **252**, 1409–1412.

HUMPHRIES, F. J. & Cliff, R. A. 1982. Sm-Nd dating and cooling history of Scourian granulites, Sutherland. *Nature,* **295**, 515–517.

JACOBSEN, S. R. & HEIER, K. S. 1978. Rb-Sr systematics in metamorphic rocks, Kongsberg sector, South Norway. *Lithos,* **11**, 257–276.

JOHNSON, Y., PARK, R. G. & WINCHESTER, J. A. 1987. Geochemistry, petrogenesis and tectonic signifi-cance of the Early Proterozoic Loch Maree amphibolites. *In*: PHAROAH, T. C., BECKINSALE, R. D. & RICKARD, D. T. (eds) *Geochemistry and Mineralisation of Proterozoic Volcanic Suites.* Geological Society, London, Special Publications, **33**, 117–128.

KALSBEEK, F. 1981. The northward extent of Archaean basement in Greenland – a review of Rb-Sr whole-rock ages. *Precambrian Research,* **14**, 203–219.

—— 1989. *Geology of the Ammassalik region, South-East Greenland.* Grønlands Geologiske Undersøgelse Rapports, **146**

—— & TAYLOR, P. N. 1985. Isotopic and chemical variation in granites across a Proterozoic continental margin; the Ketilidian mobile belt of South Greenland. *Earth and Planetary Science Letters,* **73**, 65–80.

——, AUSTRHEIM, H., BRIDGWATER, D., HANSEN, B. T., PEDERSEN, S. & TAYLOR, P. N. 1993. Geochronology of the Ammassalik area, South-East Greenland, and comparisons with the Lewisian of Scotland and the Nagssugtoqidian of West Greenland. *Precambrian Research,* **62**, 239–270.

——, BRIDGWATER, D. & ZECK, H. 1987. A 1950 + 60 Ma Rb-Sr isochron age from two Kangâmiut dykes and the timing of the Nagssugtoqidian (Hudsonian) orogeny in West Greenland. *Canadian Journal of Earth Sciences,* **15**, 1122–1128.

——, TAYLOR, P. N. & HENRIKSEN, N. 1984. Age of rocks, structures and metamorphism in the Nagssugtoqidian mobile belt, West Greenland – field and Pb-isotope evidence. *Canadian Journal of Earth Sciences,* **21**, 1126–1131.

KERR, A., RYAN, B., GOWER, C., WARDLE, R. & HALL, J. 1994. The Makkovik province: geological overview, unanswered questions and unexplained reflectors. *In*: WARDLE, R. J. & HALL, J. (compilers) *Eastern Canadian Shield Onshore-Offshore Transect (ECSOOT) Meeting (December 10–11, 1993).* Université du Quebec a Montreal, Report No. 34, 35–52.

KORSTGÅRD, J., RYAN, B. & WARDLE, R. J. 1987. The boundary between Proterozoic and Archaean crustal blocks in central West Greenland and northern Labrador. *In*: PARK, R. G. & TARNEY, J. (eds) *Evolution of the Lewisian and Comparable Precambrian High-grade Terrains.* Geological Society, London, Special Publications, **27**, 247–259.

LYON, T. P. B., PIDGEON, R. T., BOWES, D. R. & HOPGOOD, A. R. 1973. Geochronological investigation of the quartzo-feldspathic rocks of the Lewisian of Rona, Inner Hebrides. *Journal of the Geological Society, London,* **129**, 389–402.

MACHADO, N. 1990. Timing of major tectonic events in the Ungava segment of the Trans-Hudson orogen. *In: Recent advances in the geology of the eastern Churchill Province (New Quebec and Torngat orogens).* Abstracts, Wakefield Conference, Quebec, 11.

MARCANTONIO, F., DICKIN, A. P., McNUTT, R. H. & HEAMAN, L. M. 1988. A 1,800-million-year-old Proterozoic gneiss terrane in Islay with implications for the crustal structure and evolution of Britain. *Nature,* **335**, 62–64.

MARKER, M. 1990. Tectonic interpretation and new tectonic modelling along the POLAR profile, northern baltic shield. *In:* FREEMAN, R. & Mueller, St. (eds) *Data compilations and synoptic interpretations, Proceedings of the 6th Workshop, European Geotraverse (EGT) Project.* European Science Foundation, Strasbourg, 67–76.

MOORBATH, S. & PARK, R. G. 1972. The Lewisian chronology of the southern region of the Scottish Mainland. *Scottish Journal of Geology,* **8**, 51–74.

MOORES, E. M. 1991. Southwest U.S.–East Antarctic (SWEAT) connection: a hypothesis. *Geology,* **19**, 425–428.

MORTON, A. C. & TAYLOR, P. N. 1991. Geochemical and isotopic constraints on the nature and age of basement rocks from Rockall Bank, NE Atlantic. *Journal of the Geological Society, London,* **148**, 631–634.

MUIR, R. I., FITCHES, W. R. & MALTMAN, A. J. 1989. An Early Proterozoic link between Greenland and Scandinavia in the Inner Hebrides of Scotland. *Terra Abstracts,* **1**, 5.

——, ——, —— & BENTLEY, M. R. 1994. Southern Inner Hebrides-Malin Sea region. *In:* GIBBONS, W. & HARRIS, A. I. (eds) *A revised correlation of Precambrian rocks in the British Isles.* Geological Society, London, Special Reports, **22**, 54–58.

MYERS, J. S. 1987. The East Greenland Nagssugtoqidian mobile belt compared with the Lewisian complex. *In:* PARK, R. G. & Tarney, J. (eds) *Evolution of the Lewisian and comparable Precambrian high grade terrains.* Geological Society, London, Special Publications, **27**, 235–246.

OFTEN, M. 1985. The Early Proterozoic Karasjok greenstone belt, Norway; a preliminary description of lithology, stratigraphy and mineralisation. *Bulletin Norges Geologiske Undersokelse,* **403**, 75–88.

O'NIONS, R. K., HAMILTON, P. J. & HOOKER, P. J. 1983. A Nd isotope investigation of sediments related to crustal development in the British Isles. *Earth and Planetary Science Letters,* **63**, 229–240.

PARK, A. F. 1985. Accretion tectonism in the Proterozoic Svecokarelides of the Baltic shield. *Geology,* **13**, 725–729.

PARK, R. G. 1991. The Lewisian complex. *In:* CRAIG, G. Y. (ed.) *Geology of Scotland.* Geological Society, London, 25–64.

—— 1994. Early Proterozoic tectonic overview of the northern British Isles and neighbouring terrains in Laurentia and Baltica. *Precambrian Research,* **68**, 65–79.

—— & TARNEY, J. (eds) 1987. *Evolution of the Lewisian and comparable Precambrian high-grade terrains.*

Geological Society, London, Special Publications, **27**.

——, ÅHÄLL, K.-I., CRANE, A. & DALY, J. S. 1987. *The structure and kinematic evolution of the Lysekil–Marstrand area, Östfold-Marstrand belt, SW Sweden.* Sveriges Geologiska Undersökelse, **C816**.

——, CLIFF, R. A., FETTES, D. G. & STEWART, A. D. 1994. Lewisian and Torridonian. *In:* GIBBONS, W. & HARRIS, A. L. (eds) *A revised correlation of Precambrian rocks in the British Isles.* Geological Society, London, Special Reports, **22**, 6–22.

PATCHETT, P. J. & BRIDGWATER, D. 1984. Origin of continental crust of 1.7–1.9 Ga age defined by Nd isotopes in the Ketilidian terrain of South Greenland. *Contributions to Mineralogy and Petrology,* **87**, 311–318.

—— & BYLUND, G. 1977. Age of Grenville belt magnetisation: Rb-Sr and palaeomagnetic evidence from Swedish dolerites. *Earth and Planetary Science Letters,* **35**, 92–104.

——, —— & UPTON, B. G. J. 1978. Palaeomagnetism and the Grenville orogeny: new Rb-Sr ages from dolerites in Canada and Greenland. *Earth and Planetary Science Letters,* **40**, 349–364.

——, GORBATSCHEV, R. & Todt, W. 1987. Origin of continental crust of 1.9–1.7 Ga age: Nd isotopes in the Svecofennian orogenic terrains of Sweden. *Precambrian Research,* **35**, 145–160.

PETROV, V. P. 1988. On the problem of supracrustal rocks at the turn of the Archaean-proterozoic. *Geological Survey of Finland, Special Paper,* **4**, 123–129.

PIPER, J. D. A. 1976. Palaeomagnetic evidence for a Proterozoic supercontinent. *Philosophical Transactions of the Royal Society, London,* **A280**, 469–490.

—— 1982. The Precambrian palaeomagnetic record: the case for the Proterozoic supercontinent. *Earth and Planetary Science Letters,* **59**, 61–89.

—— 1992. Palaeomagnetic properties of a Precambrian metamorphic terrane: the Lewisian complex of the Outer Hebrides, NW Scotland. *Tectonophysics,* **201**, 17–48.

RODDICK, C. & Max, M. D. 1983. A Laxfordian age from the Inishtrahull platform, Co. Donegal, Ireland. *Scottish Journal of Geology,* **19**, 97–102.

SCHÄRER, U. & GOWER, C. F. 1988. Crustal evolution in eastern Labrador; contraints from U-Pb systematics in accessory minerals. *Precambrian Research,* **38**, 405–421.

SCOTT, D. & MACHADO, N. 1994a. U–Pb geochronology of the northern Torngat Orogen: results from work in 1993. *In:* WARDLE, R. J. & HALL, J. (compilers) *Eastern Canadian Shield Onshore-Offshore Transect (ECSOOT) Meeting (December 10–11, 1993).* Université du Québec a Montréal, Reports, **34**, 141–155.

—— & —— 1994b. U–Pb geochronology of northern Torngat Orogen: implications for the evolution of NE Laurentia. *Terra Abstracts,* **6**, 2, 17.

SHACKLETON, R. M. & RIES, A. C. 1984. The relation between regionally consistent stretching lineations and plate motions. *Journal of Structural Geology,* **6**, 111–117.

SKJÖLD, T. 1984. Geokronologi inom Duobblonomradet, preliminara dateringsresultat. *Meddelanolen fran Stockholms Universitets Geologiska Institut*, **255**, 205.

STARMER, I. C. 1991. The Proterozoic evolution of the Bamble sector shear belt, southern Norway: correlations across southern Scandinavia and the Grenvillian controversy. *Precambrian Research*, **49**, 107–139.

STEARN, J. E. F. & PIPER, J. D. A. 1984. Palaeomagnetism of the Sveconorwegian mobile belt of the Fennoscandian shield. *Precambrian Research*, **23**, 201–246.

SUTTON, J. & WATSON, J. 1951. The pre-Torridonian metamorphic history of the Loch Torridon and Scourie areas in the North-west Highlands and its bearing on the chronological classification of the Lewisian. *Quarterly Journal of the Geological Society of London*, **106**, 241–307.

SØRENSEN, K. 1983. Growth and dynamics of the Nordre Strømfjord shear zone. *Journal of Geophysical Research*, **88**, 3419–3437.

TAYLOR, P. N., JONES, N. W. & MOORBATH, S. 1984. Isotopic assessment of relative contributions from clast and mantle sources to the magma genesis of Precambrian granitoid rocks. *Philosophical Transactions of the Royal Society, London*, **A310**, 605–625.

VALBRACHT, P. J., OEN, I. S. & BEUNK, F. F. 1994. Sm-Nd systematics of 1.9–1.8 Ga granites from western Bergslagen, Sweden: inferences on a 2.1–2.0 Ga crustal precursor. *Chemical Geology*, **112**, 21–37.

VAN BREEMEN, O., AFTALION, M. A. & PIDGEON, R. T. 1971. The age of the granite injection-complex of Harris, Outer Hebrides. *Scottish Journal of Geology*, **5**, 269–285. VAN KRANENDONK, M. J. & WARDLE, R. J. 1994. Promontory indentation, transpression and disharmonic folding in the formation of the Palaeoproterozoic Torngat orogen, Northeastern Canada. *Terra Abstracts*, **6**, 2, 20.

VERSCHURE, R. H. 1985. Geochronological framework for the late Proterozoic evolution of the Baltic shield in South Scandinavia. *In*: TOBI, A. C. & TOURET, J. L. R. (eds) *The deep Proterozoic crust in the North Atlantic provinces*. D. Riedel, Dordrecht, 381–410.

WARDLE, R. J., GOWER, C. F. & KERR, A. 1990a. The southeastern margin of Laurentia c. 1.7 Ga: the case of the missing crust. *Programme with Abstracts, Annual Meeting, Geological Association of Canada, Mineralogical Association of Canada*, Vancouver 1990, **15**, A137.

——, RYAN, B. & ERMANOVICS, I. F. 1990b. The eastern Churchill province, Torngat and New Quebec orogens: an overview. *Geoscience, Canada*, **17**, 217–222.

——, VAN KRANENDONK, M. J., MENGEL, F. & SCOTT, D. 1992. Geological mapping in the Torngat orogen northernmost Labrador preliminary results. *In*: WARDLE, R. J. & HALL, J. (eds) *Eastern Canadian Shield onshore-offshore transect (ECSOOT)*, Lithoprobe Report, **27**, Memorial University, Newfoundland, 112–122.

WATTERSON, J. 1978. Proterozoic intraplate deformation in the light of South-east Asian neotectonics. *Nature*, **273**, 636–640.

WELIN, E., GORBATSCHEV, R. & KAHR, A.-M. 1982. *Zircon dating of polymetamorphic rocks in southwestern Sweden*. Sveriges Geologiska Undersökelse, **C797**.

WIKSTROM, A. 1991. Structural features of some younger granitoids in central Sweden and implications for the tectonic subdivision of granitoids. *Precambrian Research*, **51**, 151–159.

WILSON, M. R., HAMILTON, P. J., FALLICK, A. E., AFTALION, M. & MICHARD, A. 1985. Granites and early crustal evolution in Sweden: evidence from Sm-Nd, U-Pb, and O isotope systematics. *Earth and Planetary Science Letters*, **72**, 376–388.

WRIGHT, A. E., TARNEY, J., PALMER, K. F., MOORLOCK, B. S. P. & SKINNER, A. C. 1973. The geology of the Angmassalik area, East Greenland and possible relationships with the Lewisian of Scotland. *In*: PARK, R. G. & TARNEY, J. (eds) *The early Precambrian of Scotland and related rocks of Greenland*, University of Keele, 157–177.

WYNN, T. J. 1995. Deformation in the Mid to Lower continental crust: analogues from Proterozoic shear zones in NW Scotland. *This volume*.

Deformation in the mid to lower continental crust: analogues from Proterozoic shear zones in NW Scotland

TIMOTHY JAMES WYNN

Department. of Geology, Imperial College, Prince Consort Road, London SW7 2BP, UK
Present address: GeoScience Ltd, Falmouth Business Park, Bickland Water Road, Falmouth, Cornwall TR11 4SZ, UK

Abstract: A suite of 1.8–1.7 Ga (Laxfordian), E–W striking, amphibolite-facies shears in the Lewisian of NW Scotland have been mapped across a pre-existing rheological boundary in the Laxford–Scourie area. The rheological contrast is provided by the southern boundary of the NW striking Laxford shear zone (LSZ), which displays complex Proterozoic folding and shearing of relict Archaean granulites. The more competent Archaean gneisses lie to the SW of the LSZ which has been a focus of shearing, amphibolitization and metasomatism from 2.6 to 1.4 Ga. The shears of interest display a second-order, extensional shear band relationship to the LSZ and formed during a phase of sinistral transtensional shearing and granite intrusion at 1.8–1.7 Ga. The second-order shears show displacement variations consistent with their propagation from the strongly foliated, less competent gneisses in the LSZ into the Archaean granulites to the SW.

A map of the Laxford–Scourie area can be regarded as a section through the mid to lower crust as observed on some deep seismic lines such as the BIRPS DRUM line from the north of Scotland. In the Laxford area, the change in geometry of the second order shears from SW to NE, as they curve into the LSZ, is comparable to the curvature of extensional shears in the mid to lower continental crust. Displacement variations indicate that the shear zones may nucleate at the base of the upper/mid crust and propagate up and down dip. Alternatively the shears may nucleate in the mid/lower crust and propagate up dip, becoming steeper as they do so.

The Lewisian of NW Scotland provides an excellent opportunity to study a variety of rocks that have undergone several phases of deformation throughout the spectrum of crustal levels. Peach *et al.* (1907) established the principal features of the mainland Lewisian by excellent systematic mapping in the latter part of the last century. An important unit was recognized at Loch Laxford separating highly deformed gneisses to the north from the 'fundamental Complex' to the south.

This work was built upon by Sutton & Watson (1951) who recognized the importance of the Scourie dyke swarm as markers separating tectonic and metamorphic events throughout the Lewisian. The area between Loch Laxford and Badcall Bay was studied in some detail by Sutton & Watson (1951) and the principal metamorphic and structural changes recorded from S to N. These show a progressive increase in amphibolitization and shearing into the core of an intense zone of deformation occupied by syn-tectonic granites, later termed the Laxford Shear Zone (Beach *et al.* 1974). A variety of rock types exist in this area from ultramafic to trondhjemitic and from

orthogneiss to paragneiss and provide a good opportunity to study the effects of progressive deformation and metamorphism on rocks of contrasting rheology.

Figure 1 shows the extent of Lewisian outcrop in NW Scotland and the variety of gneissic types mapped. This paper is concerned with the evolution of the northern part of the relict Archaean Central Region (CR) where it becomes progressively sheared into the Laxford shear zone (LSZ) (Sutton & Watson 1951; Beach *et al.* 1974; Davies 1978; Coward & Park 1987; Park 1991). This area is shown in more detail in Fig. 2 where a suite of subvertical sinistral shears can be seen to merge with the main LSZ; the northern boundary of the LSZ is marked by the zone of Laxford granite sheets which dip steeply southwest. This suite of sinistral shears has been mapped in detail and their geometrical, kinematic and petrological changes recorded from E to W as they merge with the LSZ. In this paper these changes are compared with the possible kinematic and geometric evolution of crustal scale extensional faults formed in the continental crust during extensional thinning.

From COWARD, M. P. & RIES, A. C. (eds), 1995, *Early Precambrian Processes*, Geological Society Special Publication No. 95, pp. 225–241.

225

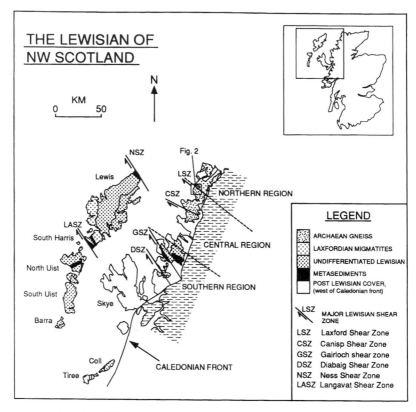

Fig. 1. Map of the Lewisian in NW Scotland. After Park & Tarney (1987).

Comparisons are provided by deep seismic lines and areas of exhumed crustal cross sections.

Geological history

Scourian

Table 1 shows the main chronological subdivisions of the Lewisian and the associated tectono-metamorphic events. One of the important events in the Lewisian was the late Scourian or Inverian tectonometamorphism which marked the change from predominantly pervasive, subhorizontal shearing (compression) to subvertical oblique-slip shearing and overthrusting on the LSZ and many of the other NW trending shears shown in Fig. 1 (Evans & Tarney 1964; Park 1964; Evans & Lambert 1974; Sheraton et al. 1973). On the southern margin of the LSZ this resulted in the formation of upright WNW-trending folds and stretching lineations, plunging 50–60° to the SE (Davies 1976, 1978). This predominantly dextral shearing was accompanied by extensive

amphibolitization within both the shears and the surrounding gneisses (Evans & Lambert 1974; Beach & Tarney 1978).

Scourie dyke swarm

Scourian tectonometamorphism was followed by the emplacement of the Scourie dyke swarm which produced the volumetrically important basic and ultrabasic dykes used so effectively by Sutton & Watson (1951) in unravelling the chronology of the Lewisian. It is now recognized that the dykes were intruded between 2.4 and 2.0 Ga, probably in several pulses (Tarney 1973; Park & Tarney 1987; Heaman & Tarney 1989). However this concept of polyphase intrusion does not alter their fundamental importance as chronological markers. The dykes are predominantly quartz dolerites in the Laxford area and are mainly subvertical with NW trends. Most dykes were metamorphosed shortly after intrusion so that hornblende completely replaces the igneous pyroxene although some original ophitic textures are visible in places, (Beach & Tarney 1978; Park & Tarney 1987).

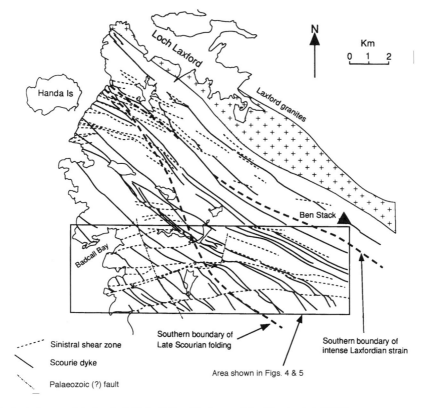

Fig. 2. Geological map of the Laxford–Badcall–Ben Stack area. After Beach *et al.* (1974).

Laxfordian

In the early Laxfordian at 1.9–1.8 Ga the Lewisian complex was subjected to the intense Laxfordian deformation and amphibolitization. Within the LSZ this resulted in continued dextral transpression along a movement vector plunging 30–40° SE. The late Scourian folds were tightened and overturned to the north and the Scourie dykes were also strongly sheared in places; (Beach *et al.* 1974). The Northern and Southern regions were intensely reworked at this time, largely obliterating the Scourian structures. Deformation in the Northern region resulted in strongly foliated gneisses dipping gently south, possibly as part of a shear zone 'flat' to the steeper LSZ which may have formed as a mid-crustal ramp (Beach 1974; Beach *et al.* 1974).

The sinistral shears of interest may have formed as second order antithetic shears to this early Laxfordian transpression (Davies 1978). However the sinistral shears at Tarbet appear to cut the late Scourian folds, so it is more likely they may have formed during a later phase of distinct sinistral transtension affecting the LSZ at 1.8–1.7 Ga

(Beach 1974, 1976; Coward & Park 1987; Coward 1990). It is possible that the Laxford granites were intruded during the transtensional shearing. The Laxford granites may have formed by partial melting of the enriched Northern region which was emplaced beneath the Central region by early Laxfordian overthrusting (Beach 1974).

Late Laxfordian to Grenvillian deformation was restricted to greenschist-facies folding and re-activation of earlier foliations and shears, mainly in the Northern and Southern regions.

P–T–t **history**

Figure 3 shows the *P–t* and *T–t* paths for the Central Region based on geochronological, palaeo-thermal and palaeobaric studies by Giletti *et al.* (1961), Evans (1965), Evans & Tarney (1964), Moorbath (1969), Lambert & Holland (1972), Beach (1973), Tarney (1973), Wood (1977), O'Hara (1977), Rollinson (1980*a*, 1980*b*, 1981), Pride & Muecke (1980), Savage & Sills (1980), Humphries & Cliff (1982), Newton & Perkins (1982),

Table 1. *Geological events in the Central Region of the Lewisian*

Archaean (Scourian)

2900 Ma   Scourian tonalites and trondhjemites produced from partial melts of subducted oceanic crust. Subhorizontal gneissic layering produced.

2700 Ma   Badcallian grasnulites facies metamorphism. Isoclinal, recumbent folding of layering, fold axial surfaces dip gently NW.

2600 Ma   Late Scourian (Inverian) tectonometamorphism, widespread static amphibolitisation. Formation of steep NW striking lineaments at Laxford, Canisp and Gruinard Bay.

Proterozoic

2400–2000 Ma   Intrusion of the Scourie dyke swarm, 4 suites from ultrabasic to basic derived from 2 separate mantle sources. The dykes mark the boundary between the Scourian and the Laxfordian (Sutton & Watson 1951).

(Laxfordian)

1900–1800 Ma   Laxfordian amphibolite-facies deformation. Central region thrust over the northern region during dextral transpression on the Laxford Shear Zone.

1800–1700 Ma   Peak of Laxfordian metamorphism, intrusion of the Laxford granites. Sinistral transtension on the main Laxford Shear Zone with the minor sinistral shears formed on the SW margin.

1700–1400 Ma   Mid–Late Laxfordian retrogression to greenschist facies, associated with kilometre-scale folding in the Southern region. Development of some pseudotachylite belts throughout the Lewisian.

1150 Ma   Minor chlorite and muscovite growth and the formation of later pseudotachylite belts. Both possibly related to the Grenvillian–Sveconorwegian orogeny.

1100–1040 Ma   Exposure at the Torridonian land surface, deposition of the Stoer and Torridon groups.

After Park & Tarney (1987).

Barnicoat (1983), Sills (1983), Sills & Rollinson (1987), Cartwright & Barnicoat (1986, 1987) and Heaman & Tarney (1989).

The main features of the curves shown in Fig. 3 are relatively rapid cooling and uplift from peak Badcallian conditions at 2.7–2.6 Ga, fairly stable

P and *T* conditions during dyke intrusion, a slight thermal peak during the early to mid-Laxfordian orogeny (and collapse) and slow cooling and uplift from 1.6–1.0 Ga when the Lewisian was exposed at the Torridonian land surface. This extremely slow exhumation history was a major factor in allowing

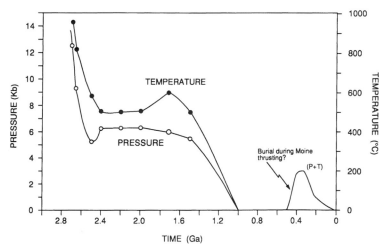

Fig. 3. Pressure–time and temperature–time plots for the Central Region (see text for references).

repeated phases of movement on the LSZ which resulted in the complex, medium to high grade deformation visible in the Lewisian today.

Sinistral shear zones

Figure 2 shows the positions of the sinistral shears with respect to the main LSZ and the boundaries of late Scourian and early Laxfordian strain. These boundaries indicate a progressive northeastwards migration of deformation on the LSZ, corresponding to an increase in strain localization. All the fabrics appear to be amphibolite facies, so the decrease in the width of the zone of penetrative shearing is unlikely to be due to cooling and is probably a result of the availability of fluids for metasomatism and subsequent weakening of the gneisses (Beach 1976; Beach & Tarney 1978).

On the basis of the similarity in structural, metamorphic and kinematic style, the Laxfordian sinistral shears in the Tarbet area (Beach 1974;

Coward 1990) are regarded as coeval with the E–W sinistral shears studied in this project. The Tarbet shears occur within, and partly define, the main Laxford shear zone and the foliation and lineation orientations have been used as reference orientations for the LSZ shear plane and the LSZ movement vector, respectively during structural analysis of the E–W sinistral shears. The geometrical and kinematic relationships between the two sets of sinistral shears indicates that the E–W set has a second order, extensional, synthetic relationship to the Tarbet shears (cf. Beach 1974; Coward & Park 1987).

Figure 4 shows the E–W sinistral shears in detail. The area mapped has been subdivided into four zones, where the zone boundaries join up inflection points on the shears. The overall change is a clockwise rotation in strike from W to E. The zone boundaries are approximately parallel to the main LSZ. Thin section analysis of the sinistral shear fabrics has shown that they were formed in amphibolite-facies conditions in all the zones

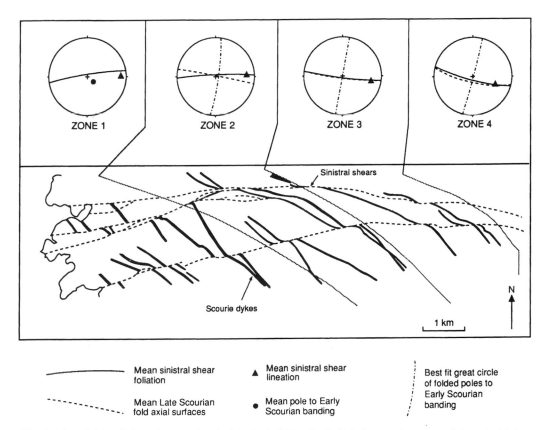

Fig. 4. Map of sinistral shears in zones 1 to 4 of the Badcall–Ben Stack–Kylesku area. Stereonets of the main fabrics in each zone are also shown.

although later, localized greenschist facies retrogression has occurred. This indicates that the kinematic and geometrical changes of the sinistral shears from W to E is not a direct function of a lateral change in metamorphic grade.

Characteristics of the gneisses surrounding the sinistral shears

Structural analysis of the older fold fabrics in each zone from the gneisses surrounding the shear zones shows progressive changes from W to E, as shown in Fig. 4. In zone 1 the Scourie gneiss banding dips gently NW, in the centre of zone 2 tight, WNW-striking folds occur which become progressively tighter and overturned to the NE in zones 3 and 4. The changes in structures from SW to NE in the gneisses surrounding the shears (as seen in Figs 2 and 4) can be summarized as follows:

(i)   increase in intensity of steep, penetrative (axial planar) foliations;
(ii) clockwise rotation of axial planar foliations and tightening of folds;
(iii) increase in retrogression of the Scourie gneisses, amphibolitization and biotite growth in the foliation zones, probably as a result of increased metasomatic activity;
(iv) Increase in effective anisotropy of the gneisses, i.e. change from sub-horizontal layering perpendicular to sinistral shears, to sub-vertical foliations, sub-parallel to shears.

These features indicate that the development of the sinistral shears in zones 1 to 4 was closely linked to changes in the structure of the surrounding gneisses. The general trend is one of decreasing obliquity between the sinistral shear fabrics and the late Scourian axial surfaces (shown in zones 2 and 3). In zone 4, all fabrics including the Scourie dykes are sub-parallel.

It is possible that these changes are equivalent to an increase in ductility of the gneisses from SW to NE, although this will depend to an extent on the orientation of the stress and strain fields during any subsequent deformation. Beach (1973, 1976) indicated that the growth of the sinistral shears was controlled by the availability of fluids to propagate fractures and metasomatize the gneisses, allowing the growth of biotite. These factors were believed to have contributed to the strain localization of these shears. However the shears only appear to be very localized in zones 1 and 2 and in zones 3 and 4 the shearing appears to be more diffuse.

This characteristic is highlighted by Fig. 5 which shows the positions of correlatable Scourie dykes across the two largest shears and the approximate positions of diffuse zones of deformation either side of the discrete parts of the shears in zones 3 and 4. These diffuse zones can be recognized by patchy fabric development within the Scourie dykes and their gradual anti-clockwise deflection as the discrete parts of the sinistral shears are approached. It is suggested that these diffuse zones of shear have formed by the reactivation of late Scourian fold axial surfaces during sinistral shearing in zones 2 and 3 (see Fig. 4). Penetrative shearing probably occurred throughout zone 4, coeval with the more discrete shearing in zones 1 to 3.

Displacement plots

Figure 6 shows the geometrical construction used to calculate the true displacements on the two

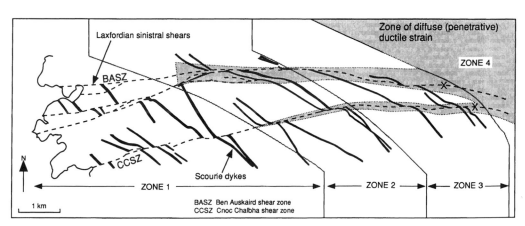

Fig. 5. Map of the sinistral shears in zones 1 to 4. The zones of diffuse deformation on the margins of the shears are also shown.

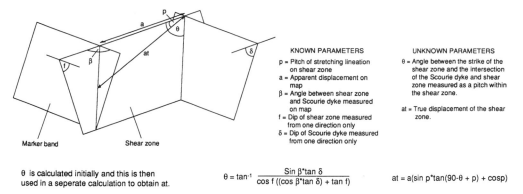

KNOWN PARAMETERS

p = Pitch of stretching lineation
    on shear zone
a = Apparent displacement on
    map
β = Angle between shear zone
    and Scourie dyke measured
    on map
f = Dip of shear zone measured
    from one direction only
δ = Dip of Scourie dyke measured
    from one direction only

UNKNOWN PARAMETERS

θ = Angle between the strike of the
    shear zone and the intersection
    of the Scourie dyke and shear
    zone measured as a pitch within
    the shear zone.

at = True displacement of the shear
     zone.

θ is calculated initially and this is then
used in a seperate calculation to obtain at.

$$\theta = \tan^{-1} \frac{\sin \beta * \tan \delta}{\cos f \left((\cos \beta * \tan \delta) + \tan f \right)}$$

$$at = a(\sin p * \tan(90 - \theta + p) + \cos p)$$

Fig. 6. Geometrical construction showing how the true displacements on the displacement plots in Fig. 7 were calculated.

sinistral shears shown in Fig. 5. This construction takes into account the dip of the shear zone, the dip of the markers (Scourie dykes) and the plunge of the movement vector. Because of the large strains accumulated on the sinistral shears, the movement vector was taken to be represented by the stretching lineations in the shears. The true displacement values obtained were then used in the displacement plots shown in Fig. 7, these displacement plots are for the Ben Auskaird and Cnoc Chalbha shear zones (BASZ and CCSZ, respectively, see Fig. 5).

Ben Auskaird shear zone

For the BASZ it can be seen that the displacements are relatively high in zone 1 (1400 m) and the western half of zone 2, between 4000 and 5000 m from the coast (1900 m displacement); this rapidly decreases in 1000 m of strike length at the boundary of zones 2 and 3 to 500 m. The rapid decrease in displacement can only be realistically explained by the partitioning of strain into the surrounding gneisses in the eastern part of zone 2 and all of zone 3, where the favourably oriented late Scourian axial surfaces could take up the deformation. This idea is further supported by the two dykes on Ben Auskaird in the centre of zone 2. Here the displacement can be separated into a total component, indicated by the deflection of the dykes across the Ben Auskaird shear zone and a discrete component, where the dykes are actually cut by a discrete shear zone. The total component is similar to the displacement values for zone 1 and the discrete component similar to the values for zone 3 (see linked points in Fig. 7a).

The large total displacement on the BASZ has shifted the southern boundary of late Scourian folding approximately 2 km to the east, so that the CCSZ cuts it at the boundary of zones 2 and 3.

Cnoc Chalbha shear zone

The displacements for the CCSZ can also be sub-divided into diffuse and discrete shearing. The plot shown in Fig. 7b contains the total displacement estimates and the discrete component in the eastern part of zone 2 and in zone 3. The total displacement systematically increases from zone 1 to zone 3, although the best fit curve changes in gradient in zone 3. The discrete component decreases in value eastwards from equal to the total displacement at the southern boundary of late Scourian folding, to virtually zero at the eastern margin of zone 3. Again this seems to indicate a partitioning of strain into the late Scourian fold axial surfaces. However, the strain partitioning is less pronounced in the CCSZ compared with the BASZ where the total displacements are much higher.

The total displacement decreases from E to W which can possibly be attributed to propagation of the CCSZ from zone 3 into zones 2 and 1. It is possible that this may be due to the relative weakness of the Scourie gneisses in zone 3 compared to the Scourie gneisses in zones 1 and 2 which contain relict granulite-facies minerals with granoblastic textures and where the compositional banding is sub-horizontal and unavailable for reactivation. This would therefore mean that the shears had to grow as new fractures in zones 1 and 2, possibly driven by fluids emanating from the Laxford granites, northeast of zone 4 (Beach 1976).

Orientations of the sinistral shears with respect to the LSZ

Figure 8 shows an equal area stereonet showing the change in orientation of the sinistral shear segments from zone 1 to zone 4 with respect to the LSZ (Tarbet) mean foliation and lineation. It is clear

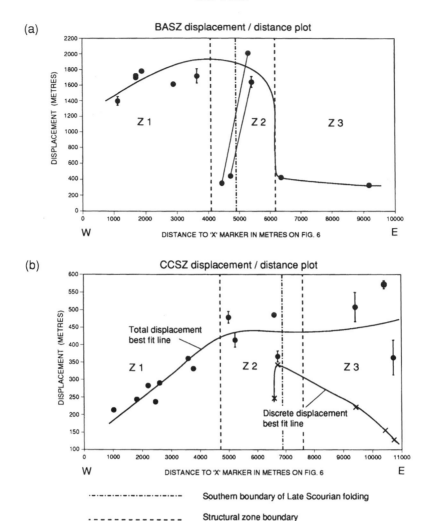

Fig. 7. Displacement plots for the Ben Auskaird shear zone (**a**) and the Cnoc Chalbha shear zone (**b**).

from Fig. 8 that the shears become closer in orientation to the LSZ as the LSZ is approached from SW to NE. Figure 9 is a 3D block diagram of the sinistral shears and the LSZ; Fig. 9 also shows how the true thicknesses of the structural zones were calculated, these thicknesses are used on the vertical axes of the plots in Fig. 10.

Figure 10a shows the angles of the mean sinistral shear foliations and the late Scourian axial surfaces with respect to the LSZ, as measured in the profile plane of the LSZ. The BASZ segments in zones 3 to 4 and the CCSZ segment in zone 3, have similar orientations to the late Scourian axial surfaces in these zones. However the BASZ segment in zone 2 and the CCSZ segment in zone 3 are oriented at slightly higher angles than the axial surfaces. The

larger obliquity between these shear segments and the late Scourian axial surfaces may be a function of the lower strains (displacements) for the western parts of the shears. Therefore, the late Scourian axial surfaces have been rotated by a smaller amount and still retain an appreciable obliquity to the discrete shear segments.

In addition to the changes in angle of the shear zone segments, the movement vectors (lineations) in the sinistral shears in zones 1 to 4 also change orientation. The best fit great circle for these lineations, shown in Fig. 8, does not have an orientation that is consistent with the sinistral shears forming as plane strain, simple shear splays. It is more likely that the sinistral shears contain a finite component of rotational movement, defined by an

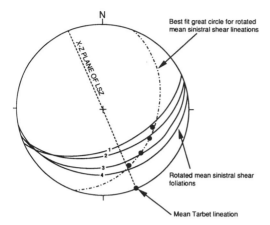

Fig. 8. Stereonet of rotated mean foliations and lineations from zones 1 to 4 of the Badcall–Ben Stack–Kylesku area.

best fit great circle to the sinistral shear lineations and the mean foliations from each zone (see Fig. 8). The angles between the shear segments and the LSZ are more consistent than those obtained in the profile plane (X–Z section) of the LSZ (see Fig. 10a). In addition, there appears to be smaller obliquity between the sinistral shear segments and the late Scourian axial surfaces in zone 2 for the BASZ and zone 3 for the CCSZ (see Fig. 10b).

If the shear zone segments propagated in a similar direction as the movement vectors of the shears, then this feature indicates that the shears propagated in a direction that had a smaller change in curvature from zone 4 to zone 1 than would be the case if they had propagated in a direction contained within the profile plane of the LSZ. The reason for this is not clear, although it is probably a function of local stress field variations and the strong control of pre-existing anisotropies in the Scourian gneisses.

Model for the formation of the sinistral shears

Figure 11 shows two models for the evolution of the sinistral shears in the Scourie–Laxford area. The main feature of the models is that the Central Region was being displaced down to the SE during sinistral transtension on the LSZ, whilst the Central

arc contained within a plane at an angle to the profile plane to the LSZ.

Figure 10b shows the angles of the CCSZ and BASZ with respect to the LSZ as measured within the best fit great circle to the mean lineations from zones 1 to 4. The angles were obtained by measuring the plunges of the intersections of the

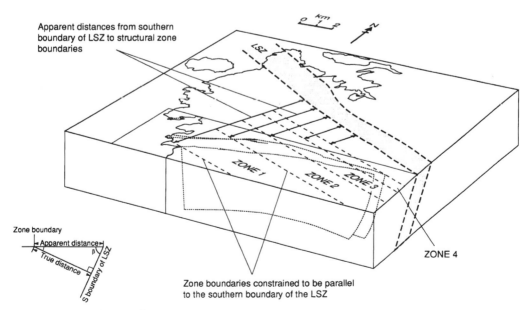

True distance = sin β * Apparent distance

Fig. 9. Block diagram of structural zones 1 to 4 to show how the true thicknesses of the zones were calculated. These thicknesses are used on the vertical axes of the plots in Fig. 10.

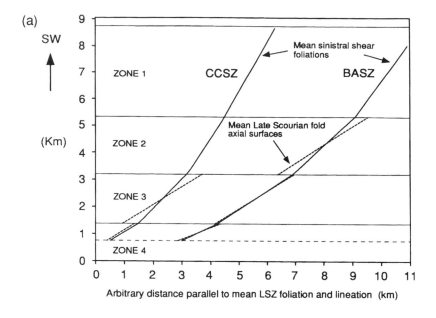

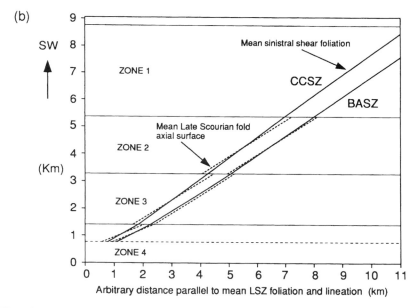

Fig. 10. Plots of angles between mean fabrics in zones 1 to 4 and the mean LSZ fabrics. (**a**) Plot of angles measured within the profile plane of the LSZ. (**b**) Plot of angles measured within the best fit great circle to the rotated mean lineations from zones 1 to 4.

Region was also extended parallel to the movement vector on the LSZ.

Figure 11a shows a model of the sinistral shears initiated as accommodation structures by a combination of simple shear and block rotation on the northern margin of the Central Region within the more penetratively deformed gneisses (Fig. 11a(i)). Continued deformation propagated the shears into the more competent Scourie gneisses and also allowed at least one discrete shear strand (CCSZ) to propagate from the southern boundary of Scourian folding eastwards back along the more diffuse zone of shearing (Fig. 11a(ii)).

Figure 11b shows an alternative model where the

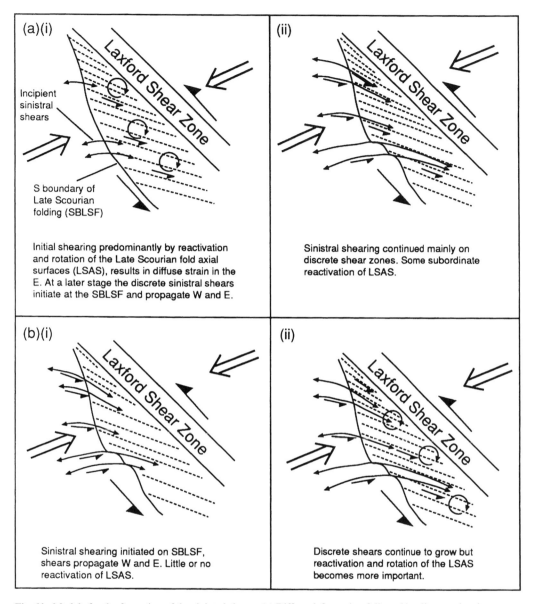

Fig. 11. Models for the formation of the sinistral shears. (**a**) Diffuse deformation followed by discrete shearing. (**b**) Discrete shearing followed by diffuse deformation.

sinistral shears nucleate at the boundary between the late Scourian folds and the early Scourian granulites and propagate WSW and ENE coevally as discrete shear zones (Fig. 11b(i)). Either during or after this discrete shearing, the gneisses affected by the late Scourian folding deformed by more diffuse shearing and rotation which produced the higher total displacements in zone 3 for the CCSZ (Fig. 11b(ii)).

The overall deformation of discrete shearing and block rotation is approximately equivalent to pure shear in a horizontal plane in this part of the Central Region (Jackson 1987). This deformation can also be equated to the 'stretching faults' described by Means (1989) but instead of the wall rocks to the shears deforming by penetrative pure shear, the strain is partitioned predominantly into simple shear and rotation. This is further complicated by the partitioning in time and space from bulk pure shear in zone 4 to very localized simple shearing with rotation in zone 1. Because there is also a decrease in displacement from zone 4 to zone 1, there is an equivalent decrease in the wall rock extension parallel to the LSZ from zone 4 to zone 1.

It is proposed that this zone of simple shear and block rotation and penetrative pure shear on the northern margin of the Central Region is equivalent to the change from simple shear and block rotation to penetrative pure shear in the middle to lower continental crust in areas of continental extension.

Mid to lower continental crust

Before a direct comparison can be made between NW Scotland and areas of continental extension it is necessary to describe the various features of the mid to lower crust via the study of exhumed orogens and extensional terranes (Fountain & Salisbury 1981; Percival et al. 1992) or deep seismic data from areas of continental extension. A large amount of deep seismic data from the continental crust has been collected in the last ten years, mainly from the Northern hemisphere and marine, rifted passive margin situations (e.g. BIRPS) but also onshore orogenic belts (e.g. DEKORP, ECORS and COCORP) (McGeary et al. 1985; Meissner et al. 1990; Pfiffner et al. 1990; Hauser et al. 1987).

Many of the surveys from Northwest Europe and North America have traversed extensional terranes (Mooney & Meissner 1992) which may be a result of the extensional collapse of orogenic belts or true continental rifting (McGeary et al. 1987; Meissner & Tanner 1993; Rey 1993). A number of features are present on many of the lines which are believed to be a direct result of the extensional deformation (Serpa & de Voogt 1987; Warner 1990; Reston 1988, 1993; Hyndmann et al. 1991; Mooney & Meissner 1992). These are listed below:

(i)   there is generally a zone of seismically transparent upper to mid crust, also with upper crustal half-grabens with sediment infill;
(ii)   there is a strongly reflective mid to lower crust where reflectors vary in length;
(iii) there is a reflection Moho at the base of the lower crustal reflectors which is usually sub-horizontal and laterally continuous;
(iv)   there is a transparent upper mantle, with occasional dipping reflectors.

The most consistent feature of many areas is that of lower crustal reflectivity which appears to be a feature of approximately 50% of the continental crust imaged to date, (McGeary et al. 1987). The possible origins of these reflectors are uncertain. Warner (1990) outlined the three most likely probabilities:

(i)   basaltic intrusions;
(ii)   shear zones;
(iii) Free aqueous fluids.

In some areas the upper and lower boundaries of the zone or layer of mid to lower crustal reflectors do not correspond to changes of velocity measured by refraction studies, i.e. the midcrustal Conrad discontinuity and the Moho, respectively. These differences indicate that the reflectivity is not always restricted to layers with particular velocities caused solely by compositional variations and indicates that the discontinuous reflectivity results in part from shearing and/or variations in metamorphic grade (Holbrook et al. 1991).

From studies of exposed 'lower crustal sections' e.g. the Ivrea zone, North Italy (Rutter & Brodie 1990; Zingg et al. 1990; Handy & Zingg 1991), it is generally agreed that the percentages of mafic and ultramafic gneisses increase with depth as does the metamorphic grade (Rutter & Brodie 1990; Percival et al. 1992). These mafic gneisses are generally interbanded with more tonalitic, granodioritic and rarely metasedimentary gneisses which would provide a strong velocity contrast and hence possible reflection surfaces in the lower crust. If this subhorizontal layering has been enhanced by pervasive pure shearing from extension or subhorizontal thrusting, then the resultant mineral alignment anisotropies can also produce strong acoustic contrasts and hence reflectivity (Mainprice & Nicolas 1989; Reston 1993; Mooney & Meissner 1992). This lithological banding may be further enhanced during extension by the intrusion of mafic sills produced by adiabatic melting of the upper mantle lithosphere. In areas of relatively recent extension, such as the Basin and Range Province, Western USA, some of these sills may still be liquid magmas (Hauser et al. 1987; Parsons et al. 1992).

Models of crustal extension from deep seismic reflection data

A number of authors have suggested models to explain the various features seen on deep seismic lines from extended terranes, particularly from BIRPS data (Reston 1990, 1993 Stein & Blundell 1990; KIemperer & Hurich 1991), but also from COCORP onshore data from the Basin and Range province (Allmendinger et al. 1987; Hauser et al. 1987). Essentially these models fall in a spectrum between lithospheric pure shear (McKenzie 1978) and simple shear (Wernicke 1985; Coward 1986). It is generally accepted that lithospheric pure shear is the most applicable model to the Mesozoic basins of the North Sea (cf. KIemperer & Hurich 1991).

Models from Stein & Blundell (1990), Klemperer & Hurich (1991) and Reston (1993) are basically similar and indicate that the lower crust acts as a ductile zone deforming by pure and/or simple shear to accommodate discrete faulting and block rotation in the upper crust and upper mantle. However the deformation averaged over the lithosphere is that of pure shear. Figure 12 shows a composite cartoon model of this idea. There may be variable amounts and types of shearing in the lower crust, as listed below:

(i) large scale, broad low angle zones of simple shear linking offset loci of upper crustal and upper mantle faulting;
(ii) distributed pure shear from anastomosing simple shears bounding deformable pods;
(iii) distributed pure shear extension forming non-rotational subhorizontal foliations.

Comparison of NW Scotland to deep reflection data

Figure 13 shows a simplified version of the map in Fig. 2 reflected about a N–S axis and rotated 135° anticlockwise to orientate the zone of Laxford granites to a horizontal position. This reorientation allows a comparison between Fig. 13 and the DRUM deep seismic line shown in Fig. 14 which shows upper crustal half-grabens, a reflective lower crust and a 6 km thick, bright upper mantle feature termed the Flannan reflector (McGeary & Warner 1985). The Flannan reflector exhibits strong positive reflection coefficients, which indicate that the composition is eclogitic (M. Warner pers. comm. 1993). Given that it is has an appreciable thickness, it may represent a relict subduction zone (M. Warner pers. comm. 1993). However the Flannan structure may have subsequently acted as a thrust during the Caledonian orogeny and then as an extensional fault during formation of the upper crustal half-grabens in Devonian to Triassic times (Coward *et al.* 1987).

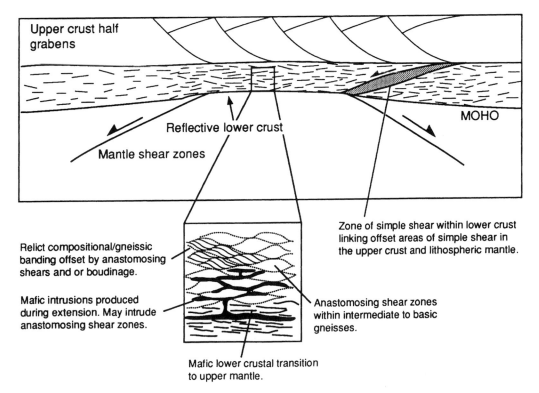

Fig. 12. Model of lower crustal shear fabrics based on BIRPS seismic reflection data. After Stein & Blundell (1990); Warner (1990); Klemperer & Hurich (1991); Reston (1993).

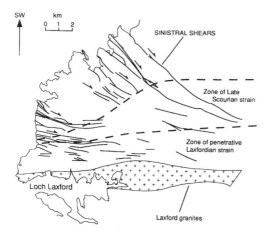

Fig. 13. Rotated and reflected version of the map in Fig. 2 to compare with Figs 12 and 14.

It is proposed that features of the sinistral shears as shown in Fig. 13 are kinematically and geo-metrically equivalent to the main features of the mid to lower crust in the DRUM line in Fig. 14 and the general model in Fig. 12. These features are discussed below.

(i)   lower crustal anastomosing shears represented by the zone of penetratively foliated Scourie gneisses, Scourie dykes and Laxford granites on the southern margin of the LSZ to zone 4;
(ii) listric upper crustal half-graben bounding faults represented by the discrete shear zones (BASZ and CCSZ) in zones 1 and 2;
(iii) 'a brittle–ductile' transition represented by the gradual change from discrete shearing in zones 1 and 2 to more penetrative, ductile strain in zones 3 and 4.

There has been considerable strain concentration into the sinistral shears in zones 1 and 2 of the Scourie–Laxford area and it is possible that this

is a function of the higher strength of the relict granulite-facies rocks causing localization of deformation, compared with the more diffuse strain in the ductile gneisses in zones 3 and 4. The higher strength relict granulite-facies rocks in zones 1 and 2 can be equated with the upper continental crust and the penetratively sheared amphibolite-facies rocks in zones 3 and 4 with the mid to lower continental crust. If the model in Fig. 11b is regarded as a cross section through the upper to lower crustal transition, then it corresponds to a situation where the shears nucleate at the brittle-plastic transition and propagate into the upper crust and mid/lower crust (cf. Jackson 1987; Jackson & White 1989; Kusznir & Park 1987). However the model in Fig. 11b indicates that later deformation is concentrated into the mid/lower crust which would result in an increase in extension with depth (Coward 1986).

An alternative model for the development of the shears is provided by Fig. 11a which can also be regarded as a cross section through the upper to lower crustal transition. In this case, the shears nucleate within the lower crust as relatively diffuse zones of shearing and propagate into the upper crust as more discrete zones. However when the brittle–plastic transition is reached the shears begin to propagate back into the lower crust as more discrete features. The later formation of discrete shears in the lower crust may be equivalent to the effective downward migration of the brittle-plastic transition during extension (England & Jackson 1987; LePichon & Chamot-Rooke 1991).

The model in Fig. 11a, as described above, implies that the extension nucleated in the lower crust is at odds with accepted models of litho-spheric failure. In such models, the faults or shears nucleate at, or just above, the brittle ductile transition which is believed to be the region of highest strength and stress concentration (Kusznir & Park 1987). Several factors may be invoked to explain this possibility.

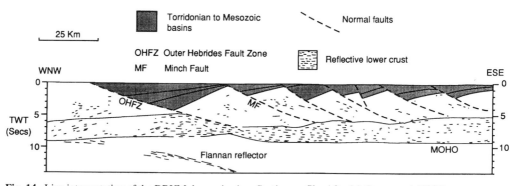

Fig. 14. Line interpretation of the DRUM deep seismic reflection profile. After McGeary *et al.* (1987).

(i) The brittle upper crust does not equate to Byerlees friction law with depth and may be weaker than previously thought. Sub-critical (sub acoustic) crack growth may allow the initiation and growth of faults whilst also supporting geologically significant stresses (Atkinson 1987). That is, stable frictional sliding controls the deformation rather than pressure sensitive cataclasis (acoustic failure).

(ii) The plastic regime of the mid to lower crust can support higher stresses than previously thought, therefore allowing stress and strain concentration and propagation of shears upwards.

(iii) Lithospheric extension is not equal with depth and the lower crust may extend locally by a larger amount compared with the upper crust especially during the initial stages of extension (Ziegler 1983; Coward 1986). If this lower crustal extension initiates before the upper crustal extension then the shear zones may nucleate within the lower crust and propagate upwards.

Conclusions

The suite of shears studied from NW Scotland formed during flattening of a relatively competent block against the LSZ to produced shears that have cut across and modified a pre-existing rheological transition. The shears may have propagated from a region of macroscopic ductile behaviour near the LSZ into the more competent, relict Scourie granulites to the southwest. These features are kinematically and geometrically equivalent to lithospheric scale extensional shears as imaged on some deep seismic reflection profiles. It is possible that large extensional faults may propagate from the plastic lower crust up into the more competent, brittle upper crust. At present, there is no evidence to suggest that the plastic regime of the lower crust can support higher stresses than the mid to upper crust. However, shearing may initiate within the lower crust if a combination of a weak upper crust and increased extension with depth can be invoked.

I would like to thank M. P. Coward for initiating and supervising this project and for providing stimulating discussions on the complexities of mid to lower crustal deformation as well as valuable comments on a draft of this paper. I would also like to thank J. Cosgrove, M. R. Warner and P. Garrard for useful discussions on the lower continental crust and structural geology in general. This work was undertaken during a NERC studentship at Imperial College.

References

ALLMENDINGER, R. W., HAUGE, T. A., HAUSER, E. C., POTTER, C. J. & OLIVER, J. 1987. Tectonic heredity and the layered lower crust in the Basin and Range province, Western United States. *In*: COWARD, M. P., DEWEY, J. F. & HANCOCK, P. L. (eds) *Continental Extensional Tectonics*. Geological Society, London, Special Publications, **28**, 223–246.

ATKINSON, B. K. 1987. *Fracture Mechanics of Rock*. Academic Press, London.

BARNICOAT, A. C. 1983. Metamorphism of the Scourian Complex, NW Scotland. *Journal of Metamorphic Geology*, **1**, 162–182.

BEACH, A. 1973. The Mineralogy of High Temperature Shear Zones at Scourie N.W. Scotland. *Journal of Petrology*, **14**, 231–248.

—— 1974. The Measurement and Significance of Displacements on Laxfordian Shear Zones, North-West Scotland. *Proceedings of the Geologists' Association*, **85**, 13–21.

—— 1976. The inter-relations of fluid transport, deformation, geochemistry and heat flow in early Proterozoic shear zones in the Lewisian complex. *Philosophic Transactions of the Royal Society of London*, **A280**, 569–604.

—— 1986. A deep seismic reflection profile across the northern North sea. *Nature*, **323**, 53–55.

——, & TARNEY, J. 1978. Major and trace element patterns established during retrogressive metamorphism of Granulite facies gneisses NW Scotland. *Precambrian Research*, **7**, 325–348.

——, COWARD, M. P. & GRAHAM, R. H. 1974. An interpretation of the structural evolution of the Laxford Front, north-west Scotland. *Scottish Journal of Geology*, **9**, 297–308.

CARTWRIGHT, I. & BARNICOAT, A. C. 1986. The generation of silica saturated melts and corundum bearing restites by crustal anatexis: Petrogenetic modelling based on an example from the Lewisian of NW Scotland. *Journal of Metamorphic Geology*, **4**, 77–79.

—— & —— 1987. Petrology of Scourian supracrustal rocks and orthogneisses from Stoer, NW Scotland: implications for the geological evolution of the Lewisian complex. *In*: PARK, R. G. & TARNEY, J. (eds) *Evolution of the Lewisian and Comparable High Grade Terrains*. Geological Society, London, Special Publications, **27**, 93–107.

COWARD, M. P. 1986. Heterogeneous stretching, simple shear and basin development. *Earth and Planetary Science Letters*, **80**, 325–336.

—— 1990. Shear zones at the Laxford front NW Scotland and their significance in the interpretation of lower crustal structure. *Journal of the Geological Society, London*, **147**, 279–286.

—— & PARK, J. 1987. The role of mid crustal shear zones in the early Proterozoic evolution of the Lewisian. *In*: PARK, R.G. & TARNEY, J. (eds) *Evolution of the Lewisian and Comparable Precambrian High Grade Terrains*. Geological Society, London, Special Publications, **27**, 127–138.

——, ENFIELD, M. A. & FISCHER, M. W. 1987. Devonian basins of Northern Scotland: extension and inversion related to late Caledonian-Variscan tectonics. *In*: COOPER, M. A. & WILLIAMS, G. D.

(eds). *Inversion Tectonics*. Geological Society, London, Special Publications, **44**, 275–308.

DAVIES, F. B. 1976. Early Scourian structures in the Scourie-Laxford region and their bearing on the evolution of the Laxford Front. *Journal of the Geological Society of London*, **132**, 543–554.

—— 1978. Progressive simple shear deformation on the Laxford shear zone, Sutherland. *Proceedings of the Geologists' Association*, **89**, 177–196.

ENGLAND, P. & JACKSON, J. 1987. Migration of the seismic-aseismic transition during uniform and non-uniform extension of the continental lithosphere. *Geology*, **15**, 291–294.

EVANS, C. R. 1965. Geochronology of the Lewisian basement near Lochinver, Sutherland. *Nature*, **207**, 54–6.

—— & LAMBERT, R. ST. J. 1974. The Lewisian of Lochinver, Sutherland; the type area for the Inverian metamorphism, *Journal of the Geological Society, London*, **130**, 125–150.

—— & TARNEY, J. 1964. Isotopic ages of the Assynt dykes. *Nature*, **204**, 638–641.

FOUNTAIN, D. M. & SALISBURY, M. H. 1981. Exposed cross sections through the continental crust: implications for crustal structure, petrology and evolution. *Earth and Planetary Science Letters*, **56**, 263–277.

GILETTI, B.J., MOORBATH, S., & LAMBERT, R. ST. J. 1961. A geochronological study of the metamorphic complexes of the Scottish Highlands. *Quarterly Journal of the Geological Society of London*, **117**, 233–264.

HANDY, M. R. & ZINGG, A. 1991. The tectonic and rheological evolution of an attenuated cross section of the continental crust: Ivrea crustal section, Southern Alps, northwestern Italy and southern Switzerland. *Geological Society of America Bulletin*, **103**, 236–253.

HAUSER, E., POTTER, C., HAUGE, T., BURGESS, S., BURTCH, S., MUTSCHLER, J., ALLMENDINGER, R R., BROWN, L., KAUFMAN, S. & OLIVER, J. 1987. Crustal structure of eastern Nevada from COCORP deep seismic reflection data. *Geological Society of America Bulletin*, **99**, 833–844.

HEAMAN, L. M. & TARNEY, J. 1989. U-Pb baddelyite ages for the Scourie dyke swarm, Scotland: evidence for two distinct intrusion events. *Nature*, **340**, 705–708.

HOLBROOK, W. S., CATCHINGS, R. D. & JARCHOW, C M. 1991. Origin of deep crustal reflections: Implications of coincident seismic refraction and reflection data in Nevada. *Geology*, **19**, 175-9.

HUMPHRIES, F. J. & CLIFF, R. A. 1982. Sm-Nd dating and cooling history of Scourian granulites, Sutherland. *Nature*, **295**, 515–517.

HYNDMAN, R. D., LEWIS, T. J. & MARQUIS, G. / HOLBROOK, W. S., CATCHINGS, R. D. & JARCHOW, C. M. 1991. Comment and reply on: Origin of deep crustal reflections: Implications of coincident seismic refraction and reflection data in Nevada. *Geology*, **19**, 1243–1244.

JACKSON, J. A. 1987. Active normal faulting and crustal extension. *In*: COWARD, M. P., DEWEY, J. F. & HANCOCK, P. C. (eds) *Continental Extensional Tectonics*. Geological Society, London, Special Publications, **28**, 3–17

—— & WHITE, N. J. 1989. Normal faulting in the upper continental crust: observations from regions of active extension. *Journal of Structural Geology*, **11**, 15–36.

KLEMPERER, S. L. & HURICH, C. A. 1991. Lithospheric structure of the North Sea from deep seismic reflection profiling. *In*: BLUNDELL, D. J. & GIBBS, A. D. (eds) *Tectonic Evolution of the North Sea rifts*. Oxford Science Publications, **181**, International Lithosphere Program, 37–59.

KUSZNIR, N. J. & PARK, R. G. 1987. The extensional strength of the continental lithosphere: its dependence on geothermal gradient, and crustal composition and thickness. *In*: COWARD, M. P., DEWEY, J. F. & HANCOCK, P. L. (eds) *Continental Extensional Tectonics*. Geological Society of London Special Publication, **28**, 35–52.

LAMBERT, R. ST J. & HOLLAND, J. G. 1972. A geochronological study of the Lewisian from Loch Laxford to Durness, Sutherland, NW Scotland. *Journal of the Geological Society of London*, **128**, 3–19.

LEPICHON, X. & CHAMOT-ROOKE, N. 1991. Extension of continental crust. *In*: LEPICHON, X. (ed.) *Controversies in Modern Geology*. Academic Press, 313–337.

MAINPRICE, D. & NICOLAS, A. 1989. Development of shape and lattice preferred orientations: application to seismic anisotropy of the lower crust. *Journal of Structural Geology*, **11**, 175–189.

MEANS, W. D. 1989. Stretching Faults. *Geology*, **17**, 893–896.

MEISSNER, R. & TANNER, B. 1993. From Collision to Collapse: phases of lithospheric evolution as monitored by seismic records. *Physics of the Earth and Planetary Interiors*, **79**, 75–86.

——, WEVER, T. H. & SADOWIAK, P. 1990. Reflectivity patterns in the Variscan mountain belts and adjacent areas: an attempt for a pattern recognition and correlation to tectonic units. *Tectonophysics*, **173**, 361–378.

MOONEY, W. D. & MEISSNER, R. 1992. Multigenetic origin of crustal reflectivity: a review of seismic reflection profiling of the lower continental crust and Moho. *In*: FOUNTAIN, D. M., ARCULUS, R. & KAY, R.W. (eds) *Continental Lower Crust*. Elsevier,.45–79.

MOORBATH, S. 1969. Evidence of the age of deposition of the Torridonian sediments of northwest Scotland. *Scottish Journal of Geology*, **5**, 154–170.

MCGEARY, S. E. & WARNER, M. R. 1985. Seismic profiling the continental lithosphere. *Nature*, **317**, 795-797.

——, ——, CHEADLE, M. J. & BLUNDELL, D. J. 1987. Crustal structure of the continental shelf around Britain derived from BIRPS deep seismic profiling. *In*: BROOKS, J. & GLENNIE, K. (eds) *Petroleum Geology of North West Europe*. Graham & Trotman, 33–41.

MCKENZIE, D. P. 1978. Some remarks on the development of sedimentary basins. *Earth and Planetary Science Letters*, **40**, 25–32.

NEWTON, R. C. & PERKINS, D. 1982. Thermodynamic calibration of geothermometers based on the assemblages garnet–plagioclase–orthopyroxene–(clinopyroxene)–quartz. *American Mineralogist*, **67**, 203–222.

O'HARA, M. J. 1977. Thermal history of excavation of Archaean gneisses from the base of the continental crust. *Journal of the Geological Society, London*, **134**, 185–200.

PARK, R. G. 1964. The structural history of the Lewisian rocks Gairloch, Wester Ross. *Quarterly Journal of the Geological Society of London*, **120**, 397–434.

—— 1991. The Lewisian Complex. *In*: CRAIG, G. Y. (ed.) *Geology of Scotland*. 3rd ed. Geological Society, London, 25-64.

—— & TARNEY, J. 1987. The Lewisian Complex: a typical Precambrian high-grade terrain? *In*: PARK, R. G. & TARNEY, J. (eds) *Evolution of the Lewisian and Comparable Precambrian High Grade Terrains*. Geological Society, London, Special Publications, **27**, 13–25.

PARSONS, T., HOWIE, J. M. & THOMPSON, G. A. 1992. Seismic constraints on the nature of lower crustal reflectors beneath the extending transition zone of the Colorado Plateau, Arizona. *Journal of Geophysical Research*, **97B**, 12 391–12 407.

PEACH, B. N., HORNE, J., GUNN, W., CLOUGH, C. T., HINXMAN, C. W. & TEALL, J. J. H. 1907. *The Geological Structure of the North West Highlands of Scotland*. Memoirs of the Geological Survey of Great Britain.

PERCIVAL, J. A., FOUNTAIN, D. M. & SALISBURY, M. H. 1992. Exposed crustal cross sections as windows on the lower crust. *In*: FOUNTAIN, D. M., ARCULUS, R. & KAY, R. W. (eds). Continental Lower Crust. Elsevier, 317–362.

PFIFFNER, O. A., FREI, W., VALASEK, P., STAUBLE, E, M., LEVATO, L., DUBOIS, L., SCHMID, S. M. & SMITHSON, S. B. 1990. Crustal shortening in the Alpine orogen: Results from deep seismic reflection profiling in the Eastern Swiss Alps, line NFP 20-EAST. *Tectonics*, **9**, 1327–1355.

PRIDE, C. & MUECKE, G. K. 1980. Rare Earth element geochemistry of the Scourian Complex, NW Scotland: evidence for the granite granulite link. *Contributions to Mineralogy and Petrology,*. **73**, 403–412.

RESTON, T. J. 1988. Evidence for shear zones in the lower crust offshore Britain. *Tectonics*, **7**, 929–945.

—— 1990. The lower crust and the extension of the continental lithosphere: Kinematic analysis of BIRPS deep seismic data. *Tectonics*, **9**, 1235–1248.

—— 1993. Evidence for extensional shear zones in the mantle offshore Britain and their implications for extension of the continental lithosphere. *Tectonics*, **12**, 492–506.

REY, P. 1993. Seismic and tectonometamorphic characters of the lower continental crust in Phanerozoic areas: A consequence of post thickening extension. *Tectonics*, **12**, 580–590.

ROLLINSON, H. R. 1980a. Iron-Titanium oxides as an indicator of the role of the fluid phase during the cooling of granites metamorphosed to granulite grade. *Mineralogical Magazine*, **43**, 623–631.

—— 1980b. Mineral reactions in a calc-silicate rock from Scourie. *Scottish Journal of Geology*, **16**, 153–64.

—— 1981. Garnet-pyroxene thermometry and barometry in the Scourie granulites, NW Scotland. *Lithos*, **14**, 225–238.

RUTTER, E. H. & BRODIE, K. H. 1990. Some Geophysical implications of the deformation and metamorphism of the Ivrea zone, Northern Italy. *Tectonophysics*, **182**, 147–160.

SAVAGE, D. & SILLS, J. D. 1980. High pressure metamorphism in the Scourian of NW Scotland: evidence from garnet granulites. *Contributions to Mineralogy and Petrology*, **74**, 153–163.

SERPA, L. & DE VOOGT, B. 1987. Deep seismic reflection evidence for the role of extension in the evolution of the continental crust. Geophysical Journal of the *Royal Astronomical Society*, **89**, 55–60.

SHERATON, J. W., TARNEY, J., WHEATLEY, T. J. & WRIGHT, A. E. 1973. The structural history of the Assynt district. *In*: PARK, R. G. & TARNEY, J. (eds) *The Early Precambrian of Scotland and related rocks of Greenland*. Univ of Keele, 31–44.

SILLS, J. D. 1983. Mineralogical changes occurring during the retrogression of Archaean gneisses from the Lewisian complex of NW Scotland. *Lithos*, **16**, 113–124.

—— & ROLLINSON, H. R. 1987 Metamorphic evolution of the mainland Lewisian complex. *In*: PARK, R. G. & TARNEY, J. (eds) *Evolution of the Lewisian and Comparable Precambrian High Grade Terrains*. Geological Society, London, Special Publications, **27**, 81–92.

STEIN, A. M. & BLUNDELL, D. J. 1990. Geological inheritance and crustal dynamics of the northwest Scottish continental shelf. *Tectonophysics*, **173**, 455–467.

SUTTON, J. & WATSON, J. 1951. The Pre-Torridonian metamorphic history of the Loch Torridon and Scourie areas in the North West Highlands and its bearing on the chronological classification of the Lewisian. *Quarterly Journal of the Geological Society of London*, **106**, 241–307.

TARNEY, J. 1973. The Scourie dyke suite and the nature of the Inverian event in Assynt. *In*: PARK, R. G. & TARNEY, J. (eds) The Early Precambrian of Scotland and related rocks of Greenland. Univ of Keele, 105–118.

WARNER, M. R. 1990. Basalts, water or shear zones in the lower continental crust? *Tectonophysics*, **173**, 163-174.

WERNICKE, B. 1985. Uniform sense normal simple shear of the continental lithosphere. *Canadian Journal of Earth Sciences*, **22**, 108-125.

WOOD, B. J. 1977. The activities of components in the clinopyroxene and garnet solutions and their applications to rocks. *Philosophical Transactions of the Royal Society of London*, **A286**, 331–342.

ZEIGLER, P. 1983. Crustal thinning and subsidence in the North Sea. *Nature*, **304**, 561–564.

Development of the Witwatersrand Basin, South Africa

MIKE P. COWARD[1], RICHARD M. SPENCER[2] & CAMILLE E. SPENCER[2]

[1] Department of Geology, Imperial College, Prince Consort Road, London SW7 2BP, UK

[2] Gemsa, Borja Larayen y Juan Pablo Sanz, Quito, Ecuador

Abstract: The Witwatersrand basin is a large hinterland basin of Archaean age developed on already consolidated continental crust south of the Limpopo plate margin. It is a flexural basin formed southeast of a major thick-skinned thrust system, modified by a second thrust system to the southeast. Studies of several thousand kilometres of seismic data, together with geophysical modelling and new field data, have allowed the thrusts and their associated hanging-wall uplifts to be mapped and dated, relative to sediments of Central Rand age. The basin subsequently underwent extension on large detachment faults. Associated half-grabens are infilled with growth sequences of Platberg sediments and intermediate volcanics. These grabens were inverted during post-Platberg and post-Transvaal times. Studies of the Witwatersrand Basin allow tectonic models to be developed for the Archaean which can be used to help interpret Cenozoic mountain belts.

The Witwatersrand basin in South Africa (Figs 1 & 2) is important in the study of Early Precambrian tectonics because of the following.

(i) It forms one of the largest concentrations of gold in the world, occurring as sediment-hosted gold within the Central Rand Group of the Witwatersrand Supergroup (see Fig. 2 for distribution of the main goldfields).

(ii) It forms one of the best preserved and most detailed investigated Archaean basins in the world, and possibly the most investigated Precambrian basin. Data are available from many hundreds of boreholes, numerous mines and several tens of thousands of kilometres of seismic data, which help determine its deep structure.

(iii) It developed as a deep intracratonic basin at approximately the same time as the greenstone belts of Zimbabwe and the Limpopo orogenic belt of Zimbabwe–Northern Transvaal. Therefore it gives an indication of intracratonic processes during the Late Archaean–Early Proterozoic and not just the processes occurring at constructive or destructive plate margins. Furthermore, this paper will argue that the tectonic processes which led to the development of the Witwatersrand basin are essentially similar to those of present day hinterland basins, i.e. basins on the opposite side of the mountain belt to the main collision zone. Present day analogues can be taken from the hinterland basins of northern China (see paper by Graham, this volume) or the sub-andean basins of South America.

This paper aims to provide a stratigraphic and structural synthesis of the Witwatersrand basin, based on collaborative work between Imperial College and Gencor Mineral Resources of South Africa. New field mapping has been carried out in all the areas where the Late Archaean Witwatersrand Supergroup is exposed. Data have been collected and assimilated from mines and boreholes throughout the basin. Over 7000 km of seismic data have been examined. These seismic data, combined with borehole, mine and field data, have allowed maps to be produced showing the detailed structure of all the principal horizons in the Witwatersrand Supergroup, together with maps showing details of all the principal structures. Over 50 cross sections have been produced through the basin. Each has been checked with seismic, gravity and magnetic data, restored and checked for compatibility with its neighbour. The results of this study are summarized in this paper.

This investigation also initiated detailed microstructural work and geochemistry of the sediments to determine the relation of gold mineralization to sedimentation, diagenesis/ metamorphism and deformation. This aspect of the work is still confidential but will form the basis of a future scientific contribution.

It is important to note the value of new field studies in determining the history of the Witwatersrand basin. Fault kinematics are readily obtainable from outcrops along the Witwatersrand (White Water ridge) both west and east of Johannesburg (Fig. 2). This ridge lies north of the original mining area opened at the end of the last century. Kinematic data are also obtainable from around the Vredefort dome (Fig. 2), the Klerksdorp area and from the area east of Heidelberg. These studies have proved invaluable for determining the structural and tectonic history of the basin and the interpretation of borehole and seismic data.

From COWARD, M. P. & RIES, A. C. (eds), 1995, *Early Precambrian Processes*, Geological Society Special Publication No. 95, pp. 243–269.

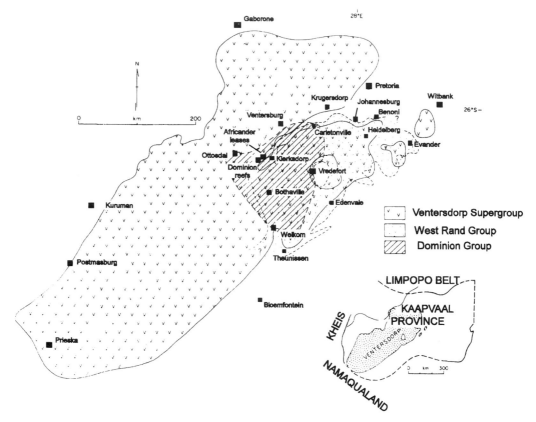

Fig. 1. Map showing the distribution of the Dominion, West Rand and Ventersdorp sequences on the Kaapvaal craton of South Africa. Inset map shows the orogenic belts bounding the craton. From Tankard *et al.* (1982). The edge of the West Rand Group essentially defines the edge of the Witwatersrand basin as commonly defined.

However the main step forward in the understanding of Witwatersrand geology has come from the use of seismic data. These have allowed all the stratigraphic units to be mapped and completely new structures identified. Many of the mysteries of Witwatersrand geology have now been solved.

Regional geology

The Dominion Group, Witwatersrand Supergroup and Ventersdorp Supergroup, often grouped as the Witwatersrand Triad, form the volcanosedimentary sequence of the Witwatersrand basin. The Witwatersrand Supergroup is further subdivided into a lower West Rand group and an upper Central Rand Group and these are divided into several Subgroups (Fig. 3). The present-day basin boundaries (Figs 1 & 2) are defined by the outcrops or subcrops of the basal part of the West Rand Group. The basin forms an elongate saucer-shaped structure, approximately 360 km by 200 km. The long

axis of the basin trends NE–SW. Basement rocks and their cover of Dominion or West Rand Group crop out, (i) in the Johannesburg Dome, north of Johannesburg, (ii) in the Devon Dome and adjacent domal outcrops, east of Heidelberg in the eastern part of the basin, (iii) along the western and north-western flanks of the basin, NW of Klerksdorp, and (iv) within an almost circular body in the central part of the basin, known as the Vredefort Dome (Fig. 2). However, throughout much of the basin the sediments of the Witwatersrand Triad are covered by the later Transvaal Supergroup or by a thin veneer of Karoo sediments. Seismic data are required to correlate the structures and map lithologies at depth.

The Dominion, Witwatersrand and Ventersdorp sediments were deposited over the centre of the Kaapvaal Archaean craton. The area of preservation decreases with age. Thus the Dominion Group is confined to the western part of the present day basin, while the Ventersdorp Supergroup is

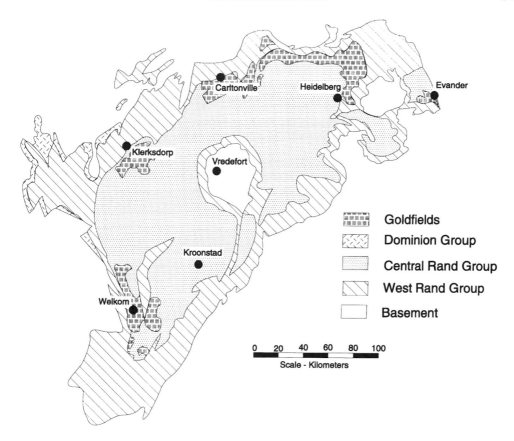

Fig. 2. Map showing the sub-Transvaal Supergroup geology showing outcrops of basement and the main stratigraphic units of the Witwatersrand Triad. The main goldfields are shown. Map redrawn from Myers *et al.* (1990).

estimated to have occupied an area of over 300 000 sq km, covering much of the western part of the Transvaal (Fig. 1).

Radiometric age determinations on Late Archaean basement granitoids and lavas in the Ventersdorp near the top of the Witwatersrand sediments, bracket the age of the sediments between 3100 Ma and 2700 Ma. Granites underlying the Witwatersrand and Ventersdorp Supergroups in the western Transvaal yield ages of *c.* 3100 Ga, while igneous rocks in the Dominion Group give ages of 3070 Ma (Armstrong *et al.* 1990). Sedimentation commenced at *c.* 3000 Ma (Barton *et al.* 1989; Robb *et al.* 1990) and continued until *c.* 2800 Ma (Robb *et al.* 1990). The radiometric data are summarized in Fig. 4.

The sediments of the Witwatersrand Triad are overlain by the Transvaal Supergroup which was deposited in a large basin covering much of the Transvaal. The terrestrial and shallow marine Wolkberg Group, the lowest group of the Transvaal Supergroup, is developed only along the northern margin of the basin, in the northern Transvaal, infilling an irregular palaeo-topography. However throughout much of the province the basal part of this supergroup is represented by quartzitic sandstones and conglomerates of the Black Reef, which was deposited across a peneplained surface.

The Black Reef forms a prominent marker horizon on seismic data and allows the identification of pre- and post-Transvaal age structures. Overlying the Black Reef is a thick sequence of dolomites, shales and ironstones, the Chuniespoort Group, overlain by shales, sandstones and volcanics of the Pretoria Group. The Hekpoort volcanics in the Pretoria Group give ages of *c.* 2.3 Ga (Van Niekerk & Burger 1964). However an age of 2557 Ma (Jahn *et al.* 1990) from Transvaal dolomites suggests that the base of the Transvaal Supergroup may be much earlier and the Black Reef may possibly be of Late Archaean age.

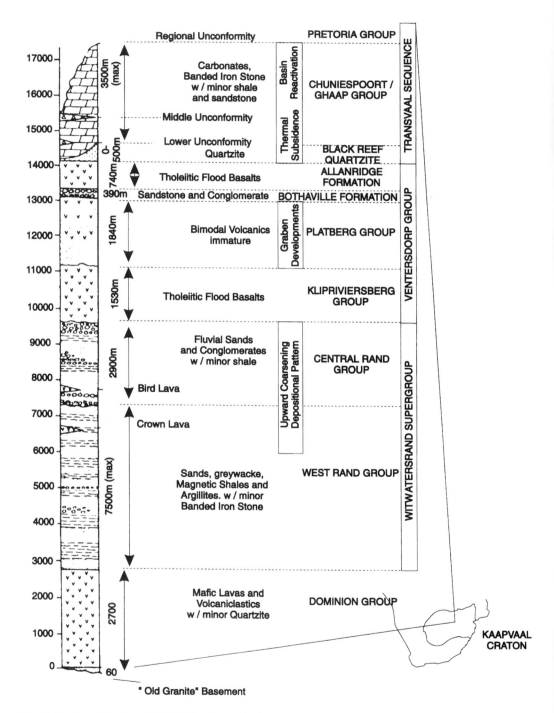

Fig. 3. Stratigraphic column for the Witwatersrand basin.

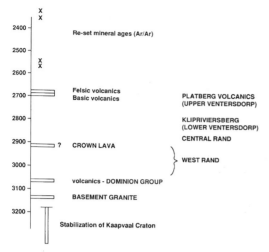

Fig. 4. Summary diagram illustrating the geocronological data for the Witwatersrand basin.

The Witwatersrand sedimentary rocks have been metamorphosed only to lower greenschist facies (Phillips & Law 1994). Deformation is heterogeneous and hence the sedimentary fabrics and textures are often very well preserved. Metamorphic mineral ages generally cluster around 2.3–2.5 Ga suggesting either a mixed age related to later thermal events, or that the rocks were unroofed and cooled at this time.

The Witwatersrand basin has long been considered to be an intracratonic basin, but its origin has remained speculative. Pretorius (1976) envisaged the basin as a large graben or half-graben with a progressively more active fault scarp along the northwestern margin and a relatively passive southeast margin. Continued uplift in the northwest caused the sedimentary prism to offlap across the shrinking basin.

Hutchinson (1975) and Minter (1976) suggested that the basin structure can be accounted for by epeirogenic tilting and warping, followed by basin closure as a result of diapiric granite doming. Similarly, according to Brock & Pretorius (1964) and Pretorius (1981), the main control on the basin shape and structure was the rise of the basement granitic domes. In the original paper (Brock & Pretorius 1964) this uplift was considered to be related to some form of gravitational movements rather than horizontal shortening or extension. Their model involved essentially vertical movements and argued for a different type of tectonic process in the Archaean from that of the present day. However Pretorius (1981) observed that the pattern and growth of the basement domes con-

trolled the palaeocurrents in the West and Central Rand Groups and produced local unconformities in the basin.

Early plate tectonic models, including that of Van Biljon (1980), considered that the Witwatersrand basin was the site of continent–continent collision. According to Van Biljon (1980) the suture zone is marked by the Swaziland greenstone belts in the adjacent craton. A more refined model of Precambrian plate accretion has been presented by De Wit et al. (1992). According to these authors the Archaean Kaapvaal craton formed by the SE–NW accretion of numerous terranes. Much of the crustal fabric seen on the deeper seismic data (Roering et al. 1990; Treloar et al. 1992) was probably derived from these accretion events.

Bickle & Eriksson (1982) concentrated on a plate tectonic extensional model. They proposed that an intracratonic Witwatersrand basin was followed by Ventersdorp rifting. Papers concentrating on the compressional tectonics include those of Burke et al. (1985, 1986), Winter (1987) and Treloar et al. (1992). Burke et al. (1985, 1986) considered the Witwatersrand basin to be foreland basin at the southeast edge of a zone of crustal thickening associated with the Limpopo belt. Treloar et al. (1992) argued that the Witwatersrand basin was a hinterland basin as neither the edge of the basin nor the facies boundaries migrated significantly during upper West Rand and Central Rand sedimentation. Burke et al. (1986) and Treloar et al. (1992) used an analogy with the Central Asian basins, such as the Tarim and Qaidam basins.

Winter (1987) favoured a continental Andean-type back-arc basin model. He argued that Phanerozoic back-arc basins are essentially similar to the Witwatersrand basin in their style of development and subsequent deformation. According to Winter (1987) the important characteristics are the alternating extensional and compressional episodes together with arc-related volcanism. Myers et al. (1990) further refined the Winter model and suggested that a present day analogue could be the Maracaibo basin of South America. They noted the importance of compressional tectonics around the edges of the basin. The different episodes of compression and extension were further documented by Treloar et al. (1992) and related to episodes of compression and mountain building in the Limpopo belt.

In the next section the main aspects of the stratigraphic development of the basin will be described, based on data collected during the current study. The stratigraphy of the basin is outlined in Fig. 3. The basin history can be simplified into different episodes of rifting, subsidence and subsequent basin inversion and uplift (Figs 5 & 6).

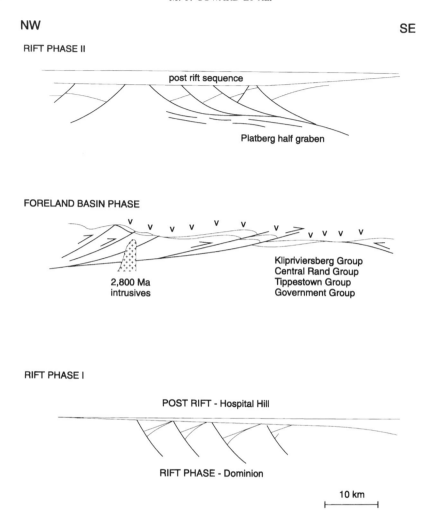

Fig. 5. Schematic NW–SE cross sections across the Witwatersrand basin showing its development during sedimentation of the Witwatersrand Triad.

Basin development

Rift sequence 1

Basin development during the Dominion Group. The Dominion Group (Figs 2 & 3) is best preserved in the western Orange Free State and southwestern Transvaal. The group is over 2700 m thick near Klerksdorp, and seismic data suggest it is locally much thicker. The group consists of clastic sediments overlain by volcanoclastic sediments, basaltic andesites and tuffs. No structures of Dominion Group age have been detected with certainty. There are local dramatic variations in thickness and the present interpretation is that the group infills a number of extensional basins,

although the position and polarity of the basin-bounding faults are speculative. Volcanics within the Dominion Group have been dated at 3.07 Ga (Armstrong *et al.* 1991).

Early Witwatersrand sedimentation and basin development. The laterally extensive Hospital Hill Subgroup unconformably overlies the Dominion basin and forms the base of the West Rand Group. It is considered to represent the thermal sag phase of rifting; no normal faults of this age have been detected. The Hospital Hill basin was marine, dominated by distal shelf facies with minor intertidal components. The basin appears to have been open to the south and west (De Wit *et al.* 1992).

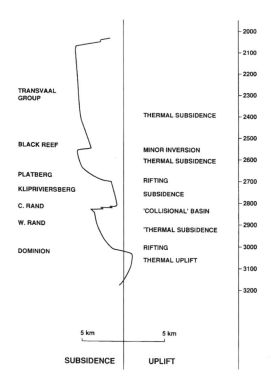

Within the diagram:
TRANSVAAL
GROUP

BLACK REEF

PLATBERG

KLIPRIVIERSBERG

C. RAND

W. RAND

DOMINION

THERMAL SUBSIDENCE

MINOR INVERSION
THERMAL SUBSIDENCE

RIFTING

SUBSIDENCE

'COLLISIONAL' BASIN

'THERMAL SUBSIDENCE

RIFTING
THERMAL UPLIFT

2000
2100
2200
2300
2400
2500
2600
2700
2800
2900
3000
3100
3200

5 km   5 km

SUBSIDENCE | UPLIFT

Fig. 6. Highly simplified subsidence diagram showing the changes in uplift and subsidence across the Witwatersrand basin with time.

The Pongola basin in the eastern part of the Transvaal is contemporaneous with the West Rand Group (De Beer & Eglington 1990; Beukes & Cairncross 1991) and may form part of the same rift and thermal subsidence basin.

The Witwatersrand foreland basin sequence

Stratigraphy and structural history. The main change in tectonic development occurs at the top of the Hospital Hill Subgroup-base of the Government Subgroup. The upper part of the West Rand Group and all of the Central Rand Group comprise syn-tectonic sediments, associated with pulses of uplift in the region northwest of the Witwatersrand basin. These uplifts may be the result of the following.

(i) Pulses of regional uplift along the southern margin of an orogenic belt, situated northwest of the Witwatersrand basin.

(ii) SE-directed thrusts, locally propagating into the basin. These structures have been recognized in the Klerksdorp–Welkom regions and in the East Rand, South Rand and Evander mining areas where they appear as growth faults during upper Central Rand sedimentation. The thrusts controlled the

shape of the basin, the distribution of alluvial fans and the local facies.

Lower Government Subgroup strata consist of immature sediments which were shed off uplifts which modified the earlier Hospital Hill basin. Fragments within the government Subgroup were derived from eroded Hospital Hill sediments and also granite–greenstone basement. The stratigraphy of the Government Subgroup is arranged in upward fining units (Winter & Brink 1991).

It is near the base of the Jeppestown Subgroup that the distribution of alluvial fans can be tied to southeasterly verging thrusts in the Klerksdorp goldfield (e.g. Antrobus *et al.* 1986); possibly some of the earlier fining upward units of the Government Subgroup were derived from the same thrust front. The Jeppestown Subgroup consists largely of fluvial sands in the Klerksdorp area which fine upwards into transgressive turbiditic mudstones. Upward coarsening sequences from these distal facies show that the Jeppestown Subgroup comprises a series of transgressive and regressive cycles (Winter & Brink 1991). These are considered to be the far-field effects of compressional tectonics occurring to the W, on the Kaapvaal Craton.

The effects of this early thrusting were largely confined to the Klerksdorp area. However an upward coarsening sequence at the top of the Jeppestown Group occurs in the Johannesburg area, probably in response to local tectonic uplift. The Welkom goldfield, in the southwest, had a SE-dipping slope at this time, possibly related to uplift of the northwest margin. In the East Rand area, stratigraphic relationships suggest the localized lateral uplift (the early precursor of the Springs monocline).

The overlying Central Rand Group consists of four genetic depositional packages which can be correlated across the basin (Fig. 7). A fifth depositional package, or megasequence, occurs in the lower part of the Ventersdorp Supergroup. These megasequences represent the main transgressions and regressions across the basin, generated by the combined effects of tectonic uplift, load-induced subsidence and sea-level changes. Figure 7 represents an attempt to construct a sequence stratigraphy diagram for the basin.

The lowest of these packages, termed megasequence 1, fines upwards and then coarsens upwards in a sequence of progressively coarser upward-fining sequences, related to uplift, alluvial fan sedimentation and subsequent erosion of the uplifted region (Winter & Brink 1991). Megasequence 2 oversteps the older stratigraphy and fines upwards to the Booysens shale. This shale represents the peak of transgression after which progradation commenced.

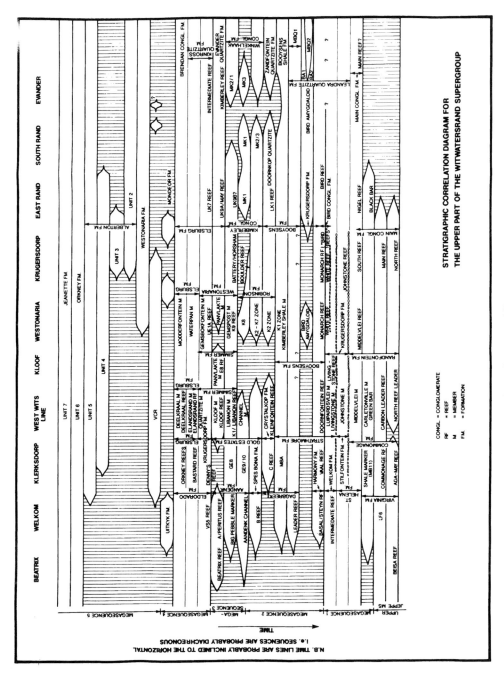

Fig. 7. Stratigraphic correlation diagram for the upper part of the Witwatersrand Supergroup.

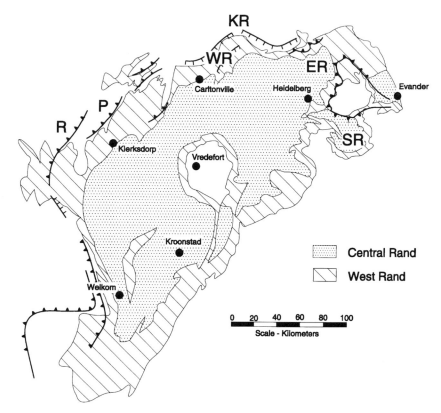

Fig. 8. Map of thrust activity during the later part of Central Rand sedimentation, i.e. thrusts which were active synchronous with deposition of the Kimberley Reef. Thrusts are shown with teeth on the hanging-wall, hanging-wall monoclines formed as fault bend or fault propagation folds above buried thrusts are shown as hatchured lines. WR, West Rand; ER, East Rand; KR, Krugersdorp; P, Palmietfontein. R, Rietkuil.

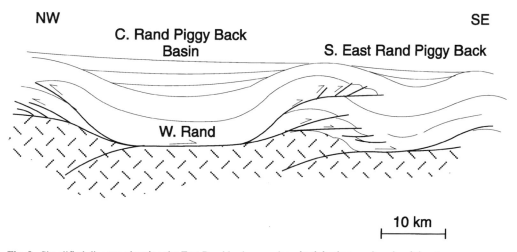

Fig. 9. Simplified diagram showing the East Rand basin as a piggy-back basin on a deep level thrust.

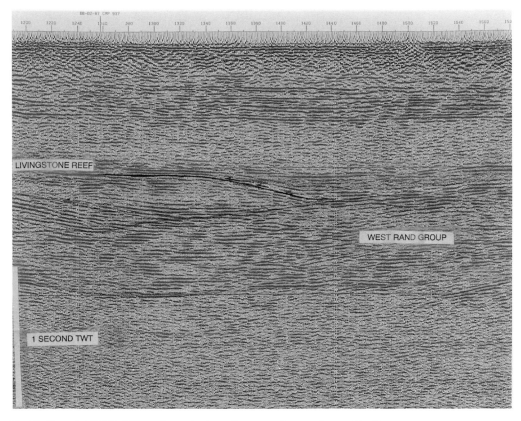

Fig. 10. Part of a WNW–ESE-trending seismic section across the central part of the Evander basin showing the onlap relationships of the central Rand sediments to syn-sedimentation thrusting.

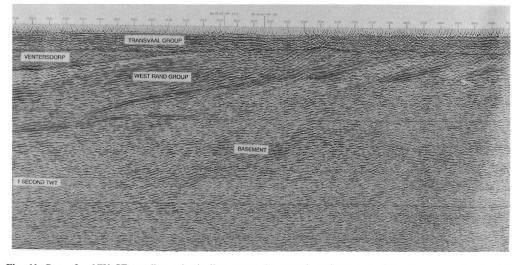

Fig. 11. Part of a NW–SE-trending seismic line across the central-southern part of the Evander basin showing imbrication of the basal part of the West Rand Group. the blue marker at the base of the Transvaal sediments is the Black Reef. Central Rand sediments are uncoloured above the West Rand Group. The red lines indicate thrusts.

Megasequence 3 consists of a basal marine facies followed by rapid progradation by alluvial deposits including coarse conglomeratic fans. A transgression resulted in extensive peneplanation of this sequence and the development of a mature marginal marine facies at the base of Megasequence 4. In Megasequence 4 the sediments coarsen upwards, culminating in the deposition of the coarsest conglomerates in the Central Rand Group at the top of the sequence. This upward coarsening is attributed to the southeastward advance of the thrust package. In the Welkom area much of the spectacular steeping of strata occurred after deposition of these conglomerates.

The Johannesburg and East Rand areas provide examples of the sporadic but southeastward propagation of the thrust system. The basement of the Johannesburg Dome was uplifted at this time. Several sub-basins (piggy-back basins) have been recognized in the East Rand (Figs 8 & 9) related to growth of thrust ramps, north of the Devon Dome. Sediment truncations and onlaps in the East Rand and South Rand basins show the topographic shape of the uplifts associated with the thrusts. Sediments of the Main Conglomerate Formation onlap folds associated with the thrusts. These onlaps can be mapped on seismic data (e.g. Fig. 10).

The thrusts imbricate the base of the West Rand Group (Fig. 11). The thrusts generally dip to the NW, but the strikes and dips vary due to folding and tilting on younger lower-level structures. The thrusts clearly propagated in a SE direction. NW–SE-trending lateral ramps and culmination walls indicate that the thrust transport direction was essentially NW–SE. In the Central Rand Group the thrusts branch into SE-verging forethrusts and NW verging backthrusts.

Smaller components of this thrust system were active in the sediments in the upper levels of the basin at the same time as the major thrusts moved at depth. These low angle thrusts deformed Central Rand sediments into gentle folds onto which the Kimberley Reef was deposited in the South Rand and Evander areas.

A southern thrust, here termed the Devon thrust, emerges from the basement beneath the South Rand basin and the Devon dome is its hanging-wall anticline above the thrust ramp (Figs 9 & 12). The Central Rand piggy-back basin formed from a combination of thrust climb above the frontal part of the Johannesburg thrust and thrust climb related to the emergence of the Devon thrust from basement into Witwatersrand Group sediments.

The distribution of lateral and frontal ramps in the basement and West Rand Group caused topographic relief not just in the basin but also in the mountain belt to the northwest. The mountain belt relief was lower at these transfer zones. Thus the

important channels and sources of the main alluvial fans seem to be situated to the northwest of important lateral ramps and thrust windows (Fig. 13). Paleocurrent directions and changes in grain size and facies of the alluvial sediments suggest that a major southerly influx of sediment occurred from the east of the Johannesburg Dome spreading westward and eastward into the East Rand and Evander mining areas, respectively. A similar influx of coarse alluvial/fluvial sediment occurred along the western margin of the basin, northwest of Welkom, where the structures are considered to be essentially lateral to the main thrust direction. Elsewhere in the Klerksdorp and Johannesburg areas the palaeocurrent data show flow directions down the mountain front slope in a southeasterly direction into the basin.

Deposition of the Central Rand Group sediments was closely followed by volcanic activity, producing the Klipsriviersberg lavas. These volcanics suggest that the region was affected by high heat flow during the last phases of thrust tectonics. The first Klipsriviersberg lavas are komatiitic and are confined to the northeastern part of the Witwatersrand basin (Myers et al. 1990), the part of the basin that was undergoing active thrust-related deformation at this time. The overlying porphyritic mafic lavas were extruded over a larger area, but their thickest development is coincident with the area overlain by komatiites. In the South Rand mining area the early lavas were deformed into a NW-trending monoclinal fold, lateral to the main thrust trend, and lava flows in the upper part of the sequence onlap this moncline.

This contrasts with the Klerksdorp area on the western side of the basin where the Venterspost Conglomerate Formation, at the top of the Central Rand Group, and the overlying Klipsriviersberg lavas were deposited after the peneplanation of the thrust-related uplifts.

The Klipsriviersberg lavas therefore provide an important time marker across the Witwatersrand basin and show that thrusting becomes younger to the SE. In the Klerksdorp area the thrusting is clearly pre-Klipsriviersberg in age, whereas in the East Rand, South Rand and Evander areas the last components of thrusting deform the Klipsriviersberg Subgroup.

Seismic data from the Standerton area, in the southeastern part of the Witwatersrand basin, show two thrust systems, a northerly verging system deformed by a southerly verging system. A northerly verging thrust forms the southern margin of the Evander goldfield where it is associated with coarse Kimberley Reef facies. This thrust is considered to be part of the older northward verging system. The younger southeasterly verging system in the Standerton area is considered to be the frontal

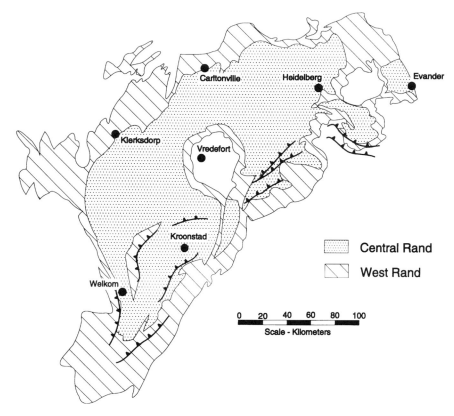

Fig. 12. Map of emergent thrust activity during early Klipriviersberg sedimentation.

part of the SE younging, syn-Central Rand–Klipsriviersberg system. This suggests that both the SE-verging and the NW-verging systems were active at the same time, advancing towards each other (e.g. Fig. 5). The two systems meet in the Evander area at the time of the Kimberley Reef.

Thrust geometry. Throughout most of the basin the thrust geometry comprises SE-verging, generally SE propagating thrusts (Figs 8 & 12). Local variations in transport directions occur, possibly associated with the development of lateral ramps. NW-verging backthrusts developed around the Johannesburg dome, generated from a buried thrust at depth. North-verging thrusts occur along the southeast margin of the basin. The basin is therefore a butterfly basin (see Graham this volume), being overridden by thrusts from opposite sides.

The southwest margin of the Witwatersrand basin, W of Welkom, is marked by a large E directed thrust (Fig. 8) which is clearly imaged on the seismic data. It is a low-angle structure, listric in section, flattening into a reflective zone in the

basement at 12–15 km below surface. It carries basement over West Rand Group strata. The trace of the fault is curvilinear, the strike changing from N–S to E–W near Welkom and then to a NNW strike, NW of the Welkom goldfield (Fig. 8).

Similar thrusts can be traced to the north into the Klerksdorp area where they and their associated folds are exposed as the Rietkuil and Palmietfontein uplifts (Fig. 8).

The northeastern extension of these thrusts was displaced in the Potchefstroon area (Fig. 8) by a large normal fault, the Buffels detachment. Another paired thrust system, consisting of the Bank and West Rand thrusts, extends north-northeastwards from the Carltonville–Kloof mining area through Krugersdorp (Fig. 8). The Bank and West Rand thrusts displace the SE verging thrusts described above, and are considered to be later, of post-Ventersdorp age. From Krugersdorp the SE-verging thrust system curves eastwards through the Johannesburg and East Rand mining areas.

South of Johannesburg two large basement faults can be recognized. The northern thrust, here termed

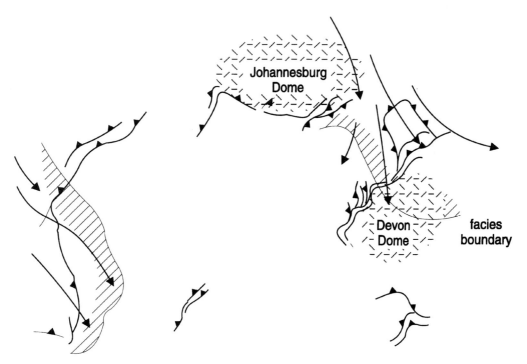

Fig. 13. Example of the influence of thrust-related topography on the sedimentology of the upper part of the Central Rand Group. The offset of thrusts and the development of lateral fold and thrust structures allows an important zone of sediment input into the eastern part of the Witwatersrand basin.

the Johannesburg thrust, climbs from basement into cover beneath the East Rand basin. The Johannesburg dome is a large hangingwall anticline related to this thrust and the SW-dipping Witwatersrand Group sediments and the Rand Ridge are part of the lateral monoclinal culmination wall above the lateral ramp in this basement thrust. The lateral ramp offsets the position of the frontal ramp where the thrust emerges into Witwatersrand Group sediments from the West Rand to the East Rand. Where the Johannesburg thrust climbs from basement into Witwatersrand cover sediments, several thrust splays occur and also important backthrusts. These backthrusts are prominent along the Rand Ridge, west of Johannesburg (e.g. Fig. 14). They detach in the lowest Witwatersrand shales or close to the top of the basement. Thin slivers of basement and cover are repeated many times in a major imbricate zone and there must be many kilometres of shortening related to these backthrusts along the Rand Escarpment. These thrusts pre-date movement on the Rietfontein fault system, which here occurs as a steeply dipping oblique structure (Fig. 14). The Rietfontein fault links extensional graben of Ventersdorp age; hence the imbricate backthrusts in the West Rand are con-

sidered to be part of the thrust sequence of Central Rand age.

Figure 14 shows a detailed map of the thrust structures between the Roodepoort and Little Falls in the Krugersdorp region of the West Rand. The most striking tectonic feature is the massive imbrication of the lower part of the West Rand Group. Along the studied section, shortening must be several kilometres (Fig. 15) and hence, as this section is typical of the Johannesburg–Krugersdorp part of the Witwatersrand basin, shortening along the entire Rand escarpment must be in the order of several tens of kilometres. In the studied area (e.g. Fig. 14), the lower West Rand quartzites, shales and Basement granite–greenstone belts are repeated in over ten thin thrust sheets, many of which can be traced across the mapped area. This repetition of quartzites causes problems in identifying stratigraphic horizons and thicknesses and indeed many of the quartzites might have been previously misidentified.

The Central Rand Group in the Johannesburg area must have been displaced several kilometres to the NW on these imbricate thrusts which are backthrusts on a hanging-wall monocline formed by a SE-verging thrust.

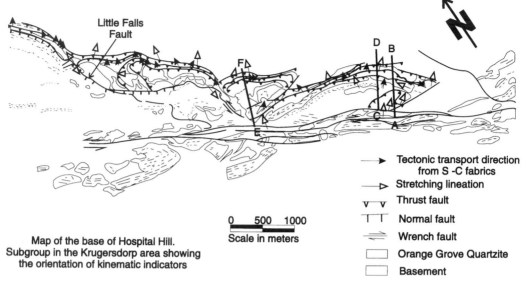

Map of the base of Hospital Hill.
Subgroup in the Krugersdorp area showing
the orientation of kinematic indicators

0   500   1000
Scale in meters

→ Tectonic transport direction
   from S -C fabrics
▷ Stretching lineation
v—v Thrust fault
⊤ ⊤ Normal fault
⇌ Wrench fault
▭ Orange Grove Quartzite
▭ Basement

Fig. 14. Map of the base of the Hospital Hill Subgroup in the Little Falls–Roodepoort district, near Krugersdorp, western Rand, showing thrusts and mapped kinematic indicators. The left lateral shear zone in the southwest part of the map forms an oblique to lateral part of the Rietfontein extensional fault system.

Along the Rand Ridge the thrust direction was dominantly towards the N or NW, splaying towards the NNE, i.e. the thrust zone appears to be a lateral to oblique structure. Kinematic indicators include

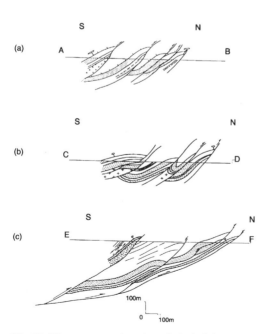

Fig. 15. Three cross sections through the imbricate zone shown in Fig. 14.

well developed lineations shown by the preferred orientations of chlorites and phyllosilicates and by stretched quartz grains. The present strike may be partially due to tilting on the southern margin of the Johannesburg dome, although some of the shears, such as the main Rietfontein shear in the southern part of the mapped area, are highly oblique to bedding and appear to be lateral ramps rather than tilted flats. Along the entire Rand escarpment there is a swing in thrust direction from NW directed in the west, to north directed in the Roodepoort–Johannesburg region. This change in trend of lineations is typical for lateral ramps/strike-slip zones.

Bedding-parallel shear zones have been mapped in the Evander mining area. These faults are characterized by ductile fabrics and verge to the SE or NW. Deformation occurred under moderate metamorphic conditions, as shown by the growth of chlorites and pyrophylite in the shear zones and the change in shape of quartz grains. This suggests that the thrusting occurred after the deposition of most of the Witwatersrand sediments and possibly following Ventersdorp sedimentation.

Rift sequence II, Ventersdorp extension

Volcanic activity was followed by extension producing the Platberg Group of syn-rift sediments, deposited in a series of half-grabens. A simplified map of the main extensional structures in the Witwatersrand basin is given in Fig. 16.

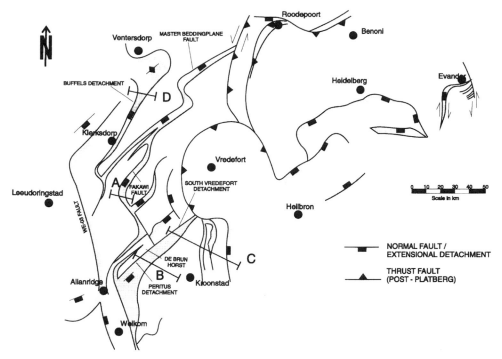

Fig. 16. Map of the main extensional structures in the Witwatersrand basin.

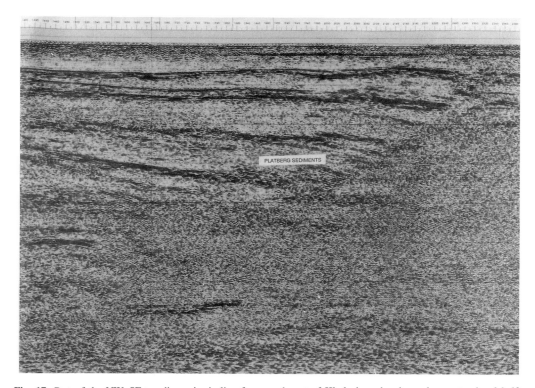

Fig. 17. Part of the NW–SE-trending seismic line from northwest of Klerksdorp showing a deep extensional half-graben infilled with Platebrg sediments. The half-graben bounding fault dips to the northwest. The vertical section is 4 seconds, two-way-travel time (equivalent to 10–11 km). The horizontal scale approximately equals the vertical scale.

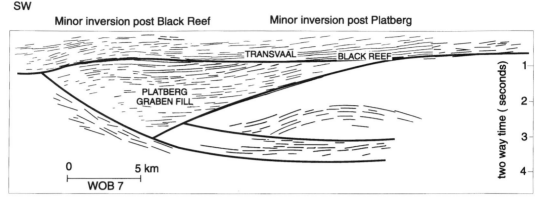

Fig. 18. Line drawing of a W–E-orientated seismic line from the western part of the Witwatersrand basin, in the Buffels area, showing the listric shape of the faults and the associated growth sequences of Platberg age. Section line shown by A on Fig. 16.

Extensional faults also occur to the northwest of the basin. Here they generally dip to the NW. They are gently listric in section, flattening slightly with depth and are infilled with a thick sequence of Platberg sediments (Fig. 17). The slightly listric shape is generally because they follow earlier Witwatersrand age thrusts.

However within the Witwatersrand basin the normal faults generally dip to the SE or E. The major faults include the following.

(i) Listric and low-angle detachment faults, as observed at Peritus, Buffels and Rietfontein, in the western part of the Witwatersrand basin, where the extensional faults are clearly listric in section (Fig. 18) and flatten onto bedding-parallel detachments, often in the Central Rand or West Rand shales. The detachment faults occur on the southest side of prominent thrust stacks, e.g. the De Bron structure, suggesting that the easy slip horizons may have been tilted during thrust evolution so as to facilitate reworking during Platberg extension (Fig. 19).

(ii) Rotational block faults where the blocks tilt most of the Central and West Rand sediments and detach on low-angle extensional detachments at depth (Fig. 20).

Half-grabens initiated during this stage were filled with clastic sediments and felsic volcanics to give rise to a syn-rift volcanosedimentary sequence, the Platberg Group. Over 7 km of Platberg sediments are preserved in some of the deeper half-grabens, west of the Witwatersrand basin. This group comprises immature clastic sediments, local stromatolitic limestones, thick quartz-porphyry flows and andesitic lavas.

The western part of the Witwatersrand basin was particularly affected by these extensional faults. The extensional system comprises low-angle detachment faults with large displacement, separated by step-over areas where the movement of one fault system is transferred to another. Some low angle detachments are linked by steeper dipping tear faults.

The WE-03 detachment (Fig. 16) is a large fault whose lateral ramp extends along the entire western edge of the Witwatersrand basin from the Welkom goldfield to the area west of Klerksdorp, where

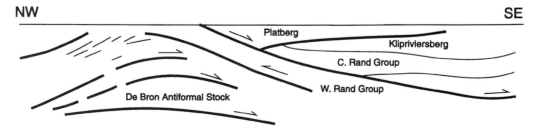

Fig. 19. Schematic section showing the development of the Peritus detachment fault on the southeast side of the antiformal stack, generally known as the 'De Bron horst'. Location B on Fig. 16.

NW SE

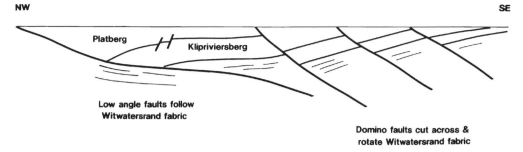

Fig. 20. Different styles of SE dipping extensional faults, based on a cross section between the Peritus area and south of the Vredefort dome (location C on Fig. 16). The main structure is a low-angle fault which appears to follow fabric in the Witwatersrand sediments. The hanging-wall of this sheet has been broken by straight rotational faults.

the strike of the detachment curves to the NE. The displacement here is *c.* 6 km and decreases steadily to the northeast. The throw is taken up progressively by the Buffels fault, located on the hangingwall of the WE-03 fault. The Buffels fault grows northeastwards where it becomes a typical detachment plane with numerous small, high-angle synthetic faults. In the Ventersdorp area the Buffels detachment defines a broad antiformal arch which is cored by basement, that is, it has the geometry of a low-angle detachment with an underlying basement core complex (Fig. 16).

The Fakawi detachment lies on the hanging wall of the WE-03 and Buffels detachments. Its southern tip in the Welkom goldfield comprises several E dipping normal faults which grow and coalesce to the northeast. The basal fault of this system develops into a flat detachment plane which underlies the De Bron horst. The De Bron horst comprises the footwalls of two relatively small faults that detach on the Fakawi fault at depth. To the northeast the Fakawi fault splits with a ramp developing from the flat and trending NW into the Klerksdorp area, parallel to the ramp on the WE-03 fault. In the Klerksdorp area the fault changes trend to strike northeastwards along the western edge of the Klerksdorp goldfield. The fault extends into the Carltonville area of the West Witwatersrand line as the Master Bedding Plane fault.

This extensional system on SE-dipping detachments is offset in the Carltonville area by a later thrust. On the hanging wall of this thrust, the Master Bedding Plane fault, or a very similar detachment, can be traced around the southern margin of the Johannesburg Dome as the Rietfontein fault system. In the southwest part of the Johannesburg Dome, the Rietfontein system strikes NW–SE, is steeply dipping and has oblique-to strike-slip displacement. In this area the Rietfontein fault is a lateral structure linking

extensional half-grabens in the West Rand with the Bez Valley graben in the East Rand.

Significant extension occurs in the southeastern part of the Witwatersrand basin, in the Evander, South Rand and Heilbron areas. In the Evander area the faults are largely rotational block faults. A large extensional fault, the Sugarbush fault, occurs along the southern edge of the Devon Dome, east of Heilbron. Myers *et al.* (1990) interpreted the normal faults in the eastern part of the Witwatersrand basin as dominantly strike-slip systems. No field evidence has been found to support this suggestion. Kinematic data suggest that the dominant movement was down to the southeast.

The large listric linked extensional faults of the Witwatersrand basin are some of the best developed and best imaged low-angle systems described. Wernicke (1985) describes gently dipping and listric extensional faults from the Basin and Range, western USA. However many of these structures are now interpreted as originally high-angle normal faults that have been rotated with increasing extension and footwall rebound (Wernicke & Axen 1988; Brun *et al.* 1994). The listric faults seen on seismic data from the Witwatersrand basin show low cut-off angles and flatten onto gently dipping reflectors at depth. They appear to be true low-angle listric faults and must owe their origin to:

(i) the presence of weak detachment planes at depth, possibly Witwatersrand shales with high fluid overpressures;

(ii) tilting and development of a SE directed surface slope due to thrust-related thickening and uplift of the hanging wall.

The extensional faults in the western part of the basin appear to cut through the West Rand sediments into basement. However the listric faults in the East Rand cannot be traced into basement. They

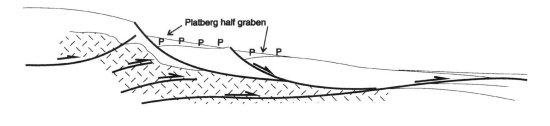

 **Base Platberg Group**

Basement

Fig. 21. Schematic section suggesting local detachment of listric faults on the basal thrust, i.e. this interpretation suggests that extension forming the Platberg grabens was co-eval with the latest thrust translation at depth.

detach within the basal shales in the West Rand Group. Figure 20 suggests that these listric faults may detach onto thrusts at depth, that is, the extension up-dip of the fault may be balanced by contraction down-dip. These structures may be considered as large scale gravity-driven collapse structures. Similar explanations have been suggested for the extensional faults in the Himalayas (Royden & Burchfiel 1987); low angle extensional faults are formed by the collapse of the mountain front into the Himalayan foreland basin. This interpretation suggests that some of the

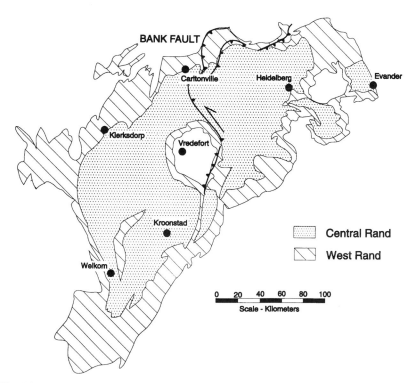

Fig. 22. Simplified map showing the distribution of major thrusts of post-Ventersdorp, pre-Transvaal age.

Platberg extensional faults were coeval with the last stages of thrusting and formed by the collapse of the mountain front into the Witwatersrand basin.

This model is supported by data from the Evander mining area where low-angle thrusts are joined by low-angle extensional shear zones. Cross-cutting relationships show that the thrusts and extensional faults were approximately contemporaneous, suggesting gravity collapse of the thrust mass.

Coward (1982) described similar gravity collapse structures from the frontal regions of the Moine thrust zone in Scotland and termed them surge zones. The Scottish surge zones are approximately 5 km by 5 km in size. The Witwatersrand collapse structures are an order of magnitude larger. They must represent large scale collapse of the mountain front with the generation of large sediment-filled basins on their hanging walls.

The NW-dipping extensional faults which occur west of the Witwatersrand basin and appear to rework earlier thrusts, presumably represent the more usual model of gravity collapse of a mountain belt (e.g. Wernicke 1985; Seranne *et al.* 1989).

Post-rift subsidence. A post-rift sag basin is represented by the Pniel Group, a volcanosedimentary package which oversteps the Platberg rift phase. In this sequence the volcanism is mainly mafic. A period of erosion locally preceeded the Pniel Group, so that the alluvial sediments were deposited on a faulted and folded paleosurface. These terrigenous clastic sediments constitute the Bothaville Formation, which were then buried beneath andesitic lavas of the Allanridge Formation.

The post-Platberg, pre-Pniel uplift may be due to combinations of:

(i) uplift of the top corners of tilted fault blocks during Platberg extension;

(ii) isostatic uplift, associated with lithospheric equilibration, following Witwatersrand thickening;

(iii) the episode of compressional tectonics responsible for uplift of the northwest edge of the basin and the gravity instability which drove the SE verging listric faults.

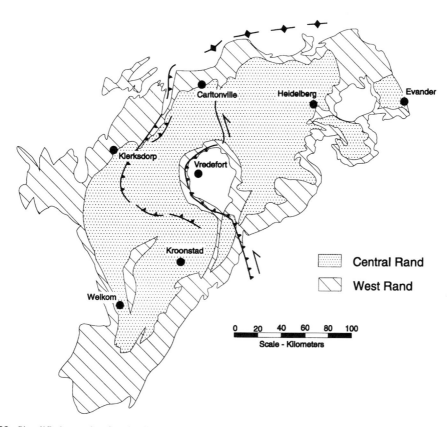

Fig. 23. Simplified map showing the distribution of major thrusts of post-Transvaal age.

Post-Ventersdorp inversion

The Ventersdorp and Witwatersrand Group sediments were locally uplifted, folded and thrust prior to Transvaal sedimentation, so that the Black Reef transgresses the earlier stratigraphy. This deformation partly pre-dates and partly post-dates the Pniel Group. An important sequence of thrusts of this age has been recognized to the west and south of the Johannesburg dome (Fig. 22). The largest structure of this age is the Bank fault in the West Rand which uplifts and offsets the earlier thrusts and extensional detachments. Some of the fault activity, southeast of Vredefort, is probably

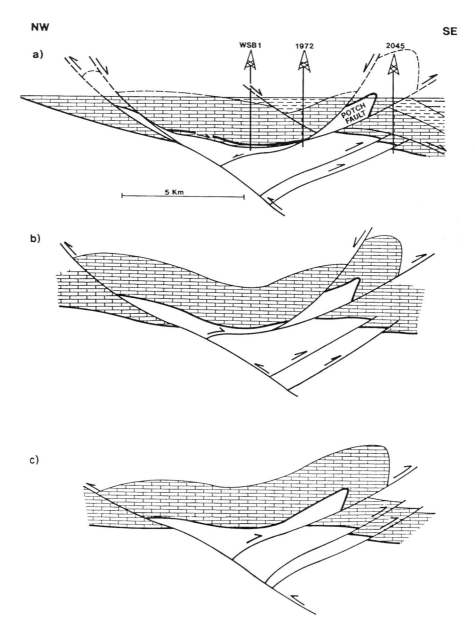

Fig. 24. Interpretation of the inversion structure near Potchefstroom. (**a**) Present section. (**b**) & (**c**) Sections sequentially restored to show the backthrusts and related hanging-wall folds associated with inversion on the Buffels detachment. Location D on Fig. 16.

of this age. Hence the renewed thrust activity is essentially confined to the eastern side of the basin, linked across the basin by a NW-SE trending transfer system. Note that this is a new interpretation of the structure of the Bank Fault, based on new seismic data, field studies and reviews of mine data.

The thrusts of late- to post-Ventersdorp age are principally confined to the eastern part of the basin. They include the area where the extensional faults detach on West Rand shales and the lowest thrust detachments (Fig. 21). They also upthrust basement along the Bank fault system and along its continuation on the thrusts, west of the Johannesburg dome. The thrusts described by Roering (1984) from Swartkops, in the western part of the Johannesburg dome, may be of this age. The Johannesburg dome must have been again tilted and uplifted at this time. The deformation may be considered as the last pulse of Archaean compressional deformation in Witwatersrand basin.

Transvaal sedimentation

Between the deposition of the Ventersdorp and Transvaal sediments there must have been sufficient time for the erosion and peneplanation of much of South Africa to allow the Black Reef quartzites to be deposited over a wide area of low relief. The lower Transvaal sediments consist of an approximately layer cake stratigraphy of shallow water carbonates and muds. Minor unconformities have been recognized within the sequence (Fig. 4). During the deposition of the Pretoria Group, in the upper Transvaal Supergroup, there must have been renewed uplift, possibly of the Limpopo belt to the north, providing a source of clastic sediments.

Post-Transvaal tectonic inversion

Following the Transvaal sedimentation there was renewed compressional tectonic activity in the Witwatersrand basin, during which many of the earlier extensional faults were reactivated as thrusts and the syn-rift Platberg Group and post-rift Pniel Group and Transvaal Supergroup were folded. Figure 23 shows the distribution of the larger of these structures. The largest post-Transvaal structure is the Vredefort uplift, sometimes referred to as the Vredefort Dome. The latter has been variously considered as (i) a meteorite impact crater, supported by the occurrence of shatter cones and pseudotachylite, or (ii) the result of some form of mantle upwelling. However new seismic data show the Vredefort dome to be largely thrust-controlled, probably formed as a push-up structure on a compressional bend to a large strike-slip zone.

Inversion geometry

During both post-Ventersdorp/pre-Transvaal and post-Transvaal age compressional deformation new thrust structures were formed, but also older extensional faults were reworked in a range in structures characteristic of tectonic basin inversion (cf. Gillcrist et al. 1987). Older normal faults were reactivated as thrusts, especially where they originally formed relatively flat-lying detachments. The inversion involved:

(i) reverse movement up the original normal faults;
(ii) back-rotation of the faults associated with basin shortening;
(iii) shortening of the hanging walls of the normal faults into a series of compressional folds;
(iv) the generation of new short-cut thrusts where the original normal faults had steeper dips; the short-cut thrusts occur on the hanging walls (e.g. in the western Rand) or footwalls (e.g. in the eastern Rand) of the earlier normal faults.

Steeply dipping faults are less likely to reactivate than gently dipping faults, particularly as they rotate away from the preferred orientation for failure (see Gillcrist et al. 1987). Hence gently dipping faults may reactivate at depth, but may stick at higher levels where they are steeper. The displacement will die out into a fold. As rotation and shortening progress, more beds are folded to allow for further strain and the zone of folding will migrate back, away from the steeper part of the fault. On the hanging wall, adjacent to the steeper-dipping portion of the fault, the layers may need to develop extra shortening strains, probably in the form of a backthrust or antithetic shear zone. Examples of these structures occur on the hanging

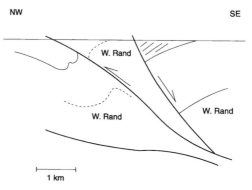

Fig. 25. Sketch section through an 'Îles flottants' structure from the western Rand.

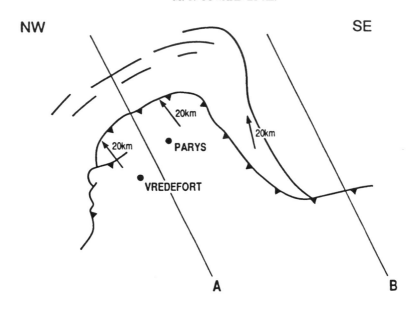

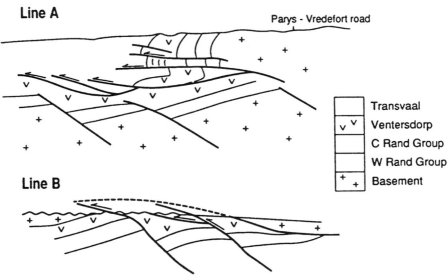

Fig. 26. Simplified map and section through the northern part of the Vredefort dome. This section may be compared with the region east of the Dome where the earlier extensional structures still show net extension.

walls of many of the detachments in the northwest part of the basin.

In the Potchefstroom area, for example, the inversion structures are most dramatic in the Transvaal sediments which are deformed by an E-verging thrust with a tight hangingwall anticline (Fig. 24). This E-verging thrust is a backthrust from a W-verging fault which thrusts Basement and Black Reef over the Malmami Subgroup of the Transvaal Supergroup. This W-verging thrust reworks the Buffels detachment fault at depth. A W-dipping normal fault, with a throw of about 2 km has collapsed the western limb of the E-verging hanging-wall fold.

Instead of folding the hanging wall of the fault, a new short-cut thrust may develop. The short-cut structure may be in the footwall or the hanging wall of the listric or kinked thrust. The footwall short-

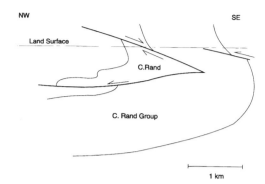

Fig. 27. Simplified section through the northwest part of the Vredefort dome showing wedge extrusion from the inner arc of the fold.

Table 1. *Limpopo belt tectonics*

Collapse of southern margin to make the Soutpansberg (related to Palala shear?)
Uplift of the Central Zone — 2000 Ma
Intrusion of the Great Dyke — 2500 ± 60 Ma
Thickening and extrusion of the Zimbabwe Craton — 2600 Ma
Rotational shears of the North Kaapvaal overthrust of the Limpopo belt and uplift of central and southern zone granulites
Collision tectonics: Limpopo belt and accretion of Zimbabwe into the Kaapvaal craton — 2850–3000 Ma
Amalgamation of main Kaapvaal craton — pre-3000 Ma

cuts commonly lead to the development of 'floating islands' of pre-rift material, bounded by the original normal fault and by the short-cut. Isolated wedges of footwall rock may be translated onto a thrust hangingwall. At small values of inversion the short-cuts may have only limited displacement and form upward fanning horsetail patterns, similar to those described experimentally by Buchanan & McClay (1991). At higher values of contraction the footwall short-cuts may be responsible for

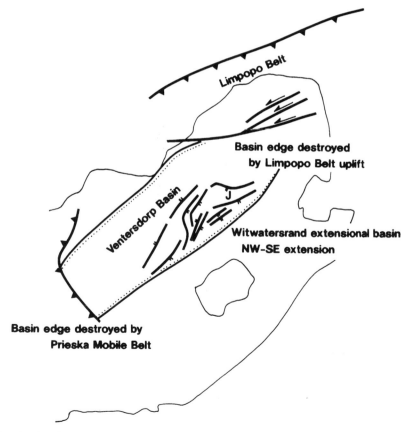

Fig. 28. Map showing the area covered by the Ventersdorp basin and the regions where the extension was dominantly down to the northwest, down-dip of thrusts and where the extension was down to the southeast on listric faults, exaggerating the shape of the Witwatersrand basin.

Fig. 29. The sub-Andean basins of South America.

generating a lower angle, more smoothly varying thrust trajectory (e.g. Gillcrist *et al.* 1987). Figure 25 shows a simplified section through an isle flottante from the West Rand.

During inversion the syn-rift fill may be squeezed out into a hangingwall fold and the post-rift sediments will show only shortening. The post-rift cover may shorten in a series of folds or thrusts, some of which may detach close to the syn-rift/post-rift boundary and, unless deep data are available, may be mistaken for thin-skinned detachment structures. The Transvaal sediments to the north and west of the Vredefort dome show both large and small wavelength folds, caused not by thin-skinned movements, but thick-skinned basement fault reactivation.

The types of folds associated with inversion are characterized by relatively planar backlimb dips and short hooked forelimbs. The faults may have extensional geometry at depth, passing through a null point (Williams *et al.* 1989) to a reverse fault at higher levels and then die out in a hooked tip beneath the forelimbs of these structures.

The largest structure associated with basin inversion is the Vredefort dome (Fig. 26). Along the northern and western margin of this dome the West Rand and Central Rand sediments are vertical to overturned. The post-Transvaal deformation involves movement along bedding parallel thrusts and the generation of fault propagation folds, which were tilted to become downward-facing during the development of the main Vredefort structure. From seismic studies around the dome, the steeply dipping to overturned beds can be shown to be a hangingwall structure to a large fault which lies a few hundred metres below the present land surface. The steeply dipping beds were then cut by late flat thrusts; the latter may branch from the main Vredefort thrust at depth or may be rootless and owe their origin to the fact that they lie on the inner arc of the tight hanging-wall fold.

These late structures include SE-dipping thrusts and normal faults which bound wedge-shaped blocks squeezed out in a NW direction from the inner arc of the inclined syncline (Fig. 27). Within the wedges the steeply dipping West and Central Rand sediments have different strikes to those above or below, suggesting that the wedges have been locally rotated during expulsion.

Beneath the northwest part of the Vredefort dome there is no clear evidence for original extensional faults. However to the east the Vredefort structures appear to pass into reactivated normal faults (Fig. 26). The amount of inversion decreases eastwards, that is, in the W the normal faults show net contraction but in the east they show only partial inversion. The biggest change in amount of inversion occurs across steep NW–SE-trending faults, which transfer the post-Transvaal compression to the north, into the region north of the Johannesburg dome (Fig. 23).

On a more regional scale, the post-Transvaal inversion appears to be confined to a NNW-trending zone across the centre of the Witwatersrand basin. The Vredefort dome appears to be a large-scale uplift in a compressional jog/offset along this zone. The presence of pseudotachylite around the Vredefort structure suggests seismic-type movements on the faults. Shatter cones suggest rapid or even explosive deformation, related to growth of the uplift, or more likely its subsequent gravitational collapse.

Discussion

The following are relevant to the origin of the Witwatersrand basin.

(1) The Witwatersrand Basin has been generated by:

(i) rift tectonics during deposition of the Platberg sediments and probably earlier, during deposition of the Dominion Group;

(ii) post-rift subsidence, possibly thermal in origin, related to asthenospheric cooling following stretching or volcanic activity;

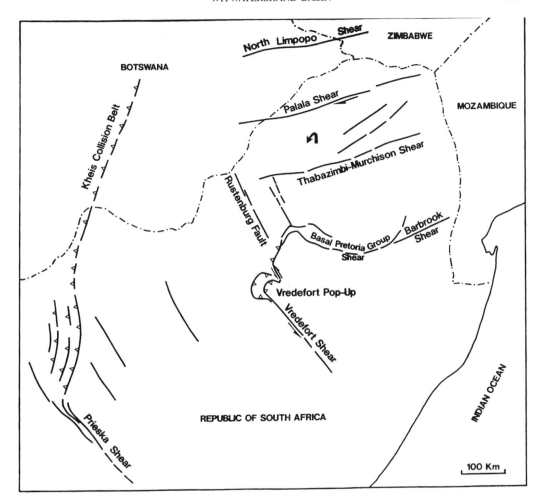

Fig. 30. the Vredefort shear system and Vredefort pop-up as a zone of intraplate deformation approximately 2000 Ma ago, related to collision tectonics of the Kheis belt in Botwana and the western Transvaal. The Palala shear and Thabazimbi–Murchison lineament were probably activated at approximately the same time.

(iii) flexural loading of the lithosphere due to thrust tectonics to the northwest and southeast of the basin.

(2) The transport direction was dominantly NW–SE throughout the extensional and compressional history. The flexural subsidence was formed by loading associated with compression, perpendicular to the strike of the basin.

(3) The most intense volcanism, stretching and compression occurred to the northwest of the basin. The thrusts of Central Rand and Klipriviersberg age propagated to the southeast into the basin. Hence the most important zone of tectonic activity lay to the northwest, in the NW Transvaal.

(4) Along the northwest edge of the Kaapvaal craton there was coeval tectonic activity associated with accretion tectonics in southern Zimbabwe and the early deformation of the Limpopo belt (Treloar *et al.* 1992). The active convergent plate margin lay along the northwest edge of the Limpopo belt, which constituted a zone of crustal thickening and high grade metamorphism at this time (Treloar *et al.* 1992). A summary of tectonic events in the Limpopo belt is given in Table 1.

(5) Limpopo belt tectonics and the Ventersdorp volcanism suggest there was a SE-dipping subduction zone beneath the Limpopo belt and the Kaapvaal craton. The area of Ventersdorp volcanism and extension is shown in Fig. 26; the Witwatersrand basin lies on the southeast margin of this volcanic province. The Platberg basin can

therefore be considered as a back-arc basin above this SE-dipping subduction zone.

(6) The Witwatersrand basin can be considered as a hinterland basin, possibly equivalent to a present day sub-Andean basin. Analogues include the Maracaibo basin in Venezuela (cf. Myers *et al.* 1990), the Magdalena basins of Colombia, or the Pampas basins of western Argentina (Fig. 29).

(7) The sediments of the Witwatersrand basin are deformed by Late Archaean structures, possibly associated with the convergent plate processes along the Limpopo margin. Tectonic thickening along the northern margin of the Limpopo belt, associated with closure of the Zimbabwe green-stone basins, may be coeval with some of the post-Ventersdorp thrusting and basin inversion. The northern edge of the Ventersdorp basin (Fig. 28) is cut by the large shear zones along the southern edge of the Limpopo belt.

(8) Proterozoic plate collision in the western Transvaal and in Botswana, generating the Kheis thrust belt at 1800–2000 Ma, is similar in age to further reworking of the Central zone of the Limpopo belt and renewed movement along shear zones in the northern Transvaal. The Vredefort uplift and the post-Transvaal thrusting and basin inversion may be coeval with these events (Fig. 30).

Conclusions

The Witwatersrand basin is an intracratonic basin, essentially similar to the sub-Andean basins of South America (cf. Winter 1987; Myers *et al.* 1990). Hence by end-Archaean times plate process-es in South Africa were not dissimilar to those of the present day at young convergent plate margins. Because of the mining exploration activity, the kinematics of the Witwatersrand basin are better understood than are those of many Cenozoic basins. The Witwatersrand basin may possibly be used as a model to help interpret some of these younger basins.

References

ANTROBUS, E. S. A., BRINK, W. C. J., BRINK, M. C., CAULKIN, J., HUTCHINSON, R. I., THOMAS, D. E., VAN GRAAN, J. A. & VILJOEN, J. J. 1986. The Klerksdorp Goldfield. *In*: ANHAEUSSER, C. R. & MASKE, S. (eds) *Mineral Deposits of Southern Africa*, I, Geological Society of South Africa, 549–598.

ARMSTRONG, R. A., COMPSTON, W., RETIEF, E. A. & WELKE, H. J. 1986. Ages and isotope evolution of the Ventersdorp volcanics. *Extended Abstract, Geocongress 86*, Johannesburg, 89–92.

——, ——, & WILLIAMS, I. S. 1990. Geochronological constraints on the evolution of the Witwatersrand Basin (South Africa), as deduced from single zirzon U/Pb ion microprobe studies. *In*: GLOVER, J. S. & HO, S. E. (eds) *Third International Archean Symposium, Extended Abstracts*, 287–288.

——, COMPSTON, W., RETIEF, E. A., WILLIAMS, I. S. & WELKE, H. J. 1991. Zircon ion microprobe studies bearing on the age and evolution of the Witwatersrand Triad. *Precambrian Research*, 53, 243–266.

BARTON, J. M., COMPSTON, W., WILLIAMS, I. S., BRISTOW, J. W., HALLBAUER, D. K. & SMITH, C. B. 1989. Provenance ages for the Witwatersrand Supergroup and the Ventersdorp contact reef: constraints from ion microprobe (U-Pb) ages of detrital zirzons. *Economic Geology*, 84, 2012–2019.

BEUKES, N. & CAIRNCROSS, B. 1991. A lithostrati-graphic–sedimentological reference profile for the Late Archean Mozaan Group, Pongola Sequence: application to sequence stratigraphy and correlation with the Witwatersrand Supergroup. *South African Journal of Geology*, 94, 44–69.

BICKLE, M. J. & ERIKSSON, K. A. 1982. Evolution and subsidence of early Precambrian sedimentary basins. *Philosophical Transactions of the Royal Society, London*, 305, 225–247.

BROCK, B. B. & PRETORIUS, D. A. 1964. Rand basin sedimentation and tectonics. *In*: HAUGHTON, S. H. (ed.) *the Geology of some ore deposits in South Africa*. Geological Society of South Africa, 1, 549–600.

BRUN, J-P., SOKOUTIS, D. & VAN DEN DRIESSCHE, J. 1994. Analogue modelling of detachment fault systems and core complexes. *Geology*, 22, 319–322.

BUCHANAN, P. G. & MCCLAY, K. R. 1991. Sandbox experiments of inverted listric and planar fault systems. *Tectonophysics*, 188, 97–115.

BURKE, K., KIDD, W. S. F. & KUSKY, T. M. 1985. Is the Ventersdorp rift system of southern Africa related to continental collision between the Kaapvaal and Zimbabwe cratons at 2,640 Ma ago? Tectonophysics, 115, 1–24.

——, —— & —— 1986. Archean foreland basin tectonics in the Witwatersrand, South Africa. *Tectonics*, 5, 439–456.

COWARD, M. P. 1982. Surge zones in the Moine thrust zone of NW Scotland. *Journal of Structural Geology*, 4, 247–256.

DE BEER, J. M. & EGLINGTON, B. M. 1991. Archean sedimentation on the Kaapvaal craton in relation to tectonics in the granite greenstone terrains: geophysical and geochronological constraints. *Journal of African Earth Sciences*, 13, 27–44.

DE WIT, M. J., ROERING, C., HART, R. J., ARMSTRONG, R. A., DE RONDE, C. E. J., GREEN, R. W. E., TREDOUX, M., PEBERDY, E. & HART, R. A. 1992. Formation of an Archean continent. *Nature*, 357, 553–562.

GILLCRIST, R., COWARD, M. P. & MUGNIER, J. L. 1987. Structural inversion and its controls: examples from the Alpine foreland and the French Alps. *Geodinamica Acta*, **1**, 5–34.

GRAHAM, R. H. 1995. Asian analogues for Precambrian tectonics? *This volume*.

HUTCHINSON, R. I. 1975. *The Witwatersrand system as a model of sedimentation in an intracratonic basin*. DSc thesis, Orange Free State, Bloemfontein.

JAHN, B.-M., BERTRAND-SATARTI, J., MORIN, N. & MACE, J. 1990. Direct dating of stromatolitic carbonates from the Schmidtsdrift Formation (Transvaal Dolomite), South Africa, with implications on the age of the Ventersdorp Supergroup. *Geology*, **18**, 1211–1214.

MINTER, W. E. L. 1976. Detrital gold, uraninite and pyrite concentrations related to sedimentology in the Precambrian Vaal Reef, Witwatersrand, South Africa. *Economic Geology*, **71**, 157–175.

MYERS, R. E., MCCARTHY, T. S. & STANISTREET, I. G. 1990. A tectono-sedimentary reconstruction of the development and evolution of the Witwatersrand Basin, with particular emphasis on the Central Rand Group. *South African Journal of Geology*, **93**, 180–201.

PHILLIPS, G. N. & LAW, J. D. M. 1994. Metamorphism of the Witwatersrand goldfields: a review. *Ore Geology Reviews*, **9**, 1–31.

PRETORIUS, D. A. 1981. Gold and uranium in quart-pebble conglomerates. *Economic Geology, 75th Anniversary Volume*, 117–138.

ROBB, L. J., DAVIS, D. W. & KAMO, S. L. 1990. U-Pb ages on single detrital zirzon grains from the Witwatersrand Basin, South Africa: constraints on the age of sedimentation and on the evolution of granites adjacent to the basin. *Journal of Geology*, **98**, 311–328.

ROERING, C. 1984. The Witwatersrand Supergroup at Swartkops: a re-examination of the structural geology. *Transactions of the Geological Society of South Africa*, **87**, 87–100.

ROERING, C. J., DE BEER, J. H., VAN REENAN, D. D. WOLHUTER, L. E., BARTON, J. M., SMIT, C. A., MCCOURT, S., DU PLESSIS, C. A., STETTLER, E. H., BRANDL, G., PRETORIUS, S. J., VAN SCHALKWYK, J. F. & GEERTSON, K. 1990. A geotransect across the Limpopo belt. *In*: BARTON, J. M. (ed.) *The Limpopo Belt: a field workshop on granulites and deep crustal tectonics*. Extended Abstracts, Rand Africaans University, Johannesburg, 100–101.

ROYDEN, L. H. & BURCHFIEL, B. C. 1987. Thin-skinned north-south extension within the convergent Himalayan region: gravitational collapse of a Miocene topographic front. *In*: COWARD, M. P., DEWEY, J. F. & HANCOCK, P. L. (eds) *Continental Extensional Tectonics*. Geological Society, Special Publications, **28**, 611–619.

SERRANE, M., CHAUVET, A., SEGURET, M. & BRUNEL, M. 1989. Tectonics of the Devonian collapse basins of western Norway. *Bulletin of the Geological Society of France*, **8**, 48–54.

TANKARD, A. J., JACKSON, M. P. A., ERIKSSON, K. A., HOBDAY, D. K., HUNTER, D. R. & MINTER, W. E. L. 1982. *The crustal evolution of Southern Africa*. Springer-Verlag, New York.

TRELOAR, P. J., COWARD, M. P. & HARRIS, N. B. W. 1992. Himalayan–tibetan analogies for the evolution of the Zimbabwe Craton and Limpopo Belt. *Precambrian Research*, **55**, 571–587.

VAN BILJON, W. J. 1980. Plate tectonics and the origin of the Witwatersrand Basin. *In*: RIDGE, J. W. (ed.) *Proceedings 5th IAGOD Symposium*, **1**, Schweizerbartsche Verlagbuchhandlung, Stuttgart, 333–384.

VAN NIEKERK, C. B. & BURGER, A. J. 1964. The age of the Ventersdorp System. *South African Geological Survey Annals*, **3**, 75–86.

—— & —— 1978. A new age for the Ventersdorp acidic lavas. *Transactions of the Geological Society of South Africa*, **4**, 99–106.

WERNICKE, B. 1985. Uniform sense simple shear of the continental lithosphere. *Canadian Journal of Earth Sciences*, **22**, 109–125.

—— & AXEN, G. J. 1988. On the role of isostacy in the evolution of normal fault systems. *Geology*, **16**, 848–851.

WILLIAMS, G. D., POWELL, C. M. & COOPER, M. A. 1989. Geometry and kinematics of inversion tectonics. *In*: COOPER, M. A. & WILLIAMS, G. D. (eds) *Inversion Tectonics*. Geological Society, London, Special Publications, **44**, 3–16.

WINTER, H. DE LA R. & BRINK, M. C. 1991. Chronostratigraphic subdivision of the Witwatersrand basin based on a Western Transvaal composite column. *South African Journal of Geology*, **94**, 191–203.

Asian analogues for Precambrian tectonics?

R. H. GRAHAM

BP Exploration Co. (Colombia) Ltd, Edifico Seguras de Conercio Carreras 9A 99-02,
Sante Fe de Bogata, Colombia

Abstract: A non-specialist view is that the Archaean crust comprises either tonalite/greenstone belt associations or gneissic terrains in granulite facies. Archaean crustal remnants lie as undeformed pods between ramifying Proterozoic mobile belts. The cratons themselves may have unconformable remnants of undeformed Proterozoic 'platform' sediments.

Coward, Windley and others have already drawn attention to similarities between the accretionary collage which makes Central Asia and the tectonic patterns of the Precambrian. These similarities are developed and amplified in this paper. Central Asian basins such as Tarim, Junggar and their satellites have developed on stretched continental fragments now accreted within the Asian collage. This stretched lithosphere, having been equilibrated and strengthened, now forms resistant kernels around which the surrounding ramifying accretionary deformation belts have been moulded.

Enclosed foreland basins have developed on the continental fragments in response to loading of the adjacent mountain belts, and great thicknesses of mostly Tertiary sediments have accumulated in them. The 'mobile belts' which surround the basins have been reactivated by later, distant, collision events with corresponding loading and sedimentation events in the basin.

The Black Sea and South Caspian basins are back-arc small ocean basins incorporated within the collage. They also form resistant kernels around which the mountain chains are moulded.

Deep erosion levels through these systems might provide broad analogues for Precambrian tectonic patterns, i.e. the Black Sea for the tonalite/greenstone belt terrains, the deep levels of Tarim for 'reworked' granulite terrains like John Sutton's homeground of the NW Highland foreland. Sutton and Watson's two orogenies are manifested as a ductile shear fabric associated with back-arc stretching (Scourian) reworked in zones after a substantial time interval (Laxfordian). Dykes like the Scourie dykes might be associated with stretching and incipient continental separation, and therefore be close in time with the ductile stretching as, arguably, the Scourie dykes are with the late Scourian (Inverian) deformation in Scotland.

John Sutton always looked for simple, unifying themes in geology. He often worked by drawing analogies from his great general geological experience to support these themes, most of which were, of course, concerned with the early history of the earth.

This paper attempts the same sort of approach. It tries to use the Phanerozoic evolution of parts of Asia as an analogue to explain some of the patterns of Precambrian tectonics, and it makes specific reference to Sutton and Watson's heartland in the Scourie–Laxford region of NW Scotland. One of the main analogues is from China where in the early 1970s, John Sutton, together with Janet Watson, was typically a torch-bearer in promoting scientific dialogue and exchange between China and the West.

There are shortcomings in this paper which I suspect John Sutton would have been the first to see. Sadly, my own first-hand knowledge of Precambrian geology came to an end 20 years ago and has not really moved on. Thanks to the work of Sutton and Watson, Windley, Bridgewater,

Shackleton and others, (e.g. Sutton & Watson 1951; Windley & Bridgewater 1971; Windley 1973; Shackleton 1976) we had at that time a generalized picture of the Precambrian crust as a collage of Archaean 'cratons' surrounded by, and partially 're-worked' in Proterozoic 'mobile belts'.

The cratons were believed to belong in two categories, either 'granite–greenstone terrains', with upright synclinal keels of low(ish) grade metasediments and metavolcanics swamped by intrusive tonalites and granites, or else areas of high grade banded gneisses locally in granulite facies and generally with sub-horizontal deformation fabrics. Both kinds of cratons are locally overlain by the remnants of once continuous, little-deformed 'platform' sedimentary sequences, e.g. Witwatersrand and younger formations in Africa, Riphean and Vendian of Eastern Siberia, Torridonian of Scotland, which may be extremely thick, e.g. 10 km of shallow water carbonates and quartzites in eastern Siberia.

The Proterozoic 'mobile belts' form linear, ramifying zones of contractional, transpressive

From COWARD, M. P. & RIES, A. C. (eds), 1995, *Early Precambrian Processes,*
Geological Society Special Publication No. 95, pp. 271–289.

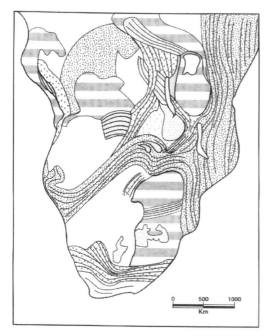

Fig. 1. Simplified tectonic map of Southern Africa showing cratons, areas of undeformed platform sediments and mobile belts (after Cahen & Snelling 1966). Trend lines show mobile belts; areas without trends are platform sediments of equivalent age. Broad strips are cratons > 2.5 Ma; fine dots 2.1–1.95 Ma; coarse dots 1.3–1.1 Ma; dashed-hatch 0.73–0.6 Ma.

or transcurrent ductile strain around the cratonic kernels. They are commonly associated with renewed metamorphism in amphibolite facies and the influx of water. Sutton and Watson's Scourian–Laxfordian story is the classic description of the 're-working' of an Archaean gneissic terrain by a Proterozoic mobile belt.

Beyond the front of 'reworking' there are localized ductile shear zones where the sub-horizontal Archaean fabric and the Scourie dykes, that cut it together become a new sub-vertical finite strain fabric. The cross-cutting relationships of the dykes are reduced or obliterated by strain.

The Lewisian complex of Scotland is only a tiny fragment of Precambrian crust, preserved on an uplifted Cenozoic passive margin, west of a Palaeozoic thrust belt. It is on the maps of the larger areas of exposed Precambrian in Africa, Australia and Canada, all areas well known to Sutton and Watson, that Proterozoic mobile belts are seen snaking round Archaean cratonic nuclei Fig. 1.

Coward (1991) and Windley (pers. comm. 1991) have already pointed to a striking superficial resemblance between these patterns, and those

visible on maps of Central Asia where Cenozoic mountain belts ramify around enclosed sedimentary basins like Junggar, Tarim, Xaidam and Fergana, and the inland marine basins of the Black Sea and the South Caspian (Fig. 2). The basins locally contain vast sedimentary thicknesses. They may be elevated, like Tarim with an average elevation of 2 km), or at or below sea level, like Turfan in Xinjaing. The orogenic belts that surround them constitute some of the greatest mountain chains on earth. Precambrian topographic expression is, of course, subdued and secondary, though perhaps there is a tendency for cratons to be higher, especially where Proterozoic platform cover remnants exist.

On a geological map, both patterns look similar to those created by garnets in a micaschist or resistant pods in a ductile shear zone system. It is difficult not to conclude that the Precambrian cratons and the Asian basins both represent strong resistant kernels around which less competent rocks have been moulded.

Central Asian basins: later history

Most of the Central Asian basins are composite, that is to say they owe their origin to more than one basin forming process. Since the beginning of the Tertiary, most of them have subsided in response to the loading of the adjacent mountain belts, episodic subsidence in the basins correlating with episodic deformation in the mountain belts.

In this sense, therefore they are foreland basins, though clearly rather special ones. Cobbold *et al.* (1993) refer to them as 'Push-down basins', Coward (1992) calls them 'Butterfly basins'. They can be seen as end-members of a family of basins associated with orogeny. The variation within this family is conveniently (non-generically) illustrated on the triangular diagram of Fig. 3 which lists the idealized characteristics of end members, and suggests where some of the world's basins might fit. It emphasizes the difference between 'classic' Alpine-type foreland basins, with their asymmetry, their association with diachronous facies, and depocentre shift in response to rolling orogenic load, and the Central Asian type basins which are relatively symmetric, relatively fixed, and commonly contain enormous thicknesses of sediment (10 km in Xaidam, 15 km in the Black Sea).

The mountain chains which surround the basins are long-lived structures which have been rejuvenated in response to every collision on the south side of Asia, most importantly the collision of India.

Although early in their histories many of these mountain belts were arc-accretionary complexes, the later part of their history, that which goes along

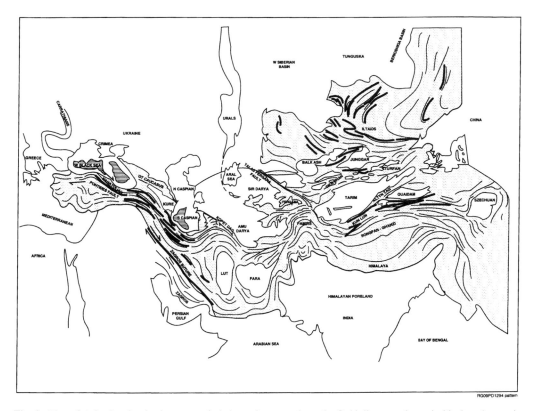

Fig. 2. Map of Asia showing basins, orogenic belts and structural trends. Ophiolites are shown in black and oceanic crust is lined.

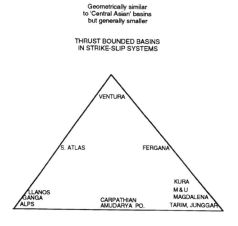

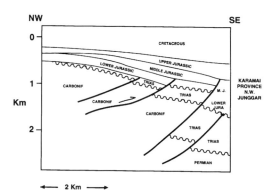

Fig. 4. 'Backstepping' deformation sequence in NW Junggar (after Lawrence *et al.* 1900).

Fig. 3. Triangular diagram illustrating the spectrum of basins associated with the loading produced by contractional or oblique-slip tectonics. The 'end members' are the apeces of the triangle, and their characteristics are noted on the figure. Many real basins fall between the end members. Examples of some real basins are shown on the diagram.

CLASSIC "ALPINE" FORELAND BASINS

Asymmetric,
Shifting depocentre
Diachronous facies,
marine to terrestrial
with time.
Piggy-back sequence
important "thin skinned"
deformation.
External drainage.

THRUST-BOUNDED CENTRAL ASIAN BASINS

More symmetric
Relatively fixed.
Either marine or terrestrial,
depending on crustal thickness.
Backstepping thrust sequence
Thick skinned, basement-
involved structures dominate
Internal drainage.

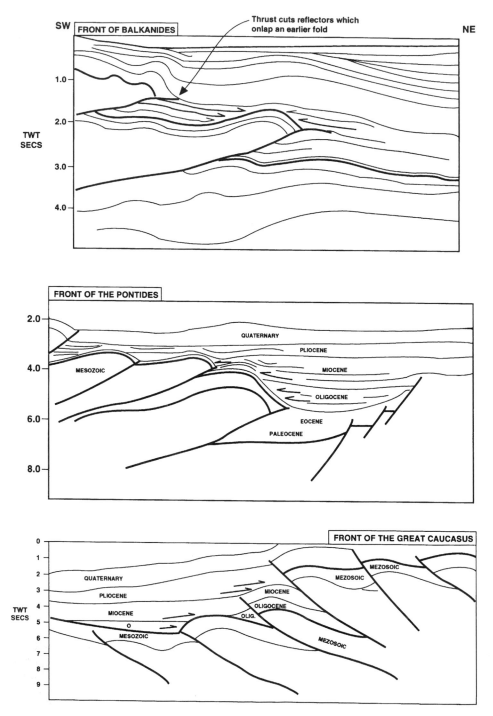

Fig. 5. Illustrations of 'backstepping' in the Western Black Sea (line drawings of interpreted seismic lines figured in Bocalletti *et al.* 1988).

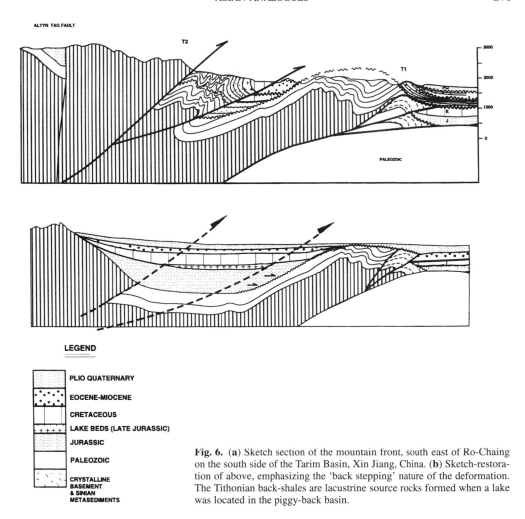

LEGEND

PLIO QUATERNARY

EOCENE-MIOCENE

CRETACEOUS

LAKE BEDS (LATE JURASSIC)

JURASSIC

PALEOZOIC

CRYSTALLINE
BASEMENT
& SINIAN
METASEDIMENTS

Fig. 6. (**a**) Sketch section of the mountain front, south east of Ro-Chaing on the south side of the Tarim Basin, Xin Jiang, China. (**b**) Sketch-restoration of above, emphasizing the 'back stepping' nature of the deformation. The Tithonian back-shales are lacustrine source rocks formed when a lake was located in the piggy-back basin.

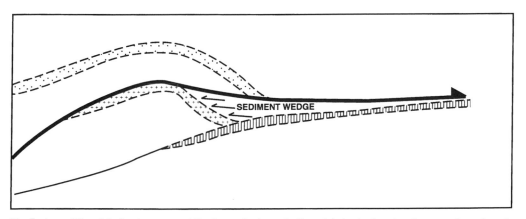

Fig. 7. A possible origin for the structural 'back stepping' seen in Central Asian basins. A sediment wedge onlaps the mountain front and reduces the critical orogenic taper. To regain the lost critical taper, the thrusts step back to increase the topographic slope. The big arrow indicates the main thrust, the small half-arrows show onlaps.

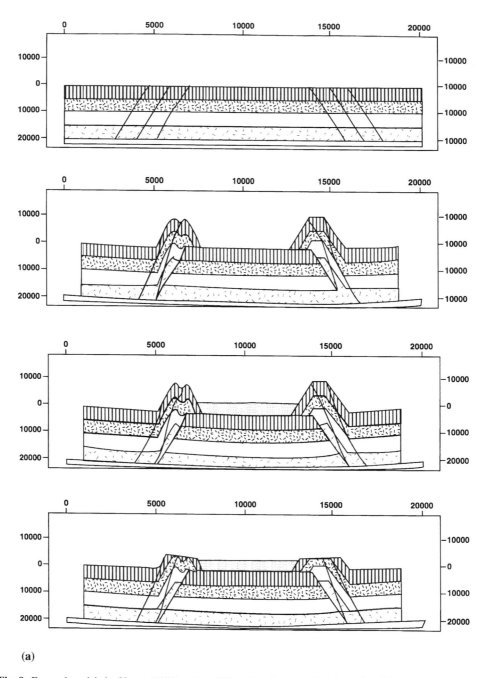

(a)

Fig. 8. Forward models by Young (1989) using a BP package incorporating thermal and isostatic responses to basin evolution) of a 'Central Asian' basin with the dimensions of Fergana, in the western Tien Shan (C.I.S.).A back stepping thrust sequences is chosen by analogy with natural examples. The area between the opposing thrusts is filled in with sediment and the thrust-related anticlines are eroded. Lower values for lithospheric rigidity, or greater distance between tht thrust, belts, would have generated a central high in the basin. Note that the end product is a basin with 10 km of sediment and 3 km of elevation (not unlike Tarim).

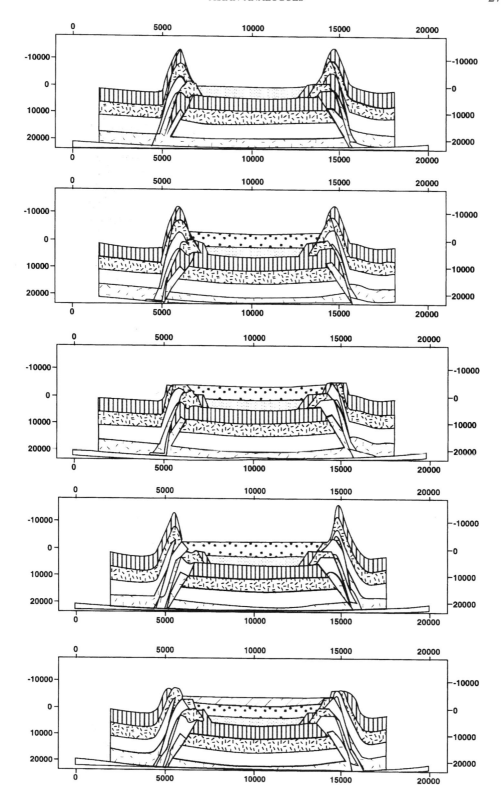

with the 'foreland basin' part of the evolution of the adjacent basins, involved a more 'thick-skinned' type of deformation, and today many of the ranges are lithospheric pop-ups more comparable with the Pyrenees, the Great Caucasus or the Eastern Cordillera of Colombia than with the Alps. Commonly they are highly transpressional and associated with great strike-slip faults like the Altyn Tag, the Talas–Fergana or the North Anatolian Fault. Also they seem to show deformation sequences which backstep, that is to say, new folds and thrusts develop above, and internal to, older ones, so thay early formed folds, deep and out in front, are onlapped by the sedimentary sequences which fill the accommodation space created by the backstepping load. We could also speak of 'out of sequence' or 'breakback'deformationn. Lawrence (1990) already illustrated this phenomenon in the Karamay hydrocarbon province of the northwestern part of the Junggar basin in Xin Jiang (Fig. 4). The hydrocarbon potential of fold structures, which become buried rather than elevated, will probably be evident to most readers.

Backstepping structural sequences are also visible on seismic sections through all the fold belts which surround the Black Sea (Fig. 5). They are as a rule on the south side of Tarim (Fig. 6), in Fergana, around the Caspian and elsewhere. Empirically they seem to be a feature of this kind of basin. Their origin is not altogether certain, but a possible explanation might be found in the internal drainage system of the basins. Most of the sediment eroded from the mountains is dumped in the basins. The sediment wedges onlap the mountains, and eventually overwhelm them. I remember standing at 4500 m on an outwash fan on the south side of the Tarim Basin. Only a few tens of metres of rock outcrop (a mountain top) peeped up through the fan. Since sedimentation in 'Central Asian type' basins was syn-tectonic, the sediment wedge presumably contributed to the overall shape of the tectonic mountain-front wedge, and reduced its taper to something below critical. In order to restore the critical taper, the thrusts must have stepped back (Fig. 7).

Young (unpublished 1989) forward modelled Central Asian type basins using a BP modelling package and series of reasonable, conventional assumptions about lithospheric behaviour. The dimensions are those of the Fergana basin of Kirghizia. The results are shown in Fig. 8. The models assume that subsidence is driven by orogenic load, and therefore contrast interestingly with Davey & Cobbold's (1991) models of lithospheric buckling.

The illustration of how vast thicknesses sediment may occur in these basins, and how the topographic height of a place like Tarim is partly a function of ther amount of sediment dumped into it, is telling.

Early history of the Central Asian basins

I have tried to make clear that the 'foreland basin', load-driven phase of the Central Asian Basins is only a part of their history. Most of them have been intra-continental basins only since the Mesozoic.

Before their incorporation into the Eurasian collage (Fig. 9), they were either back-arc rift basins or back-arc small ocean basins on the Eurasian margin, or they were microcontinents with attached trench–arc–accretion systems which had migrated across Tethys to collide with Eurasia (Fig. 10).

The passive subsidence which followed the back-arc extension or continental breakup was replaced by load driven subsidence after collision or after back-arc extension was replaced by back-arc contraction.

The mountain belts which bound the basins began their lives as trench-arc systems which became suture zones between microcontents as they accreted on a grand scale on to Eurasia during the Palaeozoic and early Mesozoic (Şengör et. al. 1993). The deep water sediments, ophiolites and volcanics in the suture zones moulded around the basins on collision, and the pattern was perpetuated and exaggerated by subsequent deformation episodes initiated by other, more distant collisions.

This is the pattern of 'garnets in a micaschist' or 'resistant pods in a ductile shear-zone' which was referred to at the begining of the paper. It obviously indicates a difference in the fundamental strength of the basins and the orogenic belts, and we must speculate on possible explanations for the origin of the pattern. One of them might follow Ziegler's (1988) arguments that new mantle lithosphere which equilibrated during, and reponsible for, post-rift thermal subsidence in 'McKenzie style' (1978) extensional basins is stronger than the material which it replaces (presumably because it is homogeneous while the old lithosphere which it replaces was heterogeneous and full of zones of weakness). If this is the case, it almost certainly follows that new mantle lithosphere would be stronger than the heterogeneous assemblage of deep water sediments, volcanics and ophiolite which would make up the arc-acretionary complexes associated with the back-arc basins and microcontinents. The latter eventually became the suture zones of the Eurasian collage, and afterwards the orogenic belts separating the basins.

This difference in the fundamental strength of the fold/thrust belts and the basins presumably applies equally to basins built on stretched microcontinents (like Tarim), to back-arc rift basins still

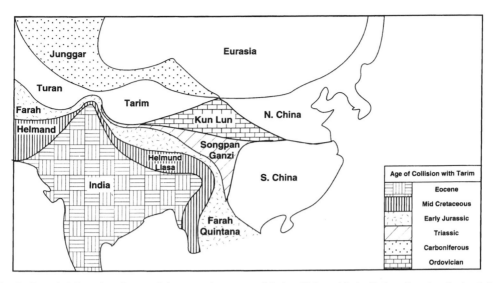

Fig. 9. Central Asian microplates and the approximate age of their collision with the Tarim microplate in the Asian collage. (After Watson *et al.* 1987; Hossack unpublished initial report BP, 1993).

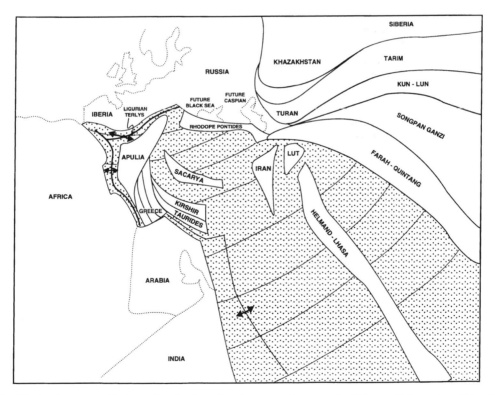

Fig. 10. A palaeogeographical snapshot showing microplates sweeping across the Tethys on their way to collide with, and accrete onto, Asia. The time slice is at about the Jurassic–Cretaceous boundary. Tarim–Turan had long since collided (Carboniferous), Helmund–Lhasa had not yet docked, the Black Sea had not yet begun to open.

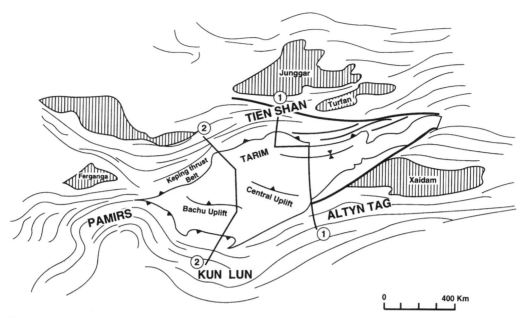

Fig. 11. Location of long sections across the Tarim basin.

floored by stretched continental lithosphere or to small ocean back-arc basins like the Black Sea or the South Caspian, since all these involve strong homogeneous mantle lithosphre thermally equilibrated after stretching. Orogenic belts wind around them all. The South Caspian is a particularly spectacular example.

The differences between back-arc basins which developed on continental crust, which did not break apart, or on continental microplates, as against those small ocean back-arc basins floored by oceanic crust, is clearly only a function of the amount of stretching, but it does allow me to develop the central argument of this paper. Small ocean basins like the Black Sea are used as analogues for Archaean granite/greentone terrains. Stretched microcontinental fragments (like Tarim) are used as analogues for Archaean granulite-facies cratonic terrains. The suture zones and orogenic belts ramifying around the basins are analogues of the Proterozoic mobile belts.

We need now to look in more detail at an example of each of these. The Western Black Sea is used as an example of a small ocean basin, and the Tarim basin as a basin built on a stretched microcontinent.

The Tarim Basin

Although Hsü (1988, 1989) has argued that Tarim is floored by oceanic crust, recent seismic inter-

pretation, field work and plate reconstruction carried out in BP strongly support the alternative view that Tarim is a continental microplate.

Where the microplate originated is uncertain, but it seems possible that it rifted away from Siberia by back-arc stretching in the Cambrian or Early Ordovician. The Eocambrian–Cambrian stratigraphy of, for example, the Berioskya basin of eastern Siberia bears an astonishingly close resemblance to that of eastern Tarim. It is contended that the back-arc rift evolved into a passive margin and that Tarim then migrated 'southwards' across Palaeozoic Tethys. Certainly the seismic stratigraphy of the Lower Ordovician sequence in eastern Tarim strongly suggests that it is a mixed carbonate and clastic turbiditic slope fan showing progradation from the south, the site of the present Altyn Tag, towards a deep sea area in the north. In the south there was an arc system partially covered by a carbonate platform which was drowned twice during the Ordovician and was re-established to disappear again in the Silurian. Cross sections through the Tarim Basin are shown in Figs 11–13.

Figure 14 shows the evolution of the Tarim basin in cartoon form. A major contractional deformation event occurred during the Ordovician, presumably in response to collision or subduction-related phenomenon within the arc. This event may have been the collision of Tarim with what is now Xaidam, or north China, which was then on the south side of Tethys.

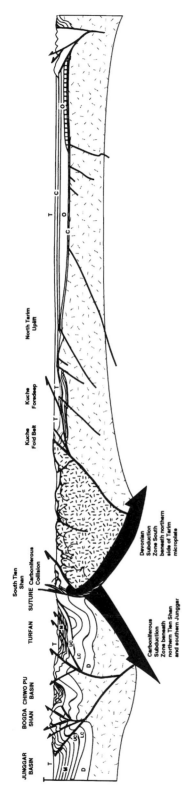

Fig. 12. Section through the eastern part of the Tarim basin from the Tien Shan to the Kun Lun (drawn by the author from field traverses, with subsurface interpretation controlled by Chinese seismic data made available to BP in 1993). Key: black is oceanic crust, light dash hatch is Proterozoic basement. Melange in the suture zone is striped, plutonic rocks (mostly Devonian and Carboniferous) are dense dash-hatched. North of the Tien Shan suture zone indicate stratigraphy from the Devonian to the Tertiary. The major unconformity is between the Upper and Lower Carboniferous.

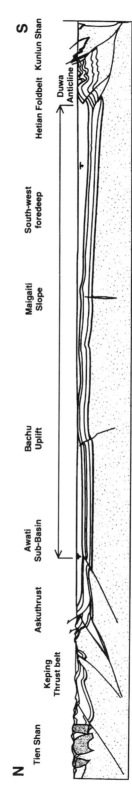

Fig. 13. Section across the western part of the Tarim basin. (Compiled from various sources by the author and J. R. Hossack.) There is reasonable subsurface control from well and seismic data made available to BP in 1993. Palaeozoic and Mesozoic strata lie beneath the asymmetrical Tertiary section (T) and the Proterozoic basement (dash-hatch).

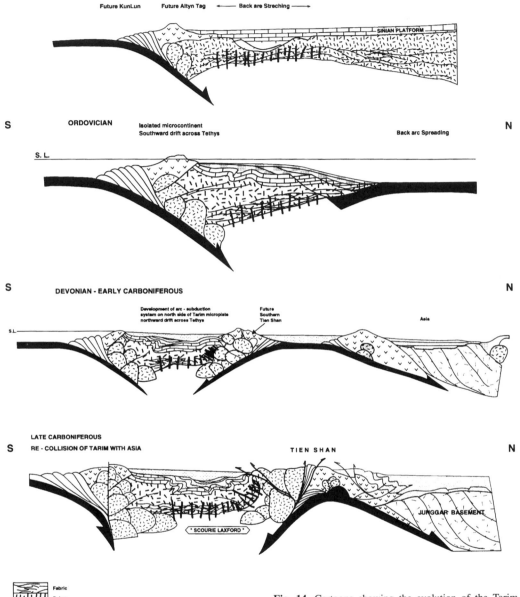

Fig. 14. Cartoons showing the evolution of the Tarim basin micro-continent as it separated from Siberia by back-arc spreading in the Early Palaeozoic, drifted across Palaeozoic Tethys, then drifted back after the establishment of an arc-subduction system on its northern margin in the Devonian. The final collision was in the Carboniferous, along another arc subduction system on the southern edge of Eurasia. The words 'Scourie–Laxford' on (d) refer to the analogy between the structures visible in that part of Scotland and those which we might imagine to see at great depth in Tarim (a stretching fabric related to back arc rifting, perhaps cut by slightly later dykes re-deformed during the later collision).

Sometime during the Devonian a trench–arc system established itself along the northern margin of the Tarim microplate. It is assumed that a southerly dipping subduction system was now established and that this resulted in a northward drift of the Tarim microplate back across the Palaeotethys. The eventual collision with Asia took place in the Carboniferous and, as Allen *et al.* (1993) have suggested, it was a collision of subduction systems with opposed dips. A massive north dipping accretion–subduction system had been growing on the south side of Asia since the Proterozoic (Şengör *et al.* 1993) with the calc alkaline volcanics of the present Tien Shan being its youngest manifestation.

The southerly dipping subduction system has been more deeply eroded than the northern one, with deep crustal granites exposed in the southern Tien Shan, but volcanics in the north. The suture separating these zones is a clearly defined belt of ophiolitic melange (Fig. 12).

After the Carboniferous collision, Tarim became a land locked basin with internal drainage. Mesozoic collisions of Triassic (Songpan Ganzi), Jurassic and Cretaceous age rejuvenated the thrust belts and encouraged coarse clastic progradation from the mountains.

Figure 6 is a sketch section through part of the Altyn Tag mountain front on the south side of Tarim. It is noteworthy for the clear rejuvenation of structure, the perched Jurassic basin with its lake, and the 'backstepping' nature of the deformation sequence in that the most internal thrust elevating Proterozoic basement is the youngest.

In most of the Tarim Basin the thickness of the Tertiary section is greater than that of the Mesozoic, presumably reflecting the fact that mid to late Tertiary contraction, and therefore loading, was more profound than Mesozoic deformation and loading. Presumably the collision of India was a more profound event than the accretion and collision of earlier terrains with more accommodation space and more sediment availability from rejuvenated mountains ranges.

A deep section through Tarim, or any of its analogue basins, would probably be something like the cartoon in Fig. 14d. The flat lying fabric shown in the lower crust and mantle on this figure is associated with ductile stretching. The back-arc stretching separated the Tarim microcontinent in the Late Precambrian or early Palaeozoic. Presumably such a fabric would be in, or near, granulite facies, and since it is associated with continental separation in a

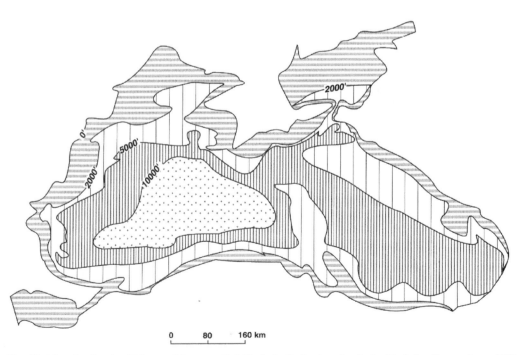

Fig. 15. Map showing the thickness of the post-rift (effectively Tertiary) section in the Black Sea (by courtesy of BP Exploration.)

back-arc setting, if can be speculated that it might be cut by dykes.

There is an analogue for this. Tricart & Lemoine (1986) have drawn attention to a remarkable deformation history in the ophiolites in the western Alps where gabbros, interpreted as subcontinental mantle, are cut by shear zones with amphibolite-facies metamorphic assemblages, and the shear zones are cut by underformed, unmetamorphosed dykes which feed pillow lavas. The shear zones are thought to be associated with continental rupture in the Jurassic and the dykes are thought to be associated with the beginnings of oceanization.

Reverse faults ahead of the mountain belts surrounding Tarim or associated with the 'uplifts' in the central part of the basin would presumably be ductile shear zones at depth and, because they are Carboniferous or younger, they would deform ('re-work') the older stretching fabric. Their association with the water-bearing accretionary rocks of the mountain belts might imply the availability of water with possible retrogression of granulite-facies assemblages to amphibolite-facies assemblages in the shear zones.

The Western Black Sea

Maps and sections through the Black Sea are shown in Figs 15–19 and a version of the tectonic evolution of the Black Sea is shown in cartoon form on Fig. 20. Fundamental to this tectonic evolution are the arguments of Guror (1988) who postulated that the Western Black Sea opened by back-arc rifting and spreading during the Mid-Late Cretaceous. This is well documented in outcrops of the Cretaceous section where the top of a shallow-water sandstone formation of Cenomanian age shows the sort of ferruginous encrustation, more or less a hard-ground, associated with deep water non-deposition. It is immediately overlain by pelagic carbonates of the Kapanbogazy Formation. The indication of rapid foundering is clear, and the most reasonable interpretation of the sequence is that it is associated with continental break up in a back-arc setting, a true 'break- up unconformity'.

The extensional fault blocks associated with the back-arc extension are visible today on the mid-Black Sea high, swamped by 15 km of Tertiary sediments (Figs 15, 16, 18 & 19).

After the formation of the back-arc basin, the

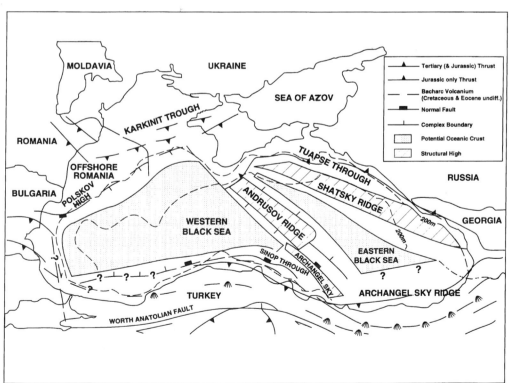

Fig. 16. Tectonic elements of the Black Sea (courtesy of BP Exploration).

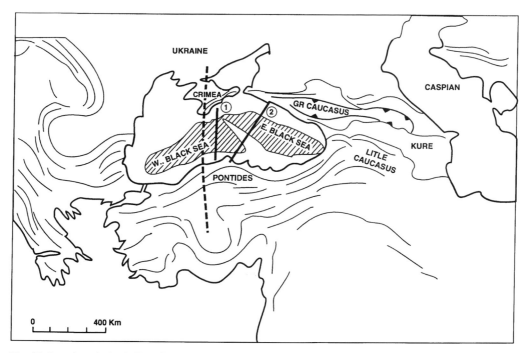

Fig. 17. Location of seismic lines figured in Figs 18 and 19.

area which is now Anatolia was subject to renewed microcontinental collisions, and the Black Sea became a 'Central Asian' type basin receiving enormous volumes of sediments from the surrounding mountain chains, in much of the area resting directly on the oceanic crust. I have noted earlier that the seismic lines crossing the mountain belts which surround the Black Sea show very clear evidence of the 'backstepping' style of deformation which is commonly associated with enclosed foreland basins. The frontal folds are always older, and onlapped by the basin fill (Figs 5, 18 & 19).

Active spreading has occurred in the Black Sea. Presumably at depth there are ultramafics, gabbros, dykes and pillow lavas of the ocean crust. Possibly ductile shear zones of Tertiary age would be deforming this oceanic crust in the vicinity of surrounding mountains belts like the Pontides.

Analogues with Precambrian crust

Analogues between oceanic crust and the volcano-sedimentary sequences of the lower stratigraphic levels of greenstone belts have been made many times before.

The Black Sea analogue might explain how the older volcanic and plutonic 'Onverwacht-type' rock of the greenstone belts could be overlain directly by large thicknesses of clastic sediments, rather than the thin sequences of the oceans.

Of course, the analogue is only relevant to the Archaean 'granite–greenstone' terrains if it can somehow be explained how continuous oceanic crust became invaded by large volumes of tonalite and related rocks, which would presumably need to be derived from the mantle, just as the tonalites of the granite terrains invaded a once more extensive greenstone belt stratigraphy. Whether it is reasonable to postulate the partial melting of tonalites from mantle on a regional scale, I do not know. Perhaps burial associated with hugely thick sediment piles, like that of the Black Sea, together with elevated Precambrian geothermal gradients, and a less differentiated mantle in the Archaean, might make such a thing possible. The temperature at 50 km depth in the Black Sea at the present day is about 1000°C (Bocaletti *et al.* 1988), close to the melting temperature of an anhydrous granodiorite or tonalite at that depth.

Central Asian drainage is mostly internal. Erosion products from adjacent mountains all end up in Central Asian basins. It has already been shown that the mountains can become overwhelmed. It is easy to imagine that with time, once the basins fill up, the drainage system might change and become a normal one with its base level as the ocean. Once

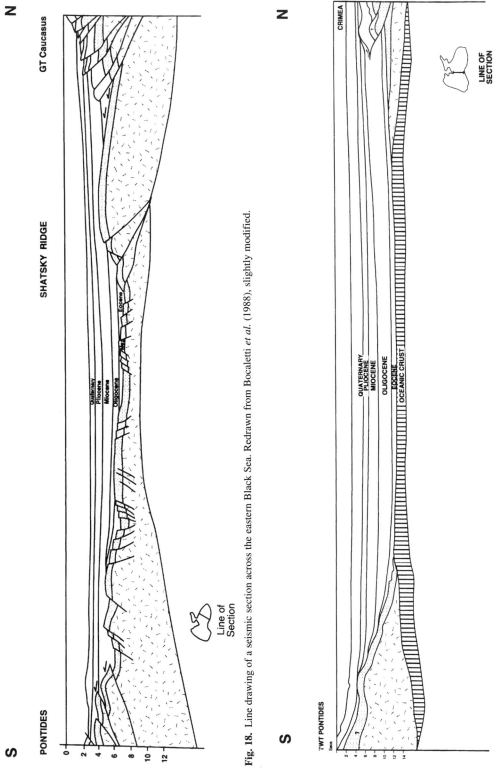

Fig. 18. Line drawing of a seismic section across the eastern Black Sea. Redrawn from Bocaletti *et al.* (1988), slightly modified.

Fig. 19. Geoseismic section across the central Black Sea, from the Crimea to the Pontides. Redrawn frm Bocalletti *et al.* (1988).

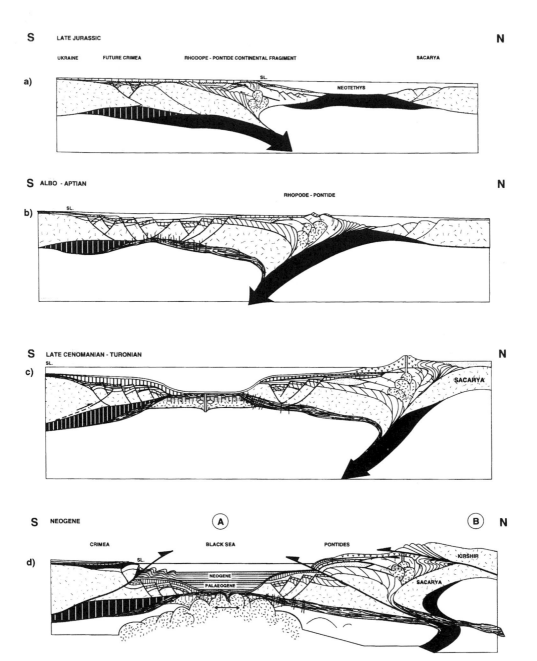

Fig. 20. Schematic evolution of the western Black Sea. In the Triassic, a major accretionary complex (called the Tavrik Formation in Crimea) accumulated during the subduction of the Palaeotethys as the Rhodope–Pontide microplate collided with Eurasia and Neotheys opened behind it. Collision occurred in the Middle Jurassic, with inversion of Middle Jurassic externsional faults in Crimea. (**a**) The Late Jurassic is a shallow water carbonate deposited above these eroded structures. It developed at this time in response to thermal subsidence on a passive margin. (**b**) Back-arc stretching, which eventually led to the formation of the Black Sea, began in the Early Cretaceous in the Crimea associated with the (northward) subduction of the Neotethys. Rifting was slightly later (Albo-Aptian) in the Pontides with the deposition of the Chaglayan Formation. Turbidites and olistostromes in the half-graben, shallow water sandstones on fault block crests and flanks. (**c**) Continental separation (the 'break up unconformity') is marked by deep water non-sequence and a pelagic drape of late Cenomanian–Turonian age (the Kapambogazi Formation). This drape is onlapped by Tertiary clastics which accumulated in the Black Sea basin Tertiary collisions (that of Kirshir is shown in (**d**) produced the contractional structures and elevations now seen on the Black Sea margin. The tonalities drawn in the central Black Sea are fancifully conceived as having melted out of the oceanic mantle. Overlaying oceanic crust and sediments are nipped into synclines between the plutons. The section A–B at depth from the Black Sea across the roots of the Pontides could be analogous with a section from the Kaapvaal craton northwards into the Limpopo mobile belt.

that had happened, with time the central part of a basin, such as Tarim, could become a regional high, with drainage radiating from it to the outside world.

It can be imagined how this could create an inversion of the existing topography with residual 'platform' sediments preserved in the one time basinal, now cratonic, areas while more deeply eroded, one time mountain belts might form lower ground. It could be argued that there are analogues with southern Africa here, with the Limpopo belt lower than the elevated Kaapvaal craton with its elevated sedimentary relics, e.g. the Witwatersrand Series.

Of course, for this simplistic hypothesis to be appropriate for the Archaean crust, some kind of 'underplating' would need to be invoked to maintain crustal thickness and isostatic equilibrium.

There can be no real doubt that one-time lower crustal rocks have been brought to the surface and are now themselves underlain by 35 km of continental crust. John Sutton often debated the mechanisms by which this might have been

achieved, and I can not add much to the discusion. It is therefore most appropriate to conclude this paper with another look at Fig. 14d and its predictions that a deep section through Tarim would show a sub-horizontal fabric in granulite facies cut by dykes, then re-deformed in steep shear zones and re-metamorphosed in anphibolite facies.

We could fondly imagine the 'Scourie–Laxford' signpost on Fig. 14d as extending the length of a walk from Scourie to Rhiconich across Laxford Bridge or across the Lewisian outcrops on the shores of Loch Torridon, the areas which Janet Watson and John Sutton described so vividly more than forty years ago, and in doing so fired their colleagues, students and followers with great enthusiasm and love for geology.

I hope John Sutton would have enjoyed the light hearted geological speculations that I have tried to present in this paper, and I believe that he might have been less worried than I about some of the scandalous generalizations. He was a bold, strong man—that was why we all respected him.

References

ALLEN, M. B., WINDLEY, B. F., ZHANG, C. & GUO, J. 1993. Palaeozoic Collision tectonics and magnatism of the Chinese Tien Shan. *Tectonophysics*, **220**, 89–115.

BOCALLETTI, M., DAINELLI, P., MANETTI, P. & MANNORI, M. R. 1988. Monograph on the Black Sea. *Bollettino di Geofisica Teorekca ed Applicata*, **30**, 117–118.

CAHEN, L., & SNELLING, N. J. 1966. *Geochronology of Equatorial Africa*. North-Holland, Amsterdam.

COBBOLD, P. R. ET AL. 1993. Sedimentary basins and crustal thickening. *Sedimentary Geology*, **86**, 77–89.

COWARD, M. P. 1976. Archean deformation patterns in southern Africa. *Philosophical Transactions of the Royal Society of London*, **A283**, 313–331.

—— 1991. *Thrust Tectonics. A bracts Volukme. Royal Holloway College, University of London.*

DAVEY, PH. & COBBOLD, P. R. 1991. Experiments on shortening of a 4-layer model of the continental lithosphere. *Tectonophysics*, **1988**, 1–25.

GUROR, N. 1988. Timing of opening of the Black Sea basins. *Tectonophysics*, **147**, 242–262.

HSÜ, K. J. 1988. Relic back-arc basins of China and their petroleum potential. *In*: KLEINSPEHN, K. (ed.) *New Perspectives in Basin Analysis. Springer*, Berlin, 245–265.

—— 1989. Origin of sedimentary basins of China. *In*: ZHU, X. (ed.) *Chinese Sedimentary Basins*. Elsevier, Amsterdam, 207–277.

LAWRENCE, S. R. 1990. Aspects of the petroleum geology of the Junggar basin, Northwest China. *In*: BROOKS, J. (ed.) *Classic Petroleum provinces*. Geological Society, London, Specia; Publications, 50, 545–557.

McKENZIE, D. P. 1978. Some remarks on the development of sedimentary basins. *Earth and Planetary Science Letters*, **40**, 25–32.

ŞENGÖR, A. M., NATALIN, B. A. & BURTMAN, V. S. 1993. Evolution of the Altaid tectonic collage and Palaeozoic crustal growth in Eurasia. *Nature*, **364**, 299–306.

SHACKLETON, R. M. 1976. Pan-African structures. *Philosophical Transactions of the Royal Society of London*, **A280**, 491–497.

SUTTON, J. & WATSON, J. V. 1951. The Pre-Torridonian metamorphic history of the Loch Torridon and Scourie areas in the North-West Highlands and its bearing on the chronological classification of the Lewisian. *Quarterly Journal of the Geological Society of London*, **106**, 241–307.

TRICART, P. & LEMOINE, M. 1986. From faulted blocks to megamullions and megaboudins Tethyan heritage in the structure of the Western Alps. *Tectonics*, **5**, 95–110.

WATSON, M. P., HAYWARD, A. B., PARKINSON, D. N. & ZHANG, ZH. M. 1987. Plate tectonic history, basin development and petroleum source rock deposition, on shore China. *Marine and Petroleum Geology*, **4**, 205–225.

WINDLEY, B. F. 1973. Crustal development in the Precambrian. *Philosophical Transactions of the Royal Society of London*, **A273**, 321–341.

—— & BRIDGEWATER, D. 1971. The evolution of Archean low- and high-grade terrains. *Geological Society of Australia Special Publication*, **3**, 33–46.

ZEIGLER, P. A. 1988. *Evolution of the Arctic-North Atlantic and the Western Tethys (with Paleogeographic maps)*. American Association of Petroleum Geologists, Memoir, **43**.

Index